Solid state physics

SOLID STATE
PHYSICS

SECOND EDITION

J. S. Blakemore

Department of Physics and Astronomy
Western Washington University

PUBLISHED BY THE PRESS SYNDICATE OF THE UNIVERSITY OF CAMBRIDGE
The Pitt Building, Trumpington Street, Cambridge CB2 1RP, United Kingdom

CAMBRIDGE UNIVERSITY PRESS
The Edinburgh Building, Cambridge CB2 2RU, UK http://www.cup.cam.ac.uk
40 West 20th Street, New York, NY 10011-4211, USA http://www.cup.org
10 Stamford Road, Oakleigh, Melbourne 3166, Australia

First published by W. B. Saunders Company 1969
second edition first published by W. B. Saunders Company 1974
This updated second edition first published by Cambridge University Press 1985
Reprinted 1986, 1988, 1989, 1991, 1993, 1995, 1998

Typeset in Bodoni Book 10/12 pt

A catalogue record for this book is available from the British Library

Library of Congress Cataloguing in Publication data

Blakemore, J. S. (John Sydney), 1927–
Solid state physics
Includes bibliographies and indexes
1. Solid state physics   I. Title
QC176.B63   1985   530.4'1   85-47879

ISBN 0 521 30932 8 hardback
ISBN 0 521 31391 0 paperback

Transferred to digital printing 2004

CONTENTS

PREFACE

This book was written as the text for a one quarter, or one semester, introductory course on the physics of solids. For an undergraduate majoring in physics, the associated course will usually be taken during the last two undergraduate years. However, the book is designed also to meet needs of those with other degree majors: in chemistry, electrical engineering, materials science, etc., who may not encounter this requirement in their education until graduate school. Some topics discussed (band theory, for example) require familiarity with the language and concepts of quantum physics; and an assumed level of preparedness is one semester of "modern physics". A reader who has taken a formal quantum mechanics course will be well prepared, but it is recognized that this is often not possible. Thus Schrödinger's equation is seen from time to time, but formal quantum mechanical proofs are side-stepped.

The aim is thus a reasonably rigorous – but not obscure – first exposition of solid state physics. The emphasis is on crystalline solids, proceeding from lattice symmetries to the ideas of reciprocal space and Brillouin zones. These ideas are then developed: for lattice vibrations, the theory of metals, and crystalline semiconductors, in Chapters 2, 3, and 4 respectively. Aspects of the consequences of atomic periodicity comprise some 75% of the book's 500 pages. In order to keep the total exposition within reasonable bounds for a first solid state course, a number of *other* aspects of condensed matter physics have been included but at a relatively brief survey level. Those topics include lattice defects, amorphous solids, superconductivity, dielectric and magnetic phenomena, and magnetic resonance.

The text now offered is on many pages unchanged from that of the 1974 second edition published by Saunders. However, the present opportunity to offer this book through the auspices of Cambridge University Press has permitted me to correct some errors, add some needed lines of explanation (such as at the end of Section 1.5), revise some figures, and update the bibliographies following this preface and at the end of each chapter. The SI system of units, adopted for the second edition, is of course retained here. Two exceptions to the SI system should be noted: retention of the *Ångstrom unit* in describing interatomic distances, and use of the *electron volt* for discussions of energy per electron of per atom. There seems no sign that crystallographers are ready to quote lattice spacings in nanometers, and the 10^{-10} conversion factor from Å to meters is an easy one. Use of the eV rather than 1.6×10^{-19} J also simplifies many descriptions of energy transformation events. Questions of units are of course important for the numerical aspects of homework problems.

These problems are grouped at the end of each chapter, and there are 125 of them altogether. Many do include a numerical part, intended to draw the student's attention to the relative magnitudes of quantities and influences more than to the importance of decimal place accuracy. The problems vary (intentionally) greatly in length and difficulty; and I have been told several times that some of these problems are too difficult for the level of the text. These can certainly provide a worthwhile challenge for one who has "graduated" from the present book to one of the advanced solid state texts cited in the General Reference list which follows this preface.

As in previous editions of this book, many more literature citation footnotes are given than are typical in an undergraduate text. These augment the bibliography at the end of each chapter in citing specific sources for *optional* additional reading. A paper so cited in a footnote may serve as the beginning of a literature search undertaken years after the owner's first exposure to this book, and the footnotes have been provided with this in mind.

The present book was written to be an account of *ideas* about the physics of solids rather than a compilation of facts and numbers. Accordingly, tables of numerically determined properties are relatively few – in contrast, for example, to nearly 60 tables of data in the fifth edition of Kittel's well-known textbook. The reader needing quantitative physical data on solids has a variety of places to turn to, with extensive data in the *American Institute of Physics Handbook* (last revised in 1972) and in the *Handbook of Chemistry and Physics* (updated annually). As noted in the list of General References on page ix, new volumes have recently been appearing in the *Landolt–Börnstein Tables* series, including data compilation for some semiconductor materials. The work of consolidating numerical information concerning solids is indeed a continuous one.

Over the years of writing and rewriting material for successive editions of this book, I have been helped by many people who have made suggestions concerning the text, worked problems, and provided illustration material. To all of those individually acknowledged in the prefaces of the first and second editions, I am still grateful. In preparing this updated second edition for Cambridge University Press, my principal acknowledgement should go to L. E. Murr of the Oregon Graduate Center for the photographs that provide a number of attractive and informative new figures in Chapter 1, and to H. K. Henisch of Pennsylvania State University for the print used as Figure 1.2.

Beaverton, Oregon   J. S. BLAKEMORE
March 1985

GENERAL REFERENCES

Solid State Physics (Introductory/Intermediate Level)

R. H. Bube, *Electrons in Solids: An Introductory Survey* (Academic Press, 3rd ed., 1992).

R. H. Bube, *Electronic Properties of Crystalline Solids* (Academic Press, 1973).

A. J. Dekker, *Solid State Physics* (Prentice-Hall, 1957). [Was never revised and is now out of print, but includes interesting discussion of several topics others omit.]

H. J. Goldsmid (ed.), *Problems in Solid State Physics* (Pion, 1968).

W. A. Harrison, *Electronic Structure and the Properties of Solids* (Freeman, 1980).

C. Kittel, *Introduction to Solid State Physics* (Wiley, 6th ed., 1986).

J. P. McKelvey, *Solid-State and Semiconductor Physics* (Krieger, 1982).

H. M. Rosenberg, *The Solid State* (Oxford Univ. Press, 3rd ed., 1988).

Solid State Physics (Advanced Level)

N. W. Ashcroft and N. D. Mermin, *Solid State Physics* (Holt, 1976).

J. Callaway, *Quantum Theory of the Solid State* (Academic Press, 2nd ed., 1991).

D. L. Goodstein, *States of Matter* (Prentice-Hall, 1975).

W. A. Harrison, *Solid State Theory* (Dover, 1980).

A. Haug, *Theoretical Solid State Physics* (Pergamon Press, 1972), 2 vols.

W. Jones and N. H. March, *Theoretical Solid State Physics* (Wiley, 1973), 2 vols.

C. Kittel, *Quantum Theory of Solids* (Wiley, 1963).

R. Kubo and T. Nagamiya (eds.), *Solid State Physics* (McGraw-Hill, 1969).

P. T. Landsberg (ed.), *Solid State Theory, Methods & Applications* (Wiley, 1969).

R. E. Peierls, *Quantum Theory of Solids* (Oxford, 1965). [Out of print, but a classic.]

F. Seitz, *Modern Theory of Solids* (Dover, 1987). [Reprint of a 1940 McGraw-Hill classic.]

J. Ziman, *Principles of the Theory of Solids* (Cambridge Univ. Press, 2nd ed., 1972).

Solid State Electronics

A. Bar-Lev, *Semiconductors and Electronic Devices* (Prentice-Hall, 1979).

N. G. Einspruch (ed.), *VLSI Electronics: Microstructure Science* (Academic Press, Vol. 1, 1981, through vol. 8, 1984, and continuing).

R. J. Elliott and A. F. Gibson, *An Introduction to Solid State Physics and its Applications* (Barnes and Noble, 1974).

A. S. Grove, *Physics and Technology of Semiconductor Devices* (Wiley, 1967).

A. G. Milnes, *Semiconductor Devices and Integrated Circuits* (Van Nostrand, 1980).

T. S. Moss, G. J. Barrett and B. Ellis, *Semiconductor Optoelectronics* (Butterworths, 1973).

B. G. Streetman, *Solid State Electronic Devices* (Prentice-Hall, 2nd ed., 1980).

S. M. Sze, *Physics of Semiconductor Devices* (Wiley, 2nd ed., 1981).

F. F. Y. Wang, *Introduction to Solid State Electronics* (North-Holland, 1980).

Quantum Phenomena

F. J. Bockhoff, *Elements of Quantum Theory* (Addison-Wesley, 2nd ed., 1976).

A. P. French and E. F. Taylor, *An Introduction to Quantum Physics* (Norton, 1978).

B. K. Ridley, *Quantum Processes in Semiconductors* (Oxford Univ. Press, 1981).

D. ter Haar (ed.), *Problems in Quantum Mechanics* (Pion, 3rd ed., 1975).

Statistical Physics

S. Fujitta, *Statistical and Thermal Physics* (Krieger, 1984).

C. Kittel, *Elementary Statistical Physics* (Wiley, 1958).

C. Kittel and H. Kroemer, *Thermal Physics* (Freeman, 2nd ed., 1980).

F. Mohling, *Statistical Mechanics* (Wiley-Halsted, 1982).

L. E. Reichl, *A Modern Course in Statistical Physics* (Univ. Texas, 1980).

R. C. Tolman, *The Principles of Statistical Mechanics* (Dover, 1979).

Wave Phenomena

L. Brillouin, *Wave Properties and Group Velocity* (Academic Press, 1960).

L. Brillouin, *Wave Propagation in Periodic Structures* (Dover, 1972).

I. G. Main, *Vibrations and Waves in Physics* (Cambridge Univ. Press, 1978).

C. F. Squire, *Waves in Physical Systems* (Prentice-Hall, 1971).

Numerical Data

American Institute of Physics Handbook (McGraw-Hill, 3rd ed., 1972).

Handbook of Chemistry and Physics (CRC Press, 66th ed., 1985).

Handbuch der Physik (S. Flügge, general editor for 54 volume series) (Springer-Verlag, 1956 through 1974).

Landolt–Börnstein Tables (Springer-Verlag). [Volumes date from the 1950s and earlier, but new ones are now appearing on solid state topics.]

chapter one

CRYSTALLINITY AND THE FORM OF SOLIDS

Solid materials can be classified according to a variety of criteria. Among the more significant of these is the description of a solid as being either *crystalline* or *amorphous*. The solid state physics community has tended during the period from the mid-1940's to the late 1960's to concentrate a much larger effort on crystalline solids than on the less tractable amorphous ones.

An amorphous solid exhibits a considerable degree of short range order in its nearest-neighbor bonds, but not the long range order of a periodic atomic lattice; examples include randomly polymerized plastics, carbon blacks, allotropic forms of elements such as selenium and antimony, and glasses. A glass may alternatively be thought of as a supercooled liquid in which the viscosity is too large to permit atomic rearrangement towards a more ordered form. Since the degree of ordering of an amorphous solid depends so much on the conditions of its preparation, it is perhaps not inappropriate to suggest that the preparation and study of amorphous solids has owed rather less to science and rather more to art than the study of crystalline materials. Intense study since the 1960s on glassy solids such as amorphous silicon (of interest for its electronic properties) is likely to create a more nearly quantitative basis for interpreting both electronic and structural features of noncrystalline materials.

In the basic theory of the solid state, it is a common practice to start with models of single crystals of complete perfection and infinite size. The effects of impurities, defects, surfaces, and grain boundaries are then added as perturbations. Such a procedure often works quite well even when the solid under study has grains of microscopic or submicroscopic size, provided that long range order extends over distances which are very large compared with the interatomic spacing. However,

1

it is particularly convenient to carry out experimental measurements on large single crystals when they are available, whether they are of natural origin or synthetically prepared.[1] Figures 1-1 and 1-2 show examples of microscopic and macroscopic synthetic crystals.

Large natural crystals of a variety of solids have been known to man for thousands of years. Typical examples are quartz (SiO_2), rocksalt (NaCl), the sulphides of metals such as lead and zinc, and of course gemstones such as ruby (Al_2O_3) and diamond (C). Some of these natural crystals exhibit a surprising degree of purity and crystalline perfection, which has been matched in the laboratory only during the past few years.[2] For many centuries the word "crystal" was applied specifically to quartz; it is based on the Greek word implying a form similar to that of ice. In current usage, a *crystalline solid* is one in which the atomic arrangement is regularly repeated, and which is likely to exhibit an external morphology of planes making characteristic angles with each other *if* the sample being studied happens to be a *single* crystal.

When two single crystals of the same solid are compared, it will usually be found that the sizes of the characteristic plane "faces" are

---

[1] For discussions of single growth techniques, see the bibliography at the end of Chapter 1.

[2] Indeed, synthetically created diamonds still do not compare in quality with the finest natural diamonds. For most other gemstones, man seems to have been able to do at least as well as nature.

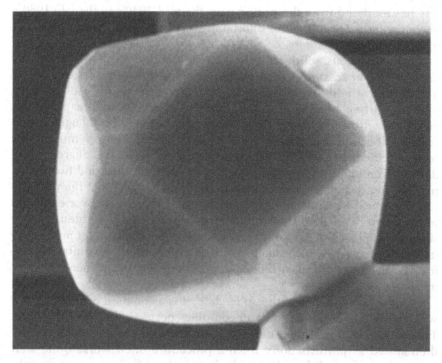

Figure 1-1 Scanning electron microscope view of small NiO crystal, with well developed facets. (Photo courtesy of L. E. Murr, Oregon Graduate Center.) At room temperature, antiferromagnetic ordering provides for NiO a trigonal distortion of the (basically rocksalt) atomic arrangement.

Figure 1-2   The growing surface of a calcium tartrate crystal, during growth in a tartrate gel infused with calcium chloride solution. From *Crystal Growth in Gels* by H. K. Henisch (Penn. State Univ. Press, 1970).

not in the same proportion (the "habit" varies from crystal to crystal). On the other hand the interfacial angles are always the same for crystals of a given material; this was noted in the sixteenth century and formed the basis of the crystallography of the next three centuries. These observations had to await the development of the atomic concept for an explanation, and it was not until Friedrich, Knipping, and Laue demonstrated in 1912 that crystals could act as three-dimensional diffraction gratings for X-rays that the concept of a regular and periodic atomic arrangement received a sound experimental foundation. More recently, the periodic arrangement of atoms has been made directly visible by field-emission microscopy.[3]

Whether we wish to study mechanical, thermal, optical, electronic, or magnetic properties of crystals—be they natural ones, synthetic single crystals (such as Ge, Si, Al_2O_3, KBr, Cu, Al), or polycrystalline aggregates—most of the results obtained will be strongly influenced by the periodic arrangement of atomic cores or by the accompanying periodic electrostatic potential. The consequences of periodicity take up a major fraction of this book, for a periodic potential has many consequences, and exact or approximate solutions are possible in many situations.

In this first chapter we shall consider how atoms are bonded together and how symmetry requirements result in the existence of a limited variety of crystal classes. There is no optimum order for consideration of the two topics of bonding and crystal symmetry, since each depends on the other for illumination; it is recommended that the

---

[3] See, for example, Figure 1-56(a) on page 79, for an ion-microscope view of atoms at the surface of an iridium crystal.

reader skim through the next two sections completely before embarking upon a detailed study of either.

The chapter continues (in Section 1.3) with an account of some of the simpler lattices in which real solids crystallize. The emphasis of the section is on the structures of elements and of the more familiar inorganic binary compounds.

Sections 1.4 (Crystal Diffraction) and 1.5 (Reciprocal Space) are closely connected, and once again it is recommended that both sections be read through before a detailed study of either is undertaken. An understanding of the reciprocal lattice helps one to see what diffraction of a wave in a crystal is all about, and *vice versa*.

Section 1.6 does little more than mention the principal types of point and line imperfection in a crystal. Bibliographic sources are cited for the reader who wishes to know more about dislocations, or about the chemical thermodynamics of defect interactions in solids.

Forms of Interatomic Binding

All of the mechanisms which cause bonding between atoms derive from electrical attraction and repulsion. The differing strengths and differing types of bond are determined by the particular electronic structures of the atoms involved. The weak van der Waals (or residual) bond provides a universal weak attraction between closely spaced atoms and its influence is overridden when the conditions necessary for ionic, covalent, or metallic bonding are also present.

The existence of a stable bonding arrangement (whether between a pair of otherwise isolated atoms, or throughout a large, three-dimensional crystalline array) implies that the spatial configuration of positive ion cores and outer electrons has less total energy than any other configuration (including infinite separation of the respective atoms). The energy deficit of the configuration compared with isolated atoms is known as the *cohesive energy*, and ranges in value from 0.1 eV/atom for solids which can muster only the weak van der Waals bond to 7 eV/atom or more in some covalent and ionic compounds and some metals.[4] The cohesive energy constitutes the reduction in potential energy of the bonded system (compared with separate atoms) *minus* the additional kinetic energy which the Heisenberg uncertainty principle tells us must result from localization of the nuclei and outer shell electrons.

In covalent bonding the angular placement of bonds is very important, while in some other types of bonding a premium is placed upon securing the largest possible coordination number (number of nearest neighbors). Such factors are clearly important in controlling the most favorable three-dimensional structure. For some solids, two or more quite different structures would result in nearly the same energy, and a change in temperature or hydrostatic pressure can then provoke a change from one allotropic form of the solid to another, as envisaged in Figure 1-3. As discussed further under the heading of the Covalent Bond, an allotropic transition to an energetically more favorable structure can sometimes be postponed, depending on the rate of conditions of cooling or warming.

---

[4] The joule is a rather large energy unit for discussion of events involving a single atom. Thus energies in this book will often be quoted in terms of electron volts per particle or per microscopic system. (It is hoped that the context will leave no doubt as to whether an energy change in eV refers to a molecule, an atom, or a single electron.) One elementary charge moved through a potential difference of one volt involves a potential energy change of 1.6022×10^{-19} joule (see the table of useful constants inside the cover). Chemists tend to cite bond energies and cohesive energies in calories per mole. 1 eV/molecule is equivalent to 23,000 calories per mole, or 9.65×10^4 joule/mole.

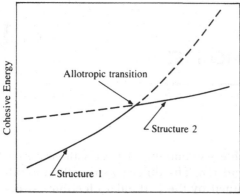

Figure 1–3 Cohesive energy versus temperature or pressure for a solid in which two different atomic arrangements are possible. An allotropic transition *may* occur at the pressure or temperature at which one structure replaces the other as having minimum energy.

THE VAN DER WAALS BOND

As previously noted, van der Waals bonding occurs universally between closely spaced atoms, but is important only when the conditions for stronger bonding mechanisms fail. It is a weak bond, with a typical strength of 0.2 eV/atom, and occurs between neutral atoms and between molecules. The name van der Waals is associated with this form of bond since it was he who suggested that weak attractive forces between molecules in a gas lead to an equation of state which represents the properties of real gases rather better than the ideal gas law does. However, an explanation of this general attractive force had to await the theoretical attentions of London (1930).

London noted that a neutral atom has zero permanent electric dipole moment, as do many molecules; yet such atoms and molecules are attracted to others by electrical forces. He pointed out that the zero-point motion, which is a consequence of the Heisenberg uncertainty principle, gives any neutral atom a fluctuating dipole moment whose amplitude and orientation vary rapidly. The field induced by a dipole falls off as the cube of the distance. Thus if the nuclei of two atoms are separated by a distance r, the instantaneous dipole of each atom creates an instantaneous field proportional to $(1/r)^3$ at the other. The potential energy of the coupling between the dipoles (which is attractive) is then

$$E_{attr} = -\left(\frac{A}{r^6}\right) \tag{1-1}$$

A quantum-mechanical calculation of the strength of this dipole-dipole attraction suggests that E_{attr} would reach 10 eV if r could be as small as 1Å. However, a spacing this small is impossible because of overlap repulsion.

As the interatomic distance decreases, the attractive tendency begins to be offset by a repulsive mechanism when the electron clouds of the atoms begin to overlap. This can be understood in terms of the Pauli exclusion principle, that two or more electrons may not occupy the same quantum state. Thus overlap of electron clouds from two atoms with quasi-closed-shell configurations is possible only by promotion of some of the electrons to higher quantum states, which requires more energy.

The variation of repulsive energy with interatomic spacing can be simulated either by a power law expression (a dependence as strong as r^{-11} or r^{-12} being necessary) or in terms of a characteristic length. The latter form is usually found to be the most satisfactory, and the total energy can then be written as

$$E = -\left(\frac{A}{r^6}\right) + B \exp\left(-\frac{r}{\rho}\right) \tag{1-2}$$

which is drawn as the solid curve in Figure 1-4. The strength of the bond formed and the equilibrium distance r_0 between the atoms so bonded depend on the magnitudes of the parameters A, B, and ρ. Since the characteristic length ρ is small compared with the interatomic spacing, the equilibrium arrangement of minimum E occurs with the repulsive term making a rather small reduction in the binding energy.[5]

We have spoken of van der Waals bonding so far as occurring between a pair of otherwise isolated atoms. Within a three-dimensional solid, the dipole-dipole attractive and overlap repulsive effects with respect to the various neighbor atoms add to give an overall cohesive energy still in accord with Equation 1-2. There are no restrictions on bond angles, and solids bound by van der Waals forces tend to form in the (close-packed) crystal structures for which an atom has the largest possible number of nearest neighbors. (This is the case, for example, in the crystals of the inert gases Ne, Ar, Kr, and Xe, all face-centered-cubic structures, in which each atom has twelve nearest neighbors.) The rapid decrease of van der Waals attraction with distance makes atoms beyond the nearest neighbors of very little importance.

---

[5] See Problem 1.1 for an exercise of this principle.

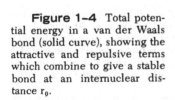

Figure 1–4 Total potential energy in a van der Waals bond (solid curve), showing the attractive and repulsive terms which combine to give a stable bond at an internuclear distance r_0.

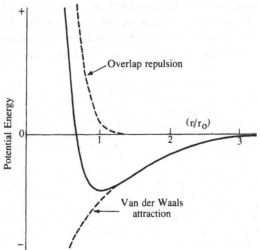

The solid inert gases[6] are fine examples of solids which are bound *solely* by van der Waals forces, because the closed-shell configurations of the atoms eliminate the possibility of other, stronger bonding mechanisms. Far more typically do we find solids in which van der Waals forces bind saturated *molecules* together, molecules within which much stronger mechanisms are at work. This is the case with crystals of many saturated organic compounds and also for solid H_2, N_2, O_2, F_2, Cl_2, Br_2, and I_2. The example of Cl_2, with a sublimation energy of 0.2 eV/molecule but a dissociation energy of 2.5 eV/molecule, shows how the van der Waals bond between diatomic molecules can be broken much more readily than the covalent Cl-Cl bond.

THE COVALENT BOND

The covalent bond, sometimes referred to as a *valence* or *homopolar* bond, is an electron-pair bond in which two atoms share two electrons. The result of this sharing is that the electron charge density[7] is high in the region between the two atoms. An atom is limited in the number of covalent bonds it can make (depending on how much the number of outer electrons differs from a closed-shell configuration), and there is a marked directionality in the bonding. Thus carbon can be involved in four bonds at tetrahedral angles (109.5°), and the characteristic tetrahedral arrangement is seen in crystalline diamond and in innumerable organic compounds. Other examples of characteristic angles between adjacent covalent bonds are 105° in plastic sulphur and 102.6° in tellurium.

The hydrogen molecule, H_2, serves as a simple example of the covalent bond. Two isolated hydrogen atoms have separate 1s states for their respective electrons. When they are brought together, the interaction between the atoms splits the 1s state into two states of differing energy, as sketched in Figure 1-5. When the two nuclei are very close together, the total energy is increased for both kinds of states by internuclear electrostatic repulsion; but for the 1s state marked[8] σ_g, which has an even (symmetric) orbital wave-function, the energy is lowered (i.e., there is an attractive tendency) for a moderate spacing.[9]

---

[6] For helium, the zero-point motion is so violent that solidification even at absolute zero can be accomplished only by applying an external pressure of 30 atmospheres.

[7] Remember that in quantum mechanics we cannot describe a specific orbit for a bound electron but only a wave-function ψ whose square is proportional to the probability of finding an electron at a location on a time-averaged basis. Then if ψ is a *normalized* wave-function (such that ψ^2 integrated over all space is unity), the average charge density at any location is the value of $-e\psi^2$.

[8] The designation of the two orbital wave-functions as σ_g and σ_u comes from the German terms "gerade" and "ungerade" for even and odd.

[9] A principal feature of the bonding attraction is the resonance energy corresponding to the exchange of the two electrons between the two atomic orbitals, as first discussed by W. Heitler and F. London, Z. Physik **44**, 455 (1927). For a recent account of this in English, see E. E. Anderson, *Modern Physics and Quantum Mechanics* (W. B. Saunders, 1971), p. 390.

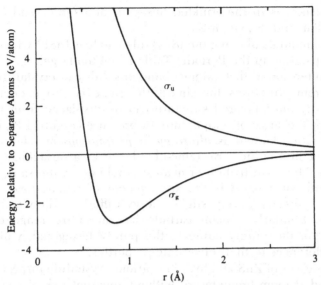

Figure 1–5 Variation of energy with internuclear spacing for the neutral hydrogen molecule, after Heitler and London (1927). The figure shows the σ_g (bonding) and σ_u (anti-bonding) states. σ_g accommodates two electrons with anti-parallel spins.

This symmetric σ_g 1s solution requires that the electron charge density $-e\psi^2$ be concentrated in the region between the two nuclei. The requirement of the Pauli principle that total wave functions combine in an anti-symmetric manner is satisfied if the σ_g 1s state is occupied by two electrons with antiparallel spins.

The alternative σ_u 1s state would have to be occupied by two electrons with parallel spins in order to conform with the Pauli principle, but as Figure 1-5 demonstrates, this state is an anti-bonding (repulsive) one at all distances. This is unimportant for H_2, since the σ_g state can accommodate the only two electrons in the system and a strong bond results.

Note that this could not happen for a double bond between two helium atoms, since the total energy would be increased by populating both of the σ_u states as well as the σ_g states. Interestingly, the molecule-ion He_2^+ *is* stable.

The wave-mechanical problem becomes much more formidable when covalent bonds are considered between multi-electron atoms, but qualitatively the picture is that sketched for the H-H bond. In all cases the closeness of approach is limited by the Coulomb repulsion of the nuclei, assisted in the heavier atoms by overlap repulsion of inner closed-shell electrons.

Some of the classes of covalently bonded materials are:

1. Most bonds within organic compounds.

2. Bonds between pairs of halogen atoms (and between pairs of atoms of hydrogen, nitrogen, or oxygen) in the solid and fluid forms of these media.

3. Elements of Group VI (such as the spiral chains of tellurium),

Group V (such as in the crinkled hexagons of arsenic), and Group IV (such as diamond, Si, Ge, α-Sn).

4. Compounds obeying the (8-N) rule (such as InSb) when the horizontal separation in the Periodic Table is not too large.

It is often found that valence-bonded solids can crystallize in several different structures for almost the same cohesive energy. The energetically most favored structure can be displaced from its prime position by a change of temperature or pressure (Figure 1-3), resulting in the situation known as *allotropy* or *polymorphism*. Thus ZnS can exist either in a cubic form (zinc-blende) or as a hexagonal structure (wurtzite). The coordination of nearest neighbors is the same for zinc-blende and wurtzite; it is the arrangement of second-nearest neighbors which creates a very slight energy difference between the two structures. Similarly, silicon carbide has an entire range of "polytypes," from the purely cubic to the purely hexagonal, which show subtle differences in their electronic properties.

In the cases of ZnS and SiC, the various crystalline forms can all be maintained at room temperature without apparent risk of spontaneous conversion to the energetically most favored form (the conditions of crystallization accounting for the various forms capable of being studied at low temperatures). With other materials, spontaneous conversion occurs quite readily.

Thus selenium cooled rapidly from its melting point (218°C) to room temperature is amorphous, but crystallization begins if the solid is warmed to 60–70°C, and the material remains crystalline on cooling back to room temperature. Another good example of allotropic conversion is provided by tin, which is stable as a gray semimetal (α-Sn) below 17°C, crystallizing in the diamond lattice with four tetrahedrally-located bonds. Temperatures above 17°C, or application of pressure even below that temperature, cause a conversion to a much more dense white metallic form (β-Sn) with a tetragonal structure in which each atom has six nearest neighbors.

COVALENT-VAN DER WAALS STRUCTURES

As previously noted, this combination of bonding mechanisms is found in materials such as solid hydrogen, in which each pair of atoms is internally covalently bonded and van der Waals bonds create a "molecular crystal." The same principles apply to most organic solids.

An example of another kind of covalent-residual bonding is provided by tellurium (Figure 1-6), in which successive atoms in each spiral chain are covalently bonded. The forces *between* chains are much weaker and are probably little more than van der Waals attraction. Consequently, tellurium has a low structural strength and is anisotropic in all its mechanical, thermal, and electronic properties.

Similarly, in graphite (Figure 1-7) carbon atoms are arranged in hexagons in each layer, so that three of the four outer shell electrons from each atom are used in valence bonds within the layer. (The fourth electron is free.) The interlayer spacing is large, with essentially only

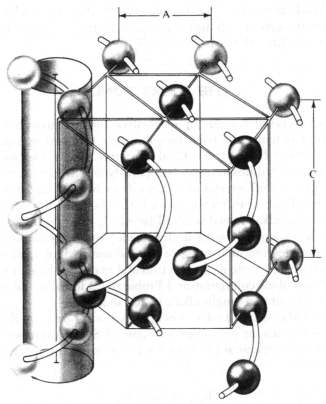

Figure 1-6 The atomic arrangement in tellurium. From Blakemore et al., *Progress in Semiconductors*, Vol. 6. (Wiley, 1962). Each atom makes covalent bonds with its nearest neighbors up and down the spiral chain. Inter-chain forces are weak. One allotropic form of selenium adopts the same structure.

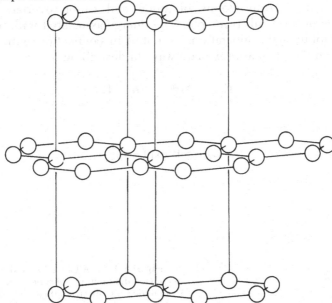

Figure 1-7 Atomic arrangement in the graphite form of carbon. Within a layer, each atom makes three strong covalent bonds ($r_0 = 1.42$ Å) in order to preserve the hexagonal array. The bonding between layers (spacing of 3.4 Å) is weak, so that the layers can slide over each other with ease.

van der Waals attraction. Thus the planes can slide over each other very easily, the property which makes graphite useful as a "solid lubricant." The same considerations apply in MoS_2.

THE IONIC BOND

An ionic crystal is made up of positive and negative ions arranged so that the Coulomb repulsion between ions of the same sign is more than compensated for by the Coulomb attraction of ions of opposite sign. The alkali halides such as NaCl are typical members of the class of ionic solids; NaCl crystallizes (almost) as Na^+Cl^-. Electron transfer from Na to Cl occurs to such a major extent because the ionization potential I_e of the alkali metal is small (work eI_e must be done to convert Na into the cation Na^+ with a closed electronic shell configuration), whereas the electron affinity E_a of the halogen is large. (Energy E_a is provided when Cl receives an electron and becomes the anion Cl^-, also with a closed shell configuration.) Problem 1.2 looks at the energetics of the ionic bond in a single alkali halide molecule.

When a Na^+ ion and a Cl^- ion approach each other in the absence of any other atoms, as envisaged in Figure 1-8, the attractive Coulomb energy at internuclear separation r relative to zero energy at infinite separation is

$$E_{coul} = -e^2/4\pi\varepsilon_0 r \tag{1-3}$$

since the (closed-shell) electronic charge distributions are spherically symmetrical. The approach distance is limited by repulsion when the closed-shell electron clouds of anion and cation overlap, in consequence of the Pauli principle. The energy associated with repulsion varies rapidly with separation, as noted in connection with van der Waals bonding; two approximate ways of describing it are

$$E_{rep} = A/r^n \qquad (n \approx 12) \tag{1-4}$$

or

$$E_{rep} = B \exp(-r/\rho) \tag{1-5}$$

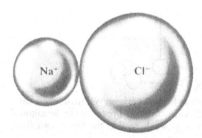

Figure 1-8 A Cl^- anion and Na^+ cation in contact, drawn in the observed ratio of sizes. In an ionic solid the size ratio plays an important part in determining the most favorable structure.

neither of which really does justice to the complicated quantum-mechanical process which constitutes repulsion.

If Equation 1-5 is adopted as at least giving some idea of how the repulsive energy varies with internuclear separation, it becomes apparent that the stable bond length between Na^+ and Cl^- will be the quantity for which

$$E_i = (E_{coul} + E_{rep}) = -e^2/4\pi\epsilon_0 r + B \exp(-r/\rho) \qquad (1\text{-}6)$$

is a minimum. This minimum is shown in Figure 1-9. Because the repulsive term is much more sensitive to changes in r than is the Coulomb term, the bond energy is only slightly smaller than $(e^2/4\pi\epsilon_0 r_0)$, while the restoring forces whenever r departs from r_0 are dictated by the values of B and ρ.

The principles noted above as being operative for a single Na^+Cl^- bond hold equally well[10] for solid NaCl, together with some additional geometric considerations. We shall be talking again about the sodium chloride structure in Section 1.3 from the viewpoint of geometry and symmetry [using Figure 1-33(a) at that time], but we need to examine all four parts of Figure 1-10 to appreciate how the particular bonding arrangement arises. In solid NaCl, each cation (i.e., each sodium ion) has six anions as its nearest neighbors [and vice versa, as we can see from Figure 1-10(b)] and the interaction with nearest neighbors involves both Coulomb attraction and overlap repulsion. As can be seen

---

[10] In the next subsection, we shall have to note that the electron transfer is not 100 per cent complete even for the most strongly "ionic" compounds, though ionic considerations are certainly the most important ones for the alkali halides.

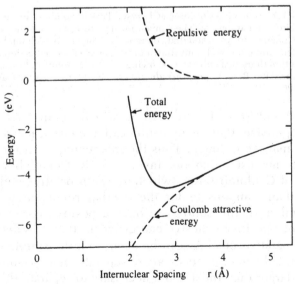

Figure 1-9 Energy of a Na^+Cl^- molecule compared with that of separate ions, according to Equation 1-6. The characteristic length of the repulsive energy is here assumed to be $\rho = 0.345$ Å and the magnitude of repulsion produces a minimum energy at $r_0 = 2.82$ Å.

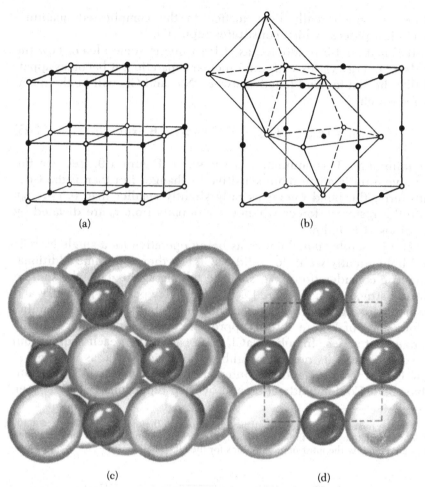

(a)　　　　　　　　　　　　(b)

(c)　　　　　　　　　　　　(d)

Figure 1–10 Four views of the NaCl crystal lattice. Solid circles are used for cations and open circles for anions. (a) The conventional picture, showing the arrangement of crystallographic planes. (b) An octahedral configuration, in which each ion has six nearest neighbors. (c) The unit cube with ions drawn to correct sizes so that cation-anion contact occurs. (d) A section through a cube face, showing that anion-anion and cation-cation contacts do *not* occur. After R. C. Evans, *Introduction to Crystal Chemistry* (Cambridge University Press, 1964).

from parts (c) and (d) of Figure 1-10, in which cations and anions are drawn to proper size, there is no cation-cation contact, *nor do the large anions even manage to touch. Thus the interaction of an ion with anything but its nearest neighbors involves only Coulomb terms. The overall sum of Coulomb terms (which are both positive and negative) must more than compensate for the overlap repulsion with the six nearest neighbors if the solid is to have a positive cohesive energy. (Possession of a positive cohesive energy means that E_i is *negative* with respect to separated ions.) As can be seen from the simple calculation in Problem 1.2, the energy necessary to separate an ionic solid into separate ions is larger than the amount necessary to separate the solid into isolated neutral atoms.

The most advantageous crystal structure for an ionic solid depends on the ratio of anion to cation radii. (Remember that an anion such as

Cl^- is considerably larger than a cation such as Na^+.) When this radius ratio permits anion-anion contact, the structure is likely to be superseded by a different ionic arrangement. Thus the NaCl structure is favored over the CsCl arrangement only if $(r_-/r_+) \geq 1.41$, and the zincblende structure is energetically even more favored if the ratio (r_-/r_+) becomes very large. Similar considerations (of many anion-cation contacts but no anion-anion contacts) dictate the most favored structures for ionic solids such as the halides of the alkaline earth elements, in which the ratio of anions to cations is not unity.

Returning to NaCl as a typical example of an ionically bound solid, we note that a sodium cation has

6	Cl^-	nearest	neighbors at a distance of	r_0
12	Na^+	next-nearest	neighbors at a distance of	$\sqrt{2}\, r_0$
8	Cl^-	further	neighbors at a distance of	$\sqrt{3}\, r_0$

and so on. The total Coulomb attractive energy *per ion pair* is thus the sum of an infinite series

$$E_{coul} = -(e^2/4\pi\varepsilon_0 r_0) \left\{ 6 - \left(\frac{12}{\sqrt{2}}\right) + \left(\frac{8}{\sqrt{3}}\right) - \ldots \right\} \qquad (1\text{-}7)$$
$$= -1.748(e^2/4\pi\varepsilon_0 r_0)$$
$$= -\alpha(e^2/4\pi\varepsilon_0 r_0)$$

The number α is the *Madelung constant* (Madelung, 1918) for the particular lattice, and α has a value controlled by the geometry of the lattice. (As a comparison with the NaCl lattice, note that α is 1.638 for the zincblende lattice, and 1.763 for the CsCl type of atomic arrangement.)

The series in Equation 1-7 which must be summed to obtain the Madelung constant has terms of alternating sign, and clearly converges rather weakly. This slowness of convergence is a problem with most lattice structures. Methods by which accurate sums may be obtained for such series have been developed by Ewald (1921) and Evjen.[11] These methods depend on dividing space extending outward from one ion into a set of zones lying between successive polyhedra. The polydedral surfaces are chosen in such a way that each zone has a total charge of zero, and an ion sitting on the boundary between two zones has its charge apportioned between the two zones. The Evjen approach produces a revised series for which the terms converge rapidly. Problem 1.3 uses this approach simplified to two dimensions.

From Equations 1-5 and 1-7 we find that the total energy per molecule of an ionic crystal relative to infinitely separated ions is

$$E_i = -\left(\frac{\alpha e^2}{4\pi\varepsilon_0 r}\right) + C \exp\left(\frac{-r}{\rho}\right) \qquad (1\text{-}8)$$

where the values of C and ρ are unknowns in the absence of a complete quantum-mechanical treatment. One of them, however, can be elimi-

---

[11] H. M. Evjen, Phys. Rev. 39, 680 (1932).

nated by using the condition that the energy passes through a minimum at the equilibrium nearest-neighbor spacing r_0. Differentiating Equation 1-8, we have

$$\left(\frac{dE_i}{dr}\right) = \left(\frac{\alpha e^2}{4\pi\varepsilon_0 r^2}\right) - \left(\frac{C}{\rho}\right) \exp\left(\frac{-r}{\rho}\right)$$

and the condition $(dE/dr)_{r_0} = 0$ requires that

$$C = \left(\frac{\alpha\rho e^2}{4\pi\varepsilon_0 r_0{}^2}\right) \exp(r_0/\rho) \qquad (1\text{-}9)$$

Then at any spacing the binding energy is

$$E_i = -\left(\frac{\alpha e^2}{4\pi\varepsilon_0 r}\right)\left[1 - \left(\frac{r\rho}{r_0{}^2}\right) \exp\left(\frac{r_0 - r}{\rho}\right)\right] \qquad (1\text{-}10)$$

At the equilibrium spacing itself, we see that the cohesive energy is

$$E_{eq} = -\left(\frac{\alpha e^2}{4\pi\varepsilon_0 r_0}\right)\left[1 - \left(\frac{\rho}{r_0}\right)\right] \qquad (1\text{-}11)$$

and since ρ is typically no more than a small percentage of r_0, the cohesive energy is dominated by the Madelung term.

The remaining unknown, ρ, can be eliminated from Equations 1-10 and 1-11 by using experimentally accessible information on the bulk compressibility, χ. This is possible since the compressibility involves the second derivative of energy with spacing at the equilibrium spacing, a quantity that is strongly influenced by the characteristic length of the short-range repulsion. The reasoning proceeds as follows. Consider an infinitesimal change dv in crystal volume (per molecule) at pressure p. For the sodium chloride structure, the volume per ion pair is $v = 2r^3$, so that $dv = 6r^2 dr$. The work done in this change is $dE = -pdv = -6pr^2 dr$, from which we must be able to express the pressure as

$$p = -\frac{1}{6r^2}\left(\frac{dE}{dr}\right)$$

whose derivative is

$$\left(\frac{dp}{dr}\right) = -\frac{1}{6r^2}\left(\frac{d^2E}{dr^2}\right) + \frac{1}{3r^3}\left(\frac{dE}{dr}\right) \qquad (1\text{-}12)$$

We remember that $(dE/dr)_{r_0}$ is zero; thus the second term on the right of Equation 1-12 vanishes at the equilibrium spacing.

Now the compressibility χ describes the pressure dependence of volume through

$$\chi = -\frac{1}{v}\left(\frac{dv}{dp}\right) = -\frac{3}{r}\left(\frac{dr}{dp}\right) \qquad (1\text{-}13)$$

This gives a value for (dp/dr) in terms of χ, which can be compared with Equation 1-12. Accordingly, at the equilibrium spacing,

$$\left(\frac{d^2E}{dr^2}\right)_{r_0} = (18r_0/\chi) \qquad (1\text{-}14)$$

Equation 1-10 can be differentiated twice to get an alternate expression for $(d^2E/dr^2)_{r_0}$, and this latter one is in terms of ρ. Comparing the result with Equation 1-14, we find that

$$(\rho/r_0) = [2 + (72\pi\varepsilon_0 r_0^4/\alpha e^2\chi)]^{-1} \qquad (1\text{-}15)$$

Thus, knowledge of the equilibrium lattice constant, the Madelung structure factor, and the compressibility, permits calculation of the cohesive energy, as utilized in Problem 1.4.

The properties of ionic crystals have been tabulated by a number of authors.[12] Among these, Mott and Gurney report that $\rho = 0.345$ Å permits a reasonable fit for all 20 alkali halides (whether of the NaCl or CsCl structures), but that the various compounds require widely varying values of C in the repulsive term $C\exp(-r/\rho)$. This happens because the closed-shell (alkali)$^+$ and (halogen)$^-$ ions behave like virtually incompressible spheres, pressed into contact at the equilibrium spacing. Mott and Gurney suggest that "basic radii" for the ions be assigned as in Table 1-1. Then if the repulsive term is written in the form

$$C\exp\left(\frac{-r}{\rho}\right) = C'\exp\left[\frac{r_a + r_c - r}{\rho}\right] \qquad (1\text{-}16)$$

the required values for C' are about the same for all 20 binary compounds, showing that the spherical ions are all deformable to about the same degree.

The semi-empirical Pauling scale of ionic radii[13] gives slightly larger sizes for alkali and halide ions, since the Pauling scale is constructed in such a way that the sum of ionic radii for nearest neighbors (in a completely ionic compound, and with six-fold coordination) should just equal the equilibrium internuclear spacing.

TABLE 1-1 BASIC RADII FOR IONS IN ALKALI HALIDE COMPOUNDS

Li$^+$	$r_c = 0.475$ Å	F$^-$	$r_a = 1.110$ Å
Na$^+$	$r_c = 0.875$ Å	Cl$^-$	$r_a = 1.475$ Å
K$^+$	$r_c = 1.185$ Å	Br$^-$	$r_a = 1.600$ Å
Rb$^+$	$r_c = 1.320$ Å	I$^-$	$r_a = 1.785$ Å
Cs$^+$	$r_c = 1.455$ Å		

[12] See for example: N. F. Mott and R. W. Gurney, *Electronic Processes in Ionic Crystals* (Oxford, Second Edition, 1948). Also, M. P. Tosi, in *Solid State Physics, Vol. 16* (Academic Press, 1964).

[13] L. Pauling, *The Nature of the Chemical Bond* (Cornell University Press, Third Edition, 1960).

How well such semi-empirical schemes work depends on how closely the bonding resembles complete electron transfer from cation to anion, and at this point we should note that this is not *fully* satisfied in *any* solid.

MIXED IONIC-COVALENT BONDS

Completely ionic binding in a compound requires the presence of an extremely electropositive component (which can be ionized with ease to form a cation) and of an extremely electronegative component (for which the electron affinity to form an anion is as large as possible). These requirements are rather well satisfied in the alkali halides, in which there is strong encouragement for electron transfer.

In compounds with less extreme electropositive and electronegative character, however, there is considerably less than 100 per cent charge transfer from cation to anion. For example, the noble metals have larger ionization energies than alkalis, and silver halides are less ionic in nature than the corresponding alkali halides. There is in fact a continuous progression from purely ionic character to purely covalent character as we consider compounds in which the electronegativity difference becomes smaller.

When there is a partial tendency towards electron sharing, the optimum binding can be considered as arising from a *resonance* between ionic and covalent charge configurations. The resulting time-averaged wave-function for a bonding electron is then

$$\psi = \psi_{cov} + \lambda\psi_{ion} \tag{1-17}$$

where ψ_{cov} and ω_{ion} are normalized wave-functions for completely covalent and ionic forms, and λ is a parameter which determines the degree of ionicity:

$$\text{per cent ionicity} = \left[\frac{100\lambda^2}{1 + \lambda^2}\right] \tag{1-18}$$

The appropriate value for λ is determined by a quantum-mechanical variational calculation, based on the premise that the most stable (and therefore, equilibrium) configuration corresponds with the value of λ for which the energy[14]

$$E = \frac{\int \psi\mathscr{H}\psi \cdot d\tau}{\int \psi^2 \cdot d\tau} \tag{1-19}$$

passes through a negative minimum.

---

[14] Equation 1-19 gives the mean energy of the system with wave-function ψ compatible with the Hamiltonian $\mathscr{H}$. An understanding of the quantum-mechanical calculation is not essential to visualize how resonance improves the bonding strength.

The quantity

$$\Delta = (E_{cov} - E) \tag{1-20}$$

is the additional binding energy resulting from the mixed character of the bond and is often referred to as the ionic-valent resonance energy. In order to estimate Δ, one must be able to evaluate the hypothetical E_{cov} of a supposedly purely covalent bond between the atoms involved. For a bond between atoms A and B, the geometric mean of the A-A and B-B bond energies is customarily used to determine E_{cov}.

Pauling[13] has developed a scale of electronegativity values from empirical information in an attempt to make semi-quantitative estimates of λ, and of Δ (which increases with λ). His resulting scale of electronegativity coefficients for the elements is shown in Figure 1-11; the values of x are such that

$$1.3 \, (x_A - x_B)^2 = \Delta_{AB} \tag{1-21}$$

measured in eV. Pauling's choice of numbers for the scale of elec-

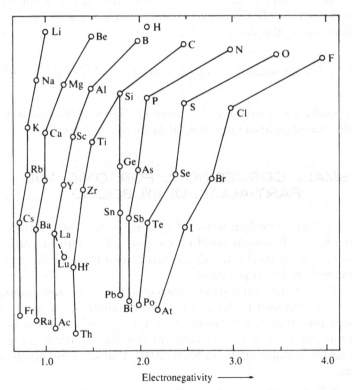

Figure 1-11 Electronegativity coefficients for some of the principal sequences of elements, after Pauling (1960). The numerical scale for the abcissa is such that Equation 1-21 should be satisfied for the resonance energy in diatomic compounds.

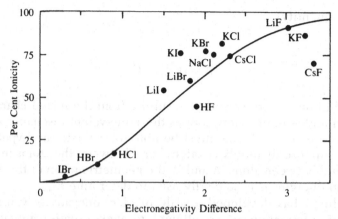

Figure 1-12 The curve follows Equation 1-22, as suggested by Pauling (1960). The points show the comparison of ionicity (as determined from dipole moments) with difference of electronegativity coefficients, using the scale of Figure 1-11.

tronegativity values was dictated by his desire that $(x_A - x_B)$ should be numerically equal to the dipole moment of the bond in the c.g.s unit, the Debye. [1 Debye = 10^{-18} esu $\cdot$ cm = 3.33×10^{-30} C $\cdot$ m]

The degree of ionicity has been defined in terms of the quantity λ of Equations 1-17 and 1-18. Pauling suggests that per cent ionicity (as determined from dipole moments) is well correlated with electronegativity difference by the equation

$$\text{per cent ionicity} = 100\left\{1 - \exp\left[-\left(\frac{x_A - x_B}{2}\right)^2\right]\right\} \quad (1\text{-}22)$$

This expression is displayed as the curve of Figure 1-12, which is compared with ionicity (dipole moment) data for various bonds.

SMALL CORRECTIONS FOR IONIC AND PARTIALLY POLAR SOLIDS

Van der Waals bonding must, of course, always be reckoned with in addition to the dominant bonding mechanism(s), but for partially or completely polar solids the van der Waals contribution to the energy is seldom more than 2 to 3 per cent.

The lattice energy must also be corrected for zero-point motion, the kinetic energy required by localization of the atoms (as a result of the Heisenberg uncertainty principle). This energy, which must be subtracted from attractive terms in determining the binding energy, is[15] ($9\hbar\omega_D/4$), and is usually of the order of 0.03 eV/atom, a relatively small correction.

---

[15] Here ω_D denotes the maximum frequency of the Debye vibrational spectrum, a topic discussed in detail in Chapter 2.

THE HYDROGEN BOND

A hydrogen atom, having but one electron, can be covalently bonded to only one atom. However, the hydrogen atom can involve itself in an additional electrostatic bond with a second atom of highly electronegative character, such as fluorine, oxygen, or to a smaller extent with nitrogen. This second bond permits a *hydrogen bond* between two atoms or structures. The hydrogen bond is found with strengths varying from 0.1 to 0.5 eV per bond.

Hydrogen bonds connect the H_2O molecules in ordinary ice (Figure 1-13), a structure similar to wurtzite in which there is a spacing of 2.76 Å between the oxygen atoms of adjacent molecules. This is much more than twice the "ordinary" O-H spacing of 0.96 Å for an isolated water molecule, and in fact neutron diffraction shows that the O-H bond is stretched only to about 1.01 A when the hydrogen atom is reaching out to extend a hydrogen bond to the next molecule. Thus in ice there is an arrangement of *almost* normal molecules. The molecules in ice can flip into a variety of arrangements, the equilibrium one depending on pressure and temperature, and numerous high pressure allotropic modifications of ice are known.

When ice is cooled to a very low temperature, it is caught in one of many possible configurations, not in general the most highly ordered one. As a result, it retains a residual entropy. For N molecules, the number of possible configurations is about $(3/2)^N$, and so a molar entropy of about $k_0 N_A \ln(3/2) \simeq 3.37$ joule/mole-kelvin is to be expected.

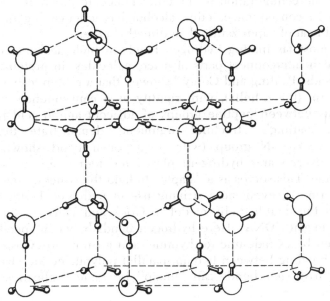

Figure 1-13 The crystalline arrangement in ice results in four hydrogen bonds for each oxygen (large circles) at the tetrahedral angles. The O-O distance is 2.76 Å and each proton (small circles) is 1.01 Å from the oxygen of its own molecule and 1.75 Å from the oxygen nucleus of the neighboring molecule. For comparison, the O-H distance is 0.96 Å in an isolated H_2O molecule. After L. Pauling, *The Nature of the Chemical Bond* (Cornell University Press, 1960).

This is in good agreement with the experimentally measured residual entropy of 3.42 joule/mole-kelvin.

Some of the hydrogen bonds persist in liquid water, as evidenced in many ways. These hydrogen bonds are responsible for the unusually high boiling point and heat of vaporization for this compound with a molecular weight of only 18. Hydrogen bonds account also for the striking dielectric properties of water and ice. The low frequency dielectric response of this material consists primarily of the rotation of polar H_2O molecules, a process in which two protons must jump to new positions which define new sets of hydrogen bonds. This rotation occurs readily in liquid water, and provides a low frequency dielectric constant $\kappa \simeq 80$. The value for ice is smaller and markedly temperature-dependent, since molecular rotation is inhibited at the lower temperatures.

In addition to their role in H_2O, hydrogen bonds are prominent in the polymerization of compounds such as HF, HCN, and NH_4F. Hydrogen bonds are responsible for the ferroelectric properties of solids such as KH_2PO_4, and we shall see in Chapter 5 that replacement of ordinary hydrogen by deuterium has a large effect on the ferroelectric transition temperature.

Hydrogen bonds are of great importance in understanding the properties of many organic compounds and biologically important materials. As a very simple example, the atomic combination C_2H_6O can exist in two well-known isomeric forms: dimethyl ether $(CH_3)_2O$, and ethyl alcohol CH_3CH_2O. The latter forms hydrogen bonds from the hydrogen of one hydroxyl group to the oxygen of a neighboring molecule, but the configuration of the ether molecule precludes hydrogen bonding. In consequence, ethyl alcohol has a much higher boiling point and heat of vaporization than dimethyl ether.

The most fascinating hydrogen bonds in biological materials are those which interconnect parts of macromolecules, in proteins and in nucleic acids. Pauling and Corey[16] showed that a protein can assume a helical form (the α-helix), with rigidity as a consequence of hydrogen bonds between peptide groups of adjacent amino acid polycondensates. Pauling's structural arguments demonstrate that each $C=O \cdots H-N$ group (with a hydrogen bond shown dotted between oxygen and hydrogen) places the four atoms in almost a straight line. This serves as a "staple" to hold the structure together in a rigid form. An even more striking use of hydrogen bonds was revealed in the double helix model of Crick and Watson[17] for deoxyribonucleic acid (DNA). Here hydrogen bonds between specific pairs of nitrogen bases (adenine to thymine, and guanine to cytosine) form the *only* bonding between two antiparallel polynucleotide chains; the series of hydrogen bonds can "unzip" for gene replication in cell

[16] L. Pauling, R. B. Corey, and H. R. Branson, Proc. Nat. Acad. Sci, U.S. **37**, 205 (1951). See also R. B. Corey and L. Pauling, Proc. Roy. Soc. **B.141**, 10 (1953).

[17] First reported in J. D. Watson and F. H. C. Crick, Nature **171**, 737 (1953). A highly entertaining and personal account is given by J. D. Watson, *The Double Helix* (Athaneum, 1968), while the current state of knowledge of nucleic acids is reviewed by J. H. Spencer, *The Physics and Chemistry of DNA and RNA* (W. B. Saunders, 1972).

division, but the "zipped up" complete DNA molecule has rigidity and can crystallize in forms which yield clear X-ray diffraction photographs.

THE METALLIC BOND

Metallic structures are typically rather empty (having large internuclear spacings) and prefer lattice arrangements in which each atom has many nearest neighbors. Three typical metallic structures are indicated in Table 1-2.

Each atom is involved in far too many bonds for each of them to be localized, and bonding must occur by resonance of the outer electrons of each atom among all possible modes. In many metals only one electron per atom is doing all of the work. This is the case for lithium, which has only one electron outside the closed K-shell configuration, and we must regard a lithium crystal as an array of Li^+ ions (spheres of radius 0.68 Å) with a surrounding *electron gas* equivalent to one electron per atom.

The weakness of the individual bonding actions in a metal is demonstrated by an enlargement of the internuclear spacing compared with that in a diatomic molecule. Thus we note in Table 1-2 that the Li-Li distance in solid lithium is 3.04 Å, whereas it is only 2.67 Å in a covalently bound Li_2 molecule. However, the total binding energy increases in a solid metal as opposed to separate molecules, because the solid has many more bonds (albeit individually weak ones). Thus the binding energy per atom increases from 0.6 eV with Li_2 to 1.8 eV in solid lithium.

In Chapters 3 and 4 we shall be repeatedly concerned with the quantum-mechanical picture of a metallic solid as a widely spaced array of positively charged "ion cores" with a superposed "electron gas" to give macroscopic charge neutrality. We now note that:

1. The wave-functions of the electrons comprising this gas overlap strongly; the wave-functions are completely delocalized (and are Bloch functions when the effect of the periodic ion potential is allowed for).

2. The electrons which perform the binding task (the valence electrons) do not all have the same energy. The binding of a metallic solid comes about because the *average* energy per valence electron is *smaller* than that of isolated atoms. This is illustrated for metallic sodium by Figure 3-39 in Section 3.4 (the band theory of solids).

TABLE 1-2 THE THREE MOST COMMON METALLIC STRUCTURES

Structure	Example	Neighbors	r_0
Body-centered cubic	Lithium	8 at distance r_0 6 at distance $2r_0/\sqrt{3}$	3.04 Å
Face-centered cubic (Cubic close-packed)	Copper	12 at distance r_0	2.56 Å
Hexagonal close-packed	Zinc	12 at distance r_0	2.66 Å

TABLE 1-3 SUMMARY OF THE CLASSIFICATION OF
BONDING TYPES IN SOLIDS

Bond Type	Typical Examples				Some Properties of the Type
	Material	Crystal Structure	Binding Energy (eV/molecule)	Nearest-Neighbor Distance (Å)	
Van der Waals (Molecular crystals)	Argon	F.C.C.	0.1	3.76	Low melting and boiling points. Easily compressible. Electrically insulating. Transparent for photon energies to far U.V.
	Chlorine	Tetragonal	0.3	4.34	
	Hydrogen	H.C.P.	0.01	3.75	
Covalent	Silicon	Cubic (Diamond)	3.7	2.35	Rigid and hard. Often high melting points. Electrical insulator or semiconductor. Strongly absorbing for photon energies above intrinsic edge, transparent (when pure) for longer wavelengths.
	InSb	Cubic (Zinc blende)	3.4	2.80	
	Mg_2Sn	Cubic (Fluorite)	1.0	2.92	
Ionic	KCl	Cubic (Rocksalt)	7.3	3.14	Rather plastic. Often dissociates on heating. Electrical insulator at low temperatures. Lattice disorder promotes ionic conduction at higher temperatures. Has reststrahlen absorption in I.R. as well as intrinsic absorption.
	AgBr	Cubic (Rocksalt)	5.4	2.88	
	BaF_2	Cubic (Fluorite)	17.3	2.69	
Hydrogen Bond	Ice	Hexagonal	0.5	1.75	Many allotropic forms. Electrically non-conducting. Dielectric activity. Optically transparent.
Metallic	Sodium	B.C.C.	1.1	3.70	Large spacing and coordination number. Electrical conductor. Opaque and highly reflecting in infrared and visible light; transparent in U.V.
	Silver	F.C.C.	3.0	2.88	
	Nickel	F.C.C.	4.4	2.48	

3. In order to be rigorous about metallic binding, allowance must be made for electrostatic repulsion of the ion cores (screened by the electron gas), van der Waals attraction of ion cores and their overlap repulsion (both very small), binding due to incomplete inner shells (important in a transition element such as iron), and electron correlations within the electron gas. Progress in the theory of solids during the 1950's and 1960's has now brought physicists to the point at which some of these topics can be included in realistic calculations.

It is not possible to prepare a single table which can show all the significant respects in which the properties of solids are influenced by the dominant bonding mechanism. An abbreviated attempt at such a comparison is shown in Table 1-3, which notes in capsule form just a few of the properties explored in later chapters.

Symmetry Operations   1.2

TRANSLATIONAL SYMMETRY

An ideally perfect single crystal is an infinite three-dimensional repetition of identical building blocks, each of the identical orientation. Each building block, called a *basis,* is an atom, a molecule, or a group of atoms or molecules. The basis is the quantity of matter contained in the *unit cell,* a volume of space in the shape of a 3-D parallelepiped, which may be translated discrete distances in three dimensions to fill all of space.

The most obvious symmetry requirement of a crystalline solid is that of *translational symmetry.* This requires that three translational vectors **a, b,** and **c** can be chosen such that the translational operation

$$\mathbf{T} = n_1\mathbf{a} + n_2\mathbf{b} + n_3\mathbf{c} \qquad (1\text{-}23)$$

(where n_1, n_2, and n_3 are arbitrary integers) connects two locations in the crystal having identical atomic environments. The translation vectors **a, b,** and **c** lie along three adjacent edges of the unit cell parallelepiped. However, translational symmetry means much more than that the atomic environment must look the same at one corner of the unit cell as at the opposite corner; it means that if *any* location in the crystal (which may or may not coincide with the position of an atom) is designated as the point **r,** then the local arrangement of atoms must be the same about **r** as it is about any of the set of points

$$\mathbf{r}' = \mathbf{r} + \mathbf{T} \qquad (1\text{-}24)$$

The set of operations **T** defines a *space lattice* or *Bravais lattice,* a purely geometric concept. A real crystal lattice results when a basis is placed around each geometric point of the Bravais lattice.

Five different Bravais lattices can be envisaged for a hypothetical two-dimensional solid, and we shall find it convenient to illustrate some features of 2-D situations before proceeding to real three-dimensional materials. Fourteen Bravais lattices are possible in three dimensions.

Only for certain kinds of lattices can the vectors **a, b,** and **c** be chosen to be equal in length, and only for certain simple lattices are they mutually perpendicular. The lattice of points **r',** and the translation vectors **a, b,** and **c,** are said to be *primitive* if *every* point equivalent to **r** is included in the set **r'.** The *primitive basis* is the minimum number of atoms or molecules which suffices to characterize the crystal structure, and is the amount of matter contained within the primitive (smallest) unit cell. For some kinds of crystalline array, there is more than one logical way in which a set of primitive vectors can be chosen (as well as an infinite number of ways in which sets can be chosen for larger, *non-primitive* unit cells). Figure 1-14 illustrates the logical (perpendicular) pair of primitive vectors for a plane rectangular lattice, along with some less useful but still valid primitive combinations, and one of the non-primitive combinations.

THE BASIS

A space lattice or Bravais lattice is not an arrangement of atoms. It is purely a geometric arrangement of points in space. In order to describe a crystal *structure*, we must cite both the lattice and the symmetry of the *basis* of atoms associated with each lattice point. The basis consists of the atoms, their spacings, and bond angles, which recur in an identical fashion about each lattice point such that every atom in the crystal is accounted for.

Fewer atoms are needed to describe the basis set for a primitive lattice than when non-primitive vectors are used, and for some simple elemental solids (such as argon or sodium) the basis consists of a single atom. Other elements crystallize in structures with several atoms per primitive basis (e.g., a basis of two atoms for silicon, four atoms for gallium), and of course for a compound the basis must comprise at least one molecule. In some complex organic compounds many thousands of atoms are required for a single basis set.

The external morphology of a crystalline substance (the angles between facets) is dictated by the Bravais lattice. However, the symmetry of the basis with respect to operations such as rotation or reflection cannot be divorced from the symmetry of the Bravais lattice. Only

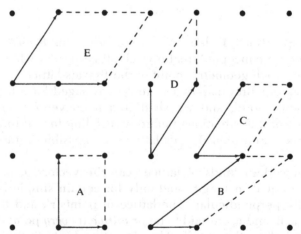

Figure 1-14 Some of the ways in which a unit cell can be defined by translational vectors for a rectangular two-dimensional lattice. Cells A through D are primitive, but E is not. For this structure, A is the obvious choice of primitive vectors and primitive cell. With structures of lower symmetry, however, there may not be a clear-cut choice of the optimum set of primitive vectors.

certain kinds of lattice are consistent with atomic groupings of any given shape and symmetry, and the highest symmetry of the lattice sets the highest symmetry which has any significance for the basis. A basis is accordingly allocated to one of the finite number of *point groups*. Each point group is characterized by the set of all rotations, reflections, or rotation plus reflection operations that take the crystal into itself with one point fixed. We shall see that for hypothetical two-dimensional solids, all possible atomic groupings can be described by ten point groups, while a total of 32 point groups exist in three dimensions.

THE UNIT CELL

In speaking of a crystal as a three-dimensional repetition of building blocks, we may think of this building block either as the basis of atoms, or as the parallelepiped defined by the vectors **a**, **b**, and **c**. The basis is the quantity of matter contained in[18] the *unit cell* of volume

$$V = \mathbf{a} \times \mathbf{b} \cdot \mathbf{c} \qquad (1\text{-}25)$$

A specification of the shape and size of a unit cell and of the distribution of matter within it gives a complete crystallographic description of a crystal. Whether or not this description is given in a particularly convenient form depends on the choice of unit cell.

This comment is made since the choice of unit cell is arbitrary to the same extent that we have many possible choices for translational vectors. Complete crystallographic information follows automatically whether or not the simplest primitive cell is chosen, but sometimes the deliberate choice of a non-primitive cell can help to emphasize some symmetries of the solid.

This is not the case for the hypothetical two-dimensional rectangular lattice of Figure 1-14, in which the primitive cell A is clearly the most useful one for any purpose. However, if we examine the face-centered-cubic (F.C.C.) lattice of Figure 1-15(a) we can see that the simplest *primitive* cell is rhombohedral, with three translational vectors each of length $(L/\sqrt{2})$, which make angles of 60° with respect to each other. The strong cubic symmetry is demonstrated if we display the lattice instead in terms of a non-primitive *unit cube*, with three mutually perpendicular translational vectors of length L. The unit cube has a volume L^3, but the volume of the rhombohedral cell is only $L^3/4$. Thus matter associated with four primitive basis units must be associated with the volume of the unit cube.

Similarly, for the body-centered-cubic (B.C.C.) lattice of Figure 1-15(b), the cubic symmetry is best demonstrated by choice of a non-primitive unit cube of side L, rather than a primitive rhombohedral cell that can be constructed with primitive vectors of length $(L\sqrt{3}/2)$. The

---

[18] An atom *inside* the unit cell we draw is clearly "contained" in this cell. An atom on the cell surface is counted partially; thus we count one-eighth of each of the eight atoms at the cube corners in Figure 1-15.

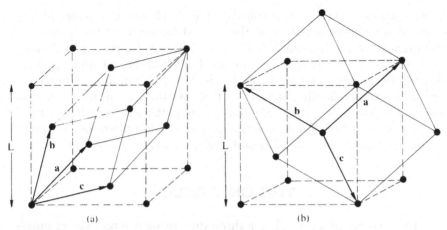

Figure 1-15 (a) The face-centered-cubic lattice, and (b) the body-centered-cubic lattice. Each part of the figure shows the conventional unit cube as dashed lines, with one possible rhombohedral form for the primitive unit cell as solid lines. If $\hat{x}$, $\hat{y}$, and $\hat{z}$ are unit vectors along three mutually perpendicular axes, then the F.C.C. primitive cell is generated by the vectors $\mathbf{a} = \frac{1}{2}L(\hat{x} + \hat{y})$, $\mathbf{b} = \frac{1}{2}L(\hat{y} + \hat{z})$, $\mathbf{c} = \frac{1}{2}L(\hat{z} + \hat{x})$. These vectors, all of length $(L/\sqrt{2})$, make 60° angles with each other. For the B.C.C. primitive cell in part (b) of the figure, the generating vectors are $\mathbf{a} = \frac{1}{2}L(\hat{x} + \hat{y} - \hat{z})$, $\mathbf{b} = \frac{1}{2}L(\hat{y} + \hat{z} - \hat{x})$, $\mathbf{c} = \frac{1}{2}L(\hat{z} + \hat{x} - \hat{y})$. These vectors, all of length $(L\sqrt{3}/2)$, make 109° angles with each other.

primitive cell has a volume of $L^3/2$, just one-half that of the cube shown with dashed lines. Were we to choose any other primitive combination of vectors to draw a B.C.C. primitive unit cell of different shape, the volume would of course still have to be $L^3/2$. There is no limit to the different unit cells of volume L^3 or more that we could construct, but the cube shown in Figure 1-15(b) is the most useful.

LATTICE SYMMETRIES

At first sight it may seem strange that every solid material does not appear in a wide range of solid forms, one corresponding with each conceivable type of Bravais lattice. The arguments of energy stability that we discussed in connection with interatomic bonding are powerful in asserting that one allotropic form of a solid should be found in practice in preference to any other *possible* allotropic forms. However, questions of crystal symmetry dictate that the number of possible allotropic forms shall be limited for any substance—and that often only one structure is at all possible. This results from the requirement that the point group (the symmetry of the basis) invariably be consistent with the symmetry of the Bravais lattice itself.

Lattice symmetry operations can be handled with great power and elegance by the use of group theory,[19] and the notations and methods of

---

[19] Group theory methods are applied to crystal structure by R. S. Knox and A. Gold, *Symmetry in the Solid State* (W. A. Benjamin, 1964), and by G. F. Koster, *Space Groups and Their Representations* (Academic Press, 1964). These books also indicate consequences of lattice symmetries in the electronic and other properties, formulated in group theory terminology.

group theory are desirable for an exhaustive study of crystallography. However, it is not necessary for us to use this algebraic notation in order to gain some idea of the physically realizable symmetry operations. The theme of this book is that an awareness of the elements of crystallography will serve as the foundation for examining those properties of solids that depend on the existence of a periodic structure. The bibliography at the end of this chapter reports on more extensive discussions of crystallography and lattice symmetry.

In addition to translational symmetry, we must consider how a lattice can be invariant with respect to (can appear identical to the original lattice following) any of the operations:

Reflection at a plane.
Rotation about an axis (1, 2, 3, 4, or 6-fold).
Inversion through a point.
Glide (=reflection + translation).
Screw (=rotation + translation).

The latter two, each requiring a combination of actions, are known as *compound operations*. Inversion can also be considered as a compound operation, since it is equivalent to a rotation of 180° followed by reflection in a plane normal to the axis of rotation. The plane at which reflection is envisaged as occurring, or the axis of a rotation, need not pass through lattice points, as will be amply demonstrated in the two-dimensional examples of Figure 2-18.

A crystal structure is said to have a center of inversion if the environment of any arbitrary location r (measured with respect to the inversion center) is identical to the environment at the inverse location − r. This is the case for far fewer real solids than one might at first anticipate. In those solids which *do* possess inversion symmetry, the inversion center is in some cases coincident with an atomic site; in others the inversion center occurs at a point in space equidistant between two atoms. As can easily be verified from the following figures, *any* Bravais lattice permits the existence of an inversion center; it is the geometry of the *basis* of atoms which rules out inversion symmetry in most solids.

As a contrasting situation, several of the major Bravais lattices are incompatible with the existence of a reflection plane (plane of mirror symmetry), no matter how simple the basis is. The discussion to follow will illustrate how a reflection plane may form one face of a unit cell, or can alternatively be placed to bisect a unit cell.

We say that a crystal possesses an n-fold axis of rotation if rotation through an angle $(2\pi/n)$ produces an atomic array identical to the original one. We have already listed rotation operations for n = 1 (a universal and trivial one which those familiar with group theory language will recognize as the identity element or operation), and for n = 2, 3, 4, and 6. Thus the reader may wonder why rotation angles other than 180°, 120°, 90° and 60° are not allowed in crystal symmetry. Other symmetry angles are certainly found elsewhere in nature, for example in a five-pointed starfish. However, n-fold rotations with n = 1, 2, 3, 4, or 6 are

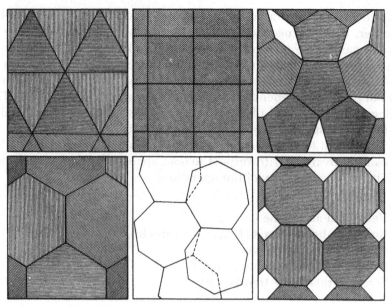

Figure 1-16 Only triangles, squares and hexagons of all the regular polygons can be fitted in a close-packed array, thus demonstrating both translational and rotational symmetry.

the only rotation symmetries which are also consistent with a crystal's requirements for translational symmetry. This can be seen in the various parts of Figure 1-16. Regular triangles, squares, and hexagons can fit together in a close-packed array, but attempts to fill an area with regular polygons of any other rotational symmetry lead either to overlapping polygons or to wasted spaces.

These statements can be verified using the simple geometric construction (Figure 1-17) of a two-dimensional "crystal" having a lattice constant a in the horizontal direction. Thus along row A, atoms 1 and m are separated by distance $(m-1)a$. Now let us assume that rotation through an angle α is a rotation allowed by the symmetry of this lattice. A counterclockwise rotation about atom 2 would move atom 1 into position formerly occupied by some atom 1'. Similarly, a clockwise rotation of angle α about atom $(m-1)$ would move atom m into the position formerly occupied by some atom m'. Evidently the atoms 1' and m' are

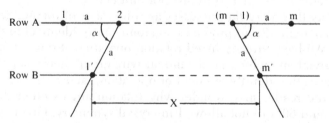

Figure 1-17 A two-dimensional construction showing what angles of rotation are also compatible with crystalline translational symmetry.

TABLE 1-4 SOLUTIONS OF
EQUATION 1-27 FOR ALLOWED
ROTATIONS IN A PERIODIC
LATTICE

$(p - m)$	$\cos \alpha$	α	Order of Rotation
-1	1	0	1-fold
-2	$1/2$	$(\pi/3)$	6-fold
-3	0	$(\pi/2)$	4-fold
-4	$-1/2$	$(2\pi/3)$	3-fold
-5	-1	π	2-fold

members of a row B, and their separation X must be an integral multiple of a if α is an allowed rotation. Setting $X = pa$, where p is an unknown integer, the difference between the two unknown integers m and
p can be described in terms of α. Some very simple trigonometry
applied to Figure 1-17 shows us that

$$X = pa = (m - 3)a + 2a \cos \alpha \qquad (1\text{-}26)$$

so that

$$\cos \alpha = \left[\frac{3 + p - m}{2} \right] \qquad (1\text{-}27)$$

Given that m and p are integers, this equation has only five solutions,
which are tabulated in Table 1-4.

We shall see, in considering the various two-dimensional and
three-dimensional lattices, that invariance with respect to an operation
may involve one of the compound operations (rotation-inversion, glide,
or screw) rather than a "simple" operation. As an introduction to a discussion of real three-dimensional lattices, we shall find it useful to
demonstrate the major concepts in terms of the rather simpler hypothetical two-dimensional lattices, point groups, and *plane groups*. A *plane
group* is the two-dimensional equivalent of what is called a *space group*
in three dimensions: the set of all symmetry elements that take a perfect infinite crystal into itself (leave the crystal looking the same as
before the operations were performed).

PLANE LATTICES AND THEIR SYMMETRIES

Linear arrays of points can be combined in five distinguishable
ways in order to create a two-dimensional Bravais lattice. These five
types of arrangement are depicted in Figure 1-18, together with the
symmetries associated with principal locations. The point symmetry
shown in each case is that of the *lattice;* the symmetry of a resulting
crystal will be lower if the basis of atoms to be associated with each lattice point has a lower symmetry than the lattice. Inversion centers are

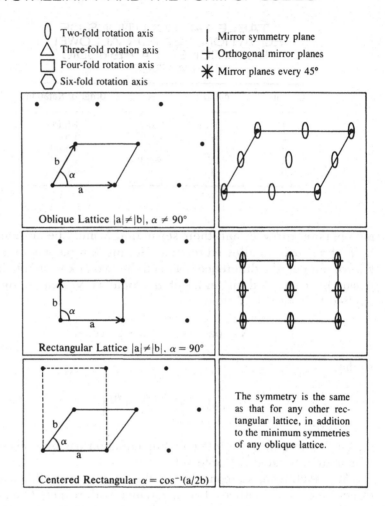

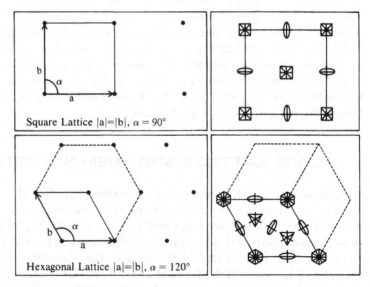

Figure 1-18 The five two-dimensional lattices and their symmetries.

possible for each of the lattices, as may be determined by inspection of the figure (see Problem 1.6).

For each of the five lattice types, rotation axes and/or mirror planes occur at the lattice points. However, there are other locations in the unit cell with comparable or lower degrees of symmetry with respect to rotation and reflection.

The most general kind of two-dimensional lattice is the *oblique* one, the first one shown in Figure 1-18. As will be recalled from the illustrations of a rectangular lattice in Figure 1-14, a unit cell (primitive or otherwise) can be selected in an infinite number of ways, but the primitive cell defined in Figure 1-18 by the unit vectors **a** and **b** is by far the most logical choice. For a general oblique lattice in which there is no special relationship between **a**, **b**, and α, the symmetries existing are only the above mentioned inversion symmetry and two-fold rotation axes at the indicated locations. Possibilities for reflection or for higher orders of rotation occur only when **a** and **b** are related in some special way, or when α has a special value.

One of these special lattice types is the *primitive rectangular* form, when $\alpha = 90°$. Since no restriction is placed on the length ratio (b/a), rotation symmetry is still no higher than two-fold; but two perpendicular sets of mirror lines now occur. Another special lattice occurs when α is not a right angle but coincides with $\cos^{-1}(a/2b)$. Even though any primitive unit cell for this lattice is an oblique parallelogram, it can be seen from Figure 1-18 that a non-primitive cell can be drawn as a centered rectangle. The *centered rectangular* lattice has the reflection symmetries and two-fold rotation symmetry associated with the lines and points we recognize for its non-primitive rectangular cell, and also has two-fold rotation symmetry with respect to the additional locations we recognize with respect to the primitive oblique unit cell.

Another of the special lattice types is the *square lattice*, which enjoys four-fold symmetry about the midpoints of the sides. The positions of highest rotation symmetry form the conjunctions of reflection lines every 45°.

The last of the special lattices which can be formed by placing restrictions on the angles and length ratios of a parallelogram is the *hexagonal* lattice. This may be thought of as a special case of a centered rectangular lattice for which a = b, and $\alpha = 60°$, but approaching the hexagonal lattice from this standpoint is liable to make us overlook many of the inherent symmetries in the hexagonal arrangement. As indicated in Figure 1-18, it is preferable that the primitive unit cell be drawn as a rhombus with $\alpha = 120°$ or 60°. Three such primitive cells make up the "hexagonal" unit cell. Reflection lines intersect every 30° at the points of highest symmetry, locations which enjoy six-fold rotation symmetry.

TWO-DIMENSIONAL POINT GROUPS AND SPACE GROUPS

In each of the two-dimensional Bravais lattices illustrated in Figure 1-18, the dot at each lattice point denotes the presence of a unit as-

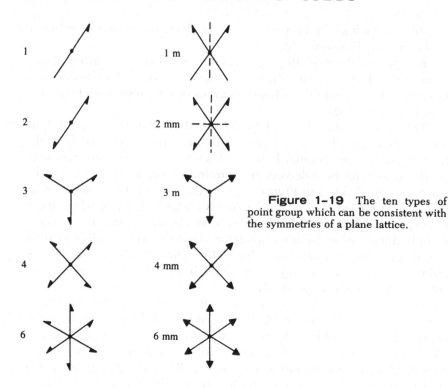

Figure 1-19 The ten types of point group which can be consistent with the symmetries of a plane lattice.

sembly or "basis" of atoms. Every basis must have the same arrangement and orientation, and the symmetry of the basis must be consistent with that of the lattice.

There are only ten possible ways in which the symmetry of the basis of atoms can be meaningfully related to the symmetry of a two-dimensional lattice. These ten *point groups* are illustrated in Figure 1-19. The numeral indicates how many positions within the basis are equivalent by rotation symmetry. A single m shows that the basis has a mirror axis of symmetry. It will be observed that when there is a two-fold rotation axis and a mirror axis, then another mirror axis *must* exist perpendicular to the first, hence the terminologies 2mm, 4mm, and 6mm.

Attempts to compare the symmetries of the ten point groups with

TABLE 1-5 SUMMARY OF TWO-DIMENSIONAL CRYSTAL SYSTEMS, LATTICES, AND PLANE GROUPS

Crystal System	Point Groups Compatible With System	Bravais Lattices Included in System	Number of Plane Groups Compatible With Lattice
Oblique	1, 2	P (primitive)	2
Rectangular	1m, 2mm	P (primitive)	5
		C (centered)	2
Square	4, 4mm	P (primitive)	3
Hexagonal	3, 3m, 6, 6mm	P (primitive)	5

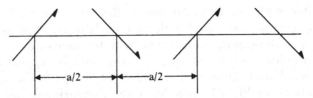

Figure 1–20 A glide operation transforms the lattice into itself by a reflection in the horizontal line through the midpoints of the arrows plus a translation of $a/2$. A second application of the glide returns the arrow to a position equivalent to its original position. Glide operation symmetries are important for three primitive rectangular structures and for one of the square structures in two dimensions.

those of the five Bravais lattices yield at first only 13 two-dimensional *plane groups* or *structures,* since each of the point groups is consistent with the symmetry of only one two-dimensional crystal system. Table 1-5 shows that the rectangular system is the only 2-D system to include more than one Bravais lattice (though we shall see that most 3-D systems include more than one lattice). Four additional plane groups can be justified by showing that each one is invariant under a glide operation (a reflection plus a translation of one-half of a principal lattice vector, as illustrated in Figure 1-20). Thus a total of 17 plane groups exists, a set illustrated by Nussbaum[20] and by Buerger[21]. Nussbaum also shows how the design of repetitive wallpaper is subjected to the same constraints as the imagination of two-dimensional crystal structures. The distribution of the 17 plane groups over the four systems is summarized in Table 1-5.

THREE-DIMENSIONAL CRYSTAL SYSTEMS, POINT GROUPS, AND SPACE GROUPS

In three dimensions, the most general lattice is the *triclinic,* for which the simplest primitive cell is a parallelepiped with all angles different from 90°, and with primitive translational vectors **a**, **b**, and **c**, all of different lengths. Consideration of special relationships between sides and angles yields 13 "special lattice types;" thus there is a total of 14 three-dimensional Bravais lattices. These are illustrated in Figure 1-21, some by their primitive cells and others by non-primitive cells which demonstrate the symmetry more clearly. Table 1-6 shows that the 14 lattices can be grouped into seven crystal systems. The conventional unit cells for these seven systems can be envisaged as arising from progressive distortions of cubic symmetry, as sketched in Figure 1-22.

A full discussion of the symmetries inherent in the seven crystal systems and in the 32 three-dimensional *point groups* would occupy a disproportionate amount of this book.[22] It is sufficient for our purposes to note that 32 point group symmetries can be described, each of which

---

[20] A. Nussbaum, *Semiconductor Device Physics* (Prentice-Hall, 1962), Chapter 5.

[21] M. J. Buerger, *Contemporary Crystallography* (McGraw-Hill, 1970), Chapter 1.

[22] See the bibliography at the end of Chapter 1 for extensive accounts (such as Megaw). Several aspects of this topic are treated in an interesting way in Chapter 5 of Nussbaum, *Semiconductor Device Physics* (Prentice-Hall, 1962).

is compatible with one and only one of the seven crystal systems. Of the 32, two can be associated with triclinic lattices, three with monoclinic, three with orthorhombic, seven with tetragonal, five with cubic, five with trigonal, and the remaining seven with hexagonal lattices. A straightforward application of these 32 point groups to the appropriate Bravais lattices yields 73 "simple" three-dimensional *space groups*. This is not the whole story, and compound operations justify the existence of an additional 157 *compound space groups*. Thus 230 three-dimensional space groups or "structures" exist in total. Despite this

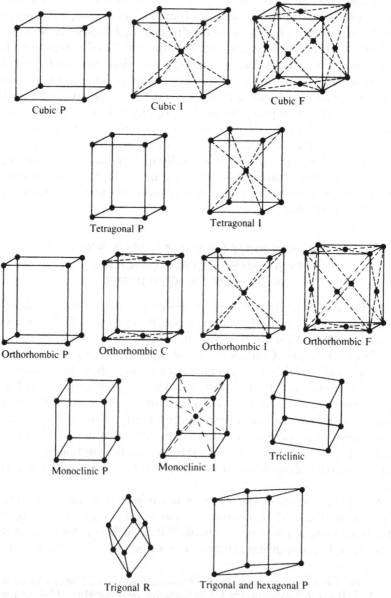

Figure 1-21 Conventional unit cells for the fourteen possible three-dimensional Bravais lattices. The cells illustrated are non-primitive for seven of the lattices. After C. Kittel, *Introduction to Solid State Physics* (Wiley, 1971).

TABLE 1-6 THE SEVEN CRYSTAL SYSTEMS
AND FOURTEEN BRAVAIS LATTICES
IN THREE DIMENSIONS

System	Restrictions on Conventional Unit Cell Dimensions and Angles	Bravais Lattices in the System
Triclinic	$a \neq b \neq c$ $\alpha \neq \beta \neq \gamma$	P (primitive)
Monoclinic	$a \neq b \neq c$ $\alpha = \gamma = 90° \neq \beta$	P (primitive) I (body-centered)
Orthorhombic	$a \neq b \neq c$ $\alpha = \beta = \gamma = 90°$	P (primitive) C (base-centered) I (body-centered) F (face-centered)
Tetragonal	$a = b \neq c$ $\alpha = \beta = \gamma = 90°$	P (primitive) I (body-centered)
Cubic	$a = b = c$ $\alpha = \beta = \gamma = 90°$	P (primitive or simple cubic) I (body-centered) F (face-centered)
Trigonal	$a = b = c$ $120° > \alpha = \beta = \gamma \neq 90°$	R (rhombohedral primitive)
Hexagonal	$a = b \neq c$ $\alpha = \beta = 90°, \gamma = 120°$	P (primitive rhombohedral)

proliferation, many of the important and widely studied solids share a few relatively simple structures.

The 157 space groups added through consideration of compound symmetry elements arise from *glide* and *screw* operations. The glide operation has already been described in connection with Figure 1-20 and the number of possible two-dimensional groups. A *screw* operation is a purely three-dimensional concept, consisting of a translation superimposed on a rotation (of 60°, 90°, 120° or 180°) in order to produce an invariance of the structure. One simple example of this operation is afforded by the tellurium crystal structure of Figure 1-6. At first glance this lattice appears to be hexagonal, but a closer look shows that the screw axis along each spiral chain violates the idea of hexagonal symmetry. A rotation of 120° superimposed upon a translation of one-third of the vector **c** does produce an invariance, and turns the crystal into itself. The same combination of translation plus 120° rotation is the required symmetry element for the crystalline synthetic polymer isotactic polystyrene, which has a helical chain of the form $(CH_2CH)_n$ with phenyl groups attached, three units per turn of the helix.[23]

---

[23] The reader may recall the previous mention of helix structures in connection with the DNA double helix and its hydrogen bonds. Does DNA have screw symmetry? The answer is: superficially yes, but in detail DNA violates the first sentence of the current section, which requires a crystal to be an infinite repetition of *identical* building blocks. Synthetic polynucleotides can be arranged to have an infinite succession of identical nitrogen bases, but in naturally occurring DNA there is a complex and aperiodic sequence of the four bases, carrying genetic information. Thus a "crystal" of DNA or of a comparable material of biological significance is apt to be a stacked assembly of similar but not identical objects. This slight inhomogeneity of a DNA crystal does not prevent it from diffracting an X-ray beam. (See footnote 17.)

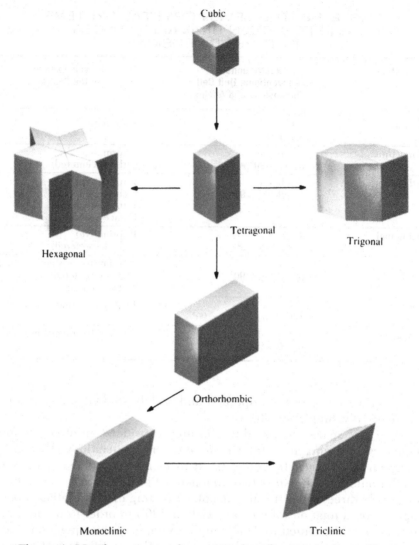

Figure 1-22 The seven crystal systems in three dimensions, obtained by successive distortions of a cube. After A. Nussbaum, *Semiconductor Device Physics* (Prentice-Hall, 1962).

It must be remembered that a "structure" or "space group" in the crystallographic sense is not a crystal; it is an array of mathematical points with a prescribed symmetry for the set of atoms which may be associated with each point. Two solids may crystallize in the same structure even though one has only two or three atoms in its basis and the other has several thousands of atoms per basis set. The gross features of X-ray diffraction will be the same for the two solids, though finer details of amplitude ratios for various diffracted lines will provide information about the atoms in the basis.

Actual Crystal Structures

1.3

MILLER INDICES AND CRYSTAL DIRECTIONS

We often find it necessary to describe a particular crystallographic plane or a particular direction within a real three-dimensional crystal. Crystal planes are usually described by their *Miller indices*, which are established as follows.

Consider a general three-dimensional lattice, for which the primitive translational vectors **a**, **b**, and **c** are not necessarily of equal length, and for which the angles between pairs of these primitive vectors are not necessarily right angles. Figure 1-23 shows a portion of this lattice. The locations O, A, B, and C in this figure are lattice points such that the vectors **OA, OB,** and **OC** are in the directions of the primitive vectors. The distances |**OA**|, |**OB**|, and |**OC**| are multiples n_1, n_2, and n_3 of the respective primitive unit vector lengths. How then can the plane ABC be most conveniently described in terms of the integers n_1, n_2, and n_3?

Description of ABC as being the (n_1|**a**|, n_2|**b**|, n_3|**c**|) plane would certainly be quite informative, but this rather clumsy notation is not used. Nor do we find ABC referred to in a more abbreviated form as the ($n_1 n_2 n_3$) plane. For reasons which will become much more apparent when the properties of reciprocal space and reciprocal lattices are explored, we shall find it useful to refer to ABC as the (hkl) plane, where hkl is a set of integers which expresses the ratio

$$h:k:l = n_1^{-1}:n_2^{-1}:n_3^{-1} \qquad (1\text{-}28)$$

A plane which satisfies Equation 1-28 is said to have *Miller indices* of

Figure 1-23 Construction for the description of the plane passing through the lattice points A, B, and C; **a**, **b**, and **c** are primitive vectors and n_1, n_2, and n_3 are arbitrary integers.

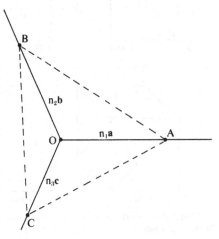

(hkl). We usually select h, k, and l to be the set of the smallest integers that satisfy Equation 1-28, but there can be an exception to this in order to provide additional information, as we shall see.

Operation of the Miller notation can be illustrated most simply for a cubic lattice (Figure 1-24), though the same principles can be applied to less symmetrical structures. A plane denoted (hk0) has an infinite intercept on the c-axis (the direction of **OC** in Figure 1-23); such a plane is evidently parallel to the c-axis. In the same fashion, a plane parallel to both the b-axis and c-axis is a (h00) plane. Following the precept of using the smallest suitable integers, any (h00) plane is called a (100) plane unless there are particular reasons for expressing additional information.

In the Miller notation, the designation ($\bar{h}$kl) uses a negative integer to show that the plane in question cuts the a-axis on the negative side of the origin. Figure 1-24 illustrates both the true (100) plane and the ($\bar{1}$00) plane, which is related to it by translational symmetry. The symbol {hkl} is used to denote all planes equivalent to (hkl) by symmetry.

Figure 1-24 shows a plane marked (200), the use of an integer other than unity for the first numeral indicating that the plane cuts the a-axis at ½a from the origin. The relation of the (222) plane to the (111) plane in Figure 1-24 is similar.

For hexagonal and trigonal crystals, the description of crystal planes by a set of three Miller indices is still entirely proper. However, for these crystal classes it is more common to use a set of *four* Miller indices, expressed in terms of the reciprocal intercepts with respect to the vectors a_1, a_2, a_3, and c shown in Figure 1-25. Thus a plane may be designated as (10$\bar{1}$0), or (0001), where the fourth index is concerned

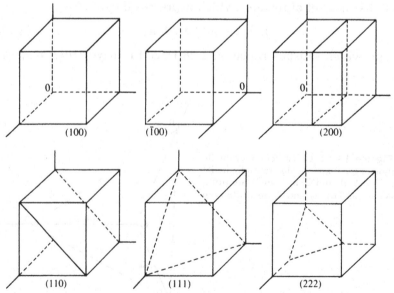

Figure 1-24 Some of the more prominent planes for cubic lattices, with their Miller notations.

Figure 1-25 Construction of the conventional unit cell for a hexagonal or trigonal crystal, showing the four vectors (of which any two of the first three are independent) used in the formulation of a four-number Miller index notation for this type of lattice.

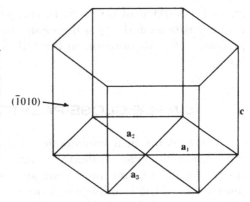

with the intercept on the c-axis. The relation between the coplanar vectors a_1, a_2, and a_3 forces a relationship between the first three Miller indices; they must add up to zero.

The identification of ($\bar{1}010$) in Figure 1-25 shows the manner in which the designation of a plane is related to the directions chosen for the vectors a_1, a_2, and a_3. A further exercise in this notation is provided by Problem 1.11.

The Miller index notation requires a little study at first, depending as it does on the measurement of the intercepts of a plane in several units of lengths for crystals of low symmetry. Once the idea has been accepted, the system is of enormous value. Without it, the description of diagonal planes in orthorhombic, monoclinic, or triclinic lattices would be much more difficult.

Curiously enough, a different approach has traditionally been taken in the description of *crystal directions*. A direction in a crystal is denoted by a set of three integers between square brackets, [hkl]. This looks so similar to the Miller notation for a crystal plane that it is fatally easy to suppose that [hkl] would be the direction of the normal to (hkl). Unfortunately, directions are defined in such a way that this supposition is correct only for cubic crystals and for a few other isolated cases.

A crystal direction is expressed as the set [hkl] of the smallest integers which are in the ratios of the cosines of angles between the direction and the unit cell axes. That is to say, h, k, and l are proportional to the magnitudes of vectors along the unit cell axes which would combine to make a vector along the specified direction.

LOCATION OF A POSITION IN THE UNIT CELL

The location of a point in the unit cell is specified by three numbers u, v, and w, for which each coordinate describes the distance from an origin in units of the cell dimension(s). Thus if one corner of the unit cell is denoted as the position 000, the opposite corner is 111, and fractional numbers are required for any intermediate point within or on the surface of the cell. In this notation $\frac{1}{2}\frac{1}{2}0$ is one of the face-center positions and $\frac{1}{2}\frac{1}{2}\frac{1}{2}$ is the center of the cell. We shall have ample practice with this notation in the following discussion of actual structures.

SIMPLE CLOSE-PACKED STRUCTURES

If the atoms in a monatomic solid are imagined to be relatively incompressible spheres, then we might well expect to see them fitting snugly together in the most compact arrangement for spheres, unless some quirk of the bonding mechanism should dictate a more spacious arrangement of different symmetry. Any casual observer of pool or snooker knows that each sphere has six touching neighbors in a close-packed plane array of spheres (Figure 1-26). Such plane arrays with hexagonal symmetry can be stacked together to make the most compact monatomic solid in two simple ways. In both of these arrangements, each atom has 12 nearest neighbors — six in its own plane, three in the plane above, and three in the plane below.

The two stacking sequences are illustrated in Figures 1-27 and 1-28. For either sequence, the second layer, B, must be placed on top of the first layer, A, with a translation which places each atom of B over one of the holes in the first layer. Layer B will obviously fit snugly into place. It is just as obvious that a third layer, identical to A, can be placed to fit on top of B. The pattern can thus repeat, ABABAB and so on, thus producing the hexagonal close-packed (H.C.P.) structure.

Figure 1-29 shows the unit cell of the H.C.P. structure with the atoms shrunken in order to display the geometric arrangement. The space lattice is *simple hexagonal* with a basis of two atoms, one at 000 and the other at $\frac{2}{3}:\frac{1}{3}:\frac{1}{2}$. Elementary geometry tells us that the ratio of the height of the cell to the nearest neighbor spacing should be

$$\frac{c}{a} = \left(\frac{8}{3}\right)^{1/2} = 1.633 \tag{1-29}$$

Figure 1–26 Spheres arranged in a close-packed layer.

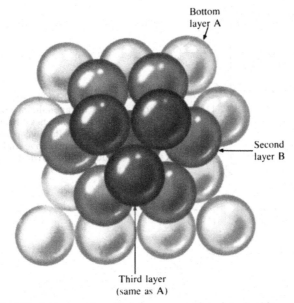

Bottom
layer A

Second
layer B

Third layer
(same as A)

Figure 1–27 Sequence in which layers are placed in building up a hexagonal close-packed structure.

for close packing of the atoms. In practice the (c/a) ratio is larger than 1.633 for most hexagonal crystals, showing that the hexagonal arrays are not in the closest possible contact. The term H.C.P. is still applied, provided that the (c/a) ratio is not more than 10 per cent away from the ideal value.

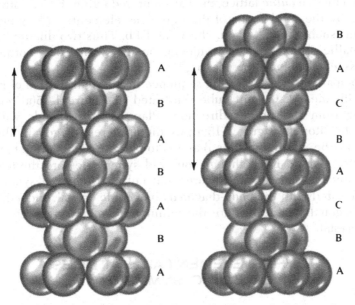

Figure 1–28 The manner in which close-packed layers of spheres can be stacked in sequence to produce (left) the hexagonal-close-packed (H.C.P.) structure and (right) the cubic-close-packed (F.C.C.) structure. After L. Pauling, *The Nature of the Chemical Bond* (Cornell, 1960).

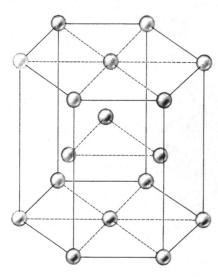

Figure 1-29 The conventional unit cell for the hexagonal-close-packed structure. Each atom has twelve nearest neighbors, six in its own plane and three each in the planes above and below. The conventional cell shown is equivalent to three primitive cells.

Examples of nominally H.C.P. crystals include the elements from Column II of the Periodic Table: Be, Mg, Zn, and Cd. H.C.P. is also the stable structure over at least part of the temperature range for several transition elements.

An alternative form of close packing is the one illustrated in the right hand portion of Figure 1-28. Here a third layer, C, is placed on top of B and is displaced so that its atoms fall on top of the spaces in layer B which were also spaces in layer A. A fourth layer identical to A can then be added and the ABCABCABC sequence continued. This procedure produces a *cubic* lattice, the face-centered-cubic (F.C.C.) structure favored by the solid forms of the noble gas elements and by metallic elements such as Cu, Ag, Au, Ni, Al, and Pb. Thus two different popular metallic structures, both enjoying many symmetries, correspond with two choices for close packing.

Comparison of the stacking sequence in the right portion of Figure 1-28 with the F.C.C. unit cube illustrated in Figure 1-15(a) requires viewing from the proper direction. Plane A corresponds with the bottom left atom in Figure 1-15(a); we then progress diagonally through the unit cube, through planes B and C to plane A again at the top right in Figure 1-15(a). Even though four-fold symmetries are the ones instinctively associated with squares and cubes, *three-fold* symmetries are associated with the body-diagonals in a cubic crystal. Indeed, these body-diagonal symmetries are the minimum symmetry aspects of any cubic crystal.

OTHER ELEMENTAL LATTICES
OF CUBIC SYMMETRY

It may at first seem surprising that F.C.C. should be the most compact type of cubic structure. Without making any calculations, one might easily assume that a body-centered-cubic (B.C.C.) structure would be at least as compact an arrangement of identical spheres as any other. That such an assumption is mistaken is asserted in Figure 1-30 and is provided as a problem in geometry (Problem 1.12). Of the filling factors quoted in Figure 1-30, we should remember that the 74 percent for F.C.C. is just the same as that for ideal H.C.P., since both are situations of maximum packing.

Each atom in the body-centered-cubic (B.C.C.) structure illustrated in Figure 1-15(b), (and Cubic I in Table 1-6 and Figure 1-21) has eight nearest neighbors. Each atom is in contact with its neighbors only along body-diagonal directions, which results in the less efficient packing. Many metals, including the alkalis and several transition elements, choose the B.C.C. structure.

Less efficient still in the packing of spheres is the *simple cubic* structure (referred to variously as S.C. or as Cubic P). Simple cubic itself is something of a rarity for simple solids. Of the chemical elements, only polonium chooses S.C., and then only as its stable form for a particular temperature range. The principal interest in S.C. lies in crystals with a multi-atom basis which have a simple cubic primitive unit cell [such as the CsCl structure, Figure 1-33(b), and the perovskite structure, Figure 1-36].

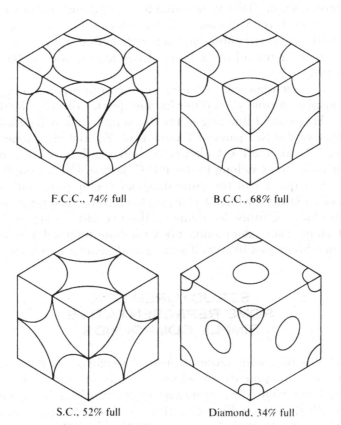

F.C.C., 74% full   B.C.C., 68% full

S.C., 52% full   Diamond, 34% full

Figure 1-30 Filling factors for identical spheres in contact, in four common cubic structures. After C. Wert and R. Thomson, *Physics of Solids* (McGraw-Hill, 2nd edition, 1970).

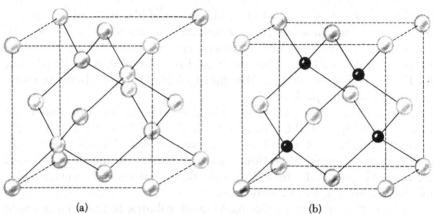

(a)   (b)

Figure 1–31   (a) The atomic arrangement in diamond, compared with (b) that for zincblende (sphalerite). In both structures, the translational symmetry is face-centered-cubic with a basis of two atoms, one at 000 and the other at $\frac{1}{4}\frac{1}{4}\frac{1}{4}$. Thus each atom has only four nearest neighbors, placed at tetrahedral angles. In zincblende, the four neighbors are all of the opposite chemical species.

Close packing is not the primary consideration in the diamond lattice [Figure 1-31(a)]. This structure is adopted by solids for which the existence of four symmetrically placed valence bonds overrides any other consideration. This is the situation in silicon, germanium, and gray tin, as well as in diamond. Diamond has the translational symmetry of face-centered-cubic, and for each lattice point of the F.C.C. Bravais lattice, diamond has a basis of two atoms, one at 000 and the other at $\frac{1}{4}\frac{1}{4}\frac{1}{4}$.

We can view diamond as the result of two interpenetrating F.C.C. lattices displaced from each other by one quarter of the cube diagonal distance. This way of thinking of the diamond lattice is facilitated by a comparison of the two parts of Figure 1-31. For the zincblende lattice of Figure 1-31(b), one F.C.C. lattice is shown with light colored atoms. The four dark atoms belong to the F.C.C. lattice which is shifted from the first by a quarter of the cube diagonal vector. We shall soon encounter other lattices which it is useful to consider as interpenetrations of simpler lattices. Since for diamond the two lattices are of the same type of atom, each atom ends up with four tetrahedrally arranged nearest neighbors just like itself and 12 next-nearest neighbors.

STRUCTURES FOR
SOME REPRESENTATIVE
SIMPLE COMPOUNDS

A tetrahedral configuration of nearest neighbors is found in many partially ionic compounds for which the ratio of anion size to cation size is not too large. The prototype of the zincblende or sphalerite structure illustrated in Figure 1-31(b) is the II-VI compound ZnS, though the same structure is found in III-V and I-VII compounds also. Sphalerite differs from diamond only in that the two F.C.C. sub-lattices are of different atomic species. Thus each zinc atom has four sulphur nearest neighbors, and vice versa. The 12 next-nearest neighbors of a zinc atom are all zinc, since the second-nearest neighbors have the 12-fold coordination of F.C.C.

The diamond structure itself already lacks many of the symmetrical features enjoyed by primitive cubic, and additional aspects of symmetry are lost in sphalerite. Since there is no center of inversion in the sphalerite structure, a (111) plane and a $(\overline{111})$ plane are not the same thing. In ZnS, (111) consists entirely of zinc atoms, while $(\overline{111})$ is a sheet of sulphur atoms.

A number of solids which crystallize as sphalerite have an alternative allotropic form in the hexagonal structure known as wurtzite. Note from Figure 1-32 that the arrangement of *nearest* neighbors in wurtzite is the same as for sphalerite; it is the disposition of the second-nearest neighbors which produces a hexagonal crystal whose cohesive energy will be only slightly different from that of the corresponding cubic crystal.

Structures which are intermediate between sphalerite (cubic) and wurtzite (hexagonal) are found in the various *polytypes* of silicon carbide.[24] These polytypes result from a variety of complicated but repeating sequences in the addition of planes of atoms perpendicular to the c-axis. Thus the well-studied 15R polytype[25] is consistent with rhombohedral symmetry, as a result of an ABCACBCABACABCB sequence of 15 planes as the repeating unit. This particular polytype has 20 atoms in the primitive basis, and five non-equivalent sites for carbon or silicon atoms in the unit cell. The electronic behavior of a SiC polytype depends on the amount of cubic versus hexagonal character in its structure.

For strongly ionic compounds with equal numbers of anions and cations, the two common structures are the NaCl and CsCl varieties illustrated in Figure 1-33. As explained earlier in connection with Figure 1-10, the energetically most favorable structure is determined

---

[24] P. Krishna and A. R. Verma, Proc. Roy. Soc. London **A272**, 490 (1963). See also *Silicon Carbide–1968*, edited by H. K. Henisch and R. Roy (Pergamon Press, 1969).

[25] L. Patrick et al., Phys. Rev. **132**, 2023 (1963).

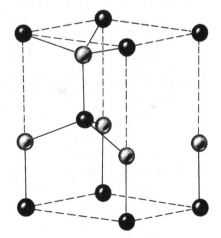

Figure 1–32 A hexagonal unit cell for the wurtzite structure.

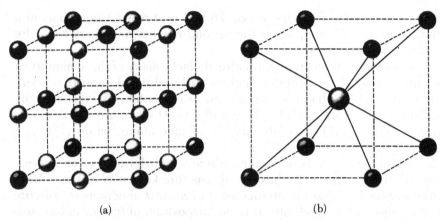

(a)   (b)

Figure 1–33 Two crystal structures commonly adopted by ionic compounds with a 1:1 ratio of anions to cations. (a) The sodium chloride (rocksalt) structure. (b) The cesium chloride structure.

by the anion-cation radius ratio, and must be a structure which obviates any risk of anion-anion contacts.

Just as cubic ZnS consists of a zinc F.C.C. lattice and a sulphur F.C.C. lattice interpenetrating so that they are displaced from each other by *one-fourth* of a cube diagonal, so cubic NaCl consists of a sodium F.C.C. lattice and a chlorine F.C.C. lattice displaced from each other by *one-half* of a cube diagonal. To put it another way, NaCl has F.C.C. translational symmetry with a basis of Na at 000 and Cl at $\frac{1}{2}\frac{1}{2}\frac{1}{2}$. In addition to many of the alkali halides and hydrides, the rocksalt structure is the favored one for silver halides, the lead chalcogenides (such as PbS, PbSe, and PbTe, etc.), and numerous II-VI compounds.

The cesium chloride structure of Figure 1-33(b) looks like body-centered-cubic but is actually simple cubic in its space lattice, with a diatomic basis of Cs at 000 and Cl at $\frac{1}{2}\frac{1}{2}\frac{1}{2}$. This structure is also referred to as the β-brass one, since in addition to some highly ionic solids it is also favored in ranges of the CuZn phase diagram and in several other metallic alloy systems.

Cubic crystals often result even for compounds in which the ratios of the chemical species are not 1:1, and Figures 1-34 through 1-36

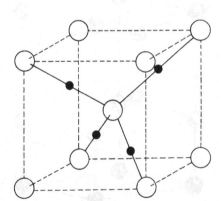

Figure 1–34 The curious cubic structure of cuprite, Cu_2O.

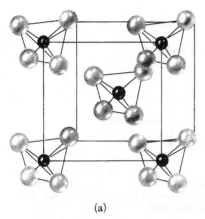 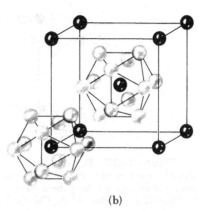

(a)   (b)

Figure 1-35 Examples of compounds which crystallize in the B.C.C. space lattice. (a) The silicon tetrafluoride structure, a fine example of the cubic requirement for tetrahedral symmetry along body-diagonal directions. (b) The structure of $MoAl_{12}$ and WAl_{12} for which each group of twelve aluminum atoms forms a regular icosahedron. After L. Pauling, *The Nature of the Chemical Bond* (Cornell, 1960).

illustrate some examples of this. The cuprite structure of Figure 1-34 is a strange one, with oxygen atoms in a B.C.C. arrangement and copper atoms bisecting just one-half of the lines that connect the body center with the corners of the cube. Problem 1.13 is an exercise about this structure.

We can see in Figure 1-34 that each oxygen atom has a tetrahedral coordination of its copper neighbors, so that there is three-fold rotational symmetry about cube diagonal directions. As we noted earlier, this constitutes the minimum set of symmetry requirements for a cubic crystal. The same requirement is obviously satisfied for the SiF_4 and $MoAl_{12}$ crystals illustrated in Figure 1-35, and each of these materials falls into the B.C.C. category.

From the preceding examples we can see that whether or not a material has a cubic structure depends much more on interbond directions within a molecule than on the simplicity or complexity of the

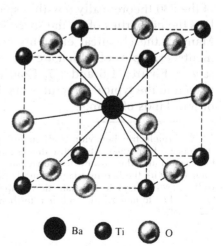

Figure 1-36 The perovskite structure as exemplified by $BaTiO_3$. The structure is cubic (as illustrated) at high temperatures, but becomes slightly tetragonal on cooling below the ferroelectric transition temperature.

Ba   Ti   O

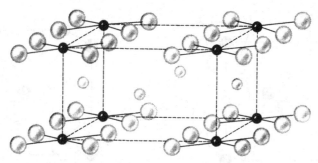

Figure 1–37 The unit cell for crystalline K_2PtCl_4. Each platinum atom (small shaded circles) is surrounded by a square of chlorine atoms (large circles). The circles of intermediate size are the potassium atoms. The space lattice is primitive tetragonal. After L. Pauling, *The Nature of the Chemical Bond* (Cornell, 1960).

chemical formula. The perovskite structure of Figure 1-36 affords another example of a material with a relatively complicated molecule which *can* be arranged in a manner consistent with cubic symmetry. For perovskite itself (calcium titanate), cubic symmetry is found at all temperatures; but for compounds such as $BaTiO_3$, $PbTiO_3$, and $KNbO_3$, there is a slight adjustment to a tetragonal form of cooling below the ferroelectric transition temperature.[26]

As a final example, the structure of K_2PtCl_4 shown in Figure 1-37 illustrates the point that for such a molecular shape, the crystal habit must be tetragonal rather than cubic. The important symmetries in this compound are those of four-fold rotations about axes parallel to the c-axis.

Crystals with more complicated and less symmetrical crystal structures have been analyzed in large numbers by X-ray diffraction and other techniques, and the bibliography at the end of the chapter provides an opportunity for the interested reader to find examples of most of the 230 theoretically possible space groups. (The book edited by Zak lists the properties of all the space groups in group theory terminology.) Many of the chemical elements crystallize in quite complex arrangements, examples of which are the graphite and tellurium structures we saw in Figures 1-6 and 1-7. Considerably *more* order, indeed, is displayed in the typical structures of compounds which involve two, three, or four kinds of atom.

---

[26] Ferroelectric behavior is discussed in Chapter 5, including some of the consequences of a permanent electric dipole moment in the lattice. The shift to a tetragonal form constitutes the upward (or downward) movement of the entire oxygen sub-lattice compared with the barium-titanium sub-lattice. Actually, when barium titanate is cooled well below the ferroelectric transition temperature, it undergoes progressive further distortions, first to monoclinic and then to rhombohedral form [W. J. Merz, Phys. Rev. **76**, 1221 (1949)].

Crystal Diffraction 1.4

DIFFRACTION AS AN INVESTIGATIVE PROCEDURE

Useful information about the structure of some crystalline solids can be deduced from macroscopic or microscopic observation of the external morphology of a sample. Moreover, techniques based on the reflection of visible light from surface defects are valuable for crystallographic *orientation* of single crystals of materials for which the structure is known. However, visible or ultraviolet light is hopelessly lacking in the resolution necessary to determine the locations and spacings of atoms and molecules. These spacings are typically of the order of 1 Å (10^{-10} m).

Microscopy in the conventional sense implies that the wavelength of the radiation used is smaller than the distances to be resolved, so that the laws of geometric optics can apply. For crystal structure determinations we far more typically work by *diffraction* techniques, using radiation of a wavelength comparable with atomic dimensions. Diffraction gives us information in Fourier form, which can be analyzed in terms of the average spacings of lines and planes of atoms in a solid, the angles between lines and planes, the symmetries of point groups, and (with suitable interpretive skill) the location of particular species of atoms. As noted in the introduction to this chapter, it is advisable to read both Sections 1.4 and 1.5 as a preliminary to a detailed study of the topics of diffraction and reciprocal space. The two topics are closely entwined.

The suggestion was made by von Laue (1912) that a crystal might be regarded as a three-dimensional diffraction grating for energetic electromagnetic waves (X-rays) of a wavelength comparable with the atomic spacing, and that a diffraction pattern should provide information about the regular arrangements of atoms. X-rays are still the principal source of new information about the crystallography of solids and are supplemented by electron and neutron diffraction. Figure 1-38 shows typical single crystal X-ray Laue spot patterns.

The energies of the three types of wave-particle used for crystallographic work must be quite different if they are to have wavelengths suitable for their respective tasks. For photons in the X-ray range,

$$\lambda = \frac{c}{\nu} = \frac{hc}{E} \qquad (1\text{-}30)$$

and for a wavelength of 1 Å the required energy is E ∼ 12,000 eV. With 　　**51**

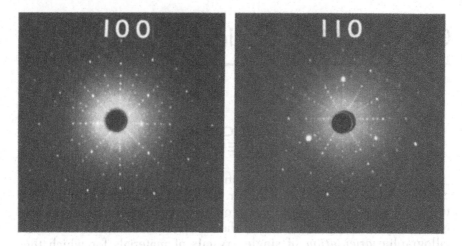

Figure 1-38 Laue X-ray diffraction spot patterns for (100) and (110) orientations with respect to pyrite (FeS$_2$) single crystals. (Photos courtesy of L. E. Murr, Oregon Graduate Center.)

electrons, less energy is necessary for a comparable de Broglie wavelength,

$$\lambda = \frac{h}{mv} = \frac{h}{(2mE)^{1/2}} \qquad (1\text{-}31)$$

Thus the energy for a one Ångstrom wavelength is only E $\sim$ 150 eV. An electron of this energy and wavelength moves at a speed v $\sim 7 \times 10^6$ m/s.

A still smaller amount of energy and speed is required for diffraction of neutrons by the ordered atomic array of a crystal,

$$\lambda = \frac{h}{M_n v} = \frac{h}{(2M_n E)^{1/2}} \qquad (1\text{-}32)$$

Thus neutrons of 1 Å wavelength move at a speed of only 4000 m/s, and have a kinetic energy of only 0.08 eV. This energy is comparable with that of a quantized unit of lattice vibrational energy, which has a lot to do with the interactions of thermal neutrons with crystals, as we shall discuss later.

USES OF THE THREE TYPES OF RADIATION

The three above-mentioned types of radiation have different, though overlapping, uses in structural studies of solids. All three types of wave-particle interact with a periodic lattice subject to the same general geometric laws, conditions which we shall consider in the next subsection.

X-rays with photon energies in the range 10 keV to 100 keV can penetrate well below the surface of a crystal, and form the basis for most of the conventional techniques for analysis of unknown three-dimensional structures. Since X-rays are scattered primarily by the extranuclear electrons, they are not too successful in determining the locations of very light atoms such as hydrogen, but do work well for the heavier elements, unaffected by the presence of several isotopes of the same element. In X-ray crystallography we are interested in the scattering performed by each electron moving in phase with the incoming electromagnetic wave, radiating *coherently* without any change of wavelength.[27] The constructive or destructive interference of the coherent scattered radiation is controlled by the geometry at the interatomic level.

The diffraction of electrons by a single crystal (Davisson and Germer, 1927) or by a crystalline film (Thompson, 1927) is an impressive demonstration both of the periodic structure of solids and of the wave-particle duality of the electron (Figure 1-39). Because an electron

---

[27] The scattering of an electromagnetic wave by an electron with no change of frequency (i.e., no energy change) was first described by J. J. Thomson, and is sometimes referred to as *Thomson* scattering. *Inelastic* scattering occurs also by the mechanism of the Compton effect, and this competes with the desired coherent scattering. Inelastic scattering provides an incoherent background of photons at all angles which have been randomly shifted to lower energies. Separation of this background from the coherently scattered photons of the original energy at the Bragg angles is an instrumental problem in structural analysis.

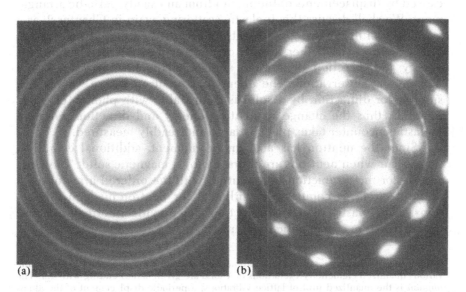

Figure 1–39 Electron diffraction patterns of thin films deposited on cleaved (001) NaCl single crystal surfaces. (a) Fine-grained polycrystalline F.C.C. Pd, rings but no preferential orientation. (b) B.C.C. vanadium, deposited on a heated substrate and largely (001) epitaxial monocrystal, as evidenced by single crystal spots as well as rings from the randomly oriented polycrystalline areas. (Photos courtesy of L. E. Murr, Oregon Graduate Center.)

is a charged particle, it interacts very strongly with matter and can penetrate only a few hundreds of Ångstrom units into a crystal before it must suffer either an elastic or an inelastic encounter. Thus electron diffraction is ill-fitted to compete with X-ray methods for study of bulk structural arrangements. It is, however, well suited for two tasks:

1. The study of surface layers and surface states on crystals. The nature of the transition from a perfectly periodic lattice to empty space outside the crystal takes many forms, depending on how the surface was prepared and what opportunities existed for contamination.

2. The study of thin films. One might well regard a sufficiently thin film as two surface layers with none of the bulk solid in between, and electron diffraction has been very useful in demonstrating some structural characteristics of matter in thin slices which deviate from the characteristics of the "ideal" single crystal.

Slow neutrons for which the de Broglie wavelength is comparable with interatomic spacings can interact with a solid in several ways. In a nonmagnetic solid, the interactions are with the nuclei only (since the neutron has no charge and has a mass much larger than extranuclear electrons). Coherent elastic scattering can take place, subject to the same geometric laws as for X-rays or electrons, and this produces a diffraction pattern which differs from an X-ray one only in its improved efficiency for light atoms.

In addition to the coherent scattering, neutrons will tend to be scattered elastically but incoherently by the presence of more than one isotope of an element, randomly distributed over the atomic sites. A more interesting process than this is the *inelastic* scattering of neutrons caused by displacements of the nuclei from an exactly periodic arrangement. We shall discuss this inelastic scattering again in Chapter 2, as a valuable tool in exploring the spectrum of lattice vibrations in a crystal. For a neutron can either absorb or emit a phonon[28] in changing its energy and direction of motion, subject to the laws of energy and momentum conservation. The energy of a neutron with a wavelength suitable for diffraction by a solid is comparable with phonon energies in solids, thus the change in neutron energy and direction when an inelastic encounter takes place is large and readily measured.

Since the neutron has a magnetic moment, additional complications occur when neutrons are diffracted by a magnetic solid. The cross section for the interaction of a neutron with the ordered array of magnetic electrons is comparable with that for neutron-nucleon interactions. Thus elastic neutron diffraction can provide information about the distribution of magnetic moments.

---

[28] Just as the particle aspect of electromagnetic radiation is called the *photon*, so the various quantized elementary excitations of a solid are given names ending with *-on*. The *phonon* is the quantized unit of lattice vibrations, a periodic displacement of the atoms from the equilibrium positions which can be characterized by its energy and momentum, as well as by a velocity, wavelength, and wave vector. The *magnon* is a magnetic spin wave, an excitation of the spins in a ferromagnetic solid which are all exactly parallel in the ground state. The *plasmon* is a collective excitation of the electron gas, a coherent oscillation of all of the electrons, which arises from the long-range character of Coulomb forces.

THE BRAGG DIFFRACTION LAW

We now consider the geometric laws which describe the *minimum* conditions for efficient diffraction by a crystal of a collimated and monochromatic beam of radiation. The condition to be discussed applies whether the radiation is of X-ray photons, electrons, or neutrons (or indeed any other species of wave-particle).

W. L. Bragg in 1913 proposed a simple formulation and expression for the geometric condition which must be satisfied if waves are to be diffracted by a parallel set of planes (of atoms). Bragg's argument was far from rigorous, since it used the laws of geometric optics for reflection from a single plane of atoms, but the laws of physical optics (Fraunhofer diffraction) for the emerging wave to experience constructive interference. Thus it is interesting that Bragg's result (Equation 1-33) is identical to that of more subtle derivations by von Laue (Equation 1-55) and by Ewald (Equation 1-52). The Ewald result uses the reciprocal lattice arguments of Section 1.5.

One important difference between diffraction of light by a plane ruled diffraction grating and X-ray diffraction by a crystal should be mentioned right away. In the former case, the angle of incidence is not equal to the angle at which the diffracted beam emerges, and there is a relationship between these two angles, the wavelength λ, and the distance between successive rulings *in the grating plane*. The Bragg diffraction condition specifies that the angles of incidence and reflection are equal (i.e., that reflection is specular) and asserts that specular reflection is efficient only when the angle of incidence is appropriate for the wavelength and for the distance between one plane and the next one parallel to it. Bragg's Law is not concerned with the arrangement of atoms within a single reflecting plane.

Consider then Figure 1-40, which supposes that monochromatic X-rays are directed towards a crystal in a parallel beam which interacts

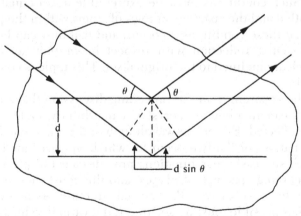

Figure 1-40 Specular reflection of a parallel beam of X-rays (or neutrons, etc.) from two successive planes of atoms in a crystal. The first plane need not coincide with the boundary of the crystal or be parallel to the boundary; our only concern is that it is a plane with a reasonably large density of atoms. The distribution of the atoms over the plane is not of concern. For constructive reinforcement, the path difference must be a multiple of the wavelength.

with all atoms in the region of the crystal to which it can penetrate. We wish to consider the interaction with a series of parallel planes of atoms which make an angle θ with the beam. Interaction with the first of these planes will produce a reflected component at the specular reflection angle θ, which will be rather feeble if the X-rays are deeply penetrating. Since this first plane has a periodic structure, it obviously acts as a two-dimensional grating for the X-rays, and feeble components will be reflected from the first plane at various angles ϕ_m corresponding with various orders of diffraction.

For the second and all subsequent planes, there will similarly be components of reflected energy at the specular angle and at angles corresponding with various diffracted orders. Since X-rays can penetrate deeply to elicit separate feeble responses from thousands of successive planes, it is apparent that destructive interference will prevent the appearance of a well-defined reflected beam for most incident angles between the X-ray beam and the set of planes being considered.

We can see from Figure 1-40, however, that the additional path traversed is 2d sin θ for a ray which suffers specular reflection from the second plane rather than the first. All specularly reflected components will be able to combine constructively in phase *if* this distance is a multiple of the wavelength. Thus the condition for *efficient* specular reflection (Bragg's Law) is

$$2d \sin\theta = n\lambda \qquad (1\text{-}33)$$

where n is an integer. As previously observed, this depends on the spacing of the planes of atoms but is independent of the atomic arrangement within each plane.

At this point one may well ask whether the conditions for constructive interference can occur at an angle of emergence which corresponds with one of the diffraction orders of the first plane, at one of the set of angles ϕ_m. Such conditions must be expressible as an equation which involves both d and the spacings of rows of atoms within the plane. The answer is that these conditions do occur, and that they can be regarded as Bragg specular reflection with respect to *another* set of parallel planes which are inclined to the original set. This topic is considered in Problem 1.14.

It will be apparent from the preceding discussion that a monochromatic X-ray beam incident on a crystal at an arbitrary angle will in general not be reflected. Either the wavelength or the angle must be varied until combinations of the two occur for which Bragg's Law is satisfied. A set of such successful combinations provides a set of pieces of information concerning spacings of planes, and this must then be analyzed to demonstrate consistency with a crystal structure. As many pieces of information as possible must be secured if the various values of d in the Bragg Law are to be assigned unambiguously to appropriate planes of Miller indices (hkl) with respect to a unique unit cell.

For a *cubic* lattice with unit cell dimension "a" in each direction, it is not difficult to show (as requested in Problem 1.15) that the spacing between successive (hkl) planes is

$$d = a[h^2 + k^2 + l^2]^{-1/2} \qquad (1\text{-}34)$$

A compact expression for a result equivalent to Equation 1-34 in a more general three-dimensional lattice requires the notation of the reciprocal lattice, a topic to which we shall turn in Section 1.5.

As noted earlier, Bragg's Law is a necessary but not a sufficient condition for efficient specular reflection. Whether or not a reflection is prominent for a particular (hkl) depends both on whether the reflection is allowed or forbidden from the standpoint of the *structure factor* and on the *atomic scattering factor* (or *form factor*) for the atoms composing the planes.

ATOMIC SCATTERING FACTOR

The diffraction of X-rays results from the interaction (Thomson scattering) of X-ray photons of wavelength λ with the charge field of all the extranuclear electrons in a solid. We now know that constructive interference of the scattered wavelets will occur only if the Bragg condition is satisfied. The intensity of the diffracted beam will be large only if the geometric structure factor is non-zero, and if the atomic scattering factor is near to its maximum value for the atoms of the solid.

The classical Thomson theory for coherent scattering of an electromagnetic wave by a free electron predicts a modest decrease of scattering efficiency as the angle 2θ between incoming and outgoing wavefronts increases. However, the efficiency of coherent X-ray scattering by the electrons of a solid falls off much more rapidly as 2θ increases. This occurs because the bound electrons of each atom are distributed over a volume with diameter comparable to the X-ray wavelength.

An atom of atomic number Z acts as a Thomson scattering point source of charge Ze for X-rays of long wavelength and/or small scattering angle. For smaller wavelength and/or larger angle, the atom scatters as a Thomson source of effective charge fe. The number f is the *atomic scattering factor* or *form factor* for the atom, and this number decreases from Z as the quantity $[(\sin\theta)/\lambda]$ increases.

Let $\psi_n(r)$ denote the radial dependence of the normalized wavefunction for the n-th electron of an atom with atomic number Z. Then the form factor for scattering X-rays of wavelength λ through angle 2θ is[29]

$$f = \sum_{n=1}^{Z} \int_0^\infty \frac{4\pi r^2 \psi_n^2(r) \cdot \sin[4\pi r(\sin\theta)/\lambda]dr}{[4\pi r(\sin\theta)/\lambda]} \qquad (1\text{-}35)$$

Each term in the summation becomes unity as $(\sin\theta)/\lambda$ goes to zero, to satisfy the requirement that $f = Z$ for this small angle limit.

[29] See, for example, Chapter 1 of *X-Ray Diffraction* by B. E. Warren (Addison-Wesley, 1969).

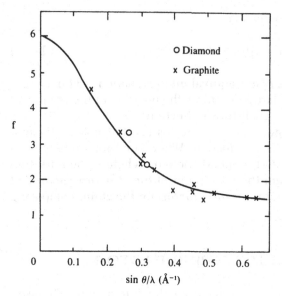

Figure 1-41 Experimental data points for the X-ray atomic scattering factor of carbon, compared with the curve generated by R. Mc-Weeny [Acta Cryst. **4**, 513 (1951)] for a self-consistent electronic distribution. Experimental data from G. E. Bacon [Acta Cryst. **5**, 492 (1952)]. The various parts of the abcissa are traced out by varying the Bragg angle, keeping λ constant. The points correspond with various low order (hkl).

The atomic scattering factor can be calculated for an isolated atom on the basis of the charge density distribution deduced from a Hartree-Fock self-consistent field distribution. Even though interatomic bonding makes some changes in the charge distributions for the outermost electrons, the scattering factors of atoms in a solid are usually rather close to the Hartree-Fock values. Figure 1-41 compares scattering factors from X-ray evidence with the curve expected from a self-consistent distribution. Note that the curve can be scanned by changing the Bragg angle rather than the X-ray wavelength. It will be seen that a carbon atom, with $Z = 6$, scatters coherently like six electronic charges at a point for long wavelengths and small angles. Since the mass of the charged scattering object appears in the denominator of the expression for Thomson scattering, the coherent X-ray scattering by the nucleus is negligible.

STRUCTURE FACTOR

In the description of the Bragg condition, it was emphasized that Equation 1-33 is only the *minimum* condition for efficient coherent reflection of X-rays by a set of parallel (hkl) planes. How much intensity (if any) is actually reflected, when the angles of incidence and reflection are equal and the Bragg condition is satisfied for some integer n, depends on whether the *structure factor* F_{hkl} is large, small, or zero for the (hkl) planes. This structure factor takes into account the form factors of the various types of atoms present in the crystal, and the places at which such atoms are situated in the unit cell. Since most solids have a polyatomic basis, it is quite possible for constructive Bragg reflection by one set of atoms to be out of phase (for certain hkl) with the waves coherently scattered from a set of atoms composed of other members of each basis.

The equation describing the structure factor for a polyatomic basis (and, indeed, for the Bragg condition itself) can be expressed to some advantage in reciprocal space language — which is not introduced until the next section. However, the principle by which a structure factor F_{hkl} is established can be seen without reciprocal space terminology. Consider a solid with N atoms in the basis, of which f_j is the form factor for the j-th atom. Imagine for a moment that all N atoms could be required to lie precisely *on* the hkl planes; none above and none below these planes, which are separated by distance d_{hkl}. Then first-order Bragg reflection of X-rays with wavelength $\lambda = 2d_{hkl}\sin\theta_{hkl}$ would have maximum efficiency. Every plane would return a reflected wave lagging in phase by exactly 2π radians compared with that of the plane above. The reflected amplitude would be proportional to the summation of all f_j over the N atoms of each basis.

Since, in practice, some atoms of the basis will lie above the respective hkl planes, then the scattered wavefronts from individual atoms will *not* be all in phase. The j-th atom of the basis, at a height Δd_j above the plane, will return a wave advanced in phase by an angle (in radians)

$$\phi_j = (2\pi\Delta d_j/d_{hkl}) \qquad (1\text{-}36)$$

compared with the wave from the level of the (hkl) plane itself. The resulting amplitude of the Bragg-reflected wave is the vector sum of the wavelets scattered from each atom per basis, and thus is proportional to the *structure factor*

$$F_{hkl} = \sum_{j=1}^{N} f_j \exp(i\phi_j) \qquad (1\text{-}37)$$

As a simple example of how two atoms in a basis can (partially or completely) oppose each other in Bragg reflection, consider the CsCl lattice of Figure 1-33(b), and the conditions under which the (100) reflection occurs. It will be seen that the cesium ions alone form a simple cubic lattice, and it is obviously possible to direct X-rays of such a wavelength and angle that components reflected from (100) planes will experience a path difference of λ (phase difference of 2π radians) for each additional depth of plane. All components of this set of reflections will add in phase. The chlorine ions in CsCl, however, form another simple cubic lattice, for which the (100) planes bisect the spaces between (100) planes of the cesium sub-lattice. Reflections from chlorine planes will be π radians out of phase with respect to the components reflected by cesium planes when the combination of λ and θ is appropriate for the (100) Bragg reflection. Then destructive interference will give a Bragg (100) reflected intensity which remains finite only because cesium and chlorine have different atomic scattering factors. Destructive interference causes a reduced intensity for X-ray diffraction of several other (hkl) also. Similar structure factor problems reduce the intensities of diffracted components in many compounds.

Looking again at the CsCl structure, we can see that if the atom in the center of the unit cell had the same atomic scattering factor as those forming the corners, the Bragg reflection for (100) and some higher orders would disappear entirely. The CsCl structure with the atom at the cube center made the same as those at the corners is just the body-centered-cubic structure of Figure 1-15(b). The rule for B.C.C. is that the *structure factor* is zero whenever the sum $(h + k + 1)$ is an odd intteger. In the F.C.C. structure, we find that the structure factor is nonzero for an (hkl) Bragg reflection only if h, k, and l are either all even or all odd (see Problem 1.16). These complexities in the rules for the structure factor in F.C.C. and B.C.C. lattices arise because we have found it convenient to characterize these lattices by non-primitive unit cells.

EXPERIMENTAL METHODS
IN X-RAY DIFFRACTION

The interested reader is referred to the bibliography at the end of this chapter for detailed descriptions of experimental methods in X-ray and neutron diffraction. There are more than a dozen ways in which X-ray diffraction can be used to study the structural features of solids, and most of these methods are applicable also in neutron diffraction, even though the equipment involved is very different. Of the X-ray methods to be briefly mentioned in this section, the Laue method is almost never used in neutron diffraction, but rotating crystal and powder methods using collimated beams of monochromatic neutrons are becoming increasingly popular. Having said this, we shall discuss experimental methods as they relate to X-rays in the following paragraphs.

In considering the Bragg diffraction of an X-ray beam by a single crystal, we must note that Equation 1-33 requires a suitable combination of d, θ, and λ for constructive interference and efficient reflection. Thus monochromatic X-rays directed on a crystal at an arbitrary angle will usually not be reflected. Either X-rays of many wavelengths must be used at a single angle of incidence (the Laue method), or monochromatic rays must be allowed to encounter the crystal at a variety of angles. This is accomplished mechanically in the rotating crystal, oscillating crystal, and Weissenberg methods, while the Debye-Scherrer method uses a powdered sample so that *every* angle of incidence is encountered for some of the crystallites.

The *Laue method* is very simple in concept and operation. A narrow pencil beam of X-rays is allowed to fall on a single crystal, as illustrated in Figure 1-42. This beam contains X-rays over a wide wavelength range, and for any wavelength satisfying the Bragg condition a diffracted beam will emerge. The figure shows two positions in which a photographic plate can be placed to record a set of diffraction "spots" which is characteristic of the structure and orientation of the crystal.

The set of diffraction spots can appear very confusing when the structure of the crystal is not known, because several wavelengths, cor-

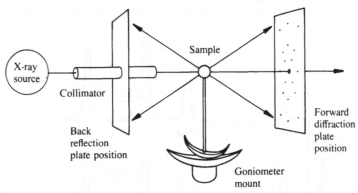

Figure 1–42 A Laue flat plane camera, using polychromatic X-rays. The pattern of Laue spots can be photographed in either the forward or the back-reflection position. The sample can be rotated about three orthogonal axes, and then X-rayed again, to confirm that a desired orientation has been obtained.

responding with different orders for the same set of planes, may be reflected to the same location on the plate. For this reason, the Laue method is rarely used for investigation of new structures. Its main use lies in the routine orientation of known crystals. Thus Figure 1-42 shows the sample crystal attached to a triple-axis goniometer; the familiar array of spots (Figure 1-38) produced after a crude attempt at orientation can be measured to determine what fine adjustment of angle is necessary for perfect alignment of a major crystal plane with the axis of the Laue camera.

As its name implies, the *rotating crystal method* is a technique in which θ is varied as a function of time while X-rays (or neutrons) of a single wavelength are presented to a single crystal. The essential aspects of the experimental arrangement are sketched in Figure 1-43. A

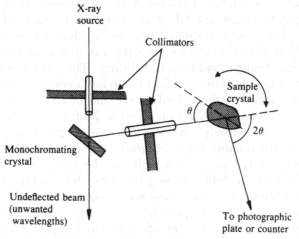

Figure 1–43 A rotating crystal arrangement, using monochromatic X-rays selected by Bragg reflection from a separate crystal. In a rotating crystal camera, the crystal is rocked back and forth while the series of diffraction images is recorded on a cylinder of photographic film placed so that its axis coincides with the rotation axis.

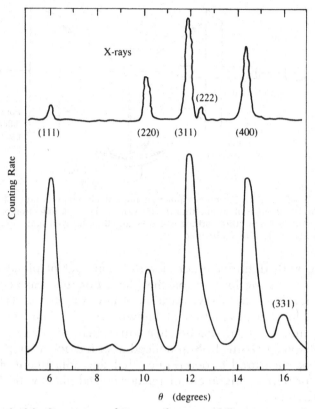

Figure 1–44 Comparison of X-ray and neutron diffraction traces for magnetite, Fe_3O_4. [After C. G. Shull et al., Phys. Rev., **84**, 912 (1951).]

single X-ray wavelength (usually the K_α line of the X-ray tube target material) can be separated from the X-ray continuum quite readily by using a separate single crystal of any convenient material as a monochromator. The sample is rotated back and forth and a cylinder of photographic film, placed coaxially with the rotation axis, records a spot whenever the Bragg condition is fulfilled. The inclination of these spots to the direction of the incident beam has a component of 2θ in the plane perpendicular to the cylinder axis, and of course most spots are also displaced along the direction of the cylinder axis. Elaborations of the rotating crystal principle to give much more information about the crystal symmetry include the *oscillating crystal* method, the *Weissenberg* method, and the *precession* method. The books by Woolfson and Buerger, cited in the bibliography, describe these methods in detail.

Instead of using a single X-ray wavelength and a time-dependent angle of incidence, we may present a crystalline sample with every θ simultaneously. This is the *Debye-Scherrer* technique of using a finely powdered crystalline sample in which the crystalline orientations are random. Rays which for one crystallite or another satisfy the Bragg condition emerge from the sample as a series of cones concentric with the incident beam direction. Thus a photographic plate records a series of concentric circles. In the analogous situation of electron diffraction (see Figure 1-39) the diffraction patterns of two polycrystalline gold films are compared, one with random crystallite orientations (rings) and one with preferential orientation (spots). The pattern of diffraction rings obtained from a powder X-ray camera is scanned with a densitometer to yield a curve with (hopefully) pronounced peaks at the various Bragg angles.

In much X-ray work with the rotating crystal method, the photographic plate is replaced with a sensitive counter, and a recorder trace is taken of the count rate while the crystal slowly rotates. (A counting and recording technique is necessary for any neutron diffraction experiment.) The information from both X-ray and neutron diffraction traces may then be compared in order to see whether one type of radiation reveals a line which is unresolved for the other. We have already noted that neutrons are useful for provoking reflections from very light atoms such as hydrogen. In many compounds, a reflection which has almost zero structure factor for either X-rays or neutrons may have a non-zero structure factor for the other. And as illustrated by Figure 1-44, the neutron diffraction trace for a magnetic material has extra lines which do not appear in the X-ray trace; lines associated with the magnetic ordering of d-shell electrons.

DIFFRACTION BY AMORPHOUS SOLIDS

In most of this chapter we have been talking about perfectly periodic lattices. It is certainly true that single crystals which have few departures from periodicity over long distances can be understood with more confidence than can amorphous and microcrystalline solids, but the problems of materials that lack long-range order cannot be ignored.

Section 1.6 discusses some of the *point* and *line* defects which degrade the perfect periodicity of an idealized crystal to the semblance of periodicity possible in real monocrystals. Here we shall concentrate on materials which are nominally amorphous, yet which produce a repeatable and meaningful X-ray diffraction pattern. Interest in amorphous solids has picked up considerably since the mid-1960's, and this has been so particularly for amorphous semiconducting materials.[30] Interest in amorphous chalcogenide glasses has been inspired in large measure by the unusual electronic properties, and by the hope that these may lead to widespread application in switching devices.[31]

A random array of isolated atoms will diffract a parallel monochromatic X-ray beam, producing an angular dispersion which is controlled by the atomic scattering factor curve for the atom, a curve shaped like the one in Figure 1-41. The diffraction pattern for a liquid or for an amorphous solid has several concentric broad and diffuse rings, superficially similar to those for a polycrystalline solid, and it was at first

[30] R. W. Douglas and B. Ellis (eds.), *Amorphous Solids* (Wiley-Interscience, 1972). See also N. F. Mott and E. A. Davis, *Electronic Processes in Non-Crystalline Solids* (Clarendon Press, Oxford, 1971).

[31] H. K. Henisch, Scientific American **221**, #5, 30 (1969); also Nature **236**, 205 (1972).

thought that the appearance of such rings was automatic evidence of crystallinity. This, however, is not the case.

On the basis of studies of vitreous silica and other amorphous solids during the 1930's, Warren[32] pointed out that the curve of the scattering factor versus $\sin\theta/\lambda$ in a solid is the sum of a monotonically decaying term (which results from the presence of isolated atoms) and an oscillatory component which describes in Fourier form the probability of various interatomic distances. Even without any long-range ordering in a solid, the mere existence of a few fairly definite interatomic distances can make the oscillatory term quite prominent for moderate values of $\sin\theta/\lambda$. Figure 1-45 shows such a curve for silica, compared with the expected monotonic decrease for purely random spacings.

The information stored in Fourier form in Figure 1-45 can be reconverted into a curve which plots the density of neighbors as a function of distance. Figure 1-46 displays the results of this reconversion for silica, using data for both X-ray diffraction and neutron diffraction, and comparing the most popular neighbor distances with those of crystalline quartz. It will be seen that there *is* correlation for the nearer neighbors, but not for more remote ones, confirming the idea that vitreous silica is a collection of SiO_4 tetrahedra linked together by their oxygen corners in a random manner. Mozzi and Warren[33] conclude that each oxygen atom is bonded to two silicon atoms, but that the Si-O-Si

[32] B. E. Warren, J. Appl. Phys. **8**, 645 (1938).
[33] R. L. Mozzi and B. E. Warren, J. Appl. Cryst. **2**, 164 (1969).

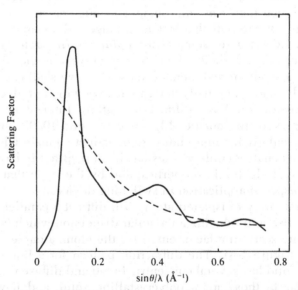

Figure 1–45 X-ray diffraction efficiency as a function of diffraction angle for vitreous silica. The dashed curve is a composite of the atomic scattering factors for isolated silicon atoms and oxygen atoms. These data are by E. H. Henninger *et al.*, J. Phys. Chem. Solids **28**, 423 (1967), and differ in only minor details from the 1938 curve of Warren. More recently, R. L. Mozzi and B. E. Warren [J. Appl. Cryst. **2**, 164 (1969)] have been able to extend this curve for vitrious silica out to $\sin\theta/\lambda = 1.6$.

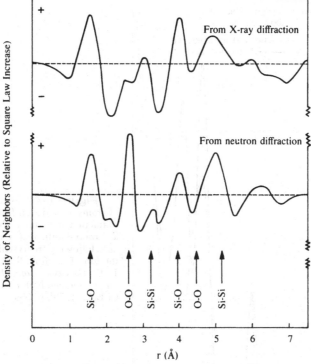

Figure 1-46 Density of atomic neighbors as a function of radius for vitreous silica (with the continuous r² increase of density subtracted). Curves are based on X-ray and neutron diffraction, the former based on the curve of Figure 1-45. From E. H. Henninger *et al.*, J. Phys. Chem. Solids **28**, 423 (1967).

bond angle can be anywhere from 120° to 180° as the SiO_4 tetrahedra link up in random fashion.

Comparable studies of other nominally amorphous materials show the extent to which short-range order can exist in a noncrystalline solid. This problem is far from straightforward with the chalcogenide glass mixtures that are of interest for electronic switching purposes,[31] and at the time of writing the best evidence seems to be available for amorphous forms of elements and of simple binary compounds. Thus diffraction evidence for amorphous selenium suggests that the spiral chain structure (Figure 1-6) exists for groups of three or four atoms, with linkages in random directions connecting these small groups. Studies have been made of the amorphous states of the well-known elemental semiconductors silicon and germanium, which have the diamond structure of Figure 1-31(a) in their better-known crystalline form. Figure 1-47 shows the variation of atomic neighbor density with distance for amorphous and recrystallized silicon, as determined from analysis of *electron* diffraction data. (The curve in Figure 1-47 differs in format from those of Figure 1-46 in that the continuous $4\pi r^2$ factor has been divided out in the latter.) The peaks in Figure 1-47 can be interpreted to show

---

[31] H. K. Henisch, Scientific American **221**, #5, 30 (1969); also Nature **236**, 205 (1972).

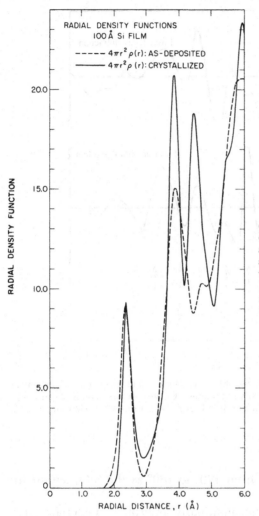

Figure 1–47 Radial density of atoms as deduced from electron diffraction data for a thin film of amorphous silicon (as deposited), and following recrystallization of the film. Data from S. C. Moss and J. F. Graczyk, *Proc. 10th Internat. Semiconductor Conference* (A.E.C., 1970), p. 658.

that each Si atom has four nearest neighbors at a distance of 2.35 Å in both amorphous and crystalline silicon, but that the arrangement of more remote neighbors is random in the amorphous state. Thus the curve for amorphous silicon shows almost no peak for the distance of third-nearest neighbors (4.5 Å), while the effects of bond angle distortion are noticeable as a line broadening at the distance for second-nearest-neighbors.

Reciprocal Space

THE RECIPROCAL LATTICE

We have reached a point in the discussion of periodic structures at which it is useful to introduce the ideas of reciprocal space and the reciprocal lattice. Just as any quantity which varies with time can be described as a sum of Fourier components in the frequency domain, so can the spatial properties of a crystal be described as the sum of components in Fourier space, or reciprocal space. For a perfect single crystal (which is a three-dimensional repetition of identical building blocks), the reciprocal lattice in Fourier space is an infinite periodic three-dimensional array of points whose spacings are inversely proportional to the distances between *planes* in the direct lattice. The reader will immediately and correctly expect to associate reciprocal lattice spacings with the Bragg diffraction condition.

Since vectors in real space have dimensions of length, those in reciprocal space have dimensions of length^{-1}. This may be compared directly with the wave-vector[34] of an excitation (such as a photon, a lattice vibration, or a moving free electron). Multiplication of each coordinate by $\hbar$ converts reciprocal space into momentum space. These brief comments are a reminder that the reciprocal lattice concept will be employed frequently in Chapters 2, 3, and 4. First, however, let us see how the reciprocal lattice formulation can give a new perspective to the Bragg diffraction criterion.

If **a**, **b**, and **c** are the primitive translational vectors of a space lattice in real space, then we *define* the *fundamental vectors* of the reciprocal lattice by

$$\left. \begin{aligned} \mathbf{a}^* &\equiv \frac{2\pi \mathbf{b} \times \mathbf{c}}{\mathbf{a} \cdot \mathbf{b} \times \mathbf{c}} \\[2mm] \mathbf{b}^* &\equiv \frac{2\pi \mathbf{c} \times \mathbf{a}}{\mathbf{a} \cdot \mathbf{b} \times \mathbf{c}} \\[2mm] \mathbf{c}^* &\equiv \frac{2\pi \mathbf{a} \times \mathbf{b}}{\mathbf{a} \cdot \mathbf{b} \times \mathbf{c}} \end{aligned} \right\} \qquad (1\text{-}38)$$

---

[34] The instantaneous amplitude of a progressive wave with speed v and frequency ν can be expressed as $A = A_0 \exp\left[2\pi i \nu\left(\frac{x}{v} - t\right)\right]$ or as $A_0 \exp\left[2\pi i\left(\frac{x}{\lambda} - \nu t\right)\right]$ in terms of the wavelength $\lambda = (v/\nu)$. We often find it more convenient to describe the wave in terms of the angular frequency $\omega = 2\pi\nu$ and the wave-vector $k = (2\pi/\lambda) = (\omega/v)$. In terms of these new parameters, the wave amplitude is $A_0 \exp[i(kx - \omega t)]$. In three dimensions, the wave-vector *is* a vector, and we should write $A = A_0 \exp[i(\mathbf{k} \cdot \mathbf{r} - \omega t)]$. For a quantized wave, the momentum associated with an excitation of wave-vector $\mathbf{k}$ is $\mathbf{p} = \hbar \mathbf{k}$.

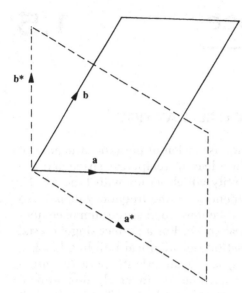

Figure 1-48 The fundamental vectors of the reciprocal lattice for a two-dimensional oblique "real" lattice. Note that a° is perpendicular to b and b° is perpendicular to a; also that $a^\circ \cdot a$ equals $b^\circ \cdot b$.

(The factor of 2π may be omitted from Equation 1-38, as we shall indicate later.) The denominator of each of the definitions in Equation 1-38 is the volume of the unit cell for the direct space lattice, which does not depend on the order in which a, b, and c are permuted.

We note from Equation 1-38 that a fundamental reciprocal vector such as a^* is not necessarily parallel to a of the direct lattice, but *is* required to be perpendicular to both b and c. We can write zeros for the various dot products

$$a^* \cdot b = a^* \cdot c = b^* \cdot a = b^* \cdot c = c^* \cdot a = c^* \cdot b = 0 \quad (1\text{-}39)$$

whereas

$$a^* \cdot a = b^* \cdot b = c^* \cdot c = 2\pi \quad (1\text{-}40)$$

These consequences are illustrated in Figure 1-48 for a two-dimensional oblique lattice.

The inverse of the transformation defined in Equation 1-38 is made in the same fashion as the first. We can easily confirm that

$$a = \frac{2\pi b^* \times c^*}{a^* \cdot b^* \times c^*} \quad (1\text{-}41)$$

and so forth.

The factor 2π which appears in Equations 1-38, 1-40, and 1-41, is placed there to make reciprocal space *numerically* as well as *dimensionally* the same as wave-vector space. The factor 2π is omitted in some discussions of reciprocal space, and in those cases a unit translation in reciprocal space can be equated with unit change of reciprocal wavelength. The confusion between the two systems is not as alarming as it might at first appear, for it will usually be apparent from the context when a location in reciprocal space described as ($\frac{1}{2}$00) is really the place otherwise referred to as (π00).

RECIPROCAL LATTICE VECTORS

At the beginning of the discussion of symmetry operations, we recognized that special significance attaches to translations in the real lattice which are members of the set

$$\mathbf{T} = n_1\mathbf{a} + n_2\mathbf{b} + n_3\mathbf{c} \qquad (1\text{-}42)$$

These translations connect pairs of points in the crystal lattice which have identical atomic environments. So also in reciprocal space is there a special interest in the set of translations that we call *reciprocal lattice vectors*. These are members of the set

$$\mathbf{G}_{hkl} = h\mathbf{a}^* + k\mathbf{b}^* + l\mathbf{c}^* \qquad (1\text{-}43)$$

where h, k, and l are integers.

We now wish to show that a reciprocal lattice vector characterized by the integers h, k, and l is perpendicular (in reciprocal space) to the (hkl) plane in real space. Recall that the (hkl) plane passes through the points (a/h), (b/k), and (c/l). Then it is easy to show (Problem 1.17) that any vector lying in the (hkl) plane must be of the form

$$\mathbf{R} = (A/h)\mathbf{a} + (B/k)\mathbf{b} + (C/l)\mathbf{c} \qquad (1\text{-}44)$$

where the coefficients A, B, and C are subject to the restriction

$$A + B + C = 0 \qquad (1\text{-}45)$$

It is then apparent from Equations 1-39, 1-40 and 1-45 that

$$\mathbf{R} \cdot \mathbf{G}_{hkl} = 0 \qquad (1\text{-}46)$$

Since $\mathbf{R}$ has been constructed in the most general way possible for a vector lying in the (hkl) plane of real space, it is apparent that $\mathbf{G}_{hkl}$ must be perpendicular to that plane.

We recognized earlier that a simple expression for the spacing between successive (hkl) planes was possible for a cubic lattice. In terms of the reciprocal lattice vector $\mathbf{G}_{hkl}$ a compact expression can be obtained for any lattice. Referring to Figure 1-49, we can see that both $\mathbf{G}_{hkl}$ and the vector from the origin of direct space to the (hkl) plane can be expressed as multiples of a unit vector $\mathbf{n}$. The equation of the (hkl) plane is that

$$d_{hkl} = \mathbf{r} \cdot \mathbf{n} = \frac{\mathbf{r} \cdot \mathbf{G}_{hkl}}{|\mathbf{G}_{hkl}|} \qquad (1\text{-}47)$$

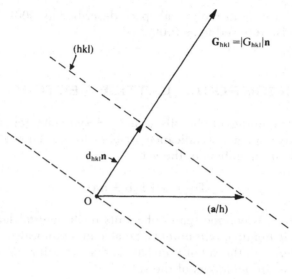

Figure 1-49 Construction for the determination of the spacing between successive (hkl) planes.

for any vector **r** whose magnitude is larger than d_{hkl}, and the vector (a/h) obviously qualifies as being of adequate length. Thus

$$d_{hkl} = \frac{\mathbf{a} \cdot \mathbf{G}_{hkl}}{h \; |\mathbf{G}_{hkl}|} \tag{1-48}$$

But from Equations 1-40 and 1-43, $\mathbf{a} \cdot \mathbf{G}_{hkl}$ is just $2\pi h$. We thus achieve the simple result that the spacing of (hkl) planes in real space is connected with the h, k, l reciprocal lattice vector length by

$$d_{hkl} = \frac{2\pi}{|\mathbf{G}_{hkl}|} \tag{1-49}$$

THE DIFFRACTION CONDITION

The reciprocal lattice permits a new and interesting perspective of the Bragg condition for specular reflection. Instead of considering just the *wavelength* of the radiation which interacts with the atoms of a set of planes (as we did in connection with Figure 1-40), we shall now concern ourselves with the initial and final *wave-vectors* **k, k′** of a reflected wave-particle. Provided that scattering is elastic, there is no change in magnitude of wave-vector:

$$|\mathbf{k}| = |\mathbf{k}'| = (2\pi/\lambda) \tag{1-50}$$

The vector triangle of Figure 1-50 demonstrates that the change $\Delta \mathbf{k}$ in **k**

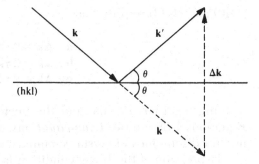

Figure 1–50 Vector triangle for the change in wave-vector when a wave-particle suffers specular reflection at an (hkl) plane.

is in the direction perpendicular to the (hkl) planes, the direction we have just been associating with G_{hkl}, or with the unit vector **n**. Thus

$$
\left.
\begin{aligned}
\Delta \mathbf{k} = (\mathbf{k}' - \mathbf{k}) &= 2\sin\theta |\mathbf{k}| \mathbf{n} \\
&= \left[\frac{4\pi \sin\theta}{\lambda}\right] \mathbf{n} \\
&= \left[\frac{4\pi \sin\theta}{\lambda |G_{hkl}|}\right] G_{hkl} \\
&= \left[\frac{2d_{hkl} \sin\theta}{\lambda}\right] G_{hkl}
\end{aligned}
\right\}
\qquad (1\text{-}51)
$$

When the combination of λ, θ, and d_{hkl} is appropriate for satisfaction of the Bragg condition,

$$\Delta \mathbf{k} = G_{hkl} \qquad (1\text{-}52)$$

Thus the set of points generated by G_{hkl} is the array of Laue spots for diffraction by a crystal. The spacings between those points are inversely proportional to the spacings of planes in the real lattice.

In view of Equation 1-52, the vector relation between the initial and final wave-vectors of a Bragg-reflected wave-particle can be written as

$$\mathbf{k}' = G_{hkl} + \mathbf{k} \qquad (1\text{-}53)$$

When each side of this equation is squared and the quantity $|\mathbf{k}|^2 = |\mathbf{k}'|^2$ is subtracted from each side, the Bragg condition appears in the form

$$G^2 + 2\mathbf{k} \cdot G = 0 \qquad (1\text{-}54)$$

We shall find this version of the diffraction condition useful in discussing the band theory of electrons in solids when we reach Chapter 3.

Since a wave-particle will experience a change in wave-vector $\Delta \mathbf{k} = G_{hkl}$ in being reflected from a set of (hkl) planes, then Equations

(1-40) and (1-43) require that

$$\left. \begin{array}{l} \mathbf{a} \cdot \Delta\mathbf{k} = 2\pi h \\ \mathbf{b} \cdot \Delta\mathbf{k} = 2\pi k \\ \mathbf{c} \cdot \Delta\mathbf{k} = 2\pi l \end{array} \right\} \qquad (1\text{-}55)$$

in tying together $\Delta\mathbf{k}$, hkl, and the vectors of the real (direct) lattice. Equations 1-55 are the *Laue equations*, which provide one useful avenue for discussions of crystal symmetry and structure.

Expression of the Bragg condition in the form of Equation 1-52 led Ewald in 1921 to suggest an interesting geometric interpretation of the diffraction requirement. This is illustrated in Figure 1-51. The planes which can participate in any resulting diffraction will of course be the ones which (in real space) are perpendicular to the direction of any **G** that can connect A with another point on the surface of the sphere (in reciprocal space).

BRILLOUIN ZONES

When the direct lattice is exactly periodic, the reciprocal lattice is periodic and infinite in extent, being a set of points generated by Equation 1-43. However, the problems in solid state physics for which the reciprocal lattice is most useful only require the use of a limited volume of reciprocal space, or k-space. These are problems associated with the dispersion relations for electrons or excitations.

It is conventional to define the *first Brillouin zone* as a volume in reciprocal space centered on one reciprocal lattice point and bounded by the set of planes which bisect reciprocal lattice vectors connecting the point with its neighbors in reciprocal space. Figure 1-52 illustrates this for a two-dimensional oblique lattice, and similar principles operate with three-dimensional lattices also (Problem 1.19). For any location $A'(\mathbf{k}')$ in reciprocal space there is a corresponding location $A(\mathbf{k})$ within the first zone, related by

$$\mathbf{k}' = \mathbf{k} + \mathbf{G} \qquad (1\text{-}56)$$

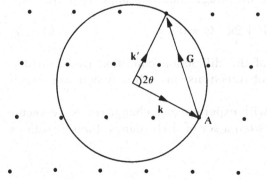

Figure 1–51 The Ewald construction for the existence of a diffracted beam. Suppose the wave-vector **k** of each incident wave-particle is drawn as a vector which terminates at a point A of the reciprocal lattice. From the other end of the vector **k**, draw a sphere of radius $|\mathbf{k}|$. Then the Bragg condition $\Delta\mathbf{k} = (\mathbf{k}' - \mathbf{k}) = \mathbf{G}$ is satisfied whenever another point of the reciprocal lattice intersects the surface of the sphere.

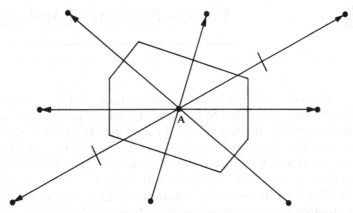

Figure 1–52 A portion of reciprocal space for a two-dimensional oblique lattice, showing the lines bisecting the midpoints of some reciprocal lattice vectors from A. The six shortest of these vectors can be bisected to produce the first Brillouin zone about A.

and we shall see in later chapters that a wave in a solid characterized by k' is indistinguishable from one characterized by the "reduced wave-vector" k. The band theory of solids, which will concern us extensively in Chapters 3 and 4, is a direct consequence of the ability of a periodic lattice to cause Bragg reflections of electrons at critical velocities (electrons whose wave-vectors coincide with Brillouin zone boundaries).

Accordingly, the Brillouin zones for (hypothetical) 1-D and 2-D solids, and for "real" 3-D lattices are encountered at several places in the book. Those for 1-D and for the 3-D F.C.C. lattice are used in discussing vibrational spectra of such solids, in Sections 2.2 and 2.3. In Section 2.4, the polyhedral shape of a 3-D Brillouin zone is approximated by a sphere of the "Debye radius" to have the same capacity. In Section 2.5, the dimensions of a (supposedly 2-D) zone become important in setting the temperature below which an "umklapp" scattering event among phonons becomes unlikely.

Brillouin zones are reexamined in Chapter 3 apropos the filling of allowed electronic states. Figures 3-23 to 3-27 deal with the zone for a 1-D solid, then Figures 3-30 to 3-33 for 2-D zones, and Figures 3-34 to 3-36 for some 3-D possibilities. References to Brillouin zones continue through Chapters 3 and 4. Of frequent interest is the zone for a solid of F.C.C. translational symmetry (true for F.C.C. itself, for the NaCl structure, and for the diamond and zincblende lattices). This important zone is used in figures such as 3-46, 3-47, 3-50, 3-51, and in Figures 4-45 and 4-50.

1.6   Crystalline Defects

We have been paying a great deal of attention to the symmetry of a perfectly periodic lattice and to the infinite array of points in reciprocal space that describes this lattice in Fourier form. But no crystal *is* perfect. The very fact that a real crystal has a surface sets an upper limit on its perfection, and indeed macroscopic samples of most solids comprise many crystallites, randomly oriented, with grain boundaries separating one crystallite from the next. Such grain boundaries act as repositories for many mobile forms of microscopic garbage. Within each monocrystalline particle, there will certainly be finite concentrations of *point defects*, and there is apt to be a finite density of *line defects*, or *dislocations*.

POINT DEFECTS

The point defects (imperfections, or flaws) in a solid can be grouped into two principal categories; impurities (foreign atoms) and native point defects. A so-called "point defect" may be a *complex* that includes several kinds of foreign atom in close proximity, or foreign atoms closely associated with some kinds of native defect. This sounds at first like a contradiction of the term "point" defect. The essential idea, however, is that the maximum dimension of a point defect in every direction is no more than a few atomic spacings. In contrast, a *line* or *surface* defect may extend for millions of atomic spacings, in one or two directions.

Impurity atoms have many consequences for the electrical, optical, magnetic, thermal, and mechanical properties of solids. We shall consider them particularly in Chapter 4, since impurities exert a dominating influence on nonmetallic conductors. It should be noted too that through electrostatic considerations and the workings of the thermodynamic laws of mass action, the concentration of an impurity can affect the equilibrium concentrations of *native* defects. Kroger (see bibliography) gives an exhaustive survey of the interactions between concentrations of native and foreign defects.

The simplest kind of native defect is the *vacancy*, the simple omission of an atom from a regular atomic site and the placement of an additional atom on the surface [Figure 1-53(a)], so that the perturbed crystal has as many atoms as the perfect one. If we suppose that the atomic array is completely rigid, it requires an energy E_v to remove an atom from the interior of a crystal; but energy E_s is liberated by placement of this atom on the surface of the crystal. The net formation energy of a vacancy on this rigid lattice model is accordingly $(E_v - E_s)$, a quantity which is typically some 10 eV/vacancy (see Problem 1.20). Since a lattice is not completely rigid, some of the formation energy $(E_v - E_s)$ is recovered by relaxation of the lattice; all the neighbors move in a little.

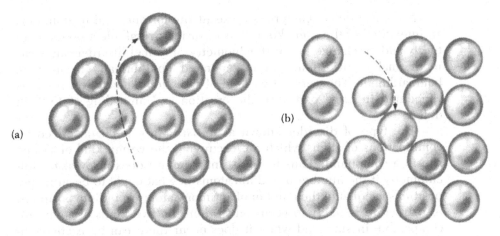

Figure 1-53 Formation in a lattice of (a), a vacancy and (b), an interstitial.

When vacancies are formed in an ionic compound, electrostatic neutrality has to be satisfied in some fashion in the volume of space which includes each vacant site.

The menagerie of defects in alkali halide crystals known as "color centers" (of which the simplest is the F-center), is created by the presence of vacancies, and of neighborhood groups of vacancies. The requirements for electrostatic equilibrium endow such defects with complex electronic and optical properties. Mott and Gurney (see p. 17, note 12) describe the early work on these defects, work which continues today.

In an ionic solid, electrostatic neutrality of the crystal as a whole is automatically preserved if there are vacancies of the various atomic species in the proportions we call *Schottky disorder*. These are the proportions which preserve *chemical stoichiometry*. Thus for Schottky disorder in RbF there are as many empty sites for Rb^+ as there are for F^-, whereas a stoichiometric balance of Cu_2O requires that empty copper sites be twice as numerous as empty oxygen sites. Schottky disorder is usually found only in binary compounds with a $1:1$ ratio, and the term Schottky defect then is taken to mean the existence of a pair of the opposite kinds of vacancy. The two members of a Schottky defect are not necessarily close together in the crystal.

An *interstitial* [Figure 1-53(b)] is an atom (or ion, as necessary) which is removed from the surface and fitted into the crystal wherever room in the lattice can be contrived by some squeezing and bulging. It is almost impossible to insert an interstitial for close-packed metals such as copper (F.C.C.) or zinc (H.C.P.), but in a relatively open lattice such as diamond there are many places where space can be made for an interstitial (either native or foreign) by a modest rearrangement of the nearer neighbors. There is no name for the simultaneous presence of electropositive and electronegative interstitial ions in stoichiometric proportions, since this is a most unlikely occurrence. For instance, Na^+ interstitials might find places to fit into a NaCl lattice, but Cl^- ions are much too large to find an interstitial home.

A *Frenkel defect* comprises a vacant site for one kind of atom plus an interstitial of the same kind (thus ensuring overall electrostatic neutrality and no change in the stoichiometry). Frenkel disorder can occur in thermal equilibrium, and is generated nonthermally by nucleon bombardment. The very active field of radiation damage[35] in solids is concerned in large part with the creation and annealing of Frenkel defects.

The type of disorder known as *antistructure* is characterized by combinations of atoms which are occupying the wrong kind of atomic site in a compound. Thus in InSb, antistructure does not violate stoichiometry if the number of indium atoms in what ought to be antimony positions is equal to the number of antimony atoms in indium positions. Antistructure is likely to occur only when the two kinds of atoms are comparable in size, and when it does occur there can be exotic complexes involving up to 50 to 100 atoms within a small region which are in an improper arrangement.

DEFECT DENSITY IN THERMAL EQUILIBRIUM

Since it takes a positive amount of energy (let us say E eV) to create a point defect such as a vacancy, it may seem unreasonable that there should be a non-zero vacancy concentration in a crystal at thermal equilibrium. However, a non-zero vacancy concentration increases the entropy of the system and permits a smaller *free energy* than for a perfect crystal, since free energy is total energy minus temperature times entropy. Suppose that we have a crystal of N atoms, and that n vacancies are randomly distributed over the available sites. The additional entropy is

$$\Delta S = k_0 \ln \left[\frac{N!}{(N-n)! \, n!} \right] \qquad (1\text{-}57)$$

and using Stirling's approximation for the factorials,

$$\Delta S = k_0 [N \ln(N) - (N-n) \ln(N-n) - n \ln(n)] \qquad (1\text{-}58)$$

Now the change in free energy due to the vacancies is

$$\begin{aligned} \Delta F &= (n \, E) - T\Delta S \\ &= (n \, E) - k_0 T [N \ln(N) - (N-n) \ln(N-n) - n \ln(n)] \end{aligned} \qquad (1\text{-}59)$$

[35] Radiation damage in solids has been extensively studied in solids since the 1940's. Military applications lie behind some of this work, but the desire to understand radiation damage also has practical backing from the peaceful uses of radiation. For example, radiation damage and degradation occur in any material used in the environment of a nuclear reactor or of a medical irradiation facility. Radiation damage occurs also in the space environment. A general review of the early work is given by D. S. Billington and J. H. Crawford, *Radiation Damage in Solids* (Princeton, 1961). The effects on semiconducting materials are reported by F. L. Vook, *Radiation Effects in Semiconductors* (Plenum, 1969), and in *Radiation Damage and Defects in Semiconductors*, edited by J. Whitehouse (Institute of Physics, 1973).

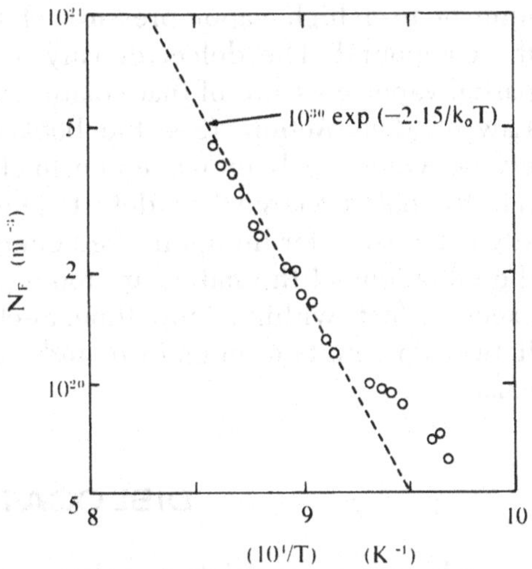

Figure 1–54 Temperature dependence of the density of Frenkel defects in germanium. As is customary for an activation-controlled population, the plot is semi-logarithmic, with $(1/T)$ as abcissa. After S. Mayburg and L. Rotondi, Phys. Rev., **91**, 1015 (1953).

and the thermodynamically most probable value for n is that for which $(\partial \Delta F/\partial n)$ is zero. From Equation 1-59 this will occur when

$$\frac{n}{(N-n)} = \exp(-E/k_0 T) \qquad (1\text{-}60)$$

Thus for all practical densities of native defects, the concentration will be an exponential function of the reciprocal temperature.

Similar expressions can be worked out for other point defects at equilibrium. In general, the free energy $\Delta F = \Delta U - T \Delta S$ must be optimized for *combinations* of various types of defects, with their various activation energies.

Figure 1-54 illustrates the curve reported by Mayburg and Rotondi (1953) for the temperature dependence of the equilibrium Frenkel defect density in germanium. As with experiments on most other solids, attempts to measure native defect densities in germanium were clouded for years with a variety of contamination effects, particularly with the effects of copper contamination. In any solid it is much easier to get a curve of a defect density which varies exponentially with $(1/T)$ than it is to be sure that the measured quantity is the desired one.

STOICHIOMETRY

We usually think of a compound AB as having equal numbers of A and B atoms. Such a crystal is said to be stoichiometric. Apparently stoichiometry is not a very highly rated virtue for many solids which happily exist with an anion-cation ratio slightly different from the nominal one. Such nonstoichiometric compounds balance their structures by the presence of vacancies, interstitials, or both.

For a compound such as NiO or PbS, the stoichiometry can be changed very easily at high temperature by exposure either to a vacuum or to a high vapor pressure of the most volatile constituent in the compound. The defect density then varies as some power of the partial vapor pressure of that volatile constituent, a consequence of the Law of Mass Action. (See the book by Kroger for details.) When a crystal which has been made nonstoichiometric by such a procedure is rapidly cooled, a large flaw (defect) density can be quenched in; the density is too large for minimum free energy at the lower temperature, but the vibrations of the lattice are too weak to permit out-diffusion of the excess defects within a finite time. Such quenched-in excess defects are liable to precipitate on grain boundaries or on dislocations where they exist.

DISLOCATIONS

Whereas a point defect such as a vacancy directly affects only one lattice point (or a few immediately adjacent lattice points), a dislocation is a line source of imperfection. This is *not* the same thing as a line of vacancies. A general dislocation may follow any curved route through a crystal, but we can understand the construction of a dislocation most easily by considering the two simplest types of straight line dislocations, the *edge* and *screw* types.

In a pure *edge dislocation*, one of the planes of atoms terminates, resembling a knife blade stuck part way into a block of cheese. This type of dislocation is illustrated in Figure 1-55(a). The displacement of the lattice (called the Burgers vector **b**) is perpendicular to the direction of the dislocation (the direction of the last line of atoms in the half-plane). The Burgers vector must be an integral multiple of the unit lattice vector.

In a *screw dislocation*, as sketched in Figure 1-55(b), part of the lattice is displaced with respect to the other part, and the displacement is *parallel* to the direction of the dislocation. We may think of accomplishing this by first cutting partly through a perfect crystal, then forcing the material on one side of the cut to move up with respect to

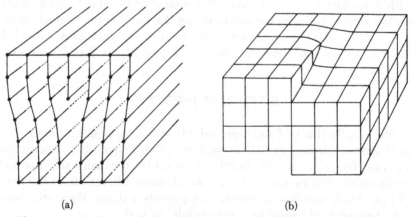

(a)   (b)

Figure 1-55 The two idealized forms of dislocation; (a) the edge type, and (b) the screw type. In general, a dislocation follows a curved path which has varying components of these two extreme characters.

the material on the other side by one or more unit lattice vectors, and finally placing the material on the two sides back into contact. The term Burgers vector is again used to describe the displacement so forced; it will always be an integral multiple of the unit lattice vector, and for a pure screw dislocation it is parallel to the dislocation axis.

A general dislocation in a real crystal is likely to have some edge and some screw character, the proportions varying as the dislocation curves in a different direction. The Burgers vector is unchanged throughout the length of the dislocation. It is not possible for a dislocation to terminate inside a crystal, thus we can have either a closed loop or an open loop which terminates at two locations on an outer surface or a grain boundary. Of course for most materials the point at which a dislocation reaches the surface of the crystal is the only place where it can actually be "seen."

When a screw dislocation runs through a crystal, we can no longer speak of planes of atoms normal to the dislocation, for all atoms lie on a surface which spirals from one end of the crystal to the other, gaining height equal to the Burgers vector for each revolution. The spiral may be either right-handed or left-handed. When a crystal is grown slowly from the vapor phase, the region along the surface step of a screw dislocation is energetically favorable for additional growth, and crystallization can proceed by the deposition of a continuous winding spiral. Two examples are illustrated below in Figure 1-56, the one at the left with the atomic-scale resolution of a field-ion microscope.

The termination point of an open dislocation loop on a crystal surface can often be revealed by chemical etching, since the etch preferentially attacks the strained region to display an *etch pit* with sides formed of characteristic lattice planes. Several such isolated etch pits are visible in Figure 1-57, while that figure also shows a line of closely spaced etch pits marking the course of a *low angle boundary* between two

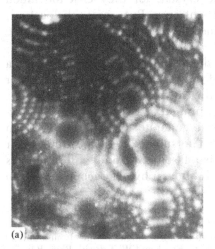 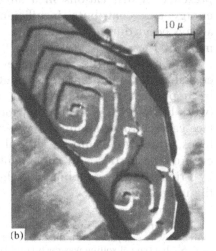

Figure 1–56 Spiral step structures on metal surfaces. (a) Atomic structure of a spiral on the surface of an iridium crystal, observed in a field-ion microscope (courtesy O. T. Inal). (b) Multilayer spiral steps on electrodeposited silver, observed by replication TEM (courtesy T. Suzuki). From L. E. Murr, *Interfacial Phenomena in Metals and Alloys* (Addison-Wesley, 1975).

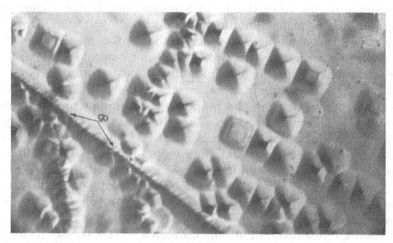

Figure 1-57 Dislocation etch pits revealed by etching a (100) surface of a LiF single crystal. In addition to randomly located pits, a low-angle boundary is marked by a closely spaced array running diagonally across the lower left of the figure. (Photo courtesy of L. E. Murr, Oregon Graduate Center.)

sections of an almost single crystal. Such photographs confirm the prediction of Burgers (1939) that a small angle change in orientation could be accommodated by a regular row of dislocations. The angle of the boundary is inversely proportional to the spacing of the dislocations (see Problem 1.22).

The *dislocation density* N_D (a quantity with dimensions of length^{-2}) is an indicator of crystal perfection. For a covalently-bonded solid such as silicon or germanium, it is possible (with great care) to grow a crystal with $N_D < 10^6$ m^{-2}. A far more typical density for a covalent solid is around 10^8 m^{-2}, while dislocation densities for metallic crystals are at least 10^{11} m^{-2}. Thermal equilibrium does not *require* the presence of dislocations in a single crystal, for they cost too much energy and contribute insufficient entropy. But some dislocations always occur during solidification, and the first ones can always create more.

Mechanical stress encourages dislocations to sweep through a crystal, generating point defects until the movement is "pinned" either by impurities or by intersection with the path of other dislocations (work hardening). A dislocation loop pinned at two points on the surface can act under strain as a repetitive generator of closed dislocation loops, which grow into the crystal.[36] The subject of dislocations, their creation, properties, and movement is a fascinating one, and the books by Friedel and by Amelinckx in the bibliography are recommended to the interested reader.

---

[36] The creation of a succession of closed dislocation loops from a doubly-pinned source may be compared with the stream of soap bubbles which emerges from a bubble pipe. Such a type of source was postulated by F. C. Frank and W. T. Read, Phys. Rev. **79**, 722 (1950) and its mode of operation has been directly demonstrated both in ionic compounds and in silicon. W. C. Dash [J. Appl. Phys. **27**, 1193 (1956)] used silicon as the host and penetrating infrared light for illumination, and precipitated copper atoms on the dislocations for contrast. He observed Frank-Read sources which had produced several complete dislocation loops, and was able to photograph them.

Problems

1.1 We are told that for a van der Waals bonded solid in which the equilibrium atomic spacing is $r_0 = 1.50$ Å, the binding energy is 10 per cent less than would be given by the attractive term alone. What is the characteristic length ρ in Equation 1-2?

1.2 In a single molecule of KF, the equilibrium internuclear separation is $r_0 = 2.67$ Å and the cohesive energy $(-E_i)$ relative to separated *ions* is 0.50 eV/molecule smaller than the Coulomb attractive energy, because of overlap repulsion. Given that the electron affinity of fluorine is 4.07 eV/electron and that the first ionization potential of potassium is 4.34 volts, show that the energy necessary to separate the molecule into neutral atoms is -0.945 E_i.

1.3 Consider the hypothetical two-dimensional square ionic lattice of Figure 1-58. Show that the Madelung constant is $\alpha = 1.61$ when neighbors as remote as fifth-nearest are considered by the Ewald/Evjen method (footnote 11), which operates as follows. Note the dashed squares, each of which is electrically neutral if an ion on one edge of a square is counted as 50 per cent inside and 50 per cent outside and one on a corner is counted as 25 per cent inside and 75 per cent outside. You should find that the contents of the first square contribute $+1.29$ towards the Madelung constant, and the area between the first and second squares an additional $+0.32$. Is the contribution of the area between any subsequent pair of squares invariably positive? Explain your answer.

1.4 Complete the derivation sketched in Equations 1-10 through 1-15. The nearest-neighbor distance is $r_0 = 3.30$ Å in KBr, a

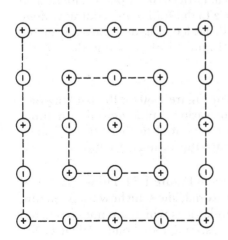

Figure 1–58 A hypothetical two-dimensional ionic lattice used for Problem 1.3, in illustration of the Evjen approach to summing the series for a Madelung constant.

material which crystallizes in the rocksalt structure (for which $\alpha = 1.748$). The compressibility of this material is $\chi = 6.8 \times 10^{-11}$ m²/N. Show that the repulsive characteristic length is $\rho = 3.37 \times 10^{-11}$ m, and that the cohesive energy relative to separate ions is $-E_{eq} = -6.85$ eV/molecule.

1.5 Derive an expression for the angle between adjacent sides of a regular n-sided polygon. Discuss this expression in terms of Figure 1-16 and in terms of the orders of rotational symmetry allowed in Table 1-4.

1.6 Sketch the five Bravais lattices which can exist in two dimensions, and show where centers of inversion exist. Now superimpose a point group of the lowest symmetry which is consistent both with the lattice and with the preservation of a center of inversion. Can this be done for each of the five types?

1.7 Consider an imaginary two-dimensional crystal for which the basis at each lattice point comprises five atoms arranged as a pentagon. Can these form a regular pentagon? What symmetries will exist, and what will be the point group?

1.8 Glide operations like those in Figure 1-20 explain the occurrence of some rectangular and square plane groups. Discuss the reasons why glide symmetry does not produce additional plane groups for the other two-dimensional Bravais lattices.

1.9 Make drawings of the primitive cells for the seven lattices illustrated in Figure 1-21 by conventional non-primitive cells.

1.10 What happens when the angles between the primitive vectors in a trigonal lattice approach either 90° or 120°?

1.11 One example of a four-numeral set of Miller indices for a trigonal or hexagonal crystal is the ($\overline{1}010$) plane labelled in Figure 1-25. Using this figure to determine the coding system, draw sketches of the H.C.P. lattice (Figure 1-29) and of the tellurium trigonal lattice (Figure 1-6) showing the (0001), ($1\overline{2}10$), and ($\overline{1}103$) planes.

1.12 Verify the numbers quoted in Figure 1-30 for the packing fractions of identical spheres in various cubic crystal structures. Show how these numbers are related to the ratio of sphere diameter to unit cube edge for the various structures.

1.13 Consider the cuprite structure of Figure 1-34. Draw a sketch of several unit cubes for this material, showing how the structure is a superposition of two sub-lattices which are not connected to each other at all by nearest-neighbor bonds. What is the structure of one of the sub-lattices alone if the copper atoms are then magically removed?

1.14 Figure 1-59 shows two parallel planes of atoms (with spacing d) in a crystal. Each plane consists of lines of atoms in the direction perpendicular to the paper, with a spacing of c between the lines. X-rays of wavelength λ have an angle of incidence θ with respect to the planes. The rays leaving at an angle ϕ_m are those of diffraction order m for the first plane considered as a plane grating. You are told that these rays are reinforced in phase by components diffracted from lower planes. Show that

$$m\lambda = c[\cos\phi_m - \cos\theta]$$

and

$$n\lambda = d[\sin\phi_m + \sin\theta]$$

where m and n are integers. Show that a crystal plane exists making an angle $\frac{1}{2}(\phi_m - \theta)$ with the original set of planes, for which the diffracted rays shown in Figure 1-59 are the result of specular reflection satisfying the Bragg condition.

1.15 Consider a lattice for which the primitive cell vectors a, b, and c have different lengths and for which the cell angles are not 90°. Consider two successive (hkl) planes, which are a distance d apart. If we denote a unit vector in the direction perpendicular to an (hkl) plane as n, show without resorting to a reciprocal lattice formulation that

$$d = \frac{a \cdot n}{h} = \frac{b \cdot n}{k} = \frac{c \cdot n}{l}$$

Demonstrate that this produces Equation 1-34 for a cubic lattice in which $|a| = |b| = |c|$ and $a \cdot b = a \cdot c = b \cdot c = 0$.

Figure 1-59 Construction for Problem 1.14.

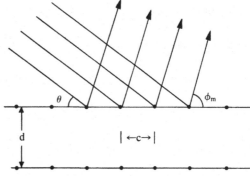

1.16 Show that the geometric structure factor of the F.C.C. lattice is zero for (100) and (110) reflections, but non-zero for the (111) reflection.

1.17 Prove that a vector lying in the (hkl) plane must satisfy Equations 1-44 and 1-45.

1.18 Show that the reciprocal lattice of simple cubic is also S.C. What is the angle between G_{100} and G_{111}? To what plane of the direct lattice is $G_{100} \times G_{111}$ perpendicular?

1.19 Set up the primitive translation vectors **a**, **b**, and **c** for a F.C.C. lattice in a convenient form and use the procedure of Equation 1-38 to get the fundamental vectors **a***, **b***, and **c*** of the reciprocal lattice. Show that formation of the planes bisecting the fourteen shortest reciprocal lattice vectors yields a Brillouin zone which is a truncated octahedron. What is the symmetry of the reciprocal lattice so produced? Without making any further analysis, can you make a rapid deduction as to the Brillouin zone for a B.C.C. lattice?

1.20 Calculate the energy $(E_v - E_s)$ for formation of a potassium vacancy in a KCl crystal, neglecting the energy recovered by lattice relaxation and using the numerical parameters from Problem 1.4. Assume that the displaced potassium ion is relocated upon a flat (100) face of the crystal.

1.21 Suppose that it takes an energy of 2 eV to create a vacancy in a certain solid. Show that the relative density of vacancies to atoms will always be less than 10^{-8} per cent unless the melting point is higher than 1000K.

1.22 A low angle boundary between two portions of a crystal is sustained by a line of dislocations, each of Burgers vector **b** and spaced a distance D apart. Show that the boundary angle is $2\sin^{-1}(b/2D)$.

Bibliography

Crystal Growth
L. Aleksandrov, *Growth of Crystalline Semiconductor Materials on Crystal Surfaces* (North-Holland, 1984).
J. C. Brice, *The Growth of Crystals from Liquids* (North-Holland, 1973).
J. J. Gilman, *The Art and Science of Growing Crystals* (Wiley, 1963).
P. Hartmann (ed.), *Crystal Growth : an Introduction* (North-Holland, 1973).
H. K. Henisch, *Crystal Growth in Gels* (Pennsylvania State Univ. Press, 1970).
R. A. Landise, *The Growth of Single Crystals* (Prentice-Hall, 1970).
B. R. Pamplin (ed.), *Crystal Growth* (Pergamon Press, 1975).
W. G. Pfann, *Zone Melting* (Wiley, 1966).

The Crystalline State
P. Gay, *The Crystalline State* (Oliver and Boyd, 1972).
A. Holden, *Nature of Solids* (Columbia Univ. Press, 1965).
D. McKie and C. McKie, *Crystalline Solids* (Wiley, 1974).
A. F. Wells, *Structural Inorganic Chemistry* (Oxford Univ. Press, 5th ed., 1983).

Non-Crystalline Solids
J. Hlaváč, *The Technology of Glass and Ceramics* (North-Holland, 1983).
N. March, R. A. Street, and M. Tosi (eds.), *Amorphous Solids and the Liquid State* (Plenum Press, 1985).
R. Zallen, *The Physics of Amorphous Solids* (Wiley, 1983).

Interatomic Binding Mechanisms
R. C. Evans, *An Introduction to Crystal Chemistry* (Cambridge Univ. Press, 1964).
N. H. Fletcher, *The Chemical Physics of Ice* (Cambridge Univ. Press, 1970).
M. O'Keefe and A. Navrotsky (eds.), *Structure and Bonding in Crystals* (Academic Press, 1981), 2 vols.
W. B. Pearson, *Crystal Chemistry and Physics of Metals and Alloys* (Wiley, 1972).
L. Pauling, *The Nature of the Chemical Bond* (Cornell Univ. Press, 3rd ed., 1960).

Crystal Symmetry Operations and Space Groups
M. J. Buerger, *Elementary Crystallography* (MIT Press, 1978).
R. S. Knox and A. Gold, *Symmetry in the Solid State* (Benjamin, 1964).
G. F. Koster, *Space Groups and Their Representations* (Academic Press, 1964).
M. Lax, *Symmetry Principles in Solid State and Molecular Physics* (Wiley, 1974).
H. D. Megaw, *Crystal Structures : a Working Approach* (Saunders, 1973).
F. C. Phillips, *An Introduction to Crystallography* (Halsted, 4th ed., 1979).
J. A. Salthouse and M. J. Ware, *Point Group Character Tables* (Cambridge Univ. Press, 1972).
J. V. Smith, *Geometric and Structural Crystallography* (Wiley, 1982).
R. W. G. Wyckoff, *Crystal Structures* (Krieger, 1982).

Crystal Diffraction and Reciprocal Lattices
G. E. Bacon, *Neutron Diffraction* (Oxford Univ. Press, 3rd ed., 1975).
L. Brillouin, *Wave Propagation in Periodic Structures* (Dover, 1972).
M. J. Buerger, *Crystal Structure Analysis* (Krieger, 1979).
H. S. Lipson, *Crystals and X-Rays* (Springer, 1970).
J. B. Pendry, *Low Energy Electron Diffraction* (Academic Press, 1974).
T. B. Rymer, *Electron Diffraction* (Methuen, 1970).
B. E. Warren, *X-Ray Diffraction* (Addison-Wesley, 1969).
M. M. Woolfson, *An Introduction to X-Ray Crystallography* (Cambridge Univ. Press, 1970).

Crystalline Defects

S. Amelinckz, *The Direct Observation of Dislocations* (Academic Press, 1964).

W. Bollmann, *Crystal Defects and Crystalline Interfaces* (Springer, 1970).

A. H. Cottrell, *Theory of Crystal Dislocations* (Gordon & Breach, 1964).

L. Eyring and M. O'Keefe (eds.), *The Chemistry of Extended Defects in Nonmetallic Solids* (North-Holland, 1970).

C. P. Flynn, *Point Defects and Diffusion* (Oxford Univ. Press, 1972).

J. Friedel, *Dislocations* (Addison-Wesley, 1964).

L. A. Girifalco, *Atomic Migration in Crystals* (Blaisdell, 1964).

J. P. Hirth and J. Lothe, *Theory of Dislocations* (McGraw-Hill, 1968).

D. Hull and D. J. Bacon, *Introduction to Dislocations* (Pergamon Press, 3rd ed., 1984).

F. A. Kröger, *Chemistry of Imperfect Crystals* (North-Holland, 2nd ed., 1973), 3 vols.

S. Mrowec, *Defects and Diffusion in Solids* (North-Holland, 1980).

N. G. Parsonage and L. A. K. Stavely, *Disorder in Crystals* (Oxford Univ. Press, 1979).

A. M. Stoneham, *Theory of Defects in Solids* (Oxford Univ. Press, 1975).

LATTICE DYNAMICS

In this chapter we are concerned with the spectrum of character-istic vibrations of a crystalline solid. This subject leads to a consideration of the conditions for wave propagation in a periodic lattice, the energy content and specific heat of lattice waves, the particle aspects of quantized lattice vibrations (phonons), and the consequences of anharmonic coupling between atoms. These topics form a significant part of solid state physics, and their discussion additionally introduces us to the concepts of permitted and forbidden frequency ranges, concepts which will be encountered again in connection with electronic spectra of solids.

The zero-point energy and thermal energy of a solid are manifest in incessant complicated vibrations of the atoms. These vibrations have Fourier components at a variety of frequencies. Additional motion is superimposed if the solid is stimulated by some external source, and we usually assume that the *principle of superposition* applies to the sum of these motions, i.e., we assume that the effect of several distur-bances is found by simply adding them together. This assumption sounds plausible, provided that we remain in the linear region (or region of elastic deformation) such that the restoring force on each atom is approximately proportional to its displacement (Hooke's Law). As we shall see in discussing thermal conductivity, there are some effects of nonlinearity or "anharmonicity" even for very modest atomic displace-ments. Anharmonic effects are also important for interactions between phonons and photons, a topic discussed in the final portion of Section 2.3.

2.1 Elastic Waves, Atomic Displacements, and Phonons

The speed with which a longitudinal wave moves through a liquid of density ρ is given by

$$v_0 = \lambda\nu = (B_S/\rho)^{1/2} \qquad (2\text{-}1)$$

where B_S is the adiabatic elastic bulk modulus, or stiffness coefficient. For solids, this equation must be modified to take into account the finite rigidity. To be more correct, we should describe the transmission of an acoustic wave through a solid in terms of the several non-vanishing elements of a fourth-order stiffness tensor. Thus in even the simplest crystalline materials, the velocity of sound is in general a function of the direction of propagation. Furthermore, we should note that solids will sustain the propagation of transverse waves, which travel more slowly than longitudinal waves. Nevertheless, we should expect Equation 2-1 to give the correct order of magnitude for the longitudinal speed of sound. The larger the bulk modulus and the smaller the density, the more rapidly can sound waves travel.

The speed v_0 from Equation 2-1, as calculated from the known crystal density and measured bulk modulus for a number of solids, is shown in Table 2-1. These values for v_0 are compared with direct observations of the speed of sound. The comparison is a tolerably favorable one for most of the solids cited, and demonstrates that sound speeds are of the order of 5000 m/s in typical metallic, covalent, and ionic solids. Some of these numbers for v_0 will be used again for the calculations in Tables 2-2 and 2-3.

The use of macroscopic concepts for wave propagation is perfectly adequate provided that the wavelength of the wave being described is very large compared with interatomic spacings. For a wave whose amplitude is not unduly large (so that Hooke's Law is obeyed), there can be no more than 21 independent coefficients in the stiffness tensor even for the most general three-dimensional lattice. This number is reduced to only three non-equivalent stiffness constants for a cubic crystal. Such constants are of course intimately connected with the restoring force exerted on each displaced atom within a periodic lattice.

It is appropriate that we should be reminded of the importance of work on the theory and practice of the continuum, or macroscopic, approach to wave propagation in solids. The frequency range of nondispersive acoustic waves, for which the product $(\lambda\nu)$ of wavelength and frequency is a constant propagation velocity, is very wide. This range has received a great deal of experimental attention. The volumes by

TABLE 2-1   ELASTIC BULK MODULUS AND SPEED OF SOUND
FOR SOME TYPICAL SOLIDS[*]

Solid	Structure Type	Nearest-Neighbor Distance r_0(Å)	Density ρ (kg/m³)	Elastic Bulk Modulus B_s (10^{10} N/m²)	Calculated Wave Speed $v_0 = (B_s/\rho)^{1/2}$ (m/s)	Observed Speed of Sound (m/s)
Sodium	B.C.C.	3.71	970	0.52	2320	2250
Copper	F.C.C.	2.55	8960	13.4	3880	3830
Zinc	H.C.P.	2.66	7130	8.3	3400	3700
Aluminum	F.C.C.	2.86	2700	7.35	5200	5110
Lead	F.C.C.	3.49	11340	4.34	1960	1320
Nickel	F.C.C.	2.49	8900	19.0	4650	4970
Germanium	Diamond	2.44	5360	7.9	3830	5400
Silicon	Diamond	2.35	2330	10.1	6600	9150
SiO$_2$	Hexagonal	1.84	2650	5.7	4650	5720
NaCl	Rocksalt	2.82	2170	2.5	3400	4730
LiF	Rocksalt	2.01	2600	6.7	5100	4950
CaF$_2$	Fluorite	2.36	3180	8.9	5300	5870

[*] Elastic data for the elements taken from K. A. Gschneidner, in *Solid State Physics, Vol. 16*, (Academic Press, 1964). Other numerical values from *American Institute of Physics Handbook* (McGraw-Hill, Third Edition, 1971).

Mason (1958) and Huntingdon (1964) cited in this chapter's bibliography include detailed studies of the relationship between the stiffness tensor and the modes of acoustic propagation.

Stiffness constants can readily be measured by sending an ultrasonic pulse through a crystal, a technique suggested by Hearmon.[1] Ultrasonic techniques in crystals have now been extended into the GHz range,[2] which as we shall soon see is still a long way from the maximum possible propagation frequency in a crystal. Since sound waves travel through a solid at a speed of some 5000 m/s, a 1 GHz wave has a wavelength of about 5 μm, or more than 20,000 atomic spacings.

There are a number of things we can learn about the complete lattice vibrational spectrum of a solid most appropriately by considering the periodic nature of a crystalline material. The continuum approach fails to alert us to the changing conditions encountered when the wavelength is *not* many times larger than an interatomic spacing, whereas a description in terms of atomic force constants is equally valid for low and high frequencies. Hence in this chapter we shall emphasize the microscopic approach.

A lattice vibrational wave in a crystal is a repetitive and systematic sequence of atomic displacements (longitudinal, transverse, or some combination of the two), which can be characterized by

$$\left. \begin{array}{l} \text{a propagation velocity} \quad v \\ \text{a wavelength } \lambda \text{ or wave vector } |k| = (2\pi/\lambda) \\ \text{a frequency } \nu \text{ or angular frequency } \omega = (2\pi\nu) = (vk) \end{array} \right\} \quad \text{(2-2)}$$

---

[1] R. F. S. Hearmon, Rev. Mod. Phys. **18**, 409 (1946).

[2] See the article by N. G. Einspruch in *Solid State Physics Vol. 17*, edited by F. Seitz and D. Turnbull (Academic Press, 1965). See also the chapter by E. H. Jacobsen in *Phonons and Phonon Interactions*, edited by T. A. Bak (Benjamin, 1964).

These are all good wave-like attributes. By considering the restoring forces on displaced atoms we can produce an equation of motion for any displacement, and generate a *dispersion relationship* between frequency and wavelength. We shall sometimes refer to frequency and wavelength, but more frequently to the companion relationship between angular frequency and wave vector.

From a classical standpoint, a wave which satisfies this dispersion relationship can be propagated with any amplitude whatsoever. At this point, however, we should note that elementary quantized lattice vibrations have a dual wave-particle nature, as do quanta of electromagnetic radiation and particles of matter. The particle aspect is called the *phonon*, and we must regard the transmission of a displacement wave in a solid as being the movement of one or many phonons, each transporting energy $h\nu = \hbar\omega$, and crystal momentum $\hbar\mathbf{k}$. Thermal conduction, electron scattering, and several other processes in solids involve the creation or annihilation of a single phonon, and in such processes the particle aspect is as important as the wave aspect. With other phenomena, our awareness of the discrete particle nature of excitation energy will be dependent on whether or not the thermally available supply of phonons is large.

Prior to the development of quantum theory, the specific heat of a solid was calculated from the classical standpoint of energy equipartition. A crystal of N atoms was conceived of as an array of 3N independent harmonic oscillators, each of energy k_0T, for a total lattice vibrational energy $U = 3Nk_0T$. At constant volume such an assembly would have a specific heat per mole of

$$C_v = (\partial U/\partial T)_v = 3N_A k_0 = 24.94 \text{ joule/mole}\cdot\text{kelvin} \qquad (2\text{-}3)$$

which is the well-known law of Dulong and Petit (1869). As typified by the data of Figure 2-1, this classical law works rather well for many solids at room temperature and above, but fails for all solids at low

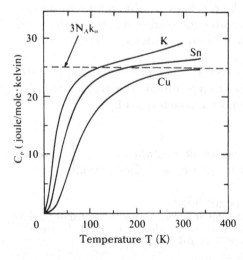

Figure 2-1 Temperature variation of the specific heat at constant pressure for typical metallic elements. Since C_p and C_v are almost the same for a solid, it is proper to compare the measured C_p with the classical law of Dulong and Petit. The departure from the classical law is described by Debye's quantized theory of specific heats. Data here are from the *American Institute of Physics Handbook* (McGraw-Hill, 1971).

temperatures. This failure is a result of the restrictiveness of quantum conditions, as discussed first by Einstein (1907), and elaborated by Debye (1912) and by Born and von Kármán (1912). We shall discuss these conditions in Section 2.4.

Quantum statistics give an answer appreciably different from that provided by classical energy equipartition for frequencies which correspond with a phonon energy $\hbar\omega$ comparable with or larger than the thermal energy $k_0 T$. Thus the *particle* aspects of vibrational energy are noticeable for a frequency in excess of a temperature-dependent threshold $\nu_{th} = (\omega_{th}/2\pi) = (k_0 T/h)$. For room temperature this threshold situation corresponds to $\nu_{th} \simeq 6 \times 10^{12}$ Hz, or $\omega_{th} \simeq 4 \times 10^{13}$ radians per second, and this is well above the frequency range explored by any extension of acoustic techniques. The corresponding threshold wavelength is

$$\lambda_{th} = (2\pi v/\omega_{th}) \qquad (2\text{-}4)$$

and since we have noted that displacement waves travel at a speed of some 5000 m/s, it would appear that the particle-like aspects of waves at room temperature are prominent only for waves with $\lambda \leqslant 10^{-9}$ m. We may notice, in addition, that the latter are waves for which the discrete atomic nature of the solid must be rather important, since interatomic spacings in solids lie in the range from 1 to 4×10^{-10} m. However, at very low temperatures quantum restrictions will limit the choice of amplitude even for waves which have a wavelength much larger than atomic dimensions.

Recognizing, then, that we may have to think in particle-like terms in order to explain many things which involve the *amplitude* of a displacement wave in a solid, it is still possible for us to begin by establishing the *spectrum* of lattice vibrational energy. For this task, a wave approach founded upon the atomic periodicity of the lattice is entirely appropriate.

2.2 Vibrational Modes of a Monatomic Lattice

THE LINEAR MONATOMIC CHAIN

In a homogeneous solid the transmission of a plane wave in the x-direction can be represented by the displacement equation

$$u = A \exp[i(kx - \omega t)] \qquad (2\text{-}5)$$

for a monochromatic wave of amplitude A, wave-vector k, and angular frequency ω. Consider a linear array of identical atoms (Figure 2-2), each of mass m, with interatomic spacing a. For this array of atoms, the displacement u of the wave in Equation 2-5 can be described only at each atomic site; u has no meaning for intermediate values of x. Thus we must make Equation 2-5 more specific. The displacement of the r th atom is

$$u_r = A \exp[i(kra - \omega t)] \qquad (2\text{-}6)$$

Differentiating this twice, we note that the acceleration of the r th atom is

$$(d^2u_r/dt^2) = -\omega^2 A \exp[i(kra - \omega t)]$$
$$= -\omega^2 u_r \qquad (2\text{-}7)$$

so that from Newton's second law, the restoring force exerted on the r th atom must be

$$F_r = m(d^2u_r/dt^2) = -m\omega^2 u_r \qquad (2\text{-}8)$$

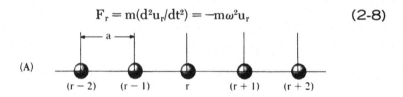

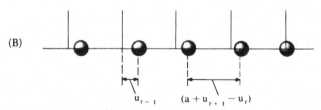

Figure 2–2 Five neighboring atoms in a linear monatomic chain lattice. (A) at their equilibrium locations, and (B) displaced by the passage of a longitudinal wave.

We should like to express this restoring force in terms of an atomic displacement force constant in order to find out how ω and k are related. In order to do so, we shall assume a simple "balls and springs" model for the atomic chain, a model for which the force on an atom is supposed to depend linearly (with constant of proportionality μ) on the extension or contraction of its *nearest-neighbor* distances and not at all on more remote neighbors. This model may at first appear to do scant justice to the sophistication of the quantum-mechanical coupling between atoms, but we should remember that small displacements in most solids *do* conform quite well with the linear Hooke's Law force dependence.

Within the framework of the Hooke's Law approach, the force on the r th atom in Figure 2-2 is

$$F_r = \mu(u_{r+1} - u_r) - \mu(u_r - u_{r-1})$$
$$= \mu(u_{r+1} + u_{r-1} - 2u_r) \qquad (2\text{-}9)$$

A comparison of this result with Equation 2-8 shows that

$$\omega^2 = (\mu/m)[2 - (u_{r+1}/u_r) - (u_{r-1}/u_r)] \qquad (2\text{-}10)$$

and from the form of Equation 2-6 we can rewrite Equation 2-10 as

$$\omega^2 = (\mu/m)\,[2 - \exp(ika) - \exp(-ika)]$$
$$= 2(\mu/m)\,[1 - \cos(ka)]$$
$$= 4(\mu/m)\,\sin^2(ka/2) \qquad (2\text{-}11)$$

Thus the dispersion relationship for allowed longitudinal waves in a monatomic chain lattice subject to nearest-neighbor restoring forces is

$$\omega = \pm 2(\mu/m)^{1/2}\,\sin(ka/2) = \pm\omega_m \sin(ka/2) \qquad (2\text{-}12)$$

The plus and minus signs in Equation 2-12 denote waves traveling either to the right or to the left. The motion at any point is periodic in time.

Figure 2-3 shows the form of the dispersion curve described by Equation 2-12. A graphical result qualitatively similar to this would have been obtained even if restoring forces resulting from more distant neighbors had been included (see Problem 2.1).

In Figure 2-3 it is convenient (as we shall find in many other aspects of solid state physics) to plot the result with wave vector k rather than wavelength λ as the independent variable. The region of *small* k is the spectral region of *long waves*, for which continuum concepts of acoustic waves are so appropriate. For if $(ka) \ll 1$, $\sin(ka/2)$ becomes essentially $(ka/2)$, and the relationship of angular frequency and wave vector is

$$\left.\begin{array}{l} \omega \approx v_0 k \\ v_0 = a(\mu/m)^{1/2} \end{array}\right\} \quad ka \ll 1 \qquad (2\text{-}13)$$

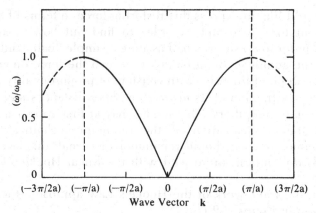

Figure 2–3 Dispersion relationship for the propagation of a longitudinal wave in a linear monatomic lattice, considering only the nearest-neighbor interactions, Equation 2-12. The range of k-space for which $|k| \leq (\pi/a)$ comprises the first Brillouin zone. The ordinate is expressed dimensionlessly in terms of the maximum angular frequency for which k is real, $\omega_m = 2(\mu/m)^{1/2} = (2v_0/a)$. Since sound waves travel through a solid at a typical speed of $v_0 \sim 5000$ m/s, this maximum angular frequency is likely to be of the order of 10^{14} radians/sec.

The phase velocity (ω/k) and group velocity ($\partial\omega/\partial k$) are the same in this non-dispersive long-wavelength, low-frequency regime, both equal to the ordinary speed of sound, v_0. The speed $v_0 = a(\mu/m)^{1/2}$ found for long waves according to this model is in full accord with Equation 2-1 for a continuous elastic medium, since (m/a) is the density and $a\mu$ the bulk modulus in one dimension. For a three-dimensional cubic solid, $\rho = (m/a^3)$ and $B_s = (\mu/a)$.

Even with the limitations of our crude one-dimensional model, we can surmise that knowledge of the speed v_0 of ordinary sound waves in a solid will permit us to estimate a value for the interatomic stiffness coefficient

$$\mu \approx m(v_0/a)^2 \qquad (2\text{-}14)$$

and hence for the bulk modulus $B_s = (\mu/a)$ or the compressibility $K = (1/B_s)$. This argument has of course been pursued in the opposite direction in Table 2-1, using a measured B_s to calculate v_0.

The "low-frequency" approximation just discussed is adequate for frequencies up to 10^{12} Hz or so, which conveniently includes the ranges of frequency normally described by the terms "acoustic" or "ultrasonic," explored experimentally by monochromatic techniques. However, as we consider shorter and shorter wavelengths, we can see from Figure 2-3 that ω reaches a limiting value $\omega_m = 2(\mu/m)^{1/2} = (2v_0/a)$ when $|k| = (\pi/a)$ and that there is considerable velocity dispersion at intervening frequencies. Expressed as functions of k, the phase and group velocities are in general given by

$$\left. \begin{aligned} v &= (\omega/k) = v_0 \left[\frac{\sin(ka/2)}{(ka/2)}\right] \\ v_g &= (\partial\omega/\partial k) = v_0 \cos(ka/2) \end{aligned} \right\} \qquad (2\text{-}15)$$

We shall see in connection with the Debye theory of specific heats (Section 2.4) what liberties are sometimes taken with the functional dependence of velocity on wave vector.

The group velocity goes to zero when nearest-neighbor atoms are moving in antiphase, which is the case for $k = \pm(\pi/a)$; that is, a wavelength $\lambda = 2a$. It makes no sense to attempt to describe the progress of a wave in terms of phase shifts of more than 180° between neighboring atoms; thus there is no physical significance to be attached to values of k for which $|k| > (\pi/a)$ (the portions of Figure 2-3 for which the curve is dotted). As we may readily verify, a set of atomic displacements consistent with Equation 2-6 for a value of k with absolute magnitude larger than (π/a) are equally consistent with any other wave vector k' for which $k' = (k + G)$, where G is a reciprocal lattice vector.[3] Thus the first Brillouin zone contains the entire spectrum.

It will be apparent from these remarks that a wave corresponding with the Brillouin zone boundary, $|k| = (\pi/a)$, is a *standing wave* rather than a traveling wave. It has a wavelength appropriate for Bragg reflection through an angle of 180°. We shall continue to observe, for situations of two or three dimensions, that an excitation (such as a phonon or an electron) suffers Bragg reflection when its wave-vector corresponds with the boundary of a Brillouin zone. For more than one dimension, a wave with k occurring at the Brillouin zone boundary must have a zero component of group velocity in real space in the direction perpendicular to the zone boundary line or surface (in k-space).

A wave of angular frequency larger than $\omega_m = (2v_0/a)$ cannot be transmitted through our imaginary one-dimensional solid, for from Equation 2-12 we see that it must correspond with a complex value for the wave-vector. An imaginary component of wave-vector results in severe attenuation; thus waves with $\omega > \omega_m$ lie in a forbidden range of the frequency spectrum. We shall find that this concept holds equally well for real three-dimensional solids. In order to discuss waves in three-dimensional solids, however, we must generalize some of the other concepts which have served us so far.

VIBRATIONS OF A
THREE-DIMENSIONAL MONATOMIC SOLID

In describing wave motion for a monatomic one-dimensional solid, we found it convenient to consider only longitudinal displacements within a linear chain. It is certainly possible to envisage a purely longitudinal or purely transverse wave in a three-dimensional solid, but only for relatively symmetrical structures (such as cubic crystals) and for those special crystallographic directions which correspond with a plane of atoms moving in unison. For a cubic crystal, a wave can be com-

---

[3] It will be recalled from Section 1.5 that a reciprocal lattice vector G connects equivalent points in two Brillouin zones. For our one-dimensional crystal, G is a multiple of $(2\pi/a)$, and the first Brillouin zone extends from $k = (-\pi/a)$ to $k = (+\pi/a)$. The maximum value of k is some 10^{10} m^{-1} in a real solid, and the maximum angular frequency ω_m is some 10^{14} rad/sec.

pletely longitudinal or completely transverse in the [100], [110] or [111] directions, and this dictates the directions in **k**-space for which experimental dispersion relations are plotted in Figures 2-4, 2-12, and 2-13.

In general, however, a wave in a three-dimensional solid has both longitudinal and transverse character, and the equations describing the transmission of a wave must contain terms appropriate for the stiffness coefficients and velocities corresponding with both longitudinal and lateral displacements. It can well be imagined that the vibrational dispersion relation for a three-dimensional solid will look more complicated than that shown in Figure 2-3. This expectation is borne out by the two experimentally determined spectra of Figure 2-4. For each of these monatomic solids there is a longitudinal branch and two transverse branches, the latter pair coinciding for the high symmetry [100] and [111] directions.

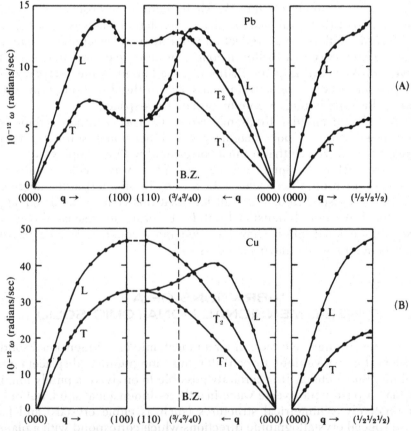

Figure 2-4 The phonon dispersion spectra for two metals which crystallize in the F.C.C. structure. Part (A) shows data for lead, and part (B) for copper. Angular frequency ω is plotted against the dimensionless vector $q = (ka/\pi)$ measured from the center of the Brillouin zone in three principal directions. For the [110] direction, the curves are extended through the Brillouin zone (B.Z.) boundary. Data for lead from Brockhouse et al., Phys. Rev. **128**, 1099 (1962). Data for copper from Svensson et al., Phys. Rev. **155**, 619 (1967), and from G. Nilsson and S. Rolandson, Phys. Rev. **B.7**, 2393 (1973). Data for both materials were obtained by inelastic scattering of monochromatic neutron beams.

Spectra such as these have been obtained by the *inelastic* scattering of slow neutrons. In Section 1.4 we were concerned primarily with the extra information neutrons could give about crystal structures when they were scattered *elastically*, subject to the Bragg condition. It was also mentioned at that time that the slow speed of thermal neutrons could make them useful for study of lattice vibrations. It is now time to return to the inelastic scattering process.

We have already noted that a lattice vibration of angular frequency ω connotes the movement of phonons, each with energy $\hbar\omega$ and crystal momentum $\hbar k$. Use of the term *crystal momentum* does not mean that "ordinary" momentum is transferred through the crystal at the corresponding rate; but "crystal momentum" is a conserved quantity in interactions which involve the creation or annihilation of crystal phonons. Thus the slowing down of a neutron in excitation of a lattice vibrational mode is subject both to energy conservation in the system and the conservation of crystal momentum, the latter amounting to a conservation law for wave-vector.

A neutron of velocity v has wave-vector

$$\mathbf{K}_n = M_n \mathbf{v}/\hbar \qquad (2\text{-}16)$$

and kinetic energy

$$E = (\hbar^2 K_n^2 / 2M_n) \qquad (2\text{-}17)$$

Suppose that in either absorbing or creating a phonon, the energy and wave vector of this neutron are changed to E' and $\mathbf{K}_n'$ respectively. Then the angular frequency ω and wave-vector k of the phonon involved will be related by the conservation equations

$$E - E' = \pm\, \hbar\omega \qquad (2\text{-}18)$$

and

$$\mathbf{K}_n - \mathbf{K}_n' = \mathbf{G} \pm\, \mathbf{k} \qquad (2\text{-}19)$$

Here the vector $\mathbf{G}$ is either zero or a reciprocal lattice vector. It can be included in Equation 2-19 since a phonon of wave-vector k is identical to a phonon of any $(k + G)$. The conservation of wave-vector is illustrated for a simple case by Figure 2-5.

The inelastic scattering process of Equations 2-18 and 2-19 can, in principle, be carried out with photons or electrons as the investigating wave-particles. However, the vastly different speeds of sound and light mean that only a small fraction of photon energy can be transferred. Thermal neutrons are well suited in speed and in energy to be inelastically scattered for an appreciable change in magnitude and direction of wave-vector.

Following the development of uranium reactors as prolific sources of slow neutrons, attempts to study lattice vibrations through inelastic

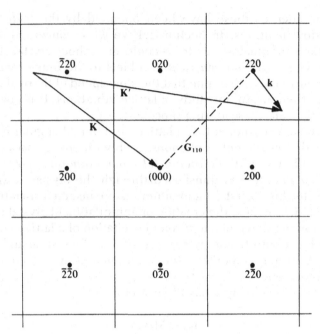

Figure 2-5 A section through k-space for a simple cubic crystal lattice, showing the first Brillouin zone and the eight surrounding zones. The coordinates for the center of each are expressed as multiples of the dimensionless variable $q = (ka/\pi)$. The vectors **K** and **K'** are shown for a neutron which suffers an inelastic collision in creating a phonon of wavevector **k**. The vector relationship between **K**, **K'**, and **k** is preserved by a reciprocal lattice vector, as in Equation 2-19.

neutron scattering were made in the early 1950's, with successes first reported[4] in 1955. The technique has proved to be very valuable, and we now have experimental data on the vibrational spectrum for many solids.[5] The change of neutron speed as a result of scattering is measured either by Bragg diffraction or by time-of-flight techniques.

Lead and copper, whose vibrational spectra are shown in Figure 2-4, both crystallize in the F.C.C. structure. Differences between their phonon spectral curves arise primarily from differing degrees of dependence on distant neighbors. With the copper data shown in the lower half of the figure, it has been suggested[6] that force constants which relate atoms more remote than eighth-nearest can be neglected, and

---

[4] The first reported successes of lattice spectra through inelastic neutron scattering were those for aluminum, by B. N. Brockhouse and A. T. Stewart, Phys. Rev. **100**, 756 (1955), and for beryllium and vanadium, by R. S. Carter, D. J. Hughes, and H. Palevsky, Phys. Rev. **99**, 611(A) (1955).

[5] For a comprehensive discussion of inelastic neutron scattering and its application to dispersion curves of lattice dynamics, see the articles by B. N. Brockhouse in either *Phonons in Perfect Lattices and in Lattices with Point Imperfections,* edited by R. W. H. Stevenson (Plenum Press, 1966), or *Phonons and Phonon Interactions,* edited by T. A. Bak (Benjamin, 1964). The emphasis in these articles is on results obtained using Bragg diffraction for momentum analysis of incoming and scattered neutrons. This requires a so-called triple-axis spectrometer.

[6] G. Nilsson and S. Rolandson, Phys. Rev. **B.7**, 2393 (1973).

that the nearest-neighbor interaction far outweighs any others for most crystal directions. In order to explain the more complex data for lead in part (a) of Figure 2-4, it is necessary[7] to take into account neighbors at least as remote as ten atomic planes away for a reasonably quantitative fit to the dispersion curves.

Additional small anomalies in the data for lead arise from a different source, the result of a difference in the coupling depending on whether the wave-vector of the phonon is larger or smaller than twice that for the most energetic free electrons in the metal. These *Kohn anomalies*[8] are seen in a number of metals for which there is a strong electron-phonon coupling.

Curves qualitatively similar to those of Figure 2-4 are found for other solids with a monatomic basis. The most useful directions to be displayed and the locations of the abscissa maxima depend on the shape of the Brillouin zone for the crystal structure involved. (See Problem 2.3 in connection with the zone appropriate to Figure 2-4.)

MODE COUNTING.
THE DENSITY OF STATES

Although any vibrational pattern is possible for an elastic continuum, there is a limit to the number of distinguishable modes of vibration for a lattice of finite size made up of discrete atoms.

In exploring the distribution of these modes with respect to frequency or wave-vector, let us first consider a linear monatomic chain of $(N + 1)$ atoms, which will have a length of Na. Suppose that the end atoms are clamped, so that $u_1 = 0$ and $u_{N+1} = 0$. Then either longitudinal or transverse vibrations can exist for which there are 1, 2, 3, . . . or N half-wavelengths in the distance Na. These permitted vibrational modes are situations for which the wave-vector is

$$k = \frac{\pi}{Na}, \frac{2\pi}{Na}, \frac{3\pi}{Na}, \ldots \text{ or } \frac{\pi}{a} \qquad (2\text{-}20)$$

We note that these are linearly spaced.

Of course the spacing will be very small if N is a large number, and it will then be appropriate to describe the number of *states* (number of distinguishable vibrational modes) lying between k and $(k + dk)$ as $(Na/\pi) \cdot dk$. Then the number of states *per unit length* of our one-dimensional crystal, in an interval dk, is simply

$$g(k) \cdot dk = (1/\pi) \cdot dk, \quad k \leq (\pi/a)$$
$$= \quad 0 \quad , \quad k > (\pi/a) \qquad (2\text{-}21)$$

We refer to g(k) as the *density of states* (per unit length, per unit in-

[7] B. N. Brockhouse *et al.*, Phys. Rev. **128**, 1099 (1962).

[8] W. Kohn, Phys. Rev. Letters **2**, 393 (1959).

terval of k) with respect to k. For the one-dimensional example chosen, we find that g(k) does not depend on k within the permitted spectral range. For most other situations we shall find that g(k) has an explicit dependence on k.

It may seem that the preceding evaluation of a density of states was weakened by our assumption that the end atoms in the chain were motionless (since such an assumption permits solutions only for standing waves rather than running waves). Alternatively, we might suppose an infinite linear chain of atoms with atomic spacing a. Now let running waves occur, subject to the restriction that any permitted mode repeats its set of displacements after the distance L = Na. Thus $u_r = u_{N+r}$, and so on.[9] Modes which can propagate subject to this restriction must have wave vectors

$$k = \pm \frac{2\pi}{L}, \pm \frac{4\pi}{L}, \pm \frac{6\pi}{L}, \ldots \text{ or } \pm \frac{\pi}{a} \qquad (2\text{-}22)$$

where the positive and negative signs refer to waves traveling to the right or to the left.

A comparison of Equations 2-20 and 2-22 makes it obvious that for either standing or running waves, the density of states g(k) in terms of unit change in |k| is $(1/\pi)$ for a one-dimensional crystal. Indeed, we can assert that the density of states for an infinite chain is independent of the boundary conditions applied. Now an infinite linear chain of atoms is not the kind of crystal we study experimentally. However, by demonstrating the unimportance of boundary conditions for a linear chain we can go on to assert that the density of states with respect to wavevector, frequency, or energy in a real three-dimensional crystal will not depend on the shape or nature of the surface, provided only that the crystal is large compared with atomic dimensions.

With problems of solid state physics it is often necessary to know the density of states with respect to variables other than k. (This applies for electrons as well as phonons.) For example, $g(\omega)$ is often needed as well as g(k). In the linear monatomic lattice we have just been considering, we can write

$$g(\omega) = g(k) \cdot (dk/d\omega) = (1/\pi) \cdot (dk/d\omega) \qquad (2\text{-}23)$$

Now since

$$\omega = 2(\mu/m)^{1/2} \sin(ka/2)$$
$$= \omega_m \sin(ka/2) \qquad (2\text{-}24)$$

---

[9] This is the method of "periodic boundary conditions," developed by M. Born and Th. von Kármán, Phys. Zeit. **13**, 297 (1912). As we shall find in later chapters, the artifice of periodic or cyclic boundary conditions is invaluable in exploring both the vibrational and the electronic spectra of solids. When we speak of cyclic rather than periodic boundary conditions, we imagine a chain of N atoms forming a circle, so that $u_r = u_{N+r}$ because the r th atom *is* the (N + r) th atom. For most of us, cyclic boundary conditions are less easy to visualize in three dimensions, although periodic boundary conditions can still be imagined. Imagination is all that is necessary, for the imagined number N disappears as soon as we express the density of states per *unit length* or per *unit volume* of crystal.

then

$$(d\omega/dk) = (a\omega_m/2)\cos(ka/2)$$
$$= (a/2)[\omega_m^2 - \omega^2]^{1/2} \qquad (2\text{-}25)$$

so that

$$g(\omega) = (2/\pi a)[\omega_m^2 - \omega^2]^{-1/2}$$
$$= (1/\pi v_0)[1 - (\omega/\omega_m)^2]^{-1/2} \qquad (2\text{-}26)$$

Thus $g(\omega)$ *does* have an explicit dependence on ω, and in fact becomes infinite at the upper limit. We shall have to return to this expression for $g(\omega)$, and to $g(\omega)$ for more complicated situations, when we consider the quantum theory of specific heats in Section 2.4.

We have seen that the allowed modes for a linear chain with required periodicity over a length L fall in a sequence which can be represented by a series of points in a one-dimensional k-space, the spacing being $(2\pi/L)$. Figure 2-6(a) shows this representation.

The same kind of representation can be used to depict the distribution of allowed modes in k-space for a three-dimensional crystal. Figure 2-6(b) shows just a few of the modes permitted when (as a matter of convenience) we specify that a permitted set of displacements must be periodic with respect to the distance L along each of the Cartesian axes. Thus we can see from the figure that the volume of k-space associated with each permitted state is $(2\pi/L)^3$. Then, since the volume of k-space in a spherical shell centered on the origin of radius $k \equiv |\mathbf{k}|$ and

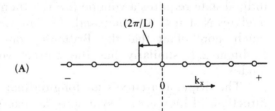

Figure 2-6 Location of allowed states of vibration in k-space when periodicity is required over a distance L. (A), for a monatomic linear lattice. (B), showing the states in the positive quadrant of the k_x k_y plane for a three-dimensional lattice.

thickness dk is $4\pi k^2 \cdot dk$, the number of allowed vibrational states *per unit volume* in a range dk must be

$$g(k)\cdot dk = \frac{4\pi k^2 \cdot dk}{L^3(2\pi/L)^3}$$
$$= (k^2/2\pi^2)\cdot dk \qquad (2\text{-}27)$$

The distribution of modes in k-space depicted in Figure 2-6(b) will extend to the Brillouin zone boundaries in the various directions.

The enumeration of $g(\omega)$ is considerably more complicated for a three-dimensional than a one-dimensional solid, though it can readily be demonstrated that $g(\omega)$ varies as ω^2 in the low-frequency nondispersive range (see Problem 2.4). Evaluation of the *total* number of modes would be a complicated calculation for most real crystals, but fortunately the answer is a simple one: any three-dimensional array of N atoms in total can vibrate in 3N distinguishably different ways. 3N is, of course, the classical number of degrees of freedom for N atoms.

Of the 3N vibrational modes, two-thirds (2N) are associated with transverse displacements and one-third (N) with longitudinal waves. In k-space language, the volume of the Brillouin zone for the crystal structure is adequate for describing one longitudinal mode for every atom, and the same volume of k-space can simultaneously accommodate twice as many points describing transverse modes. If we cut some material from the end of the crystal, the k-space volume of the first Brillouin zone is unchanged but the points in k-space representing distinguishable vibrational states are further apart.

As a justification for the above statement that N longitudinal and 2N transverse modes exist for N identical atoms in a three-dimensional array, consider a simple cubic crystal of length L on each side, so that $N = (L/a)^3$. The Brillouin zone is a cube of edge $(2\pi/a)$ and total volume $(2\pi/a)^3$. Since we know from Figure 2-6(b) that each allowed longitudinal state requires a volume $(2\pi/L)^3$, then the complete zone accommodates N states, as anticipated. The interested reader can prove with much more effort that the Brillouin zone for F.C.C. (examined in Problem 2.3) similarly has the correct volume to accommodate N modes.

The dispersion curves for longitudinal and transverse phonons in directions of high crystal symmetry (as exemplified by Figure 2-4) can be analyzed in terms of the most important interatomic force constants. This new information can then be used for a numerical calculation of the density of vibrational states as a function of frequency. As may be imagined, in this computer-oriented age the procedure can be carried out in much more detail than was possible in the pioneer calculations of density of states.[10] The result of such a numerical analysis for the

---

[10] A classic series of papers by M. Blackman [Proc. Roy. Soc. **A.148**, 365 (1935); **A.159**, 416 (1937)] showed how $g(\omega)$ should vary with ω for models of relatively simple crystal structures, taking first- and second-nearest-neighbor interactions into account. The ability to compute $g(\omega)$ has since been facilitated both by the use of computers and by employment of a numerical technique [G. Gilat and L. J. Raubenheimer, Phys. Rev. **144**, 390 (1966)] which greatly decreases the computation time. In this technique, the Brillouin zone is divided into a cubic mesh, and constant frequency surfaces inside each small cube are approximated by a set of parallel planes.

copper data of Figure 2-4 is shown in Figure 2-7, in which the small contributions of displacements with respect to neighbors as remote as sixth-nearest have been taken into account.[11] More than 75 per cent of the vibrational states lie in the part of the frequency spectrum for which $g(\omega)$ is *not* proportional to ω^2.

For solids in which the scattering of slow neutrons is primarily *incoherent* (because of nuclear spin disorder), Placzek and Van Hove[12] have pointed out that $g(\omega)$ can be measured directly from the incoherent component. This is rather interesting in that it is a technique particularly suited to those solids for which the preponderance of incoherent over coherent neutron scattering makes the measurement of dispersion curves so difficult. Figure 2-8 shows an example of $g(\omega)$ for vanadium, measured in this fashion.

The appearance of sharp peaks and abrupt changes in slope can be seen in Figures 2-7 and 2-8, particularly in the former case where the curve has a more quantitative backing. The values of ω for which $g(\omega)$ suffers an abrupt change of slope are known as *critical points,* or Van Hove singularities.[13] These are the angular frequencies for which the group velocity of phonons vanishes in some direction, because a contour of constant ω in k-space changes character with respect to the Brillouin zone boundary. Thus we see that the highest peak in Figure 2-7 occurs at the maximum allowed frequency for longitudinal phonons

---

[11] E. C. Svensson, B. N. Brockhouse, and J. M. Rowe, Phys. Rev. **155**, 619 (1967).

[12] G. Placzek and L. Van Hove, Phys. Rev. **93**, 1207 (1954).

[13] Such *critical points* in the lattice vibrational spectrum were noted in a paper by L. Van Hove, Phys. Rev. **89**, 1189 (1953). The terms "critical point" and "Van Hove singularity" are also commonly used in discussions of the density of *electron states* as a function of electron energy in a solid, since similar considerations apply.

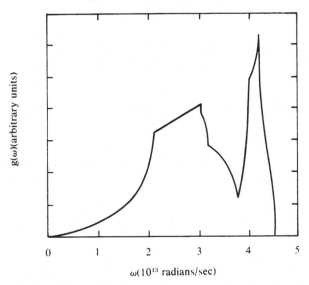

Figure 2-7 Total density of vibrational states as a function of ω, for copper. The curve shown here is derived from numerical analysis of the various branches in the experimentally measured dispersion curves in Figure 2-4(B). After Svensson et al., Phys. Rev. **155**, 619 (1967).

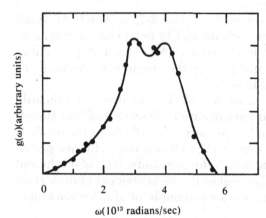

Figure 2-8 Total density of vibrational states as a function of ω, for vanadium. The determination was made in this case by the *incoherent* scattering of neutrons. Data reported by A. T. Stewart and B. N. Brockhouse, Rev. Mod. Phys. **30**, 250 (1958). This curve differs significantly from that reported by R. S. Carter et al., Phys. Rev. **104**, 271 (1956) which points up the difficulties of this kind of experiment.

in the [110] direction; also that g(ω) suffers an abrupt reduction in slope at the frequency for which transverse phonons in the [111] direction coincide with the Brillouin zone boundary.

Critical point analysis is of considerable value in the interpretation of vibrational spectra. As we shall see later, Van Hove singularities occur in the electronic spectra of solids and are equally valuable in the elucidation of the electronic energy bands.

Vibrational Spectrum For a Structure with a Basis

Up to this point, the discussion has deliberately been restricted to structures in which the basis consists of a single atom. Additional features appear for more complex structures in which there are two or more atoms per primitive basis (whether or not the two atoms are of different chemical species). The most striking feature to appear is that permitted and forbidden frequency ranges alternate.[14] We can learn how the alternation of regions with real and complex **k** arise in connection with lattice vibrations, benefiting from this experience when we encounter comparable phenomena in the electronic spectrum of a solid.

It is convenient (though not essential) to begin our consideration of these more complicated phenomena by assuming the presence of two different kinds of atom.

NORMAL VIBRATIONAL MODES OF A LINEAR DIATOMIC LATTICE

Imagine a one-dimensional crystal as depicted in Figure 2-9. This differs from Figure 2-2 only in that atoms of masses M and m are alternately placed. We shall suppose that m < M. Now if a longitudinal dis-

---

[14] Those readers whose backgrounds have involved much exposure to electrical circuit theory will recognize that the alternation of permitted and forbidden ranges of ω is a mathematical parallel to the alternation of pass bands and stop bands for transmission lines or combinations of tuned circuits. The similarity between circuit theory and the approach to frequency ranges of real or complex **k** in a solid is used extensively in Shockley's classic semiconductor monograph *Electrons and Holes in Semiconductors* (Van Nostrand, 1950). The close similarity between acoustic, electrical, and other waves which must experience periodic constraints is also fully explored in L. Brillouin, *Wave Propagation in Periodic Structures* (Dover, 1953).

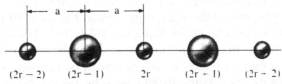

Figure 2-9 A sequence of neighboring atoms in a linear diatomic chain. The smaller atoms are assumed to have a mass m which is less than the mass M of the larger atoms. Note that the *repeat* distance is now 2a. This will affect the size of the Brillouin zone.

turbance is encouraged to travel along the crystal, the displacements of the two kinds of atom will usually have different amplitudes

$$
\left.
\begin{aligned}
u_{2r} &= A \exp\{i[2kra - \omega t]\} \\
u_{2r+1} &= B \exp\{i[(2r + 1)ka - \omega t]\}
\end{aligned}
\right\}
\tag{2-28}
$$

If we suppose that only the nearest-neighbor restoring forces are important and that displacements are in the elastic range of Hooke's Law, then the force equations for the displacements u_{2r} and u_{2r+1} can be written in the same fashion as for Equations 2-7 to 2-9 with a monatomic chain:

$$
\left.
\begin{aligned}
-m\omega^2 u_{2r} &= m(d^2 u_{2r}/dt^2) = \mu[u_{2r+1} + u_{2r-1} - 2u_{2r}] \\
-M\omega^2 u_{2r+1} &= M(d^2 u_{2r+1}/dt^2) = \mu[u_{2r+2} + u_{2r} - 2u_{2r+1}]
\end{aligned}
\right\}
\tag{2-29}
$$

Substitution of the two portions of Equation 2-28 into Equation 2-29 produces the two simultaneous equations

$$
\left.
\begin{aligned}
-m\omega^2 A &= \mu B[\exp(ika) + \exp(-ika)] - 2\mu A \\
-M\omega^2 B &= \mu A[\exp(ika) + \exp(-ika)] - 2\mu B
\end{aligned}
\right\}
\tag{2-30}
$$

which we can rearrange as

$$
\left.
\begin{aligned}
A(2\mu - m\omega^2) &= 2\mu B \cos(ka) \\
B(2\mu - M\omega^2) &= 2\mu A \cos(ka)
\end{aligned}
\right\}
\tag{2-31}
$$

These can be solved together to eliminate A and B, expressing a dispersion relationship between k and ω:

$$
(2\mu - m\omega^2)(2\mu - M\omega^2) = 4\mu^2 \cos^2(ka)
\tag{2-32}
$$

so that

$$
\omega^2 = \mu\left(\frac{1}{m} + \frac{1}{M}\right) \pm \mu\left[\left(\frac{1}{m} + \frac{1}{M}\right)^2 - \frac{4\sin^2(ka)}{mM}\right]^{1/2}
\tag{2-33}
$$

Whereas we previously found that for a monatomic linear chain there was one solution for $k > 0$ (a wave traveling to the right) and one for $k < 0$ (a wave traveling to the left), we now have *two* values of ω for a given k even in the positive quadrant. The spectrum of the result for ω as a double-valued function of k is shown in Figure 2-10.

The lower branch of the spectrum seen in Figure 2-10 is described by the negative sign choice in Equation 2-33. This branch is usually called the *acoustic branch* and corresponds fairly well with what we have already seen in monatomic lattices, with two exceptions:

1. Any set of displacements can now be described in terms of a wave-vector with an absolute value no larger than $(\pi/2a)$, compared with a Brillouin zone boundary at $\pm (\pi/a)$ for a linear monatomic chain. This reminds us that the range covered by a Brillouin zone is deter-

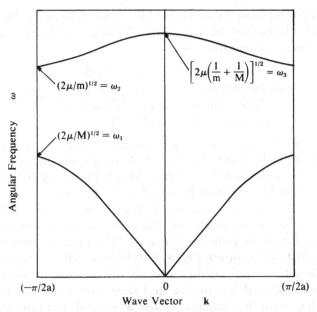

Figure 2–10 Dispersion relationship for the propagation of a longitudinal wave in a linear diatomic lattice. The first Brillouin zone comprises the range of k-space for which $|k| \leq (\pi/2a)$. The lower branch (which may be compared with that of Figure 2-3) is the *acoustic* branch, and the new upper branch is usually called the *optical* branch of the vibrational spectrum.

mined by the periodic repeat distance 2a, not by the nearest-neighbor distance.

2. The maximum possible angular frequency for acoustic mode vibrations is

$$\omega_1 = (2\mu/M)^{1/2} \tag{2-34}$$

from Equation 2-33, and this appears to be independent of the lighter atoms in the chain.

The reason for the second of these observations can be understood if we look into the amplitudes of the two types of atoms as a function of frequency. Equation 2-31 can be arranged as

$$\left[\frac{\text{Amplitude of heavy mass M}}{\text{Amplitude of light mass m}}\right] = \left[\frac{B}{A}\right] = \left[\frac{2\mu - m\omega^2}{2\mu \cos(ka)}\right]$$
$$= \left[\frac{2\mu \cos(ka)}{2\mu - M\omega^2}\right] \tag{2-35}$$

Thus the amplitude ratio is approximately unity (all atoms moving in the same way) for the long wavelength, low frequency, acoustic waves which satisfy the following set of conditions:

$$\left.\begin{array}{l} \text{wave vector} \quad |k| \ll (\pi/2a) \\[2mm] \text{speed} \quad v_0 = \left[\frac{2\mu a^2}{M+m}\right]^{1/2} \\[2mm] \text{angular frequency} \quad \omega = kv_0 \ll (2\mu/M)^{1/2} \end{array}\right\} \tag{2-36}$$

Looking at the term on the far right of Equation 2-35, it can be seen that as **k** and ω both increase in the acoustic branch, so does the ratio (B/A). In the limiting case, for the mode with

$$
\left.
\begin{array}{rl}
\text{wave vector} & \mathbf{k} = \pm(\pi/2a) \\
\text{angular frequency} & \omega = \omega_1 = (2\mu/M)^{1/2} \\
\text{phase velocity} & (\omega/k) = (8\mu a^2/\pi^2 M)^{1/2} \\
\text{group velocity} & (d\omega/dk) = 0
\end{array}
\right\} \quad (2\text{-}37)
$$

then $(B/A) \to +\infty$ as $\omega \to \omega_1$. This means that A must be zero regardless of the amplitude B. Since the lighter atoms are not in motion for the limiting mode at frequency ω_1, it is no longer surprising that the magnitude of m should not enter into Equation 2-34 for ω_1.

Note that above the angular frequency ω_1 there is a frequency range which has no solutions for real **k**, just as we found for a monatomic chain for *all* frequencies above ω_m. The range of frequencies for which **k** must be complex forms a forbidden band, since any wave initiated at a frequency in this interval will be heavily damped. The interesting additional feature we find from Equation 2-33 and Figure 2-10 is that with a diatomic chain a second permitted range of frequencies exists, the range of optical mode vibrations.

THE OPTICAL BRANCH

The optical branch of lattice vibrations (the set of optical modes) is so named because modes of this type can be excited with light of an appropriate frequency in solids which are at least partly ionic. This can happen because the amplitude ratio (B/A) is *negative* throughout the range of frequencies encompassed by the optical branch; nearest neighbors move out of phase with each other. Such motion can be excited by the transverse electric vector of a suitable e.m. wave when nearest neighbors carry electric charges of opposite sign. We shall consider this topic later on in this section.

For a linear diatomic chain the optical branch is described by the *positive* sign choice in Equation 2-33. Long wavelength optical modes (which occupy the region of **k**-space close to the origin) appear to satisfy the curious set of conditions:

$$
\left.
\begin{array}{rl}
\text{wave vector} & \mathbf{k} \to 0 \\
\\
\text{angular frequency} & \omega \to \omega_3 = \left[2\mu\left(\frac{1}{m} + \frac{1}{M}\right)\right]^{1/2} \\
\\
\text{phase velocity} & (\omega/k) \to \infty \\
\text{group velocity} & (d\omega/dk) \to 0
\end{array}
\right\} \quad (2\text{-}38)
$$

By manipulation of Equation 2-35, we can find that the amplitude ratio of the two kinds of atoms is

$$
\left[\frac{B}{A}\right] = -(m/M), \quad \text{when} \begin{cases} k = 0 \\ \omega = \omega_3 \end{cases} \quad (2\text{-}39)
$$

Thus for these long wavelength vibrations, neighboring atoms are moving in opposite directions in such a way that the common center of mass for a neighbor pair does not move.

Up to this point we have considered only the *longitudinal* vibrations of a diatomic chain, but the essential difference between the requirements of acoustic and optical modes for displacements of neighboring atoms can be seen more clearly by plotting *transverse* displacements. This has been done in Figure 2-11 for a wave of intermediate wave-vector.

Whether a vibrational mode is longitudinal or transverse, a given atomic amplitude of motion requires much more energy for a long wave optical mode than for an acoustic mode of the same wavelength. For it will be appreciated from Figure 2-11 or its longitudinal equivalent that optical modes minimize changes in second-nearest-neighbor distances by maximizing changes in the nearest-neighbor separations. Since the nearest-neighbor stiffness constant is usually considerably larger than any others, each harmonic oscillator must store a considerable amount of energy.

The amplitude ratio (B/A) remains negative throughout the optical branch. From Equation 2-35 we can determine that as k increases and as ω in consequence decreases to its short wave limit of ω_2, then (B/A) approaches zero from the negative side. Thus for the shortest possible

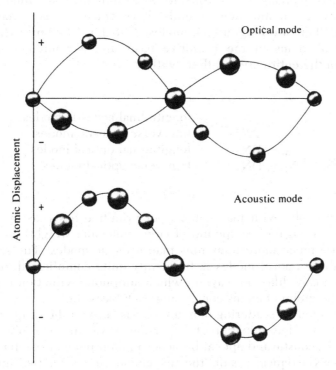

Figure 2-11 The sequence of atomic displacements involved in the transmission of a transverse wave along a diatomic chain. The difference between the acoustic and optical modes can be seen more clearly when transverse displacements are sketched rather than a longitudinal wave.

wavelength ($\lambda = 4a$) in the optical branch, the heavy atoms do not move at all and the wave involves the light atoms alone. For the frequency limit

$$\omega_2 = (2\mu/m)^{1/2} \qquad\qquad (2\text{-}40)$$

the group velocity ($d\omega/dk$) again becomes zero. Note that waves of frequency ω_1 or ω_2 are standing waves of zero group velocity which are reflected through $180°$ by Bragg diffraction.

The mass ratio (m/M) determines both the width of the forbidden frequency range and the width of the optical branch. When m and M are not too dissimilar, the forbidden range is a narrow one, and the optical branch covers a frequency range of up to $1.4:1$. However, if M is considerably larger than m, the gap is a wide one, and all of the optical branch vibrations are cramped into a narrow frequency range at $\omega \approx (2\mu/m)^{1/2}$. We shall be reminded of this when we consider theories of lattice energy and lattice specific heat.

THREE-DIMENSIONAL SOLID WITH A POLYATOMIC BASIS

The division of the vibrational modes into acoustic and optical branches is still applicable when our discussion of a solid with a polyatomic basis is broadened to consider three-dimensional structures. We must now (as in monatomic solids) have twice as many transverse modes as there are longitudinal modes. Thus for pN atoms (where p = number of atoms in the primitive basis and N = number of basis groups in the entire crystal) there will be

N	longitudinal acoustic modes
2N	transverse acoustic modes
$(p-1)N$	longitudinal optical modes
$2(p-1)N$	transverse optical modes

Then for a solid with many atoms per primitive basis (as in a single crystal of an organic compound of large molecular weight) the optical modes are much more numerous than acoustic modes. The spectrum $g(\omega)$ of these optical modes is frequently quite complicated, resulting from the many different ways in which antiphonal vibration can occur *within* the group of atoms comprising each basis set.

The coherent scattering of neutrons has (as with the simpler solids) been used for determination of dispersion curves in some solids which have both acoustic and optical branches in their phonon spectra. Figure 2-12 shows a typical result, the dispersion curves for two important directions in k-space for the ionic solid sodium iodide. Each of the "transverse" curves is doubly degenerate. Since iodine is much more massive than sodium, the forbidden gap in frequency is a wide one.

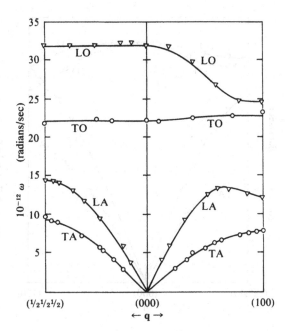

Figure 2-12 Dispersion curves for lattice vibrations in sodium iodide, plotted for reduced wave-vector along two important directions away from the center of the zone. The letters L, T, O, and A signify longitudinal, transverse, optical branch, and acoustic branch respectively. The data were taken at 100K by A. D. B. Woods et al., Phys. Rev., **131**, 1025 (1963).

Phonon dispersion curves should be independent of temperature, provided that the measurement temperature is well below the melting point, since the harmonic (Hooke's Law) approximation is valid when the amplitudes of wave motion are small. Thus room temperature can be a perfectly proper measurement temperature for relatively refractory solids. However, the authors of the paper cited in Figure 2-12 reported[15] that NaI had to be measured at 100K in order to obviate thermal deterioration of the data. The *curves* of Figure 2-12 were calculated by the same authors[16] on the basis of a theoretical "shell model" which allowed for the deformability of the nominally spherical ions, and for a less than 100 per cent transfer of charge from cation to anion. A variety of related "shell models" have been used in predicting the vibrational spectra of other ionic solids.

It will be noted that the LO and TO branches of the vibrational spectrum in Figure 2-12 have considerably different frequencies for all values of **k** (or **q**). This feature is found in any partially *ionic* material. The material of the next subsection shows how the ratio of ω_L to ω_T for long wavelengths is linked to a dispersion of dielectric constant in the infrared for an ionic or partially ionic compound.

The lattice vibrational spectrum includes both acoustic and optical branches for a completely *non-ionic* solid, provided that the crystal structure has a polyatomic basis. This is the case, for example, with the diamond structure, in which the basis comprises two carbon atoms.

[15] A. D. B. Woods, B. N. Brockhouse, R. A. Cowley, and W. Cochran, Phys. Rev. **131**, 1025 (1963).

[16] R. A. Cowley, W. Cochran, B. N. Brockhouse, and A. D. B. Woods, Phys. Rev. **131**, 1030 (1963).

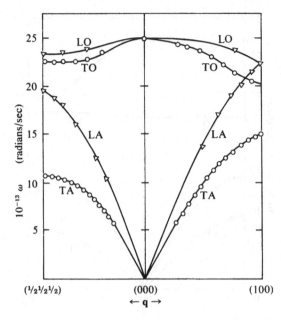

Figure 2–13 Dispersion curves for diamond. Data from J. L. Warren, R. G. Wenzel and J. L. Yarnell, *Inelastic Scattering of Neutrons* (Vienna, International Atomic Energy Agency, 1965).

Accordingly, dispersion curves for diamond (Figure 2-13) display branches for nearest-neighbor-in-phase and for nearest-neighbor-out-of-phase kinds of motion. The additional symmetries resulting from the mass equality of the two atoms in the basis mean that, for example, there is no gap between the top of the LA branch and the bottom of the LO branch at the zone boundary in the [100] direction. Moreover, the LO and TO modes are degenerate at the center of the zone, since there is no ionicity to produce a dispersion of dielectric behavior. Other nonionic solids with a polyatomic basis similarly show both acoustic and optical modes, with a common frequency for the long-wave LO and TO modes.

IONIC POLARIZABILITY AND RESTSTRAHLEN

The ratio of momentum to energy is much larger for an acoustic mode vibration than for electromagnetic radiation, which is one way of saying that light travels much faster than sound. This disparity makes it impossible to satisfy both energy conservation and momentum conservation in converting a photon into an acoustic phonon, unless some third body can provide almost all of the momentum for the phonon. However, the same difficulty does not necessarily arise if we consider annihilation of a photon in the creation of an optical mode phonon. Thus an optical mode phonon of very small momentum (corresponding, perhaps, with a wavelength of 10^4 to 10^5 atomic spacings) has a finite energy, as indicated by the maximum of Figure 2-10. The optical phonon frequencies shown in Figures 2-12 and 2-13 show that such long-wave optical modes have an energy typically of 0.01 to 0.1 eV.

Thus if any suitable coupling mechanism exists, a long-wave optical phonon can be created by annihilation of a photon (from the infrared part of the electromagnetic spectrum) with no momentum imbalance. Yet the oscillating transverse electric vector of an electromagnetic wave polarizes any ionic lattice when photons pass through; neighboring ions of opposite polarity are encouraged to move in anti-phase. It is the unavoidable presence of some non-linearity in the polarization response which provides the coupling mechanism for photon annihilation and phonon generation in an *ionic* solid. The "optical modes" derive their name from the existence of this coupling and its consequent effects, a name which is applied whether or not the solid has any ionic character to support the mechanism.

When an electromagnetic wave of relatively low frequency passes through an ionic solid, the dielectric constant κ_0 receives contributions from both the response of outer electrons and the displacements of the ion cores. When the wave being transmitted is of much higher frequency, only the electrons can respond with sufficient speed, and the dielectric constant will have a rather smaller value κ_∞, as sketched in Figure 2-14. Further dispersion will occur at some higher frequency in the ultraviolet part of the optical spectrum, beyond which even the electrons in the crystal do not have time to respond.

In any spectral region for which there is a dispersion of the dielectric constant, the dielectric response contains both real and imaginary parts of the dielectric constant, corresponding with retardation and

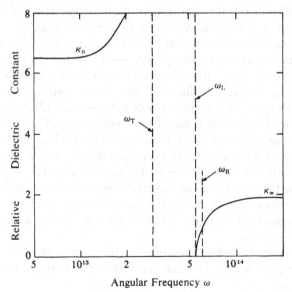

Figure 2-14 Dispersion of the dielectric constant in an ionic solid as angular frequency passes through the Reststrahlen range of long wave optical modes. The parameters used are those for rubidium fluoride, though the magnitude of the coupling between photons and phonons in this solid is large enough to round off the edges of the discontinuity at angular frequency ω_T. In this figure, ω_L is the frequency for which $\kappa = 0$, and it is also the frequency for long wave longitudinal modes. The frequency ω_R is that for which the dielectric constant is unity, to give zero reflectivity.

attenuation[17]. For the dispersion from κ_0 to κ_∞, the absorption is directly connected with the optical modes of ions.

Thus suppose that electromagnetic waves of moderate frequency pass through a crystal. The oscillating transverse electric vector[18] associated with these waves stimulates transverse oscillations of anions and cations in *opposite* directions. This is just the kind of motion required for optical modes. The more nearly the angular frequency of the electromagnetic wave approaches ω_T, the resonant frequency for transverse optical modes of long wavelength (i.e., zero crystal momentum), the more violent will be the amplitude of the ionic response. A solution of the simple differential equation

$$\frac{d^2u}{dt^2} + \gamma \frac{du}{dt} + \omega_T^2 u = \frac{-eE_0}{m} \exp(i\omega t) \qquad (2\text{-}41)$$

is the complex displacement

$$u = \frac{-(eE_0/m) \exp(i\omega t)}{[\omega_T^2 - \omega^2 + i\gamma\omega]} \qquad (2\text{-}42)$$

In these equations γ is a parameter which represents the strength of anharmonic effects in producing damping (absorption) of the electromagnetic excitation. The absorption is quite powerful, because an ionic crystal behaves as a set of Lorentzian line shape oscillators, all at the same angular frequency ω_T. Thus Figure 2-15 shows the optical transmission of a thin alkali halide crystal as photon energy in the infrared is scanned through the region of ω_T. The minimum transmission corresponds with an absorption coefficient of about 10^6 m^{-1}, or an extinction coefficient (absorption index) of about 10. This latter is appreciably larger than the refractive index for lower energies.

The frequency ω_T is known as the *Reststrahlen* frequency. This German term, meaning residual ray, refers to the fact that an ionic crystal which is strongly absorbing for frequencies comparable to ω_T is simultaneously strongly reflecting for those same frequencies. A common technique for producing monochromatic radiation at a wavelength in the far infrared has been reflection of a polychromatic beam from a succession of *Reststrahlen plates*, which are formed from an ionic solid. Only the wavelength corresponding to ω_T for the compound is reflected with enough efficiency to survive a series of such reflections. This method for securing spectral beams in the far infrared is less crucial now, since grating monochromators have essentially closed the gap between the visible region and microwaves; yet Reststrahlen plates still have many uses.

It will be shown when we come to the more extensive discussion of dielectric phenomena in Chapter 5 that the contributions of elec-

---

[17] We shall return to the subject of dielectric behavior in Chapter 5. The subject of dielectric dispersion and the interaction between electromagnetic radiation and optical mode vibrations is discussed at length in Born and Huang (see bibliography).

[18] The oscillating magnetic vector associated with the electromagnetic wave produces a much smaller effect, because the magnetic interaction is with the ion velocity rather than the ion position. Thus the effect is diminished in approximately the ratio of the speed of light to peak ion speed. How much smaller this is will be left to the reader as a problem (Problem 2.9).

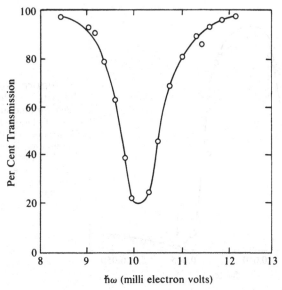

Figure 2-15 Optical transmission in the Reststrahlen region of the infrared spectrum for a thin film (thickness ~ 1μm) of rubidium iodide at 4.2K. The minimum transmission occurs for a photon energy equal to that of a transverse optical phonon of long wavelength. After G. O. Jones et al., Proc. Roy. Soc. **A.261**, 10 (1961).

tronic and ionic response in the Reststrahlen spectral region produce a complex dielectric constant

$$\kappa(\omega) = \kappa_\infty + \frac{[\kappa_0 - \kappa_\infty]\omega_T{}^2}{(\omega_T{}^2 - \omega^2 + i\gamma\omega)} \tag{2-43}$$

This equation requires that the real part of the dielectric constant become negative for a range of frequency extending upwards from ω_T. Within this range, the refractive index is imaginary, and thus a slab of finite thickness should be perfectly reflecting.

The angular frequency ω_L for long-wave longitudinal optical vibrations is higher than ω_T in an ionic solid. The reason for this difference can be seen in the different character of the polarization produced over a macroscopic volume. When an LO mode is excited, planes of ions move back and forth, and at the moments of maximum amplitude the front and rear surfaces of the crystal carry sheets of positive and negative charge. Thus there is a bulk polarization of the crystal, with an attendant amount of stored energy. No such contribution to the energy is required for a TO vibration, with charged planes oscillating in a transverse manner and setting up no macroscopic polarization.

The relation of ω_L to ω_T was examined by Lyddane, Sachs, and Teller,[19] who showed that the extra energy for creation of the bulk polarization with LO phonons was consistent with

$$(\omega_L/\omega_T)^2 = (\kappa_0/\kappa_\infty) \tag{2-44}$$

---

[19] A good account of the Lyddane-Sachs-Teller (LST) relation and a discussion of the interrelation of infrared properties and vibrational parameters are given by M. Born and K. Huang, *Dynamical Theory of Crystal Lattices* (Oxford University Press, 1954), pp. 82 et seq. These authors give a table of the related parameters for a variety of ionic solids. The original reference to the LST relation is: R. H. Lyddane. R. G. Sachs, and E. Teller, Phys. Rev. **59**, 673 (1941).

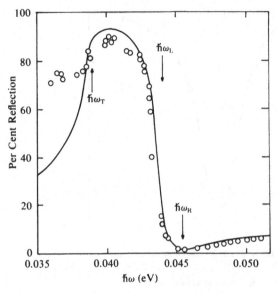

Figure 2-16 Optical reflectivity in the Reststrahlen range for zinc sulphide at 300K. After T. Deutsch, International Conference on Physics of Semiconductors, Exeter (Institute of Physics, 1962). The photon energies marked with the arrows are consistent with $\omega_T = 5.9 \times 10^{13}$ radians/sec., $\kappa_0 = 6.2$, $\kappa_\infty = 4.7$.

We may substitute the LST result into Equation 2-43 to find that the real part of the dielectric constant passes from its negative value through zero back toward positive values when $\omega = \omega_L$. Thus ω_L should be the upper frequency limit of the highly reflective Reststrahlen range.

At an angular frequency slightly higher than ω_L, the dielectric constant passes through unity. We should expect the reflectivity to be very close to zero for this frequency, increasing slowly again with a further increase of frequency. A reflectivity maximum and then minimum of this type is indeed seen with many ionic solids. Figure 2-16 shows a reflectivity spectrum for zinc sulphide, for which the dielectric constant is some 25 per cent smaller above the Reststrahlen frequency than below it. Essentially zero reflectivity is indicated for a frequency ω_R, and we expect this to be the frequency, slightly larger than ω_L, for which the real part of the dielectric constant is unity. It can be deduced from Equations 2-43 and 2-44 that

$$(\omega_R/\omega_T)^2 = \frac{\kappa_0 - 1}{\kappa_\infty - 1} \tag{2-45}$$

provided that the magnitude of γ in Equation 2-43 is considerably smaller than ω_T.

PHONON COMBINATION BANDS

We have noted that optical modes occur even in covalent solids, but that they cannot be optically excited at the Reststrahlen frequency. However, there are optical frequencies for which *multi-phonon* interactions can occur in annihilation of a photon. These are the optical frequencies of the "combination bands," absorption bands which are much less strong than the Reststrahlen band of an ionic solid. Phonon combination bands are found when the photon energy is sufficient to create two or more phonons for which the total crystal momentum is vanishingly small or for which it is possible to turn a phonon of small energy into a more energetic one of the same wave-vector. The term "two-phonon overtone band" is more appropriate than "combination band" if the result of photon annihilation is the creation of two phonons of equal energy and opposite momentum.

Phonon combination bands will have an opportunity to appear in the optical absorption spectrum if there is appreciable anharmonicity in the description of the potential energy associated with lattice vibrations,[20] or any non-vanishing higher order electric multipole moments.[21] The literature on the experimental and theoretical aspects of this field has been expanding rapidly since the first measurements of phonon combination bands were reported in germanium and silicon.[22] Figure 2-17 shows a set of data for infrared absorption in silicon, in which several combination bands are prominent.

The question of which combination bands will appear and which will be absent is closely related to the symmetry of a given crystal. Cri-

---

[20] M. Born and M. Blackman, Z. Physik **82**, 551 (1933).

[21] M. Lax and E. Burstein, Phys. Rev. **97**, 39 (1955).

[22] R. J. Collins and H. Y. Fan, Phys. Rev. **93**, 674 (1954). Since this first paper appeared, many others have followed each year on phonon combination bands in solids. The topic is a popular one in the proceedings of the biennial International Semiconductor Conferences (Exeter 1962, Paris 1964, Kyoto 1966, Moscow 1968, Cambridge 1970, Warsaw 1972, etc.).

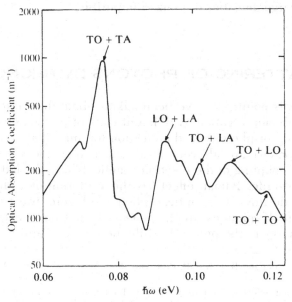

Figure 2–17 Infrared optical absorption of silicon, showing the various phonon combination bands, and the TO overtone band. After F. A. Johnson, Proc. Phys. Soc. 73, 265 (1959).

teria can be developed for the selection rules.[21] Thus a comparison of theory and experiment concerning combination modes can be helpful in improving our knowledge of the symmetry at critical points in the phonon spectra.

LOCAL PHONON MODES

Single photon-phonon reactions can occur even in covalent solids which contain impurities. Such reactions occur, for example, when the photon energy coincides with that of a *local phonon mode* associated with an impurity. It has been pointed out[23] that the presence of substitutional impurity atoms with a mass different from those of the host lattice will (1) slightly perturb the frequencies of all the normal vibrational modes, and (2) create vibrational states within the forbidden frequency range above the acoustic branch and below the optical branch. Since these latter are modes of complex **k**, the amplitude of vibration must fall off exponentially with the distance from the impurity atom.

These localized vibrational states will also appear if the impurities provided have the same mass as the host atoms, but a different charge.[23] In general, impurities will differ from the host in both mass and charge. By destroying the inherent symmetry of the lattice, their presence makes it possible for the electric vector associated with an electromagnetic field to interact with localized vibration modes. Moreover, since the symmetry is destroyed it is now possible for the electromagnetic wave to interact with the (slightly shifted) normal optical modes of the crystal. Both of these effects have been seen in semiconductor crystals containing large densities of ionized impurities.

SCATTERING OF PHOTONS BY PHONONS

Up to this point, we have been talking about the *annihilation* of a photon in phonon creation. A different class of phenomena arise from the *scattering* of photons in the creation (or annihilation) of phonons. Such phenomena are all very weak, since they too depend on the anharmonic coupling of the electromagnetic field to the lattice.

The first-order Raman effect[24] is the scattering of a photon which creates or destroys one optical phonon of zero wave-vector. The "Stokes-shifted" photons are the ones which have created phonons (and lost energy in the process), while there is also "anti-Stokes-shifted

---

[23] P. G. Dawber and R. J. Elliott, Proc. Roy. Soc. **A. 273**, 222 (1963); Proc. Phys. Soc. **81**, 453 (1963).

[24] The Raman effect was reported by C. V. Raman and K. S. Krishnan (1928) in liquids and vapors, and was soon noted also in solids by Landsberg and Mandelstam (1928). Raman phenomena in solids are reviewed by R. Loudon, Advances in Physics **13**, 423 (1964).

radiation" from phonon-destroying collisions. Raman-shifted radiation is usually observed at right angles to the incident light beam, since this is a favorable alignment condition for looking at the very small Raman intensities.

The second-order Raman effect involves the creation of (or removal of) pairs of phonons with equal and opposite wave-vector. Thus, there is a band rather than a line phenomenon, extending over the width of the optical branch. The analysis of second-order Raman scattering is closely connected with critical point analysis of the vibrational spectrum.

The scattering of a photon in the creation or annihilation of *acoustic* phonons is known as Brillouin scattering.[25] Because of the large difference between the speeds of sound and of light, the change of photon energy is extremely small in Brillouin scattering (see Problem 2.10). Thus considerable resolution is needed for study of the Brillouin-shifted photon spectrum, and a highly monochromatic source is desirable. Lasers have become very popular as photon sources for both Raman and Brillouin scattering studies.

Even though the anharmonic terms which facilitate Raman and Brillouin phenomena are quite small, and the scattering efficiencies are painfully small, an intense laser source does provide enough photons to permit a variety of interesting scattering effects. The density of ultrasonic phonons can be sufficiently increased with pulsed laser light to produce *stimulated Brillouin scattering,* the coherent amplification of an ultrasonic vibration at the expense of increased scattering of the incident photons.[26] Raman laser phenomena have also proved to be fruitful sources of data on the vibrational states of solids.

---

[25] L. Brillouin, Ann. Physique **17**, 88 (1922). See also the paper by Loudon.[24]

[26] The stimulated Brillouin scattering of laser light has been seen in crystalline solids such as quartz and sapphire. The simple experimental arrangement for such measurements is described by R. Y. Chiao, C. H. Townes, and B. P. Stoicheff, Phys. Rev. Letters **12**, 592 (1964).

2.4 Phonon Statistics and Lattice Specific Heats

In the last two sections, the particle-like aspects of lattice vibrations have been cited occasionally; as for example in describing the rationale of neutron scattering experiments or in discussing photon-phonon interactions. However, these sections were primarily concerned with dispersion laws for which it was usually convenient to adopt a wavelike nomenclature. Armed with this information about dispersion laws and densities of modes with respect to $\mathbf{k}$ or ω, we now find it appropriate to invoke the quantized particle idea of the phonon once again, in order to see how the vibrational energy of a crystal depends on temperature. If U denotes the entire vibrational energy of the crystal (per kilogram, per cubic meter, or per mole), then we are interested in $C_v = (\partial U/\partial T)_v$, the specific heat at constant volume. For purposes of actual measurement, the specific heat at constant pressure, C_p, is much more convenient, but $(C_p - C_v)$ is fortunately very small for a solid because the work performed in thermal expansion is so small.

CLASSICAL MODEL OF LATTICE ENERGY

Suppose that an atom of mass m, forming part of a solid, is performing harmonic motion of amplitude x_m at angular frequency ω. The restoring force constant is μ. At any instant, the atom's displacement from equilibrium is x, its velocity is $v = \dot{x}$, and its acceleration is $\ddot{x} = (-\mu x/m) = -\omega^2 x$. Thus the total energy associated with this motion is

$$\begin{aligned} E &= \text{(kinetic energy)} + \text{(potential energy)} \\ &= (mv^2/2) + (\mu x^2/2) \\ &= (m/2)[v^2 + \omega^2 x^2] \end{aligned} \qquad (2\text{-}46)$$

Averaged over a Boltzmann distribution, the classical expectation value is

$$\langle E \rangle = \frac{\int_{v=0}^{v_m} \int_{x=0}^{x_m} E \cdot \exp(-E/k_0 T) dv \cdot dx}{\int_{v=0}^{v_m} \int_{x=0}^{x_m} \exp(-E/k_0 T) dv \cdot dx} \qquad (2\text{-}47)$$

As we may readily determine (Problem 2.11), when we substitute the energy of Equation 2-46 into Equation 2-47, the result is

$$\langle E \rangle = k_0 T \qquad (2\text{-}48)$$

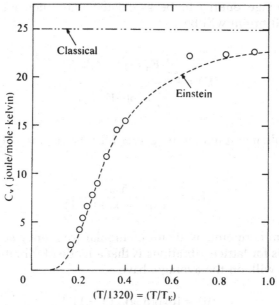

Figure 2-18 The molar specific heat of diamond, compared with the classical law of Dulong and Petit and with the curve predicted by the Einstein model for a characteristic temperature $T_E = 1320K$. This implies an angular frequency $\omega_E = (k_0 T_E/\hbar) = 1.73 \times 10^{14}$ radians sec^{-1}. After A. Einstein, Ann. Physik, **22**, 180 (1907).

Thus for a set of N atoms, each with three classical degrees of freedom, the total lattice energy would be

$$U = 3Nk_0T \qquad (2\text{-}49)$$

and the lattice specific heat would be the result previously quoted as Equation 2-3.

Our explanation of why specific heats do in fact decrease on cooling must center on why the average energy to be associated with a mode must be a function of temperature *and* of its frequency ω.

THE EINSTEIN MODEL

Einstein[27] made the first progress on this topic, based on the success of Planck (1900) in describing black body radiation by quantized rules. If it is assumed that a quantized oscillator of frequency $\nu = (\omega/2\pi)$ can exist only in the energy states

$$E_n = n\,\hbar\omega, \qquad n = 0, 1, 2, 3, \ldots \qquad (2\text{-}50)$$

and that the probability of the n th state is

$$g_n = \exp(-E_n/k_0T) \qquad (2\text{-}51)$$

---

[27] A. Einstein, Ann. Physik **22**, 180 and 800 (1907).

then the average energy to be associated with such an oscillator in thermal equilibrium will be

$$\langle E \rangle = \frac{\sum\limits_{n=0}^{\infty} E_n \exp(-E_n/k_0 T)}{\sum\limits_{n=0}^{\infty} \exp(-E_n/k_0 T)} \tag{2-52}$$

A fairly simple manipulation (suggested for the reader as Problem 2.12) results in

$$\langle E \rangle = \frac{\hbar \omega}{[\exp(\hbar \omega/k_0 T) - 1]} \tag{2-53}$$

for a quantum harmonic oscillator of angular frequency ω. Another way of stating this for lattice vibrations is that a mode of vibration at angular frequency ω will, on the average, have

$$\langle n \rangle = [\exp(\hbar \omega/k_0 T) - 1]^{-1} \tag{2-54}$$

phonons associated with it at temperature T. Then $\langle n \rangle$ is called the *phonon occupancy* of the mode.[28]

The energy $\langle E \rangle$ approaches the classical limit $k_0 T$ for temperatures much larger than $(\hbar \omega/k_0)$, but decreases much more rapidly than $k_0 T$ for low temperatures.

For simplicity, Einstein supposed that a solid with N atoms would have 3N modes of vibration all at the same angular frequency ω_E. He treated this frequency as an adjustable parameter to secure agreement between his theory for specific heat and what actually happens in a solid. Now if each vibrational mode has the same energy $\langle E \rangle$ from Equation 2-53 associated with it, the total lattice vibrational energy will be

$$U = \frac{3N\hbar\omega_E}{[\exp(\hbar\omega_E/k_0 T) - 1]} \tag{2-55}$$

(apart from the zero-point energy, which was not a part of Einstein's model). The corresponding specific heat at constant volume is

---

[28] According to the principles of "modern" quantum mechanics, which of course were developed twenty years after Einstein's formulation, the possible energies of a quantum oscillator are modified to $E_n = (n + \frac{1}{2})\hbar\omega$, so that

$$\langle E \rangle = \hbar\omega\left\{\frac{1}{2} + [\exp(\hbar\omega/k_0 T) - 1]^{-1}\right\}$$

The additional $\frac{1}{2} \hbar\omega$ added to each mode is the *zero-point energy*, so called because it exists at all temperatures, including absolute zero. The zero-point energy is an automatic part of the lattice energy, but it is a part which does not depend on temperature. Thus inclusion or exclusion of this term in the energy does not affect the specific heat.

$$C_v = \left(\frac{\partial U}{\partial T}\right)_v = 3Nk_0 \cdot F_E(\omega_E, T) \tag{2-56}$$

where the *Einstein function*, $F_E(\omega_E, T)$, is

$$F_E(\omega_E, T) = \frac{(\hbar\omega_E/k_0 T)^2 \exp(\omega_E/k_0 T)}{[\exp(\hbar\omega_E/k_0 T) - 1]^2} \tag{2-57}$$

The Einstein function approaches unity at high temperatures, thus conforming with the classical result, Equation 2-3. However, the function decreases exponentially for temperatures well below the Einstein *characteristic temperature* $T_E = (\hbar\omega_E/k_0)$:

$$F_E(\omega_E, T) \approx (\hbar\omega_E/k_0 T)^2 \exp(-\hbar\omega_E/k_0 T) \text{ if } T \ll T_E$$
$$= (T_E/T)^2 \exp(-T_E/T) \tag{2-58}$$

Figure 2-18 illustrates the example originally used by Einstein, the comparison with experimental specific heat data for diamond. The fit was not perfect, but it was a remarkable step forward from the failure of the classical model. T_E was used as an adjustable parameter, forcing a fit between theory and experiment of the *magnitude* of C_v (but not of its derivative) at a single low temperature. It will be noted that Einstein's parameter T_E corresponds with a universal angular frequency ω_E which is near to the center of gravity of the modes described for diamond in Figure 2-13.

The Einstein model is unrealistic from a physical point of view, since one would expect all vibrational modes to be at the same frequency only if all particles in a solid vibrated independently of each other, which they do not. Another approximate picture, but one which is more believable, was suggested by Debye.[29] The Debye model also happens to provide a better fit to the low temperature specific heat data of many solids, since it is found that C_v is often proportional to T^3 at low temperatures, as predicted by the Debye model but in contrast to the exponential behavior of Equation 2-58.

THE DEBYE MODEL

Debye followed Einstein in postulating that a solid of N atoms would have 3N vibrational modes, each with an energy given by Equation 2-52 or a phonon occupancy given by Equation 2-54. He observed, however, that the angular frequency ω of a mode must depend on its wave-vector k. There must be some maximum angular frequency ω_m such that

$$3N = \int_0^{\omega_m} g(\omega) \cdot d\omega \tag{2-59}$$

is the total number of distinguishable modes. Then the same ω_m should

[29] P. Debye, Ann. Physik **39**, 789 (1912).

be the upper limit for the integral describing the total vibrational energy:

$$U = \int_0^{\omega_m} \frac{(\hbar\omega) \cdot g(\omega) \cdot d\omega}{[\exp(\hbar\omega/k_0 T) - 1]} \qquad (2\text{-}60)$$

This is not a simple task for the *actual* density of states, $g(\omega)$, in a real solid. Attempts to solve the formal problem of Equation 2-60 were first made by Born and von Karman,[30] and later in the series of papers by Blackman to which we have previously referred.[10]

Debye suggested that it should be possible to obtain useful results by expressing $g(\omega)$ *as though* the phase velocity $v = (\omega/k)$ were a suitably chosen speed of sound v_0 for *all* modes of vibration. It is then necessary to choose an artificial value ω_D for the upper integration limit in Equations 2-59 and 2-60 in order to make the right side of Equation 2-59 still equal to 3N. This suggestion requires an improper treatment of the high frequency modes, but quantum restrictions ensure that these modes will have very few phonons at the low temperatures for which markedly non-classical behavior is observed. Thus the result should not be too sensitive to the change enforced for $g(\omega)$ at the upper end of the spectrum.

The Debye approximation to the density of states can be applied in a real or imaginary solid of any number of dimensions. Problem 2.13 looks at the Debye model for a one-dimensional solid. For a solid of m dimensions, the temperature dependence of the specific heat at low temperatures is predicted to be

$$C_v = A\, T^m \qquad (2\text{-}61)$$

and it is encouraging that C_v in many real (3-D) solids does vary as T^3 at low temperatures.

Important parameters for the Debye model are the speed of sound v_0 and the maximum supposed frequency, ω_D. We do not have the option to select both of these parameters independently to secure a fit between the Debye model and experiment. In the final expressions, ω_D is usually replaced by the *Debye characteristic temperature* $\theta_D = (\hbar\omega_D/k_0)$ as the quantity in terms of which C_v is quoted.

The three-dimensional Debye model consists of substituting an appropriate $g(\omega)$ into Equation 2-60, with a limit consistent with Equation 2-59. For this purpose, we recall from Equation 2-27 that $g(k) = (k^2/2\pi^2)$ for longitudinal modes of vibration, plus twice as many again for transverse modes. As is covered in Problem 2.4, $g(k)$ converts into

$$g(\omega) = \frac{\omega^2}{2\pi^2}\left[\frac{1}{v_L^3} + \frac{2}{v_T^3}\right], \qquad \omega \ll (v_L/a) \qquad (2\text{-}62)$$

[30] M. Born and T. von Kármán, Z. Physik **13**, 297 (1912).

as the density of states with respect to frequency for the long-wave acoustic limit (supposing that longitudinal and transverse waves are likely to be transmitted at different speeds). The Debye approximation then consists of writing

$$g(\omega) \approx (3\omega^2/2\pi^2 v_0^3), \qquad 0 < \omega < \omega_D \qquad (2\text{-}63)$$

for *all* the vibrational modes. In Equation 2-63 the parameter v_0 and the integration limit ω_D are related by the requirement that a solid comprising N atoms in a volume V must have altogether some (3N/V) modes per unit volume. Thus

$$(3N/V) = \int_0^{\omega_D} g(\omega) \cdot d\omega = (\omega_D^3/2\pi^2 v_0^3) \qquad (2\text{-}64)$$

This artificially imposed upper limit can be expressed in terms of a Debye characteristic temperature:

$$\theta_D = (\hbar\omega_D/k_0) = (\hbar v_0/k_0)(6\pi^2 N/V)^{1/3} \qquad (2\text{-}65)$$

It just happens to be convenient to express the operations of the Debye model in terms of the parameter θ_D with dimensions of temperature, rather than in terms of the limiting frequency $\omega_D = (k_0\theta_D/\hbar)$ or a maximum radius $k_D = (\omega_D/v_0)$ in reciprocal space. Our thinking concerning this model will be assisted by remembering that a sphere of radius k_D occupies the same volume of k-space as the true Brillouin zone. Thus a phonon with wave-vector comparable with k_D (frequency comparable with ω_D, energy comparable with $k_0\theta_D$) is a phonon near to the zone boundaries. Phonons of such large energy and wave-vector are not present in appreciable numbers except at high temperatures.

The process of replacing a highly complex $g(\omega)$ with one which increases quadratically with ω up to an artificial upper limit is illustrated for copper in Figure 2-19. It should be remembered that a crude treatment of the high frequency modes is not likely to have very serious results. Of more interest is the fact that $g(\omega)$ as deduced from neutron scattering increases more rapidly at the lower end of the spectrum than does the Debye curve shown here. Yet the Debye curve is based on a Debye temperature θ_D [with its accompanying commitment on ω_D, v_0, and $g(\omega)$] derived from specific heat data with copper. This should serve to remind us that θ_D is used as an *adjustable parameter* to enforce the most favorable fit between calculated and measured *specific heats*.

At any rate, the Debye procedure calls for the density of states $g(\omega)$ from Equation 2-63 to be substituted into Equation 2-60. The resulting lattice vibrational energy per unit volume is then

$$U = \left(\frac{3\hbar}{2\pi^2 v_0^3}\right) \int_0^{\omega_D} \frac{\omega^3 \cdot d\omega}{[\exp(\hbar\omega/k_0 T) - 1]} \qquad (2\text{-}66)$$

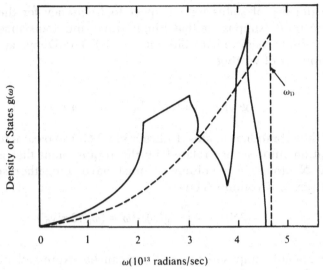

Figure 2–19 Density of states for phonons in copper. The solid curve is deduced from experiments on neutron scattering by Svensson *et al.* (1967) and by Nilsson and Rolandson (1973), and is the same as the curve in Figure 2-7. The broken curve is the three-dimensional Debye approximation, scaled so that the areas under the two curves are the same. This requires that $\omega_D = 4.5 \times 10^{13}$ radians sec^{-1}, or a Debye characteristic temperature $\theta_D = 344K$.

which, by a change to a dimensionless variable $x = (\hbar\omega/k_0 T)$, can be written as

$$U = \left[\frac{9N\, k_0 T^4}{V\theta_D{}^3}\right] \int_0^{(\theta_D/T)} \frac{x^3 \cdot dx}{(e^x - 1)} \qquad (2\text{-}67)$$

Differentiating with respect to temperature, we obtain a specific heat

$$C_V = \left(\frac{\partial U}{\partial T}\right)_V = \left[\frac{9N\, k_0 T^3}{V\theta_D{}^3}\right] \int_0^{(\theta_D/T)} \frac{x^4 \cdot e^x \cdot dx}{(e^x - 1)^2} \qquad (2\text{-}68)$$

The thermal properties U and C_v involve integrals which can be expressed analytically only for the extremes of high or low temperatures. However, C_v can be evaluated numerically at any temperature[31] and Figure 2-20 shows the Debye curve compared with typical experimental data.

It comes as no surprise that for temperatures much larger than θ_D the integral of Equation 2-67 reduces to $\frac{1}{3}(\theta_D/T)^3$ to give the classical energy $U \approx (3Nk_0 T/V)$ and specific heat $C_v \approx (3Nk_0/V)$. It is more to the point for us to concentrate on the low temperature regime, for which the measured decrease of C_v is less severe than the exponential predicted by an Einstein model.

---

[31] Extensive tables of the Einstein and Debye functions are given in E. S. R. Gopal, *Specific Heats at Low Temperatures* (Plenum Press, 1966). Tabulations of these functions can also be found in *The American Institute of Physics Handbook* (McGraw-Hill, 1971) and in the *Handbook of Mathematical Functions*, edited by M. Abramowitz and I. A. Stegun (Dover, 1965).

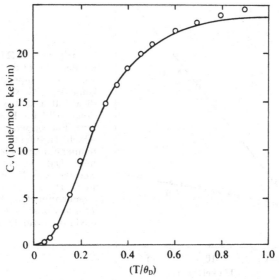

Figure 2–20 The curve of the specific heat of a solid (per mole) as a function of temperature, according to the Debye model in three dimensions. The experimental points are the data for yttrium reported by L. D. Jennings, R. E. Miller, and F. H. Spedding, J. Chem. Phys. **33**, 1849 (1960); they have been plotted here with an abscissa scale appropriate for θ_D of 200K. (An optimum fit for the lowest temperature points would require a slightly higher Debye temperature.) Note that the experimental points are above the Debye curve at high temperatures because C_p is the quantity actually measured, not C_v.

For temperatures less than about $(\theta_D/10)$, no appreciable error is involved if we choose to evaluate the integrals of Equations 2-67 and 2-68 over the range zero to infinity. The integral of Equation 2-67 is then $(\pi^4/15)$; thus

$$U = \left[\frac{3\pi^4 N k_0 T^4}{5V\theta_D{}^3}\right], \qquad T < (\theta_D/10) \tag{2-69}$$

and

$$\left.\begin{aligned} C_v = \left(\frac{\partial U}{\partial T}\right)_v &= (12\pi^4 N k_0/5V)(T/\theta_D)^3 \\ &= 1944\,(T/\theta_D)^3 \text{ joule/mole} \cdot \text{kelvin} \end{aligned}\right\} T < (\theta_D/10) \tag{2-70}$$

Equation 2-70 shows the cubic dependence of C_v on T which occurs so often in practice for solids at low temperatures, and which was not accounted for in the Einstein model. At intermediate temperatures, the rise of C_v with T can often be described with tolerable accuracy by either an Einstein or a Debye model.

When C_v at low temperatures is in accordance with Equation 2-70, a measurement at a single temperature determines θ_D and permits us to predict C_v at other temperatures. A check on this procedure is shown in Figure 2-21, both for a non-metal and for a metallic conductor. For KCl a plot of (C_v/T) against T^2 is linear and passes through the origin; thus

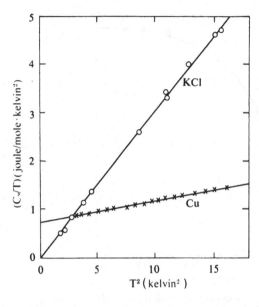

Figure 2-21 Temperature dependence of the specific heat at very low temperatures. A linear slope with these coordinates indicates that C_V has a term proportional to T^3. For KCl this is the only term. For copper, C_V also has a term which is linearly dependent on temperature, the electronic specific heat. Data for KCl after P. H. Keesom and N. Pearlman, Phys. Rev. **91**, 1354 (1953). Data for copper after H. M. Rosenberg, *Low Temperature Solid State Physics* (Oxford University Press, 1963).

the Debye T^3 law is obeyed and the value of θ_D can be determined from the slope of the experimental line.

For the copper data of Figure 2-21, a linear dependence of (C_v/T) on T^2 is also found, but with a non-zero intercept. This shows that

$$C_v = AT + BT^3 \qquad (2\text{-}71)$$

where the second term is the lattice specific heat (Debye law) and the first term arises from the specific heat of the free electron gas. We shall consider the energy and other properties of a metallic electron gas in detail in Chapter 3. For the present it is sufficient to remember that electronic terms (if any) must be allowed for before we attempt to compare specific heat data with more complicated models.

ELABORATIONS OF THE DEBYE MODEL

Since the Debye model involves a rather drastic simplification of the actual $g(\omega)$, we may expect that the value of θ_D which fits C_v at a certain temperature will not necessarily be the optimum value for *all* temperatures. However, for any pair of measurements (C_v *and* T), it is possible to calculate the equivalent parameter θ_D. A series of such measurements then permits us to plot a curve of θ_D against T. There is always *some* variation of θ_D with temperature, though for most solids the variation is smaller in magnitude than that shown by the curve of Figure 2-22.

The principal characteristics of the curve in Figure 2-22 (that θ_D should first decrease and then increase again on warming) are accounted for by a number of theoretical models based on the Einstein and Debye approaches. As we have noted, Blackman[10] made some of the early attempts at calculating $g(\omega)$ for simple lattices and deducing the corresponding curves of C_v versus T and θ_D versus temperature. Blackman[32] also showed that θ_D has a minimum at an intermediate temperature if one represents all the acoustic modes in a diatomic lattice by a Debye distribution and all the optical modes by an Einstein function.[33] (See Problem 2.15.)

Much more complicated curves for $g(\omega)$ are now available, thanks to computer-aided theoretical calculations of vibrational modes, and thanks to neutron scattering and other new experimental techniques. Such work is unlikely to add in large measure to what we know about lattice specific heats, however, for C_v and C_p are indifferent to these small details. When we use temperature as a variable in scanning lattice energy, this is spectroscopy with very coarse resolution.

The more recent and complicated curves for $g(\omega)$ can, however, tell us significant things about important groups of phonons: their energy, wave-vector, directionality, polarization, and ability to interact with other phonons and other excitations.

---

[32] M. Blackman, Repts. Progr. Phys. **8**, 11 (1941); Handbuch der Physik **7**, 325 (Springer, 1956).

[33] It is intriguing that Einstein should have chosen to compare his simple model with specific heat data for diamond. With the advantage of hindsight, we can look at the optical modes of diamond clustered in a narrow frequency range (Figure 2-13) and have good reason to expect that an Einstein model would do reasonable justice to the behavior of C_V.

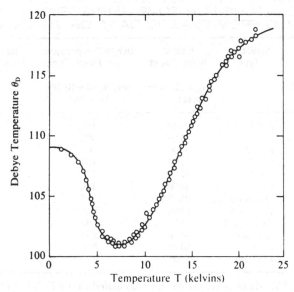

Figure 2-22 Temperature dependence of the Debye characteristic temperature θ_D (as measured from specific heat) for metallic indium. Data from J. R. Clement and E. H. Quinnell, Phys. Rev. **92**, 258 (1953). The electronic component of the thermal capacity has been subtracted before conversion of specific heat data into the corresponding θ_D.

THE DEBYE TEMPERATURE

The concept of the Debye temperature θ_D is useful in connection with many topics in solid state physics other than specific heats, a fact which may come as a surprise. It may have seemed in our discussion up to this point that the Debye approach to specific heat ignored the periodicity of a crystal lattice and the Brillouin zone limitation on the physically realizable ranges of wave vector and frequency. This is not entirely the case, since the ω_D of Debye's model is automatically comparable with the angular frequencies of phonons whose wave-vectors are close to the zone boundaries. These are the *majority* of phonons which are excited at temperatures higher than θ_D. Conversely, at temperatures well below the Debye temperature, the only phonons excited are those for **k** quite close to the center of the zone and well away from zone boundaries. Thus phenomena such as thermal conduction (controlled by the anharmonic coupling of phonons to other phonons) and electrical conduction (controlled by the scattering of electrons by phonons) take different forms above and below the Debye temperature.

Above the Debye temperature, most phonons have wavelengths of only a few atomic spacings. However, at temperatures well below the Debye temperature the most probable phonon wavelength is of the order of $(a\theta_D/T)$ (see Problem 2.16). This wavelength may be hundreds, or indeed thousands, of atomic spacings at sufficiently low temperatures.

The magnitude of the elastic constants determines the Debye temperature, and thus dictates how small T must be before the condition

TABLE 2-2 THE DEBYE CHARACTERISTIC TEMPERATURE FOR SOME REPRESENTATIVE SOLIDS, DERIVED FROM ELASTIC DATA OR THERMAL DATA[°]

Solid	Structure Type	Elastic Wave Speed $v_0 = (B_S/\rho)^{1/2}$ (m/sec)	Debye Temperature From Elastic Data $(\hbar v_0/k_0)(6\pi^2 N/V)^{1/3}$ (kelvins)	Debye Temperature From Low Temperature Specific Heat θ_D (kelvins)
Sodium	B.C.C.	2320	164	157
Copper	F.C.C.	3880	365	342
Zinc	H.C.P.	3400	307	316
Aluminum	F.C.C.	5200	438	423
Lead	F.C.C.	1960	135	102
Nickel	F.C.C.	4650	446	427
Germanium	Diamond	3830	377	378
Silicon	Diamond	6600	674	647
SiO_2	Hexagonal	4650	602	470
NaCl	Rocksalt	3400	289	321
LiF	Rocksalt	5100	610	732
CaF_2	Fluorite	5300	538	510

[°] Note: The elastic wave speed is as calculated in Table 2-1. Debye temperatures from specific heat data (with the electronic specific heat subtracted) are as listed by K. A. Gschneidner in *Solid State Physics, Vol. 16* (Academic Press, 1964) for the elements, and in *American Institute of Physics Handbook* (McGraw-Hill, Third Edition, 1971) for the four compounds.

$T \ll \theta_D$ is well satisfied. The strong interatomic forces in a solid such as diamond or sapphire dictate a large θ_D, and in such a solid a relatively modest amount of cooling is sufficient to render negligible the phonon occupancy number for all modes except those of wavelength very large compared with unit cell dimensions.

Values for the Debye temperature are cited for some representative solids in the final column of Table 2-2. These values have all been derived from the lattice contribution to the low temperature specific heat. As we can see from the table, the characteristic temperature deduced from the elastic constants may be either larger or smaller than the "thermal" Debye temperature.

2.5   Thermal Conduction

The particle — or phonon — aspect of lattice vibrations is particularly appropriate when we are concerned with energy transformation. Such energy transformation processes include the *creation* of and *annihilation* of phonons. Thermal conduction is most conveniently described in terms of the *scattering* of phonons; by other phonons, by static imperfections, or by electrons.

LATTICE THERMAL CONDUCTION AND PHONON FREE PATH

Heat energy can be transmitted through a crystal via the motion of

> phonons,
> photons,
> free electrons (or free holes),
> electron-hole pairs,
> or excitons (bound electron-hole pairs).

The electronic components of heat conduction are usually the largest components in a metal, but almost all of the thermal current in a non-metal is carried by the lattice vibrations (phonons), except at the highest temperatures when photons may become dominant. We now wish to consider the magnitude of this phonon current when there is a finite temperature gradient.

We recall from Equation 2-54 that at temperature T the number of phonons excited to the mode of wave-vector $\mathbf{k}$ and angular frequency ω_k is

$$\langle n_k \rangle = [\exp(\hbar\omega_k/k_0 T) - 1]^{-1} \qquad (2\text{-}72)$$

In thermal equilibrium, when there is no temperature gradient, $\langle n_k \rangle$ is equal to $\langle n_{-k} \rangle$; that is, there is a perfect balance between the rates of phonon flow in any pair of opposite directions. Thus the net heat flow is zero.

When a finite temperature gradient ∇T is established, the thermal conductivity κ_ℓ can be defined in terms of ∇T and the rate of energy flow per unit area normal to the gradient:

$$\mathbf{Q} = -\kappa_\ell \nabla T \qquad (2\text{-}73)$$

This thermal conductivity can be expressed in terms which relate to the microscopic properties of phonons by appealing to the analogy between conduction in a phonon "gas" and in a conventional molecular gas. The thermal conductivity result of the kinetic theory of gases simplifies for the case of all particles having the same speed (as is true for phonons well below the Debye energy) to

$$\kappa_\ell = \frac{1}{3} C_v v_0 \Lambda \tag{2-74}$$

Here C_v is the lattice specific heat per unit *volume* (which is a measure of the phonon density), v_0 is the phonon velocity (speed of sound), and Λ is the *phonon mean free path*.

Now the path length traveled by a phonon from the instant it is created until the instant it is annihilated or otherwise transformed is often a rather drastic function of phonon energy. It may be quite large for the phonons of small energy but become very much smaller for those phonons whose energy exceeds the Umklapp cut-off energy $k_0\theta_u$ (of which we shall say more shortly). Nevertheless, for any distribution of phonons it is always possible to *define* a mean free path Λ by means of Equation 2-74.

For temperatures near the melting point of a solid, Λ may be as small as six to ten interatomic spacings. At very low temperatures, Λ may be as large as 1 mm. The general form of temperature dependence for Λ and the consequent variation of κ_ℓ are shown in Figure 2-23. We must understand the behavior of Λ before we can make much sense of what happens to the thermal conductivity.

Table 2-3 lists the thermal conductivity of six crystalline nonmetallic solids at three temperatures and notes the mean free path for each situation from Equation 2-74, using the speed of sound from Table 2-1 and the Debye temperature of Table 2-2 in the determination of C_v. We must now see how such a range of phonon free path is possible.

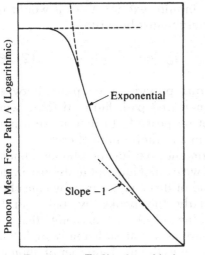
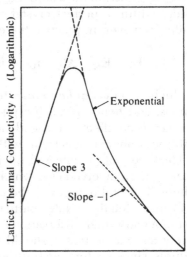

Figure 2-23 Typical variation of phonon mean free path and phonon thermal conductivity with temperature, plotted in double logarithmic fashion. The increase of mean free path on cooling (as Umklapp-processes become less likely) is eventually ended when defect and surface scattering control the thermal path.

TABLE 2-3 LATTICE THERMAL CONDUCTIVITY AND PHONON MEAN FREE PATH IN TYPICAL NONMETALLIC SOLIDS[*]

Solid	T = 273K		T = 77K		T = 20K	
	Thermal Conductivity κ_ℓ (W/m · K)	Phonon Mean Free Path $\Lambda = (3\kappa_\ell/v_0 C_v)$ (m)	Thermal Conductivity κ_ℓ (W/m · K)	Phonon Mean Free Path $\Lambda = (\kappa_\ell/v_0 C_v)$ (m)	Thermal Conductivity κ_ℓ (W/m · K)	Phonon Mean Free Path $\Lambda = (3\kappa_\ell/v_0 C_v)$ (m)
Silicon	150	4.3×10^{-8}	1500	2.7×10^{-6}	4200	4.1×10^{-4}
Germanium	70	3.3×10^{-8}	300	3.3×10^{-7}	1300	4.5×10^{-5}
Crystalline Quartz (SiO_2)	14	9.7×10^{-9}	66	1.5×10^{-7}	760	7.5×10^{-5}
CaF_2	11	7.2×10^{-9}	39	1.0×10^{-7}	85	1.0×10^{-5}
NaCl	6.4	6.7×10^{-9}	27	5.0×10^{-8}	45	2.3×10^{-6}
^{7}LiF (isotopically pure)	10	3.3×10^{-9}	150	4.0×10^{-7}	8000	1.2×10^{-3}

[*] Data from R. Berman, Cryogenics **5**, 297 (1965) and references therein, among other sources. For each conductivity value, the mean free path is calculated using v_0 from Table 2-1, and C_v (per unit volume) chosen to be consistent with the "thermal" Debye temperature of Table 2-2.

ANHARMONIC EFFECTS

If Hooke's law were rigorously obeyed in a solid, and the energy of an atom at location r (whose equilibrium position was r_0) could be expressed as

$$E_r = E_{r_0} + A(r - r_0)^2 \qquad (2\text{-}75)$$

then the principle of superposition for lattice vibrations would be exact. In such a situation it would be impossible for phonons to interact (collide) within a perfect crystal of infinite extent; Λ itself would be infinite. However, in practice the energy must be written

$$E_r = E_{r_0} + A(r - r_0)^2 + B(r - r_0)^3 + \cdots + \cdots \qquad (2\text{-}76)$$

The higher order *anharmonic* terms provide a coupling between phonons, and hence make the phonon free path finite. If there were no anharmonic effects, it would not be possible to thermalize a non-equilibrium phonon spectrum (within an infinite perfect crystal)!

The theory of the effect of anharmonic coupling on phonon-phonon scattering is an involved one, which we cannot attempt to discuss in detail here. The bibliography at the end of this chapter may be consulted for further details. (Note particularly the books by Bak and by Ziman.) As an initial observation on the subject, we may note that it is entirely reasonable that Λ should vary as T^{-1} at sufficiently high temperatures, since for temperatures above θ_D the total number of phonons excited (and available to interact with a given phonon in decreasing its free path) is proportional to T. In this same high temperature range, Λ and κ_ℓ should have the same temperature dependence, since C_v should be almost independent of temperature.

When we consider lower temperatures, an explanation of what happens to Λ must take into account two different types of phonon-phonon interaction, one of which is effective and the other ineffective in keeping Λ (as defined through Equation 2-74) finite.

NORMAL PROCESSES AND UMKLAPP PROCESSES

Suppose that two phonons collide and form a third phonon. The probability of such a collision process is controlled by the magnitude of the anharmonic terms in the description of the energy. The properties of the resulting phonon are controlled by laws of energy conservation and of crystal momentum conservation.

We remarked earlier in connection with the inelastic scattering of slow neutrons by the vibrations of a crystal that the crystal momentum $\hbar\mathbf{k}$ of a phonon is not the same thing as "ordinary" momentum. Thus a vibrational mode with $\mathbf{k}$ corresponding to a Brillouin zone boundary is a standing wave, and this has zero momentum with respect to the crystal as a whole. Yet ordinary momentum *is* transferred to a crystal when phonons are created or destroyed by external stimulation such as the scattering of an incident neutron or β-particle. The question of whether crystal momentum $\hbar\mathbf{k}$ must be conserved or not depends on the detailed circumstances of the transition being considered.

At any rate, crystal momentum is conserved when two phonons collide to produce a single phonon such that

$$\hbar\omega_1 + \hbar\omega_2 = \hbar\omega_3 \qquad (2\text{-}77)$$

and

$$\hbar\mathbf{k}_1 + \hbar\mathbf{k}_2 = \hbar\mathbf{k}_3 \qquad (2\text{-}78)$$

in the conservation of energy and momentum. Such a process is known as a *normal-process*, or N-process. As we can see if we sketch the vector diagram of Equation 2-78, an N-process does not alter the direction of energy flow; thus it makes no contribution towards the thermal resistance. If N-processes were the only possible phonon-phonon interactions inside a perfect crystal, the lattice thermal conductivity would be infinite.[34]

---

[34] There appears to be an inconsistency here. How can we have infinite lattice conductivity if there are scattering processes which provide for finite free paths? The answer is that the *mean free path* Λ is infinite, for we have defined it through Equation 2-74 to be influenced only by the spectrum of processes which can usefully work towards thermalizing a non-equilibrium phonon population. Even though N-processes appear to provide for a finite free path length of an individual phonon, this is an illusion. Whether the phonon distribution is an equilibrium distribution or not, N-processes are creating phonons of a given ω_k at the same average rate that they are annihilating them. Only in a very specialized situation (such as that of solid ^{4}He below 1K, to be discussed later in this section) is it found that N-processes have any influence on the *rate* at which phonons can diffuse.

However, Peierls[35] showed that thermalization of a phonon population could proceed by processes which satisfied Equation 2-77 for energy conservation, and also the vector relationship

$$\mathbf{k}_1 + \mathbf{k}_2 = \mathbf{k}_3 + \mathbf{G} \qquad (2\text{-}79)$$

Here $\mathbf{G}$ is a reciprocal lattice vector. Since a phonon of wave-vector $(\mathbf{k}_3 + \mathbf{G})$ is indistinguishable from a phonon of wave-vector $\mathbf{k}_3$ in a periodic lattice, Equation 2-79 is evidently a valid conservation law. Peierls called such events *Umklapp-processes*, or U-processes. The intriguing aspect of a U-process is that it destroys momentum and changes the direction of energy flow (as illustrated in Figure 2-24 in which energy is carried to the right by $\mathbf{k}_1$ and $\mathbf{k}_2$, but to the left by the final state $\mathbf{k}_3$). Thus U-processes provide thermal resistance to phonon flow and can thermalize a phonon distribution.

Peierls noted that the mean free path set by U-processes would vary as T^{-1} at high temperatures, since the excitation of all lattice modes is proportional to T for temperatures larger than θ_D. Interesting things happen when we consider the chances for U-processes at lower temperatures. For at low temperatures, only the regions of k-space close to the center of the Brillouin zone remain heavily populated with phonons; yet a U-process cannot occur unless $(\mathbf{k}_1 + \mathbf{k}_2)$ extends beyond the zone boundary. This requires that $\mathbf{k}_1$ and $\mathbf{k}_2$ should each be comparable with $(\mathbf{G}/2)$. Peierls anticipated that the probability of U-processes would fall off as $\exp(-\theta_u/T)$ at low temperatures, where the parameter θ_u was expected to be comparable to $(\theta_D/2)$. This is quite different from the low temperature probability of N-processes, which should fall off as T^5 well below the Debye temperature.[36]

---

[35] R. Peierls, Ann. Physik **3**, 1055 (1929). See also R. Peierls. *Quantum Theory of Solids* (Oxford University Press, 1955).

[36] Most of the phonons excited at a temperature well below θ_D have long wavelengths (see Problem 2.16), and these are effective in scattering only through small angles. Averaged over a cosine law, the scattering efficiency for N-processes varies as the square of the temperature. Then since the total number of phonons excited varies as T^3 in that same range, we expect to find $A_N \propto (T/\theta_D)^{-5}$. Similar considerations lead to a phonon-limited mean free path for *electrons* at low temperatures which also varies as T^{-5}, as we shall find in Chapter 3.

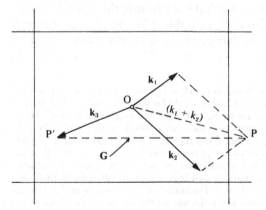

Figure 2-24 A section through k-space, showing the vector relationships of wave-vectors for an Umklapp process. The vectorial sum of $\mathbf{k}_1$ and $\mathbf{k}_2$ must extend beyond the boundaries of the first Brillouin zone in order for a U-process to occur. P and P' are equivalent points in neighboring Brillouin zones.

A phonon free path with the temperature dependence

$$\Lambda \propto \exp(\theta_u/T) \qquad (2\text{-}80)$$

is indeed seen in numerous solids, though the value of θ_u required for a fit is usually smaller than $(\theta_D/2)$. Thus in sapphire (Al_2O_3) the Debye temperature $\theta_D \sim 1000K$ while the *Umklapp temperature* θ_u is only some 250K. The phonon mean free path in sapphire at 50K is about 100 μm, some 30 times larger than would be given by an extrapolation of the high temperature T^{-1} line.

Phonon-phonon processes more complicated than the ones we have considered so far also occur. Both N-processes and U-processes involving larger numbers of phonons have to be considered in more complete discussions of thermal resistance. These higher order processes affect the behavior to an appreciable extent only at rather high temperatures.

DEFECT-CONTROLLED PHONON SCATTERING

In addition to the scattering of phonons by each other, phonons may also experience scattering by

(1) point defects, such as impurities, and vacancies;
(2) line defects (dislocations);
(3) grain boundaries in a polycrystal; the outer surface of a mono-crystal;
(4) disorder in an alloy, both short range and long range;
(5) random distribution of different isotopes of a given chemical species.

All of these can readily absorb both phonon energy and crystal momentum. Thus in a crystal of poor quality, Λ is doomed to remain small at all temperatures. In a very good crystal, Λ will rise on cooling (as U-processes become improbable) and will saturate at low temperatures at a value set by the distribution of unavoidable defects.

In Figure 2-23 we showed Λ as leveling off at low temperatures, at a magnitude set by the defect distribution. Since this is the temperature range for which C_v varies as T^3 (the Debye law), we can now see that $\kappa_\ell = (C_v v_0 \Lambda/3)$ should similarly vary as T^3 for temperatures so low that U-processes are ineffective compared with other scattering mechanisms.

Some beautiful data illustrating this point are shown in Figure 2-25. This figure plots the thermal conductivity of lithium fluoride crystals of high crystalline perfection, for which the isotopic abundance of ^7Li has been enriched to 99.9 per cent in order to minimize isotope scattering. As a result, the conductivity of both samples rises on cooling in accordance with the Peierls law for U-scattering, then follows a course determined by phonon scattering from the sides of the respective samples.

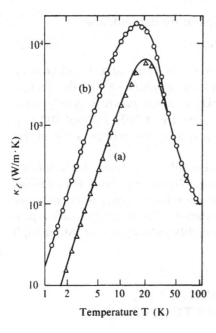

Temperature T (K)

Figure 2-25 Lattice thermal conductivity as a function of temperature for single crystal bars of lithium fluoride. Sample (a) has a cross-section 1.33×0.91 mm, while the cross-section of sample (b) is 7.55×6.97 mm. The conductivity is controlled by U-processes at the higher temperatures, and by scattering from the sides of the crystal at low temperatures. The lithium was isotopically enriched to 99.9 per cent ^{7}Li in order that isotope scattering should not interfere with observation of the "size effect." Data are from R. Berman, Cryogenics **5**, 297 (1965).

The larger the sample diameter, the larger the low temperature mean free path (provided that defect scattering within the bulk is kept small enough). The possibility that phonon transport could be limited by scattering from the crystal walls was suggested by Casimir;[37] the mechanism has been likened to "Knudsen flow" or "ballistic transport" in the kinetic theory of gases.

The curves of Figure 2-25 provide good evidence that the Casimir condition is satisfied at low temperatures; the phonon mean free path can be identified with the diameter of the sample (Problem 2.17). Moreover, the reduction of mean free path at higher temperatures corresponds with the onset of U-processes. For most samples of most materials, however, the sharp maximum of κ_{ℓ} is softened by the presence of rival phonon scattering mechanisms. Among these is the isotope scattering mechanism which is essentially independent of temperature. Figure 2-26 shows that for LiF in which there is a mixture of ^{6}Li and ^{7}Li, the thermal conductivity rises less steeply on cooling than if only U-processes were active, and that a lower, flatter maximum is reached. Germanium, also illustrated in Figure 2-26, is a material for which isotope scattering ordinarily contributes 10 per cent of the thermal resistance at room temperature, and which overshadows the Umklapp mechanism at liquid hydrogen temperatures.[38] An increased low temperature maximum can be obtained by isotopic enrichment of ^{74}Ge, but the upper germanium curve of Figure 2-26 is still far from being Umklapp-limited. A massive rise of Λ at low temperatures is seen only in those materials (such as diamond or alumina) for which isotopic

---

[37] H. B. G. Casimir, Physica **5**, 495 (1938).

[38] C. J. Glassbrenner and G. A. Slack, Phys. Rev. **134**, A1058 (1964).

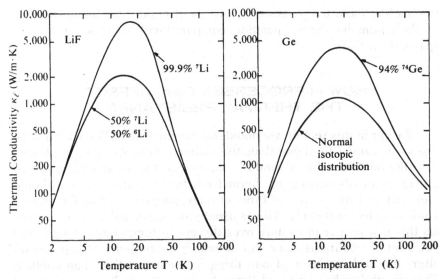

Figure 2–26 Two examples of the influence of isotope scattering on the maximum lattice thermal conductivity. At the left, curves for LiF either of 50:50 isotopic distribution or with lithium enriched to 99.9 per cent of the heavy isotope, after R. Berman, Cryogenics **5**, 297 (1965). At the right, curves for "normal" germanium, and for partially enriched germanium, after T. H. Geballe and G. W. Hull, Phys. Rev. **110**, 773 (1958).

competition is naturally absent or for which complete isotopic enrichment is possible.

The lattice periodicity seen by a phonon is similarly impaired in an alloy within which atoms of two species are randomly distributed over equivalent crystallographic sites. Thus the lattice thermal conductivity for an alloy will be appreciably smaller than for material of either of the terminal compositions. Figure 2-27 illustrates this point for alloys of composition intermediate between GaAs and GaP. The phonon mean

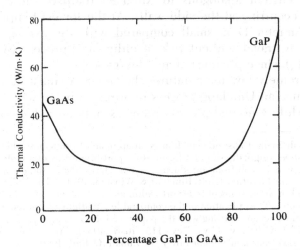

Figure 2–27 The effect of alloy scattering on the lattice thermal conductivity. The curve shows the thermal conductivity at 300K for alloys of the composition $GaAs_xP_{1-x}$. After P. D. Maycock, Solid State Electronics **10**, 161 (1967).

free path at room temperature is only about eight interatomic spacings for the equimolecular composition, compared with some 30 interatomic spacings for GaP.

HOW N-PROCESSES CAN AFFECT THE THERMAL RESISTANCE

Earlier in this section we noted the conclusions of Peierls,[35] that N-processes cannot by themselves thermalize a non-equilibrium phonon distribution. From this it would seem logical that we should ignore N-processes in discussions of thermal conduction, and that we should look only at U-processes (of three or more phonons) and at the various kinds of defect scattering. This is usually the case, and for most materials this is a proper procedure over the entire temperature range. And yet, a set of conditions *can* occur for which the *presence* of N-processes alters the odds in favor of one thing happening rather than another. This possibility has emerged through use of the Boltzmann transport equation approach[39] for the description of phonon transport. We shall return to the Boltzmann equation in Chapter 3 in connection with electron transport.

Detailed examinations of the Boltzmann transport equation approach have shown[40] that N-processes do indeed play no role when the average distance traveled by a phonon before it suffers a U-collision is small compared with the dimensions of the sample or crystallite. At the reasonably high temperatures for which this is the case, we may think of phonon transport as resembling the flow of a rather dense fluid along a pipe, with N-processes as elastic collisions between molecules and U-processes as the resistive collisions which produce a viscous drag.

For the opposite extreme of very low temperatures, we have seen that the flow resembles that of a very dilute fluid; and that if there is not much scattering by static defects within the medium, there will be a phonon current analogous to Knudsen (ballistic) flow, with most collisions occurring at the side walls. At the lowest temperatures, the sample diameter D is small compared with the average distance a phonon may travel without risk of either a U-process (Λ_U) or an N-process (Λ_N), since phonons of any energy are rare.

At extremely low temperatures, however, Λ_U must be much larger than Λ_N since an Umklapp-process requires collision with an *energetic* phonon, while a normal-process requires only an anharmonic interac-

[39] The Boltzmann transport equation is an approach to transport phenomena in any statistical system, originally devised for the transport properties of a molecular gas. [For an exposition which is far from introductory, see S. Harris, *An Introduction to the Theory of the Boltzmann Equation* (Holt, Rinehart & Winston, 1971).] The Boltzmann equation has been used in discussions of transport within an electron gas since the Lorentz model of 1905 (Section 3.2). Application to the comparable situations of conduction in a phonon gas is much more recent, starting with the papers of J. M. Ziman, Canad. J. Phys. **34**, 1256 (1956) and of J. Callaway, Phys. Rev. **113**, 1046 (1959). The subject is reviewed in Chapter 8 of *Electrons and Phonons* by J. M. Ziman (Oxford, 1960).

[40] R. N. Gurzhi, Soviet Physics J.E.T.P. **19**, 490 (1964). See also R. A. Guyer and J. A. Krumhansl, Phys. Rev. **148**, 778 (1966).

tion with a phonon of any energy. As noted previously, the probability of a U-process varies as $\exp(-\theta_u/T)$ at low temperatures, which is much more drastic than the T^5 dependence for probability of an N-process. Thus there must be some limited range of temperature for which

$$\Lambda_N \ll D \ll \Lambda_U \qquad (2\text{-}81)$$

Nothing exciting happens in this temperature range if there is any appreciable amount of defect scattering. But if defect scattering can be minimized, the only *resistive* collisions are the ones with the walls, but *non-resistive* collisions (N-processes) occur much more frequently. Under these conditions, the flow of phonons is analogous with Poiseuille flow of a fluid; the progression of a phonon is a random walk process, as in Brownian motion. Development of the problem as a random walk process (Problem 2.18) shows that the average phonon will traverse a path length of the order of (D^2/Λ_N) between successive collisions with the side walls. Thus the mean free path *as measured in a thermal transport experiment* will appear to be larger than the sample diameter, and will increase up to a point as N-processes become more numerous! An upper limit to the mean free path (and to the thermal conductance) will occur when the temperature is high enough to make U-processes competitive with boundary scattering; with further warming the conductivity rapidly drops to join the usual Umklapp line.

The set of inequalities in Equation 2-81 must be met over a limited range of temperature for any material. However, Poiseuille flow will be seen only if defect scattering is sufficiently small in that same temperature range. The latter condition is not met for the LiF data of Figure 2-25, and as a result κ_ℓ rises smoothly as a cubic function of T until the Umklapp mechanism takes over at about 20K. We could expect to see such behavior with LiF (or any other dielectric material which can be made in monocrystal form) of extremely high isotopic perfection, high crystalline perfection, and very low impurity content.

The necessary set of conditions is met more readily in solid helium, since the thermodynamic properties of helium make elimination of impurities rather simple, and the existence of a superfluid phase for ^{4}He makes isotope separation possible to a high degree. Figure 2-28 shows an experimental curve for κ_ℓ in a crystal of B.C.C. ^{4}He, crystallized and maintained at a pressure of 85 atmospheres. The behavior at the lowest temperatures conforms with the Casimir model for boundary scattering: a mean free path of about the sample diameter and a thermal conductivity which varies (as does C_v) as T^3. This is the range of Knudsen flow. On warming to above 0.6K, however, κ_ℓ rises more steeply, to a peak at about 1.0K. The range from 0.6 to 1.0K behaves in the manner expected for Poiseuille flow, with phonon transport *facilitated* by the presence of N-processes. It will be noted that κ_ℓ varies as T^8 in that short temperature range, as expected from $C_v \cong (T/\theta_D)^3$ and $\Lambda_N \cong (T/\theta_D)^{-5}$. At 1K the inferred "mean free path" is three times larger than the crystal diameter, but with further warming the mean free path (and conductivity) decrease as the Umklapp processes take control.

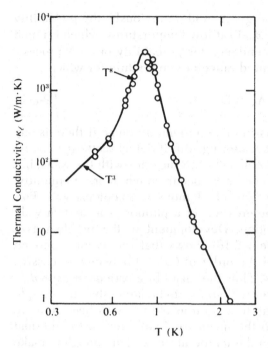

Thermal Conductivity κ_ℓ (W/m·K)

T (K)

Figure 2-28 A double-logarithmic plot (with the vertical scale compressed by a factor of three) of the temperature dependence of thermal conductivity for hexagonal-close-packed solid ⁴He, maintained at a pressure of 85 atmospheres. After L. P. Mezhov-Deglin, Soviet Physics, J.E.T.P. **22**, 47 (1966). The maximum conductivity corresponds with a phonon mean free path of 7 mm, compared with a crystal diameter of 2.5 mm.

This intriguing but rather obscure influence of N-processes on conduction has been sketched at some length, since we can expect it to have a variety of ramifications in low temperature experimentation. The possibility of Poiseuille flow is connected with the possibility of transmission of a "second sound" heat pulse through a solid without excessive attenuation. Second sound is well known in superfluid liquid helium and has been observed[41] in solid ⁴He. It can be expected in other solids for which the ratios of the various scattering probabilities are favorable.

ELECTRONIC THERMAL CONDUCTION IN METALS AND SEMICONDUCTORS

Heat is conducted in a metallic medium both by phonons (which are scattered by other phonons, defects, and electrons) and by electrons (which are scattered by phonons and defects). In a pure metal the ratio of the rate of electron-phonon collisions to that of U-processes results in an electronic contribution which is usually more than 90 per cent of the total conductivity. Only in very impure metals or disordered alloys is it likely that the electron mean free path would be so diminished that phonon conduction could dominate at ordinary temperatures.

A discussion of the theory underlying electronic thermal conduction would be inappropriate at this point. The discussions of electronic trans-

[41] C. C. Ackerman, B. Bertman, H. A. Fairbank, and R. A. Guyer, Phys. Rev. Letters **16**, 789 (1966).

port in metals and semiconductors (Chapters 3 and 4) automatically involve us with the electronic transport of energy as well as that of charge. An extensive discussion of experimental information on the electronic thermal conductivity in a metal is given by Rosenberg.[42]

The electron population in a semiconducting material is usually too small to produce a very large electronic thermal conductivity, though such conduction can be measured in some semiconductors at fairly high temperatures (when thermally excited electrons are numerous, and Λ_U is very small). An appreciable amount of energy can be transported through an intrinsic semiconductor by the diffusion of hole-electron pairs; each carries energy comparable with the intrinsic gap of the semi-conductor, and diffuses as a neutral complex. Reference may be made to Drabble and Goldsmid[43] for extensive literature on energy transport through semiconductors.

---

[42] H. M. Rosenberg, *Low Temperature Solid State Physics* (Oxford University Press, 1963), Chapter 5.

[43] J. R. Drabble and H. J. Goldsmid, *Thermal Conduction in Semiconductors* (Pergamon Press, 1962).

Problems

2.1 Consider the propagation of longitudinal waves through a linear monatomic lattice. The mass of each atom is m, the equilibrium atomic spacing is a, and the stiffness constant with respect to the j th nearest neighbor is μ_j. Recall that Equation 2-11 was obtained by considering only the nearest-neighbor restoring forces, $\mu_j = 0$ except for $j = 1$. Show that when more remote neighbors cannot be ignored Equation 2-11 is replaced by a dispersion relation

$$\omega^2 = (2/m) \sum_{j=1}^{\infty} \mu_j [1 - \cos(jka)]$$

Plot the shape of the curve relating ω to k if $\mu_2 = 0.25\mu_1$ and all $\mu_j = 0$ for $j > 2$. How is this further affected if we now introduce a third stiffness constant $\mu_3 = -0.2\mu_1$?

2.2 Consider a three-dimensional monatomic solid which crystallizes in the simple cubic structure, with lattice constant a. Show that for the low frequency region of the vibrational spectrum, the density of states with respect to frequency is of the form

$$g(\omega) = (\omega^2/2\pi^2) \sum_{l} (m/a^2\mu_l)^{3/2}$$

where the summation extends over the longitudinal and transverse branches. What will happen if $\omega_i = \omega_{mi} \sin(ka/2)$ for each branch over the entire spectral range? Derive an expression for $g(\omega)$, and plot your results for the case in which $v_L = 2v_T$.

2.3 Note that the metals for which Figure 2-4 displays the phonon spectra both crystallize in F.C.C. Deduce the array of points in k-space which is reciprocal to the F.C.C. lattice, and show the size and shape of the first Brillouin zone. Use this to confirm that the locations $q = [100]$, $q = [\frac{3}{4}\frac{3}{4}0]$, and $q = [\frac{1}{2}\frac{1}{2}\frac{1}{2}]$ all lie on the boundaries of this zone. Note some other locations in q-space which lie on the zone boundary and are of high symmetry.

2.4 Consider an isotropic three-dimensional solid for which the dispersion relation for lattice vibrations is $\omega = (2v_0/a)\sin(ka/2)$ $= \omega_m \sin(ka/2)$ for all directions. You are told that the propagation speeds of longitudinal and transverse waves are negligibly different. Show that the total density of states with respect to frequency must be

$$g(\omega) = \frac{6[\sin^{-1}(\omega/\omega_m)]^2}{\pi^2 a^2 v_0 [1 - (\omega/\omega_m)^2]^{1/2}}$$

so that $g(\omega)$ is proportional to ω^2 in the nondispersive range. Plot the entire course of $g(\omega)$ as expressed here, and discuss why the equation fails at the upper end of the range.

2.5   Draw the (110) plane of the reciprocal lattice for F.C.C. and mark the zone boundaries. Show that the distances to the boundary in [111], [100], and [110] directions are in the ratio 0.87 : 1.00 : 1.06, as a justification of the scales used in the abscissa of Figure 2-4. Sketch in contours of constant ω for longitudinal waves in copper, based on the curves of Figure 2-4. Relate what you produce to some aspects of Figure 2-7.

2.6   Consider a linear diatomic chain, composed of atoms with alternate masses m_1 and m_2. The repeat distance is $2a$ and the nearest-neighbor separation is just half of this. Describe the spectrum of acoustic and optical phonons if there is a stiffness constant $\mu_{12} = \mu_{21}$ constraining deviations of the nearest-neighbor distance, and additional stiffness coefficients μ_{11} and μ_{22} for deviations relative to the nearest neighbor of the same mass. Plot the results for $m_1 = 2m_2$, $\mu_{12} = 2\mu_{11} = 4\mu_{22}$.

2.7   Derive an expression for the group velocity of longitudinal waves in a linear diatomic lattice, and confirm that this is zero for the angular frequencies ω_1, ω_2, and ω_3 shown in Figure 2-10. Describe the variation of amplitude ratio and group velocity with frequency for both branches of the vibrational spectrum. Illustrate your remarks with curves for a diatomic chain in which $M = 3m$ and for which interactions beyond the first nearest neighbors are zero.

2.8   Discuss the reason why Bragg reflection occurs when the wave vector coincides with a Brillouin zone boundary. Why cannot a wave be transmitted for which the wave-vector is complex or pure imaginary?

2.9   An electromagnetic wave passes through a solid with a refractive index of 5.0 for frequencies well below the Reststrahlen frequency $\omega_T = 10^{14}$ radians/sec. The solid consists of singly charged positive and negative ions, all of mass 50 a.m.u. Suppose that the wave has an intensity of 10^6 watt/m^2 at angular frequency $\omega = 10^{13}$ radians/sec. Calculate the R.M.S. amplitudes of the transverse electric and magnetic vectors associated with the wave. Set up the equation of motion for an ion driven at this frequency by the electric vector, and determine the R.M.S. amplitude of its motion. From the Lorentz interaction of the ion velocity with the oscillating magnetic vector, determine how much influence the magnetic component has on the ionic motion.

2.10   Visible photons, each with an energy of 2.0 eV, pass through a solid of refractive index 2.5, in which the speed of sound is 5000 m/s. Determine the energy of phonons which can be created by first order Brillouin scattering, as a function of the angle between the incident light and the created phonons. Plot your results as curves of the sound frequency and of the shift in light wavelength versus angle.

2.11   Verify that the energy $\langle E \rangle$ of Equation 2-47 is the sum of two ratios of integrals, each of which is equal to $(k_0 T/2)$; one representing the average kinetic energy of a classical harmonic oscillator and the other representing the average potential energy.

2.12   Verify that Equation 2-53 is the result of Equations 2-50 and 2-52. *Hint:* Create a geometric progression infinite series of which you can take the natural logarithm, and then differentiate the logarithmic series to produce the numerator and denominator of Equation 2-52. The series you started with can be summed to demonstrate the desired result.

2.13   Consider the Debye approximation for the lattice energy and specific heat of a one-dimensional monatomic lattice with atomic spacing a, and speed of sound v_0 from Equation 2-13. Show that the correct total number of states is preserved for a Debye frequency and temperature $\omega_D = (\pi v_0/a)$ and $\theta_D = (\hbar \omega_D/k_0)$. Derive integral expressions for the lattice energy and specific heat, and show that $C_v = (\pi^2 k_0 T/3a\theta_D)$ per unit length at low temperatures, while C_v has the classical value (k_0/a) per unit length at high temperatures.

2.14   Derive an expression for the temperature at which the thermal lattice energy is equal to the zero-point energy for an Einstein model, all modes degenerate at frequency ω_E. Now write down a condition for the equality of zero-point energy and thermal energy for a Debye solid with a speed of sound v_0.

2.15   In order to get some insight into how the parameter used to characterize the thermal properties of a solid varies with temperature, suppose a solid in which we have N quantum oscillators at frequency ω_1, 2N oscillators at frequency $2\omega_1$, and 3N oscillators at frequency $3\omega_1$. Show how the vibrational energy (neglecting zero-point energy) and specific heat will vary with temperature. Compare this with tabulated information for a Debye function to show how the Debye θ_D of our model solid varies with temperature.

2.16   Consider a three-dimensional solid which has a density of states in conformity with the Debye approximation. Show that at temperatures lower than θ_D the most probable phonon energy is $\hbar\omega_p$, where

$$\exp(\hbar\omega_p/k_0 T)[1 - (\hbar\omega_p/2k_0 T)] = 1$$

From this you should find that the most probable phonon wavelength is $\lambda_p = (a\theta_D/T)$.

2.17   Show that for temperatures well below the Debye temperature, the mean free path of a phonon can be expressed in the form

$$\left.\begin{array}{l}\Lambda = (A\kappa_\ell\theta_D^3/v_0 T^3) \\ = (B\kappa_\ell v_0^2/T^3)\end{array}\right\}$$

Evaluate this for LiF which uses exclusively the ^{7}Li isotope for a molecular weight of 27. Lithium fluoride crystallizes in the NaCl structure, with a nearest-neighbor distance of 2.014 Å. Show that if we identify Λ for the lowest temperatures in Figure 2-25 with the sample diameter, the rising curves of this figure are consistent with $\theta_D \sim 720$K, or $v_0 \sim 5000$ m/sec.

2.18   Discuss Brownian motion for phonons which can experience only N-processes with each other or inelastic collisions with the walls of a crystal. Show that a description in terms of Brownian motion is consistent with a phonon traveling some (D^2/Λ_N) along a crystal of diameter D between successive collisions with the walls, if the inequalities of Equation 2-81 are satisfied.

Bibliography

Elastic Properties

P. D. Edmonds (ed.), *Ultrasonics : Methods of Experimental Physics, Vol. 19* (Academic Press, 1981).

A. K. Ghatak and L. S. Kothari, *An Introduction to Lattice Dynamics* (Addison-Wesley, 1972).

H. B. Huntingdon, *Elastic Constants of Crystals* (Academic Press, 1964).

W. P. Mason and R. N. Thurston (eds.), *Physical Acoustics*, from Vol. 1 (1964) through Vol. 17 (1984) (Academic Press).

M. J. P. Musgrave, *Crystal Acoustics* (Holden-Day, 1970).

T. S. Narasimhamurty, *Photoelastic and Electro-Optic Properties of Crystals* (Academic Press, 1981).

Lattice Dynamics

H. Bilz and W. Kress, *Phonon Dispersion Relations in Insulators* (Springer, 1979).

M. Born and K. Huang, *Dynamical Theory of Crystal Lattices* (Oxford Univ. Press, 1954).

W. Cochran, *Dynamics of Atoms in Crystals* (Arnold, 1973).

B. Donovan and J. F. Angress, *Lattice Vibrations* (Chapman and Hall, 1971).

G. K. Horton and A. A. Maradudin (eds.), *Dynamical Properties of Solids*, from Vol. 1 (1974) through Vol. 5 (1984) (North-Holland).

A. A. Maradudin, E. W. Montroll, G. H. Weiss, and I. P. Ipatova, *Theory of Lattice Dynamics in the Harmonic Approximation* (Academic Press, 2nd ed., 1971).

G. Venkataraman, L. A. Feldkamp, and V. C. Sahni, *Dynamics of Perfect Crystals* (MIT Press, 1975).

B. T. M. Willis and A. W. Pryor, *Thermal Vibrations in Crystallography* (Cambridge Univ. Press, 1975).

Phonons and Phonon Interactions

B. Di Bartolo and R. C. Powell, *Phonons and Resonances in Solids* (Wiley, 1976).

S. W. Lovesey, *Theory of Neutron Scattering from Condensed Matter* (Oxford Univ. Press, 1984), 2 vols.

H. J. Maris (ed.), *Phonon Scattering in Condensed Matter* (Plenum Press, 1980).

J. A. Reissland, *Physics of Phonons* (Wiley, 1973).

J. M. Ziman, *Electrons and Phonons* (Oxford Univ. Press, 1960).

Thermal Properties

R. Berman, *Thermal Conduction in Solids* (Oxford Univ. Press, 1976).

M. Blackman, Handbuch der Physik **7**, 325 (Springer, 1955). [On specific heats.]

G. Caglioti and A. F. Milone (eds.), *Mechanical and Thermal Behavior of Metallic Materials* (North-Holland, 1982).

J. R. Drabble and H. J. Goldsmid, *Thermal Conduction in Semiconductors* (Pergamon Press, 1962).

E. S. R. Gopal, *Specific Heats at Low Temperatures* (Plenum Press, 1966).

C. Y. Ho, R. W. Powell, and P. E. Liley, *Thermal Conductivity of the Elements* (National Bureau of Standards, 1975).

P. H. Keesom and N. Pearlman, Handbuch der Physik, **14**, 282 (Springer, 1956). [On specific heats at low temperatures.]

P. G. Klemens, Handbuch der Physik, **14**, 198 (Springer, 1956). [On thermal conductivity.]

J. E. Parrott and A. D. Stuckes, *Thermal Conductivity of Solids* (Pion, 1975).

B. Yates, *Thermal Expansion* (Plenum Press, 1972).

chapter three

ELECTRONS IN METALS

While metallic conduction has attracted considerable interest for a long time, a clear knowledge of the character of electron dynamics in metals has emerged surprisingly recently. In this chapter we shall follow the historical sequence in first examining simple free electron models, for which it was supposed that atoms in metals could liberate their outer electrons to produce an electron "gas" for random thermal motion and contributions to conduction. These models explained a number of important metallic properties, but raised a new set of questions, which remained unanswerable until it was realized that this electron gas moves through space also occupied by a periodic array of positively charged atomic cores.

The periodic nature of a crystal lattice has often been emphasized in the last two chapters. In this chapter we shall see that the periodicity of the ion core array produces an electrostatic field distribution which profoundly affects the relationship between energy and momentum for a mobile electron. We call this relationship the "band theory of solids." In addition to permitting a more realistic picture of metallic conduction, band theory explains why many solids have insulating or semiconducting properties.

As a curious consequence, advances in the understanding of semiconducting and insulating solids were very rapid during the 1940's and 1950's, and we might well say that by the mid 1950's some simple semiconductors were better understood than any metals. The pendulum has started to swing back since that time, and with the development of highly sophisticated experimental techniques there has been a lively resurgence of interest in metals. Part of this interest has taken the form of geometric studies of the variation of momentum with direction for an electron at constant energy; this topology of the "Fermi surface" in k-space has given many new insights into the conduction processes of metals.

A mobile electron in a metal interacts with the periodic array of ion cores, with crystal defects and impurities, with thermal displacements of the ion core array (i.e., phonons), and with other electrons. We shall be much concerned with understanding how perfect periodicity permits an electron to keep moving in a straight line and how defects and phonons encourage a change in the motion. In this chapter we shall refer briefly to collective electron effects in metals (including "plasma oscillations" and superconductivity), but most of the standard theory of metals is based on the assumption that correlation of the motion of one electron with that of another can be ignored. Thus the standard theory of metals is the theory of single electrons which do not interact with each other and which obey Fermi-Dirac statistics. The simplicity of Fermi-Dirac statistics is lost once it becomes necessary to deal with assemblies of interacting particles.

The discussion in this chapter of "free electron theories" (both classical and quantized) is rather extensive. Our purpose in treating the free electron models is not primarily historical; these models are worthy of attention in that they incorporate basic ideas about conduction which are adopted with little change in the band theory of solids. The end result of band theory is usually a validation of the "nearly free electron" approach, once masses have been renormalized and some scalars have been generalized into tensors.

Some Features of the Metallic State

TYPICAL OBSERVED PROPERTIES OF METALS

Before we talk about the various theoretical models for the metallic state, we shall find it useful to remind ourselves of the more important properties of a metal. These are properties which should emerge as natural consequences of a successful theory. With such a goal in mind, we must admit that no completely satisfactory theory now exists. One by one, successive theoretical models have been able to *approach* a qualitative and quantitative understanding of the various types of property.

We should note the following properties of metals:

1. Under isothermal conditions a metal obeys Ohm's Law quite well. We often express this law as

$$\mathbf{J} = \sigma \mathbf{E} \tag{3-1}$$

where the scalar electrical conductivity σ (in units of ohm^{-1} m^{-1}) relates the current density $\mathbf{J}$ (in amp/m²) to the electric field gradient $\mathbf{E}$ (volt/m).

2. A metal is a very good conductor of electricity. Whereas the conductivity of an insulator may be as small as 10^{-16} ohm^{-1} m^{-1} (see Figure 3-19) and that of a semiconducting material is typically within the range 10^{-4} to 10^5 ohm^{-1} m^{-1}, most metals have a room temperature conductivity of from 10^6 to 10^8 ohm^{-1} m^{-1}. The conductivity for a sampling of metals is given in Table 3-1. This table progresses from monovalent metals through multivalent elements to transition elements.

3. A metal also has a large electronic thermal conductivity κ_e. Wiedemann and Franz (1853) observed that a good thermal conductor is also a good electrical conductor, and the consistency of (κ_e/σ) among metals at a given temperature is still referred to as the *Wiedemann-Franz law*. At ordinary temperatures κ_e is usually relatively temperature-independent, whereas σ is likely to display a T^{-1} behavior. (This is the case for the copper data of Figure 3-1). In 1881 Lorenz noted that $(\kappa_e/\sigma T)$ has a temperature-independent value shared in common by many metals, and the value of $L \equiv (\kappa_e/\sigma T)$ is known as the *Lorenz number*. Some values for the Lorenz number L are listed in Table 3-1.

4. An increase of κ_e, and a faster increase of σ, are observed when a metal is cooled well below a characteristic temperature (which is related to the characteristic temperature θ_D for thermal capacity). For copper (Figure 3-1) and some other monovalent metals, σ varies as T^{-5} in the steepest part of the range.

TABLE 3-1 ELECTRICAL CONDUCTIVITY AND LORENZ NUMBER FOR SOME METALLIC ELEMENTS[*]

Metal	T = 100K		T = 273K	
	Electrical Conductivity (ohm^{-1} m^{-1})	Lorenz Number (volt/kelvin)^2	Electrical Conductivity (ohm^{-1} m^{-1})	Lorenz Number (volt/kelvin)^2
	σ	$L = (\kappa_e/\sigma T)$	σ	$L = (\kappa_e/\sigma T)$
Copper	2.9×10^8	1.9×10^{-8}	6.5×10^7	2.3×10^{-8}
Gold	1.6×10^8	2.0×10^{-8}	5.0×10^7	2.4×10^{-8}
Zinc	6.2×10^7	1.8×10^{-8}	1.8×10^7	2.3×10^{-8}
Cadmium	4.3×10^7	2.1×10^{-8}	1.5×10^7	2.4×10^{-8}
Aluminum	2.1×10^8	1.5×10^{-8}	4.0×10^7	2.2×10^{-8}
Lead	1.5×10^7	2.0×10^{-8}	5.2×10^6	2.5×10^{-8}
Tungsten	9.8×10^7	2.8×10^{-8}	2.1×10^7	3.0×10^{-8}
Iron	8.0×10^7	3.1×10^{-8}	1.1×10^7	2.8×10^{-8}

[*] Values for σ are taken from G. T. Meaden, *Electrical Resistance of Metals* (Plenum Press, 1965). Thermal conductivity data from *American Institute of Physics Handbook* (McGraw-Hill, 3rd edition, 1971) then permits calculation of the Lorenz number.

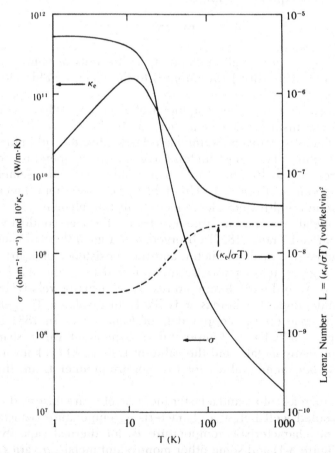

Figure 3-1 Temperature dependence of the electrical conductivity σ and the electronic thermal conductivity κ_e for highly pure copper.

5. At a sufficiently low temperature, σ reaches a plateau value controlled by impurities and lattice imperfections. The electrical *resistivity* (the inverse of conductivity, measured in ohm-meters) is often found to approximate the behavior of *Matthiessen's Rule*, that the contribution of impurities and imperfections to the resistivity is the same at all temperatures:

$$[1/\sigma(T)] = [1/\sigma_{imp}] + [1/\sigma_{pure}(T)] \qquad (3\text{-}2)$$

Of course, the contribution of σ_{imp} is seen most readily at very low temperatures when σ_{pure} is heading for infinity.

6. Magnetic effects in ferromagnetic metals and alloys also contribute to the electrical resistivity.

7. About half of the metallic elements become superconducting at sufficiently low temperatures. These elements are so marked in the Periodic Table on the inside cover.

8. The free electron gas has a very small electronic specific heat, one which is proportional to the absolute temperature. The electron gas also has a very small paramagnetic susceptibility, which is temperature-independent.

9. In the presence of a combination of electrical, magnetic, and thermal gradients, a variety of galvano-thermo-magnetic effects arise, where, for example, a temperature gradient may produce electrical currents or potential differences. Under such circumstances the relation between a given field and the resultant phenomena must be described by a magneto-thermo-conductivity tensor. Most of these higher order effects are quite small in a metal and are not easy to measure with precision.

10. For very pure samples of single crystals, a variety of orientation-dependent phenomena can be observed in the presence of a very large magnetic field. Several of these phenomena are oscillatory functions of magnetic field strength.

FREE ELECTRONS AND POSITIVE ION CORES

In Section 1.1 we were preoccupied with the various forms of interatomic binding in solids. At that time we noted that the valence electrons in an ionic or covalent solid were associated with specific atoms, but that bonding was a collective affair in a metal. Each atom in a typical metallic structure has many nearest neighbors and many bonds. These bonds are individually very weak, and a metal has considerable strength only because its bonds are so numerous.

According to the theory of metals, a metallic crystal may be envisaged as the superposition of:

1. A periodic array of positive ion cores. Each core consists of an atomic nucleus and the electrons which make up closed shells. From the Pauli exclusion principle, we know that an electron in a closed-shell

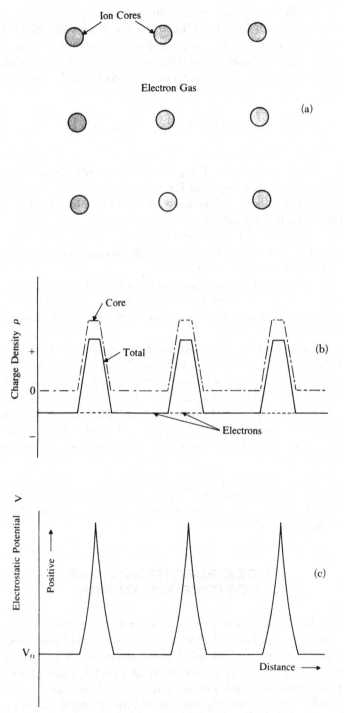

Figure 3-2 (a), A two-dimensional slice through a metallic crystal, indicating a periodic array of relatively compact positive ion cores. The "gas" of free electrons is envisaged as being uniformly distributed over all the space. (b), The variation of charge density with distance for a line chosen to pass through a sequence of positive ion cores. Along such a line, ρ alternates from positive to negative and back again. If we choose to examine ρ along a parallel line which does not intersect any cores, we shall find that ρ is uniformly negative along the entire line. (c), The periodic fluctuation of electrostatic potential required by the charge density curve of part (b).

orbit around an atom is to be associated exclusively with that atom.

2. A quasi-uniform negative charge density due to all outer elec-trons. These electrons form an "electron gas," of which the particles are moving at thermal speeds. By their collective action the electrons in the gas bind the solid together. The free electron density may range from one electron per atom ($2 \times 10^{28}\,\mathrm{m}^{-3}$) upwards, depending on the valence of the atoms.

This combination is shown in part (a) of Figure 3-2, and the other parts of the figure show the conclusions we may draw from this picture. The *average* charge density of the combination in Figure 3-2(a) must be zero for electrical neutrality of the solid. Yet if we look at the *local* situa-tion, we can surmise that there must be a periodic sequence of positive humps in the charge density ρ, superimposed upon an essentially uni-form negative sea of charge from the electron gas. This oscillation of ρ is sketched in Figure 3-2(b). Since the closed-shell core of an atom is usually considerably smaller than the interatomic distance, the cores will customarily occupy only a small fraction of the total volume,[1] and the positive excursions of ρ will be of large magnitude over a small vol-ume.

From Poisson's equation

$$\nabla^2 V = -\rho/\varepsilon_0 \qquad (3\text{-}3)$$

there must be a three-dimensional periodic variation of electrostatic po-tential in a metallic crystal, as sketched in Figure 3-2(c). The shape of

---

[1] See Problem 3.1 for an examination of the volume available for free electrons in the simplest metals — the alkali metals of Group I A.

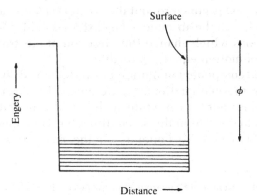

Figure 3-3 The energy of an electron in a solid is determined by the sum of its po-tential and kinetic energies; it has an appreciable amount of the latter only if it is moving rapidly. Even the most energetic electron in a solid has much less *total* energy than one at rest in free space outside the crystal, for its potential energy is large and negative. An elec-tron, with its negative charge, can be pulled from a crystal into a surrounding vacuum only if the surface potential barrier is surmounted. The height of this barrier is called the *work function* ϕ, and is typically a quantity of a few eV per electron. Electrons may be en-couraged to pass over this external barrier thermally (thermionic emission) or by absorp-tion of an energetic photon (the external photoelectric or photoemissive effect).

each hump in potential, and the height of each potential maximum, will depend on the details of the charge configuration in the ion core.

These periodic humps in the electrostatic potential are treated as being of *negligible* effect in the *free electron model*. Calculations are made on the assumption that "free" electrons move in a kinetic manner with complete freedom throughout a metallic body, subject only to a potential barrier at the surface of the solid. This external barrier (Figure 3-3) is the work function ϕ, which enters into phenomena such as thermionic and photoelectric emission.

The free electron density may range from some 2×10^{28} m^{-3} upwards, depending on the location of the metallic atoms in the Periodic Table. A statistical treatment of this electron population with total neglect of the interatomic potential is most successful (in quantum form) for the monovalent alkali metals, though it can also give us a useful first idea about the behavior of multivalent metals such as aluminum or lead.

RANDOM MOTION
AND SYSTEMATIC
ELECTRON MOTION

The various theoretical models for metallic conduction share the idea that electron motion can be viewed as the sum of two components. The first component is the random motion resulting from thermal and zero-point energy. (In classical models only thermal motion is recognized. In quantized models of a "degenerate" electron gas, the zero-point motion is predominant). The instantaneous speed of any electron is quite large, but the word "random" implies that at thermodynamic equilibrium in the absence of any external influence the ensemble average velocity is zero.[2] For any specified electron, the scalar *speed* averaged over a long period of time (comprising many free paths between collisions) is non-zero, but the (vector) *velocity* averaged over a long period is zero in the absence of any external field. This comes about because collisions which change the direction of electron motion make all directions of motion equally probable.

The second component of motion is a systematic drift motion caused by electrical or thermal gradients in the metal. The average velocity of the assembly does not vanish while such a field is operating, and every electron tends to migrate in the same direction. How much of an observable effect this can have depends on the *mean free time* between collisions, τ_m.

---

[2] For any finite assembly of electrons, the average drift velocity $v_D = \dfrac{1}{N} \sum\limits_{i=1}^{N} v_i$ oscillates about zero as a function of time in thermodynamic equilibrium. In consequence, a fluctuating open-circuit voltage may be observed between any two points on a conductor, even in the absence of any macroscopic currents. This *thermal noise* has components of the same amplitude for all frequencies from zero up to $(k_0 T/h)$ at temperature T. Thermal noise, discussed first by J. B. Johnson [Phys. Rev. **32**, 97 (1928)] and H. Nyquist [Phys. Rev. **32**, 110 (1928)], is described in relation to other fluctuation phenomena in *Elementary Statistical Physics* by C. Kittel (Wiley, 1958) and in *Noise and Fluctuations: an Introduction* by D. K. C. MacDonald (Wiley, 1962).

Classical Free Electron Theory   3.2

THE MAXWELL-BOLTZMANN
VELOCITY DISTRIBUTION

We now know that quantum restrictions dictate the distribution of electron speeds. However, several aspects of metallic behavior were accounted for in the classical models which preceded the development of Fermi-Dirac statistics. These were models based on the classical Maxwell-Boltzmann velocity distribution. It is assumed that the reader recalls from some prior exposure to statistical physics that in a Boltzmann distribution with n electrons per unit volume, the number

$$f_0 dv_x dv_y dv_z = n \left(\frac{m}{2\pi k_0 T} \right)^{3/2} \exp\left[\frac{-m(v_x^2 + v_y^2 + v_z^2)}{2k_0 T} \right] dv_x dv_y dv_z \quad (3\text{-}4)$$

per unit volume lie within the velocity range $dv_x dv_y dv_z$ centered on $v_x v_y v_z$ *at thermodynamic equilibrium*. In Equation 3-4, f_0 is known as the equilibrium *distribution function*.

The equilibrium Boltzmann distribution is spherically symmetrical in velocity space, as indicated by the shading of Figure 3-4(a). Thus the probability of a scalar speed in the range from s to (s + ds) is

$$\frac{dn}{n} = 4\pi s^2 \left(\frac{m}{2\pi k_0 T} \right)^{3/2} \exp\left[\frac{-ms^2}{2k_0 T} \right] ds \quad (3\text{-}5)$$

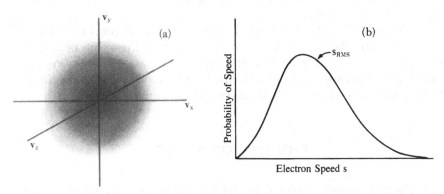

Figure 3-4 The Maxwell-Boltzmann distribution of Equation 3-4. (a), The distribution of vector velocities in velocity space. (b), The equilibrium probability curve for scalar speed s, without regard for the direction of that speed, as in Equation 3-5. The most probable speed is $(2k_0 T/m)^{1/2}$, but the speed for the average energy is $s_{RMS} = (3k_0 T/m)^{1/2}$.

regardless of the direction of that speed [as seen in Figure 3-4(b)]. The average kinetic energy per electron in a Boltzmann distribution is

$$\tfrac{3}{2}k_0T = \tfrac{1}{2}\, m(s_{RMS})^2 \tag{3-6}$$

for an RMS speed of $s_{RMS} = (3k_0T/m)^{1/2}$. At ordinary temperatures this speed is some 10^5 m/s.

ELASTIC SCATTERING AND THE MEAN FREE PATH

In the development of a classical free electron model for conduction, it seemed entirely natural to assume that the mean free path of an electron would be controlled primarily by elastic collisions with the positive ion cores. In a metal containing N ion cores per unit volume, each of effective scattering radius R, the mean free path of an electron should be some

$$\lambda \sim (\pi R^2 N)^{-1} \tag{3-7}$$

from purely geometric considerations (Problem 3.2), regardless of the incident speed. Such collisions can produce a complete spectrum of scattering angles with very little change in electron energy because of the much larger ion mass. The mean free path λ of Equation 3-7 would then seem to be the appropriate quantity to be substituted into

$$\tau_m = \lambda/s_{th} \tag{3-8}$$

where s_{th} denotes the (thermal) speed with which an electron moves from one scattering center to another, and τ_m is the mean free time between collisions. Any external field has to exert its influence in changing the electron velocity during a succession of short intervals, each comparable with τ_m.

Discussions of the field-aided drift of electrons proceed without mention of the thermal motion (except in setting the value of τ_m). We should always remember that drift velocities are *small* but systematic velocities, superimposed on a thermal motion of much larger amplitude, which cancels out only as an ensemble average.

THE DRUDE MODEL

The first simple classical model for a free electron gas in a metal was described by Drude (1900). A description of Drude's model is useful here, for despite its limitations it did introduce some ideas which are still incorporated in the more elaborate treatments.

Drude supposed that each atom in the metal would contribute one or more electrons to the "electron gas" for a total electron density n per unit volume. He worked out the consequences of a highly simplified model in which it was supposed that *every* electron had the kinetic energy corresponding with three classical degrees of translational freedom; i.e., that each electron had a kinetic energy $\frac{3}{2}k_0T$. Thus in Drude's model all electrons moved with the RMS speed of a Boltzmann distribution (Equation 3-6).

Such an assumption would endow the electron gas with a total energy of $\frac{3}{2}nk_0T$ and an electronic specific heat of $\frac{3}{2}nk_0$. Then in a metal (such as sodium or copper) for which the electron and atom densities were the same the electrons should boost the measured specific heat by about 50 per cent, compared with a nonconductor. This anticipated extra 50 per cent of thermal capacity is not found in actual metals. The specific heat of the electrons in a metal is actually very small and is linearly dependent on temperature, not independent of it. Thus at the outset we can point to a grave shortcoming of Drude's theory (and of all subsequent classical theories). However, we can still derive some benefit from looking at Drude's formulations of drift motion and electrical conductivity.

Drude supposed that moving electrons are scattered by random collisions with the ion cores. The term "random" is used here to indicate that the *average* velocity is zero immediately after any scattering collision. This means that any drift velocity built up in the direction of an external field is wiped out by a scattering collision. A field can exert its influence to change the magnitude and/or direction of an electron's motion during the interval from one collision to the next, but the effect is eradicated each time a collision takes place. Thus the larger the mean free time τ_m between collisions, the larger the influence of an external field.

We can define the mean free time τ_m as the reciprocal of the collision probability per unit time. Let us consider a group of n_0 electrons at time $t = 0$. Then the number of electrons which have survived without a collision until time t is

$$n_t = n_0 \exp(-t/\tau_m) \qquad (3\text{-}9)$$

The rate at which collisions are then removing electrons from the ranks of survivors is

$$\frac{-dn}{dt} = \left(\frac{n_t}{\tau_m}\right) = \left(\frac{n_0}{\tau_m}\right) \exp\left(\frac{-t}{\tau_m}\right) \qquad (3\text{-}10)$$

Now suppose that an electric field **E** has been present while these electrons have been moving and colliding. After a time t, an electron which has not yet been scattered has achieved a drift velocity

$$\Delta v_t = \left(\frac{-eEt}{m}\right) \qquad (3\text{-}11)$$

over and above its thermal motion. Hence the distance traveled in the direction of the field is

$$x_t = \left(\frac{-eEt^2}{2m}\right) \tag{3-12}$$

superimposed upon the random thermal motion. Equation 3-12 is the ordinary Newtonian result for distance covered in a situation of uniform acceleration.

The total electronic transport along the field direction for n_0 electrons in one free path is then

$$
\begin{aligned}
\int_0^\infty x_t\left(\frac{dn}{dt}\right) dt &= \left(\frac{-eEn_0}{2m\tau_m}\right) \int_0^\infty t^2 \exp(-t/\tau_m)\cdot dt \\
&= \left(\frac{-eEn_0\tau_m^2}{m}\right) \int_0^\infty \frac{1}{2}\, y^2 e^{-y}\cdot dy \\
&= \left(\frac{-eEn_0\tau_m^2}{m}\right)
\end{aligned}
\tag{3-13}
$$

This is transport equivalent to that of n_0 particles, all having the same free time τ_m and a *mean drift velocity*

$$\Delta\bar{v} = \left(\frac{-eE\tau_m}{m}\right) \tag{3-14}$$

If we suppose that our metal has a total of n electrons/m³, all with constant drift velocity $\Delta\bar{v}$ in an electric field **E**, then the electrical current density is

$$\mathbf{J} = (-en\Delta\bar{v}) = (ne^2\tau_m\mathbf{E}/m) = \sigma\mathbf{E} \tag{3-15}$$

where the positive scalar quantity

$$\sigma = (ne^2\tau_m/m) \tag{3-16}$$

is the electrical conductivity. This *form* for the conductivity is preserved in more complicated quantized models and in band theory; the differences lie only in how n, m, and τ_m are defined.

Equation 3-15 conforms with the empirical statement known as Ohm's Law, that applied voltage and resultant current are linearly related in a conducting solid. As we proceed to look at more complicated situations of conduction in a solid, it will sometimes be found that **J** can become non-linear with respect to **E**. We shall also find that σ may be a tensor quantity, relating a field in one direction to a current density in another.

Electrical conductivity is often expressed in terms of quantities other than the scattering mean free time. A common expression is

$$\sigma = n\, e\mu \tag{3-17}$$

where

$$\mu = (-\Delta\bar{v}/E) = (e\tau_m/m) \tag{3-18}$$

is the *drift mobility* of electrons (the drift velocity achieved in unit field strength). The conductivity is also often cited in terms of the electronic *mean free path*, which is the distance $\lambda = s_{th}\tau_m$ that any electron moves by virtue of its *thermal* speed during the mean free time τ_m. Thus in the Drude model we can write σ as

$$\sigma = \left(\frac{ne^2\tau_m}{m}\right) = \left(\frac{ne^2\lambda}{ms_{th}}\right) = \frac{n\,e^2\lambda}{(3mk_0T)^{1/2}} \tag{3-19}$$

Since the electrical conductivity of almost every common metal varies as T^{-1} over a wide range, it was necessary for Drude to argue that collisions of electrons with the fixed array of positive ion cores must somehow result in a mean free path which varied as $T^{-1/2}$. It was never at all clear why this should be so, a failing which helped to stimulate further thought on kinetic models for an electron gas. On further cooling, the conductivity of a typical metal rises as T^{-5} before reaching a low temperature plateau (Figure 3-1), and these features were even less palatable for classical free electron models.

The Drude kinetic arguments we have used here for σ can similarly be applied to predict an electronic thermal conductivity

$$\kappa_e = \tfrac{2}{3}\tau_m s^2 C_{cl} \tag{3-20}$$

The quantity C_{cl} in Equation 3-20 denotes the classical specific heat at constant volume for the *electron gas*. As we have already noted, this specific heat should be $\tfrac{3}{2}k_0n$ for Drude's model of an electron gas, so that

$$\kappa_e = nk_0\tau_m s^2 = (3n\tau_m k_0{}^2 T/m) \tag{3-21}$$

This may be combined with Equation 3-16 for the electrical conductivity to give a Lorenz number

$$L \equiv (\kappa_e/\sigma T) = 3(k_0/e)^2$$
$$= 2.2 \times 10^{-8} \text{ (volt/kelvin)}^2 \tag{3-22}$$

Drude's result for the Lorenz number was in surprisingly good agreement with the conductivity ratios measured for many metallic solids around room temperature, as may be seen from Table 3-1. This was one of the most celebrated successes of Drude's model, and encouraged efforts towards the development of more complicated models.

The ratio of thermal to electrical conductivities in most real metals changes significantly from the Wiedemann-Franz number upon cooling below 100K. However, it is found for many metals that $(\kappa_e/\sigma T)$ becomes stable again at the very lowest temperatures. (Needless to say, the

Drude model says nothing about that.) Figure 3-1 shows that $(\kappa_e/\sigma T)$ for copper becomes constant again at temperatures so low that scattering by impurities and defects controls τ_m for both electrical and thermal conductivity.

THE LORENTZ MODEL

A gross simplification made by Drude was the assumption that every electron has the same thermal speed. This assumption was abandoned by H. A. Lorentz (1905) (not to be confused with Lorenz, who had published 25 years earlier). Lorentz[3] investigated the properties of a classical Maxwell-Boltzmann distribution for electron velocities. He considered the perturbation of the Maxwellian distribution in the presence of an electrical or thermal gradient, either of which will tend both to displace and to distort the symmetry of the equilibrium distribution function of Equation 3-4.

The approach taken by Lorentz was that of the *Boltzmann transport equation*, which we previously encountered in Section 2.5, in connection with heat conduction by the diffusion of phonons through a solid. It was noted at that time that the Boltzmann equation approach is used in several areas of physics to describe transport by a statistical distribution of mobile entities. Now in the Lorentz model we are concerned with transport of the electrical charges associated with an electron gas (electrical conductivity), and also with transport of the kinetic energy of these same electrons (electronic thermal conductivity).

A completely rigorous treatment of electrical conduction by the Boltzmann approach is quite lengthy and complicated, but the simplified treatment sketched here shows the important principles involved. The interested reader may consult the accounts of Seitz[4] or Ziman[5] for more extensive discussions.

Suppose that the electron gas in a metal has velocities distributed according to the equilibrium distribution f_0 of Equation 3-4 in the absence of any external fields or influences. This distribution was illustrated in Figure 3-4(a), and is repeated in part (a) of Figure 3-5. To simplify the Boltzmann equation, suppose that the metal is homogeneous, so that f_0 is independent of spatial coordinates.

Now let an electric field be applied to the metal, resulting in a systematic electron drift. The velocity distribution f, while the field is present, will in general differ from f_0. Once again, for simplicity we shall consider the consequences of a uniform field $\mathbf{E}$, so that the spatial derivative of $(f-f_0)$ is zero. Then anywhere within the metal crystal, and at any instant while the field is acting, the rate at which f changes with time is the sum of two different kinds of contribution:

$$\left(\frac{df}{dt}\right) = \left(\frac{\partial f}{\partial t}\right)_{\text{field}} + \left(\frac{\partial f}{\partial t}\right)_{\text{scat}} \tag{3-23}$$

---

[3] The lectures of Lorentz are available in a reprinted English translation, under the title *The Theory of Electrons* (Dover, Second Edition, 1952).

[4] F. Seitz, *Modern Theory of Solids* (McGraw-Hill, 1940).

[5] J. M. Ziman, *Principles of the Theory of Solids* (Cambridge, 2nd edition, 1972).

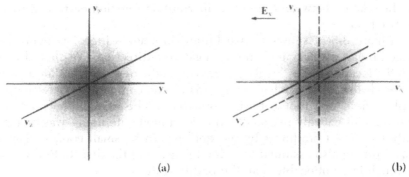

Figure 3-5 (a), The distribution of velocities in velocity space for a Boltzmann electron gas at thermodynamic equilibrium. (b), The distribution for the same assembly of electrons when an electric field is applied in the x-direction. The center of gravity of the distribution now coincides with the intersection point of the two dashed lines. Also, the distribution has almost certainly been perturbed in shape as well as being displaced.

The first term on the right is the influence of the field in changing the distribution, while the second is the effect of scattering in the attempt to restore the equilibrium distribution. Equation 3-23 must be satisfied for each point in velocity space.

Since force is rate of change of momentum, and an electron of speed v has momentum mv and experiences a force $-e\mathbf{E}$ in an electric field $\mathbf{E}$, then the first term on the right of Equation 3-23 is

$$\left(\frac{\partial f}{\partial t}\right)_{\text{field}} = \left(\frac{\partial \mathbf{v}}{\partial t}\right) \cdot \left(\frac{\partial f}{\partial \mathbf{v}}\right) = \left(\frac{-e\mathbf{E}}{m}\right) \cdot \left(\frac{\partial f}{\partial \mathbf{v}}\right) \qquad (3\text{-}24)$$

Thus the effect of an electric field in changing the distribution depends on the derivative of the distribution function with respect to velocity (which is related to the derivative with respect to energy).

Scattering collisions are supposed to erase any acceleration of the preceding free path, and to recreate a velocity distribution f_0. Thus in considering the influence of collisions in minimizing $(f-f_0)$, Lorentz assumed that $(\partial f/\partial t)_{\text{scat}}$ would be directly proportional to (f_0-f). The proportionality can be expressed in terms of a relaxation time τ_r:

$$(\partial f/\partial t)_{\text{scat}} = (f_0-f)/\tau_r \qquad (3\text{-}25)$$

Note that the quantity τ_r is defined for each location in velocity space. Thus the relaxation time is certainly related to the mean free time τ_m of the Drude model (Equations 3-8 and 3-16), but the relation is an integral one.

With the aid of Equations 3-24 and 3-25, we can rewrite Equation 3-23 in the form of a *continuity equation*

$$\left(\frac{df}{dt}\right) + \frac{e\mathbf{E}}{m} \cdot \left(\frac{\partial f}{\partial \mathbf{v}}\right) + \frac{f-f_0}{\tau_r} = 0 \qquad (3\text{-}26)$$

for the relation between f_0 and the distribution function perturbed by an electric field.

Figure 3-5(b) shows a distribution function which has been displaced by an electric field in the negative x-direction. We should note that the location of the "center of gravity" of the new distribution, as well as any distortion of its shape, depends on the competition between the applied field and the scattering processes. The assumption made by Lorentz—an entirely reasonable one for a moderate field—was that the bodily shift of f produced by the field would be small compared with s_{RMS}. Lorentz also assumed that the distortion of the distribution would be much less noticeable than the bodily shift.

When a steady state is reached by applying a constant electric field for a time that is long compared with τ_r, the first term of Equation 3-26 must become zero. Continuity then requires that the second and third terms be equal and opposite for every location in velocity space, so that

$$f = f_0 + (\tau_r eE/m) \cdot (\partial f/\partial v) \tag{3-27}$$

is the steady-state perturbed distribution. An integration over f yields a nonvanishing drift velocity for the entire electron gas; in short, the metal has a finite electrical conductivity. If, as in Figure 3-5(b), we suppose the field to lie along the x-direction, then the current density is

$$J_x = -\int ev_x f \cdot dv_x dv_y dv_z \tag{3-28}$$

and since this integral performed with respect to f_0 vanishes, then

$$J_x = \sigma E_x = -\int (E_x e^2/m)\tau_r v_x \cdot (\partial f/\partial v) \cdot dv_x dv_y dv_z \tag{3-29}$$

In order to carry out this integration, we shall simplify the treatment further with the plausible assumption that τ_r depends only on the *magnitude* of electron speed, not on the *direction* of that motion. A common assumption in solving the Boltzmann equation is that τ_r varies as a power law of electron speed:

$$\tau_r = As^j \tag{3-30}$$

The exponent j depends on the nature of the scattering mechanism. Now Lorentz assumed that electrons were elastically scattered by the fixed array of ion cores. For this mechanism, the mean free path λ is independent of electron speed, as noted in connection with Equation 3-7 and Problem 3.2. This requires us to identify λ as the quantity A in Equation 3-30, with $j = -1$, i.e.,

$$\tau_r = \lambda/s \tag{3-31}$$

If this *is* the operative scattering mechanism, then the more slowly moving members of the distribution are the ones whose motion will be most severely affected by an electric field.

Substitution of Equation 3-31 into 3-29 yields an integral

$$\sigma = -\int (\lambda e^2 v_x/ms) \cdot (\partial f/\partial v) \cdot dv_x dv_y dv_z \qquad (3\text{-}32)$$

for the electrical conductivity, which we can transform into an integral with respect to the scalar speed s, noting that $v_x{}^2 = (s^2/3)$ averaged over all electrons, and that a shell in velocity-space of radius s and thickness ds has volume $4\pi s^2 \cdot ds$. Thus Equation 3-32 becomes

$$\sigma = (4\pi e^2/3m) \int_0^\infty \lambda s^2(-\partial f_0/\partial s)ds \qquad (3\text{-}33)$$

as an integral with respect to scalar electron speed. The conductivity can equally easily be expressed as an integral with respect to electronic kinetic energy $\epsilon = \tfrac{1}{2}ms^2$.

It will be noted that the major contribution to the conductivity of Equation 3-33 comes from that range of electron speed for which the distribution function has the maximum negative derivative with respect to speed.[6] Problem 3.4 suggests that the integral of Equation 3-33 (and of the companion integral with respect to energy) be carried out using f_0 from Equation 3-4 to verify that

$$\sigma = \frac{4ne^2\lambda}{3(2\pi mk_0 T)^{1/2}} \qquad (3\text{-}34)$$

Note that Equation 3-34 has the same general form as for the Drude model. When the mean free path is defined in the same manner, the expressions of Equations 3-19 and 3-34 differ only by a factor of $(3\pi/8)^{1/2} = 1.09$. In a similar way, the Boltzmann transport equation approach of Lorentz produces a result for electronic thermal conductivity which is similar to Drude's result but about one-third smaller.[7]

OTHER EFFECTS PREDICTED
BY THE LORENTZ MODEL

The Lorentz model, just as for the Drude model, requires an electronic kinetic energy of $\tfrac{3}{2} nk_0 T$. This in turn would require a large elec-

---

[6] This may not appear too exciting for f_0 of Equation 3-4, but it assumes a new importance when the Boltzmann equation is applied to quantized models of conduction, for then the Fermi-Dirac distribution function has a zero derivative *except* for electrons of approximately the Fermi speed and energy.

[7] The Lorentz results for both electrical and thermal conductivities may be combined to give a predicted Lorenz number $L = (\kappa_e/\sigma T) = 2(k_0/e)^2$. Recall that Drude's result for the Lorenz number (which fitted experimental results rather well) was 50 per cent larger than the number we have just quoted. It may seem disturbing that the more realistic Boltzmann distribution should be less accurate than Drude's model in this respect, but this should not be taken too seriously. Neither Drude's nor Lorentz's model uses the *correct* (Fermi-Dirac) distribution; our interest in Lorentz's approach is that the solution of the Boltzmann transport equation is at the heart of the modern one-electron theories of transport. The various second-order effects (such as magnetoresistance, Nernst effect, and the like) appear as natural consequences in the Lorentz model and all succeeding models.

tronic specific heat, which is not seen in practice. Since we now know that paramagnetism arises from the magnetic moment of spinning electrons, a Boltzmann gas of free electrons should also have a large magnetic susceptibility, with a magnitude inversely proportional to the absolute temperature and a field-dependence controlled by a Langevin function.[8] The non-appearance of this phenomenon did not distress Lorentz, since the idea of electron spin came much later.

Several interesting phenomena do follow as natural consequences of the Lorentz solution of the Boltzmann transport equation in a combination of external influences. Thus an electromotive force is developed in a conductor within which temperature varies from end to end (the Thomson effect), and an expression for the magnitude of the Thomson effect can be calculated by solving the Boltzmann transport equation with a velocity distribution of electrons which varies along the temperature gradient. More complicated effects arise when electrical, thermal, and magnetic fields are combined.[9]

When electrons in a metal are subjected to a magnetic induction **B** as well as an electric field **E**, then from Maxwell's equations the total force on an electron of velocity **v** is

$$\mathbf{F} = -e[\mathbf{E} + (\mathbf{v} \times \mathbf{B})] \qquad (3\text{-}35)$$

rather than just $-e\mathbf{E}$. Accordingly, $(\partial f/\partial t)_{\text{field}}$ of Equation 3-24 must be generalized to

$$(\partial f/\partial t)_{\text{field}} = (-e/m)[\mathbf{E} + (\mathbf{v} \times \mathbf{B})] \cdot (\partial f/\partial \mathbf{v}) \qquad (3\text{-}36)$$

and the continuity equation (Equation 3-26) must be similarly generalized.

Note that the effect of a magnetic field is constantly to change the direction of motion of an electron, since the magnetic Lorentz force is in the direction of $\mathbf{v} \times \mathbf{B}$. The effect averaged over the distribution of velocities for thermodynamic equilibrium is zero, but a net effect results from the interaction of a magnetic field with the small systematic drift velocities superimposed by an electric field. This interaction has the effect that electrons drifting down a metallic rod will be encouraged to drift toward one side, in the direction of $\mathbf{B} \times \mathbf{v}_D$. (Problems 3.5 and 3.6 are addressed to this topic.) The small resulting departure from electrostatic neutrality sets up an opposing electric field, so that in steady state the current density **J** is directly along the rod, but the total electric field **E** has components parallel to and perpendicular to **J**.

---

[8] In Chapter 5 we shall discuss the Langevin function for partial alignment of a set of spins in a magnetic field, when thermal fluctuations oppose ordering.

[9] A monograph which deals in detail with the variety of effects that can be observed in the presence of electrical, thermal, and magnetic fields in metals is *Galvanomagnetic and Thermomagnetic Effects* by L. L. Campbell (Longmans, 1923, reprinted by Johnson Reprints, 1960). The theoretical discussion in this early work conforms with classical theories. For a more recent account based on quantized models of an electron gas, see J-P Jan, "Galvanomagnetic and Thermomagnetic Effects in Metals" in *Solid State Physics Vol. 5*, edited by F. Seitz and D. Turnbull (Academic Press, 1957).

Under these combined influences, the steady-state solution of the Boltzmann equation to first order in **B** is that

$$\sigma \mathbf{E} = \{\mathbf{J} + [\pi e\lambda/2(2\pi mk_0T)^{1/2}]\mathbf{J} \times \mathbf{B}\}$$

$$= \{\mathbf{J} - [\sigma R_H]\mathbf{J} \times \mathbf{B}\} \qquad (3\text{-}37)$$

Here σ is the conductivity given in Equation 3-34. The parameter R_H is called the *Hall coefficient*, and this sets the magnitude of the transverse electric field in the presence of a magnetic induction. E. H. Hall first reported such an effect experimentally in 1879.

It will be seen by comparison of Equations 3-34 and 3-37 that the Lorentz model predicts a Hall coefficient

$$R_H = -(3\pi/8ne) \qquad (3\text{-}38)$$

which depends on the electron density but not on the mean free path. We should be able to work backwards from experimental Hall data to a deduced electron density, as attempted in Table 3-2.

One column in Table 3-2 lists the combination $(-\sigma R_H)$, which is called the *Hall mobility* of the carriers. This mobility is measured in units of $m^2/V \cdot s$ (identical with $tesla^{-1}$), and differs only by a numerical factor close to unity from the drift mobility of the Drude model (Equation 3-17). From the Hall mobilities of Table 3-2, the corresponding mean free paths are shown in the final column. Table 3-2 is arranged to progress from simple monovalent metals to multivalent metals and transition elements. Note that the sign of the Hall effect is positive for some metals. For these "anomalous" metals (which are not anomalous when described in terms of band theory), the entries for Hall mobility and mean free path are omitted.

Table 3-2 shows that the electron density deduced from the Hall effect agrees pretty well with the density of atoms for monovalent sodium, and not quite so well for monovalent silver. For trivalent aluminum, the Hall data are indicative of four electrons per atom rather than

TABLE 3-2 COMPARISONS OF HALL AND CONDUCTIVITY DATA FOR METALS WITH THE LORENTZ CLASSICAL MODEL FOR ELECTRONIC TRANSPORT

Metal	Hall Coefficient* R_H (m^3/C)	Electrical Conductivity† For T = 100K σ_{100} $(ohm^{-1} m^{-1})$	Electron Density Deduced From Hall Effect $n = (-3\pi/8eR_H)$ (m^{-3})	Atomic Concentration Deduced From Lattice Constants (m^{-3})	Hall Mobility For T = 100K $(-\sigma R_H)_{100}$ $(m^2/V \cdot s)$	Mean Free Path Deduced From Hall Mobility at T = 100K λ (Å)
Sodium	-2.5×10^{-10}	8.7×10^7	2.9×10^{28}	2.5×10^{28}	0.0218	76
Silver	-8.4×10^{-11}	2.4×10^8	8.8×10^{28}	5.9×10^{28}	0.0202	70
Cadmium	$+6.0 \times 10^{-11}$	4.3×10^7	Negative	4.6×10^{28}	Negative	Negative
Aluminum	-3.0×10^{-11}	2.1×10^8	2.5×10^{29}	6.0×10^{28}	0.0063	22
Iron	$+2.5 \times 10^{-11}$	8.0×10^7	Negative	8.5×10^{28}	Negative	Negative
Nickel	-6.0×10^{-10}	1.0×10^8	1.2×10^{28}	9.2×10^{28}	0.0600	210

* Hall coefficient data from *Handbook of Physics* (edited by E. U. Condon and H. Odishaw) (McGraw-Hill, 1958).
† Conductivity data from G. T. Meaden, *Electrical Resistance of Metals* (Plenum Press, 1965).

three. More serious troubles occur with the other elements listed in the table.

From the first term on the right of Equation 3-37 it appears that the conduction of an electron gas *along* a sample should be unimpaired by the presence of a magnetic field with a component perpendicular to **J**. In practice the conductance of any metal decreases in a magnetic field. This is the *magnetoresistance effect* discovered by W. Thomson (1856). In the Lorentz model, when the Boltzmann equation is solved to powers of **B** beyond the first, it is found that there should be a decrease of conductance dependent on the square of the magnetic field (and with smaller terms depending on higher powers of **B**). This square law dependence is in accord with experiment.

No magnetoresistance is predicted for a model (such as the Drude model) in which all electrons are presumed to have the same thermal velocity. However, since electrons *do* have a distribution of speeds, only for an electron of some "average" speed is the magnetic deflection of the Lorentz force exactly compensated by the electrostatic Hall field. An electron moving either faster or slower than this average speed must follow a curved (and hence elongated) path from one end of the sample to the other. Curvature of many electronic paths gives a lower macroscopic conductance:

$$\sigma_B = \frac{J^2\sigma}{J^2 + (\sigma R_H)^2 |J \times B|^2} \tag{3-39}$$

Thus magnetoresistance is maximized when the magnetic field is wholly perpendicular to the direction of current flow, and for a free electron model it should vanish when **B** is parallel to **J**.

As a result of complications which are explicable only in terms of the band theory of solids, longitudinal magnetoresistance (**B** parallel to **J**) is *not* zero for some metals. Moreover, the Hall coefficient of some metals is measured to be a positive quantity, which cannot be accounted for by the free electron expression of Equation 3-38. Nevertheless, the Lorentz results for the various magnetoconductive effects were a major step forward. Typical values for the Hall coefficient (Table 3-2) were indeed consistent with the picture of one or more free electrons per atom.

THE FAILURES OF THE CLASSICAL MODELS

We have spent some time discussing the classical models of Drude and Lorentz in order to lay the groundwork for the quantum models of today. The most disturbing failures of the classical models are not the anomalous sign of the Hall coefficient for a few metals or magnetoresistive behavior of a complicated type. More disturbing is the fact that a metal does not exhibit the large specific heat and susceptibility expected for completely free charge carriers. It is also distressing that the mean free path deduced from Hall effect and/or conductivity data is very large compared with interatomic spacings (see the last column of Table 3-2) and increases on cooling. Such behavior is not predicted by Equation 3-7.

Several papers were addressed to these various problems in the two decades following Lorentz's work. Wien (1913) speculated that the electrons involved in conduction must have the same speed down to absolute zero and must not be impeded in their motion by the array of positive ion cores, though it was not clear to him how such conditions could be justified. J. J. Thomson (1915) suggested that electrical charge might travel through a crystal in the form of "electrical doublets," an intriguing anticipation of the "Cooper pairs," suggested forty years later, which lie at the heart of the BCS theory of superconductivity. Lindemann (1915) postulated that the electrons might form a medium more like a rigid lattice than a free gas, in order to explain the absence of the many degrees of freedom expected from classical specific heat theory. Lindemann imagined that the rigidity of this electron lattice would endow it with a high wave speed, and hence good conduction properties despite a very small electronic specific heat.

All these and other suggestions failed to get to the heart of the problems of a dense electron gas. These failures were largely resolved by the application of the Pauli exclusion principle, embodied in the Fermi-Dirac[10] statistics for particles (such as electrons) with half-integral spin.

---

[10] E. Fermi, Z. Physik **36,** 902 (1926); P. A. M. Dirac, Proc. Roy. Soc. **A112, 661** (1926). These papers were the first to explore the consequences of the Pauli exclusion principle for a large assembly of electrons. The Fermi-Dirac distribution function is described in any of the statistical physics and atomic physics books cited in the general reference list at the beginning of this book.

3.3

The Quantized Free Electron Theory

FERMI-DIRAC STATISTICS OF AN ELECTRON GAS

We assume that the reader has encountered the ideas of electron spin and of the Pauli exclusion principle on previous occasions, and the following discussion of Fermi-Dirac statistics is accordingly brief.

The electrons in a metal customarily have both kinetic and potential energy as a function of position. For the present we shall consider a *free electron* gas in which the potential energy is arbitrarily set as zero. Then total energy and kinetic energy are synonymous, and an electron of velocity v and momentum $p = mv$ has an energy $\epsilon = (p^2/2m)$. From the de Broglie equivalence of particle aspects and wave aspects, this electron has a wave-vector $\mathbf{k} = (p/\hbar)$. This electron can be represented by a vector from the origin to a specific location in v-space, p-space, or k-space, or by just a point at that location. One such location in k-space is shown in Figure 3-6, and we should remark at once that all points in k-space at the same distance from the origin will have the same energy, since

$$\epsilon = \frac{\hbar^2}{2m}[k_x^2 + k_y^2 + k_z^2] = \frac{\hbar^2 k^2}{2m} \qquad (3\text{-}40)$$

does not depend on the direction of $\mathbf{k}$.

From a classical standpoint any wave vector would be permissible,

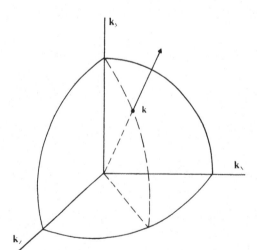

Figure 3-6 The description of the behavior of an electron by a point in k-space. The components of **k** describe the periodicity of the electron wave-function in the various directions.

and there would be no restriction on the number of electrons with the same wave-vector. The first of these classical postulates was overturned by the Heisenberg uncertainty principle (1925), and the second by the Pauli exclusion principle (1925). The combination of these leads to Fermi-Dirac statistics (1926).

In formulating rules subject to the Heisenberg uncertainty principle, we require that wave-functions for electrons must satisfy the Schrödinger equation subject to some boundary conditions.[11] The concept of periodic boundary conditions (wave-function ψ periodic with period L along each Cartesian axis) is useful here, just as it was for phonons. Then a plane wave solution

$$\psi = C \exp(i\mathbf{k}\cdot\mathbf{r}) \qquad (3\text{-}41)$$

for an electronic wave-function is acceptable only if

$$\left.\begin{aligned}
k_x &= (2\pi n_x/L)\\
k_y &= (2\pi n_y/L)\\
k_z &= (2\pi n_z/L)
\end{aligned}\right\} \qquad (3\text{-}42)$$

where n_x, n_y, and n_z are real integers. Now only certain positions in k-space correspond with acceptable wave-functions, such that k-space would be filled if each of these positions were to be surrounded with a cubical cell of volume $(2\pi/L)^3$ (see Figure 3-7).

For a real metallic crystal of macroscopic dimensions, these "cells" are very small and there are many of them. Thus we can use calculus procedures (as with vibrational states) to write the number of cells lying between spheres of radii k and (k + dk) as $(k^2 L^3/2\pi^2)dk$. Then, *per unit volume* of metal, there are $(k^2/2\pi^2)dk$ cells available within a scalar

---

[11] We assume that the reader has had some exposure to the quantum-mechanical language of modern physics. Recall that according to the Heisenberg uncertainty principle there is a relation between the energy ϵ of an electron and the spatial dependence of the wave which describes the probability of finding the electron at any location. The wave-function ψ (which may be complex) has a value such that $-e|\psi|^2$ is the electrical charge density produced from the long-term average of the electron's motion. ϵ and ψ are related by the Schrödinger equation

$$[(-\hbar^2/2m)\nabla^2 + \mathcal{V}(\mathbf{r})]\psi = \epsilon\psi$$

where $\mathcal{V}(\mathbf{r})$ is the spatially dependent potential energy of an electron. Since an electron has a charge of $-e$, an electrostatic potential $V(\mathbf{r})$ results in $\mathcal{V}(\mathbf{r}) = -eV(\mathbf{r})$. In band theory, $\mathcal{V}(\mathbf{r})$ is finite and has the periodicity of the lattice. We shall consider this later. In quantized free electron theory it is assumed that $\mathcal{V}(\mathbf{r}) = 0$ throughout, so that ϵ and ψ must be related by

$$(-\hbar^2/2m)\nabla^2\psi = \epsilon\psi$$

Solutions of this wave equation exist in the form $\psi = A \sin(\mathbf{k}\cdot\mathbf{r}) + B \cos(\mathbf{k}\cdot\mathbf{r})$, or $\psi = C \exp(i\mathbf{k}\cdot\mathbf{r})$ provided that $\epsilon = (\hbar^2 k^2/2m)$ and subject to appropriate boundary conditions. These boundary conditions could take the form that ψ must vanish over a closed surface in real space (the particle-in-a-box problem). Allowed waves are then standing waves. The same density of states, but with physically more meaningful running waves, results if we use the Born-von Kármán idea of cyclic boundary conditions, just as we did for lattice vibrational modes.

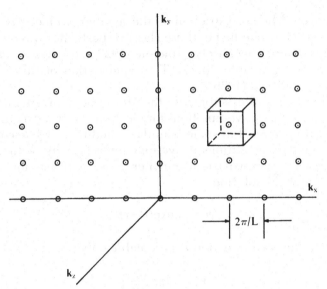

Figure 3-7 Some of the permitted locations in k-space for cyclical boundary conditions with a periodicity of L in each direction of real space. All k-space would be filled if each permitted point were to be surrounded with a cube of volume $(2\pi/L)^3$. One such cube is illustrated.

range dk of wave-vector. The density of available *electron states* is twice as large as this because of the two possibilities for electron spin. Wavefunctions for electrons with spin $s = \pm\frac{1}{2}$ are regarded as separate permitted states according to the Pauli principle, even though their k_x, k_y, and k_z are identical. Thus the number of *electron states* per unit volume within an infinitesimal range of energy or wave-vector is

$$g(\epsilon)d\epsilon = g(k)dk = (k/\pi)^2 dk \qquad (3\text{-}43)$$

Since energy and wave-vector are related by Equation 3-40 [as sketched in part (a) of Figure 3-8], k can be eliminated to give a relationship between electron energy and density of electron states:

$$g(\epsilon) = \frac{1}{2\pi^2}\left(\frac{2m}{\hbar^2}\right)^{3/2} \epsilon^{1/2} \qquad (3\text{-}44)$$

This parabolic density of states law is shown in Figure 3-8(b).

The argument leading to $g(k)$ for vibrational states in Chapter 2 was similar to the preceding, and we can note a similarity between Equations 2-27 and 3-43. Since the dispersion law for electrons is different from that for phonons, there is less similarity between $g(\omega)$ for phonons (Problem 2.4) and $g(\epsilon)$ for electrons (Equation 3-44).

However, the significant difference between quantized phonon and electron distributions is that phonons obey Bose-Einstein statistics (Equation 2-54), for which there is no limit on the number of phonons excited per mode; while electrons, as fermions, must obey the Pauli exclusion principle. This limits the occupancy of any allowed state to either one or zero electrons. In thermodynamic equilibrium at temperature T, the Fermi-Dirac expression for the expectation of occupancy of a state with energy ϵ is

$$f(\epsilon) = \frac{1}{1 + B \, \exp(\epsilon/k_0 T)} = \frac{1}{1 + \exp\left(\dfrac{\epsilon - \epsilon_F}{k_0 T}\right)} \qquad (3\text{-}45)$$

Here, B is a dimensionless normalizing parameter. The normalization can also be carried out in terms of the parameter $\epsilon_F = -k_0 T \ln (B)$, which has the dimensions of energy. Normalization is required so that

$$n = \int_0^\infty f(\epsilon) \, g(\epsilon) d\epsilon = \frac{1}{2\pi^2} \left(\frac{2m}{\hbar^2}\right)^{3/2} \int_0^\infty \frac{\epsilon^{1/2} \cdot d\epsilon}{1 + \exp\left(\dfrac{\epsilon - \epsilon_F}{k_0 T}\right)} \qquad (3\text{-}46)$$

is equal to the total density of electrons per unit volume.

When an electron gas contains very few electrons per unit volume, then the normalization condition, Equation 3-46, is satisfied for ordinary temperatures with a negative value of the parameter ϵ_F. The distribution function is then that of a Boltzmann distribution for all real values of electronic kinetic energy, with $f(\epsilon) \ll 1$ even for the least energetic electrons. However, the total electronic density in a real metal is so large that Equation 3-46 can be consistent with a negative ϵ_F only for temperatures exceeding 10^5K. At ordinary temperatures in a metal, ϵ_F must be a positive quantity, and its value is known as the *Fermi energy* or *Fermi level*.

Let us imagine a metallic "hulk" which consists of a periodic array of atomic cores, denuded of the electron gas which is necessary for a real metal at electrostatic neutrality. Now suppose that electrons are added to this array of cores until electrical neutrality is restored. For simplicity we shall suppose that this restoration process occurs at absolute zero temperature.

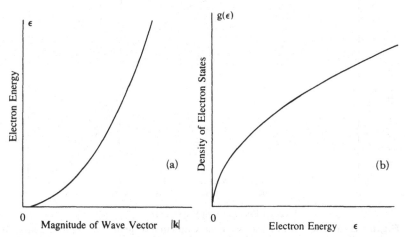

Figure 3-8 (a), The energy of an electron as a function of the magnitude of its wavevector (Equation 3-40). (b), The density of distinguishable quantized electron states as a function of energy, for free electrons (Equation 3-44).

The first electrons added will go into the states of the lowest possible kinetic energy; these are the long-wavelength states of small wavevector since $\epsilon = (\hbar^2 k^2 / 2m)$. The next electrons added must go into regions of k-space further from the origin, since those regions closest to the origin have already been taken. Further electrons must build up occupancy of successive spherical shells in k-space, each of higher energy. Finally, enough electrons will have been supplied to restore neutrality, and the last electrons will enter states at the low temperature Fermi energy ϵ_{FO} which satisfies

$$n = \frac{1}{2\pi^2}\left(\frac{2m}{\hbar^2}\right)^{3/2}\int_0^{\epsilon_{FO}} \epsilon^{1/2}\cdot d\epsilon = \frac{1}{3\pi^2}\left(\frac{2m\epsilon_{FO}}{\hbar^2}\right)^{3/2} \qquad (3\text{-}47)$$

In numerical terms,

$$n = 4.55 \times 10^{27}\epsilon_{FO}{}^{3/2}\ m^{-3} \qquad (3\text{-}48)$$

for energies expressed in eV per electron. Since metals usually contain more than 10^{28} free electrons per cubic meter, the low temperature Fermi energy

$$\epsilon_{FO} = (3\pi^2 n)^{2/3}\hbar^2/2m \qquad (3\text{-}49)$$

is several electron volts per electron.

Curves in Figures 3-9 and 3-10 show the abrupt cutoff of $f(\epsilon)$ and of $f(\epsilon)g(\epsilon)$ at absolute zero, a situation which means that the normalization parameter we call the Fermi energy is the energy of the most energetic electrons in the gas. As the other curve of Figure 3-9 shows, the situation

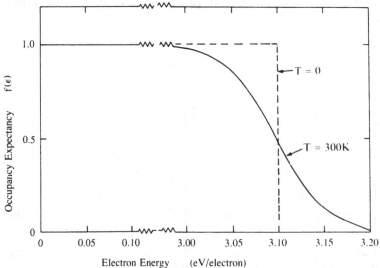

Figure 3-9   The Fermi-Dirac occupancy function (Equation 3-45), for a dense electron gas at ordinary temperatures. The Fermi energy of 3.1 eV/electron is appropriate for metallic sodium. Thus we see that for sodium at room temperature, occupancy is complete up to 0.95 ϵ_F and zero beyond 1.05 ϵ_F.

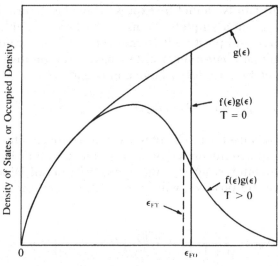

Figure 3-10 The density of electronic states as a function of energy on the basis of the free electron model, and the density of occupied states dictated by the Fermi-Dirac occupancy law. At a finite temperature, the Fermi energy moves very slightly below its position for absolute zero (see Problem 3.10). The effect shown here is an exaggerated one; the curve in the figure for $T > 0$ would with most metals require a temperature of thousands of kelvins.

is not far different even at temperatures well above absolute zero. Note that the occupancy factor $f(\epsilon)$ drops from unity to zero as energy increases by a few k_0T in the range centered on ϵ_F. Since k_0T is only some 0.026 eV at 300K, the transition range is a narrow one, and at all temperatures below the melting point of a metal there is very sparse occupancy of states above ϵ_F.

Strictly speaking, we should draw a distinction between ϵ_{FO} as a parameter which measures the total free electron density and is given by Equation 3-49, and ϵ_F as the parameter which satisfies the normalization requirements of Equation 3-46 at a finite temperature. At any temperature, ϵ_F is the energy for which the occupancy probability is one-half. This energy has significance from a thermodynamic point of view, and is sometimes referred to as the *chemical potential*, or as the *electrochemical potential*. As sketched in Figure 3-10, there is a slight downward trend of ϵ_F with increasing temperature, which we can easily see to be reasonable in a qualitative manner. As temperature rises, thermal excitation makes available for occupancy additional states higher than ϵ_{FO} and requires that some states below ϵ_{FO} become unoccupied. Since $g(\epsilon)$ increases with energy, the energy for 50 per cent occupancy must sink as thermal broadening of the transition range increases.

In order to express the temperature dependence of the Fermi energy in a quantitative manner, we should observe that Equation 3-46 involves a member of the family of *Fermi-Dirac integrals*

$$F_j(y_0) = \int_0^\infty \frac{y^j \cdot dy}{1 + \exp(y - y_0)} \qquad (3\text{-}50)$$

Integrals of this family cannot be expressed in closed form for any arbitrary value of y_0, but asymptotic forms exist when y_0 is either large and negative or large and positive.[12] Numerical tables of Fermi-Dirac integrals for integer and half-integer orders have been published.[13]

The asymptotic form for y_0 large and negative is

$$F_j(y_0) \approx \Gamma(j+1) \exp(y_0), \qquad (-y_0) > 2 \qquad (3\text{-}51)$$

We shall find this useful in Chapter 4 when we discuss the small electron densities of a semiconductor. For an electron density of metallic proportions, the interesting asymptotic form of a Fermi-Dirac integral is that for y_0 large and positive, when the solution is a series starting with the two terms

$$F_j(y_0) \approx \frac{y_0^{j+1}}{(j+1)} \left[1 + \frac{\pi^2 j(j+1)}{6 y_0^2} + \cdots \right], \quad y_0 \gg 1 \qquad (3\text{-}52)$$

Continuing terms in the series are in ascending powers of y_0^{-2}. The series terminates when j is an integer; but $F_j(y_0)$ is a converging infinite series for situations when j is not an integer. In any event, the first two terms suffice when y_0 is a sufficiently large number.

In Problem 3.10 the reader is asked to verify by comparison of Equations 3-46, 3-49, 3-50, and 3-52 that the Fermi energy ϵ_F at any finite temperature in a metal is related to that at zero temperature by

$$\epsilon_F \approx \epsilon_{FO} \left[1 - \frac{(\pi k_0 T)^2}{12 \epsilon_{FO}^2} \right] \qquad (3\text{-}53)$$

As the numerical example associated with that problem confirms, the change in ϵ_F with temperature is very small, and we shall usually refer simply to ϵ_F, without regard for temperature, in the remaining discussion.

SPECIFIC HEAT OF A DEGENERATE ELECTRON GAS

Among others intrigued by the application of quantum-mechanical ideas to macroscopic systems, Arnold Sommerfeld[14] explored the free

---

[12] The asymptotic form for y_0 large and positive was first discussed by A. Sommerfeld, Z. Physik, **47**, 1 (1928), and there have been numerous subsequent discussions of the analytic properties of the Fermi-Dirac integrals. Much of this is summarized in Appendix C of *Semiconductor Statistics* by J. S. Blakemore (Pergamon, 1962). Some analytic properties can be handled more readily in terms of the functions $\mathscr{F}_j(y_0)$ which are related to the $F_j(y_0)$ by $F_j(y_0) = \Gamma(j+1)\mathscr{F}_j(y_0)$.

[13] Tables for $F_{1/2}(y_0)$ and other half-integer values were calculated and cited by J. McDougall and E. C. Stoner, Phil. Trans. Roy. Soc. A**237**, 67 (1968). Further tables exist in a number of more recent papers. A compilation in terms of the related $\mathscr{F}_j(y_0)$ is found in Appendix B of *Semiconductor Statistics* by J. S. Blakemore (Pergamon, 1962).

[14] A. Sommerfeld, Naturwiss, **15**, 824 (1927); Z. Physik **47**, 1 (1928). For a review of this by Sommerfeld in English, see A. Sommerfeld and N. H. Frank, Rev. Mod. Phys. **3**, 1 (1931).

electron model for a metal using Fermi-Dirac rather than Maxwell-Boltzmann statistics. He recognized that at ordinarily attained temperatures, most of the members of a metallic electron gas are several k_0T or more below the Fermi energy. These electrons cannot undergo a scattering collision to a slightly different energy because all states of comparable energy are already occupied, and additional occupancy is forbidden by the Pauli principle.[15] Thus only the uppermost fraction of the electron population, a fraction of about $(4k_0T/\epsilon_F)$, has the ability to respond gradually to an externally imposed electrical or thermal gradient. Sommerfeld concluded from this that only the electron states near ϵ_F control the observable properties of a metal.

Let us consider the energy content of an electron gas, and its derivative, the electronic specific heat. At absolute zero the electron gas has a minimum total energy

$$U_0 = \int_0^{\epsilon_{FO}} \epsilon g(\epsilon) d\epsilon = \frac{\epsilon_{FO}^{5/2}}{5\pi^2} \left(\frac{2m}{\hbar^2}\right)^{3/2} \tag{3-54}$$

from Equation 3-44. We can use Equation 3-47 to express this energy as

$$U_0 = (3n\epsilon_{FO}/5) \tag{3-55}$$

Thus there is an average energy of $(3\epsilon_{FO}/5)$ per electron without any thermal stimulation. Electrons are moving with large speeds (as evaluated in Problem 3.9), and in equilibrium the motion of any electron at any moment is matched by the opposing motion of another electron somewhere in the same metallic crystal.

The electronic energy is slightly larger than U_0 at any higher temperature, because a few electrons are thermally promoted from states just below ϵ_F to states higher than ϵ_F. At any finite temperature

$$U = \int_0^\infty \epsilon g(\epsilon) \, f(\epsilon) d\epsilon$$

$$= \frac{1}{2\pi^2} \left(\frac{2m}{\hbar^2}\right)^{3/2} \int_0^\infty \frac{\epsilon^{3/2} \cdot d\epsilon}{1 + \exp\left(\frac{\epsilon - \epsilon_F}{k_0T}\right)} \tag{3-56}$$

which involves the function $F_{3/2}(y_0)$ from the family of functions described by Equation 3-50. Using both Equations 3-52 and 3-53,

---

[15] It is possible for an electron in a low-lying energy state, well below the Fermi level, to be scattered into a state of slightly different energy if the former occupant of *that* state is simultaneously scattered into any empty state it can find (such as the original state of our first electron). Of course, since electrons are indistinguishable, interchange of two electrons has no observable result. Thus we can notice the scattering of an electron from a low energy state only if it is raised to an empty state near the Fermi energy; this is not the kind of thing we can readily accomplish with an electric field. In view of the indistinguishability of electrons, the large-energy-change scattering of an electron has the same probability as a multiple process in which all intermediate electrons move up one infinitesimal step. It is easy to see that such a process has a very small statistical weight.

$$\int_0^\infty \frac{\epsilon^{3/2} \cdot d\epsilon}{1 + \exp\left(\frac{\epsilon - \epsilon_F}{k_0 T}\right)} \approx \frac{2\epsilon_F^{5/2}}{5}\left[1 + \frac{5}{8}\left(\frac{\pi k_0 T}{\epsilon_F}\right)^2\right], \qquad \epsilon_F \gg k_0 T$$

$$\approx \frac{2\epsilon_{FO}^{5/2}}{5}\left[1 - \frac{5}{24}\left(\frac{\pi k_0 T}{\epsilon_F}\right)^2\right]\left[1 + \frac{5}{8}\left(\frac{\pi k_0 T}{\epsilon_F}\right)^2\right]$$

$$\approx \frac{2\epsilon_{FO}^{5/2}}{5}\left[1 + \frac{5}{12}\left(\frac{\pi k_0 T}{\epsilon_F}\right)^2\right] \tag{3-57}$$

Thus the energy of Equation 3-56 is

$$U \approx \frac{\epsilon_{FO}^{5/2}}{5\pi^2}\left(\frac{2m}{\hbar^2}\right)^{3/2}\left[1 + \frac{5}{12}\left(\frac{\pi k_0 T}{\epsilon_F}\right)^2\right]$$

$$= \left[U_0 + \frac{n\pi^2 k_0^2 T^2}{4\epsilon_F}\right], \qquad T \ll \epsilon_F/k_0 \tag{3-58}$$

And so to a first approximation the electronic specific heat is

$$C_e = (\partial U/\partial T)_e = (\pi^2 n k_0^2 T / 2\epsilon_F) \tag{3-59}$$

which we should compare with the value $C_{cl} = (3nk_0/2)$ expected for a classical electron gas. The quantum restrictions have changed the electronic portion of the total specific heat to

$$C_e = (\pi^2 k_0 T / 3\epsilon_F) C_{cl} \tag{3-60}$$

This result is often stated in the form that the electronic heat in a metal is *degenerated* by a factor of about $(\epsilon_F/3k_0 T)$ from its classical value. An electron gas for which $k_0 T \ll \epsilon_F$ is referred to as a *degenerate* gas.[16]

Sommerfeld's result of Equation 3-59 for the electronic specific heat conforms in its magnitude and temperature dependence with the electronic component of specific heat actually seen in metals. Since C_e varies as the first power of the absolute temperature, and the lattice heat varies more or less as T^3, the electronic portion can be measured most easily at very low temperatures. Figure 3-11 shows the total specific heat as a sum of the kind

$$C_p = AT + BT^3 \tag{3-61}$$

expected from the Sommerfeld and Debye contributions. As we noted in

---

[16] When two states of a system (such as two excited states of an atom) happen to have the same energy, we often say that the two states are "degenerate." It is unfortunate that the word "degenerate" has two quite different meanings. In the present paragraph, the word is used in the sense that the electronic specific heat has been degenerated (degraded) from the large value expected in classical models. As we shall see, a number of other properties (such as magnetic susceptibility) are also degenerated as a consequence of quantum restrictions, and it is an easy step to the figure of speech that in a metal one is dealing with a "highly degenerate electron gas." In Chaper 4 we shall find that the electron gas in a semiconducting material may be described as "degenerate" or "non-degenerate," depending on whether or not mobile electrons are sufficiently numerous to make quantum restrictions on electron motion important.

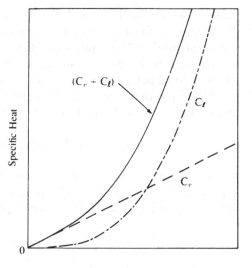

Figure 3–11 Combination of the electronic specific heat C_e and the lattice specific heat C_ℓ for a metallic solid at low temperatures. According to the Debye theory, C_ℓ varies as T^3, while for a degenerate electron gas C_e is linear in T.

connection with Figure 2-21, the two terms can be determined from the intercept and slope of a plot of (C_p/T) versus T^2.

Since only the states lying within a few k_0T of the Fermi level influence the measured properties of a degenerate electron gas, it can be quite logical to express C_e in terms which relate specifically to that part of the energy range. Thus (as suggested in Problem 3.12), the electronic specific heat can be described in terms of the *density of states* $g(\epsilon_F)$ at the Fermi energy:

$$C_e = (\pi^2/3)k_0^2T \cdot g(\epsilon_F) \qquad (3\text{-}62)$$

This is equivalent to the classical specific heat $(3k_0/2)$ per electron for all the electrons in an energy range slightly over $2k_0T$ wide, centered on the Fermi energy.

PAULI PARAMAGNETISM

Quantum statistics have been able to explain other apparent anomalies of the classical models. Thus they can explain why a large assembly of spinning electrons has a rather small and temperature-independent magnetic susceptibility.

It has been recognized (since the discovery of electron spin) that paramagnetism is produced by the alignment of the magnetic moments of extranuclear electrons. Each electron has a magnetic moment of one *Bohr magneton:*

$$\mu_B = e\hbar/2m = 9.3 \times 10^{-24} \text{ joule/tesla} \qquad (3\text{-}63)$$

in a direction anti-parallel to its angular momentum (spin). Paired electrons in inner shells cancel each other in their magnetic effect, but paramagnetic alignment occurs for unpaired inner shell electrons in a variety of solids.[17] We might expect the outer electrons of a metallic free electron gas to be capable of alignment, and of producing a magnetic moment M depending on the applied field.

For a *non-degenerate* electron gas this magnetic moment should be given by a classical Langevin function[18]

$$M = \chi_m H = n\mu_B L(\mu_B \mu_0 H/k_0 T) \qquad (3\text{-}64)$$

In a magnetic field well below that required for saturation, $L(\mu_B\mu_0 H/k_0 T) \simeq (\mu_B\mu_0 H/3k_0 T)$. As a result, the susceptibility χ_m is a field-independent but temperature-dependent quantity

$$\chi_m = (n\mu_B{}^2\mu_0/3k_0 T) \qquad (3\text{-}65)$$

This is known as *Curie Law* behavior, but it is not in fact displayed by the electrons of any real metal.

Figure 3-12 shows the features of the degenerate electron distribution which give a paramagnetism different from Curie's Law. These features were pointed out by Pauli in 1927. Pauli noted that his exclusion principle permits two electrons of opposite spin to occupy the same region of k-space (that is, to have identical translational wave functions).

---

[17] As we shall discuss in Chapter 5, paramagnetic alignment of unpaired inner shell electrons occurs in many materials which incorporate transition or rare earth elements. Cooperative alignment leads to ferromagnetic, antiferromagnetic, or ferrimagnetic behavior.

[18] The Langevin function is derived in Chapter 5.

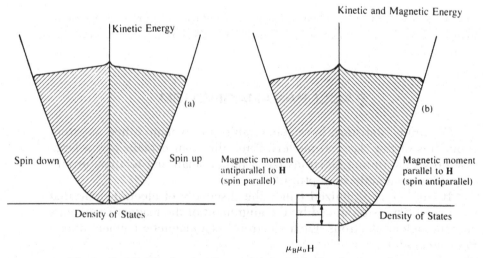

Figure 3–12 The density of states (as abscissa) plotted against electron energy (as ordinate), dividing the density of states into two components of opposing spin directions. The occupancy is shown for some finite temperature. (a), Equal occupancy in the two groups when no magnetic field is present. (b), Unequal occupancy in the presence of a magnetic field, which results in Pauli paramagnetism.

Thus we can draw the total density of states versus energy as the sum of two components, each governed by half the $g(\epsilon)$ of Equation 3-44. In the absence of a magnetic field, states in one half of Figure 3-12(a) are as likely to be occupied as those in the other half, representing an opposite spin direction.

Now if a magnetic field is applied, we can regard one direction of spin as *parallel* to H and the other as *anti-parallel*. Each set of states in Figure 3-12 is shifted by $\pm\mu_B\mu_0H$ from the energy at zero field as shown in Figure 3-12(b). Then the occupancy of the lower energy spin anti-parallel (that is, magnetic moment parallel) states will grow at the expense of decreased occupancy in the spin parallel set in order to preserve a common electrochemical potential. This results in a positive magnetic moment, or paramagnetism.

We should note that to a first order of approximation the magnetization will be proportional to the applied field. Thus, susceptibility χ_m is a meaningful parameter. Moreover, the transfer of electrons from parallel to anti-parallel spin states is not much affected by the small thermal rippling of occupancy at the Fermi energy. Then the susceptibility will be almost independent of temperature (which accords delightfully with experimental observation). And so we may simply calculate χ_m at absolute zero.

The net magnetic moment of the two groups of electrons can be seen from Figure 3-12 to be

$$M = \chi_m H = (\mu_B/2) \int_{-\infty}^{\infty} f(\epsilon)[g(\epsilon + \mu_B\mu_0H) - g(\epsilon - \mu_B\mu_0H)]d\epsilon$$

$$= (C\mu_B/2)\left[\int_0^{(\epsilon_F + \mu_B\mu_0H)} \epsilon^{1/2}d\epsilon - \int_0^{(\epsilon_F - \mu_B\mu_0 H)} \epsilon^{1/2}d\epsilon\right]$$

$$= (C\mu_B/3)[(\epsilon_F + \mu_B\mu_0H)^{3/2} - (\epsilon_F - \mu_B\mu_0H)^{3/2}] \qquad (3\text{-}66)$$

In the second and third lines of the above equation, the density of states for both spin directions combined has been expressed as $g(\epsilon) = C\epsilon^{1/2}$, where we know from Equation 3-44 that $C = (1/2\pi^2)(2m/\hbar^2)^{3/2}$. The change of integration limits and removal of $f(\epsilon)$ from the integral comes about since we are considering absolute zero, for which $f(\epsilon)$ has a step function transition from unity to zero at the Fermi energy.

Provided that $\mu_B\mu_0H \ll \epsilon_F$, which is well satisfied for any attainable magnetic field (see Problem 3.13), then the net paramagnetic magnetization of Equation 3-66 is

$$M = \chi_m H \simeq H\mu_B^2\mu_0C\epsilon_F^{1/2} = H\mu_B^2\mu_0g(\epsilon_F) \qquad (3\text{-}67)$$

Thus the paramagnetic susceptibility can be expressed in terms of $g(\epsilon_F)$ (as was done with the specific heat of Equation 3-62) as

$$\chi_m = \mu_B^2\mu_0g(\epsilon_F) \qquad (3\text{-}68)$$

It is hardly surprising to note that this equals the magnetic effect of the electrons within energy $\mu_B\mu_0H$ of the Fermi limit. Since this is a very small fraction of the total electron population, Pauli's model correctly predicts that magnetic susceptibility (like specific heat) is degenerated from the classical value.

The extent of this degeneration can be seen most clearly by expressing χ_m in terms of the ratio n/ϵ_F. From Equations 3-44 and 3-49, we know that

$$g(\epsilon_F) = (1/2\pi^2)(2m/\hbar^2)^{3/2}\epsilon_F^{1/2} = (3n/2\epsilon_F) \qquad (3\text{-}69)$$

Substituting this into Equation 3-68, we have a Pauli paramagnetic susceptibility

$$\chi_m = 3n\mu_B^2\mu_0/2\epsilon_F \qquad (3\text{-}70)$$

which is degenerated by a factor of $(2\epsilon_F/9k_0T)$ compared with the classical Langevin function result of Equation 3-65.

In the derivation of this susceptibility, we have overlooked the effect of a magnetic field on the translational motion of the free electrons. Such effects were first examined for a degenerate free electron model by Landau (1930). We shall discuss this in more detail in Section 3.5 in connection with cyclotron resonance and oscillatory magnetic phenomena, and shall note here only that a magnetic field imposes restrictions on electron motion in directions perpendicular to the field (the Lorentz force) which produces a diamagnetic susceptibility of $-\mu_B^2\mu_0 n/2\epsilon_F$. Thus diamagnetism cancels out just one-third of the susceptibility created by paramagnetism. The combined resultant susceptibility is

$$\chi_m' = n\mu_B^2\mu_0/\epsilon_F \qquad (3\text{-}71)$$

when the Pauli and Landau factors are taken into account. This is in tolerable agreement with the measured, temperature-independent susceptibility of simple metals such as the alkali elements.

For most metals, agreement between Equation 3-71 and experiment would require that electrons behave as particles with an *effective mass* different from the conventional electronic mass. Such a mass renormalization (which affects the density of states, the Fermi energy for a given n, and so forth) is usually required for a quantitative fit of electronic specific heat data in the same metals. Thus an enlarged density of states is needed to explain the relatively large specific heat and susceptibility in transition elements. Data consistent with an effective mass smaller than m are found for some other solids. These anomalies can all be accounted for (in principle) by the band theory of solids. Effective mass formalism will be discussed in Section 3.5.

SOMMERFELD'S MODEL FOR METALLIC CONDUCTION

We have already considered the contribution made by Sommerfeld to the theory of electronic specific heat by considering Fermi-Dirac statistics of an electron gas, and now examine his formulation of conduction in such a gas.[19] Sommerfeld used Equation 3-44 for the density of states, and Equation 3-45 for the occupancy factor. Thus conduction could be regarded as a responsibility of only the electrons near the Fermi level in energy. In other respects Sommerfeld used the same set of assumptions that Lorentz had in working out electronic transport from the Boltzmann transport equation. Just as in Section 3.2, this approach leads to an integral expression (compare with Equation 3-33) for the electrical conductivity

$$\sigma = \frac{-4\pi e^2}{3m} \int_0^\infty \lambda s^2 \left(\frac{\partial f_0}{\partial s}\right) ds = \frac{-8\pi e^2}{3m^2} \int_0^\infty \lambda \epsilon \left(\frac{\partial f_0}{\partial \epsilon}\right) d\epsilon \quad (3\text{-}72)$$

Now the normalized Fermi-Dirac distribution function f_0 is defined in terms of the distribution of occupied states in velocity space (as was done in Equation 3-4 for a Maxwellian distribution). The new distribution function is

$$f_0 = \frac{2(m/2\pi \hbar)^3}{1 + \exp\left(\dfrac{ms^2 - 2\epsilon_F}{2k_0 T}\right)} = 2(m/2\pi \hbar)^3 f(\epsilon) \quad (3\text{-}73)$$

where $f(\epsilon)$ is the occupancy factor of Equation 3-45. The origin of the normalizing factor in Equation 3-73 is suggested as the topic for Problem 3.14.

Insertion of Equation 3-73 into 3-72 permits us to write the conductivity as

$$\sigma = \frac{-2e^2 m}{3\pi^2 \hbar^3} \int_0^\infty \lambda \epsilon \left(\frac{\partial f(\epsilon)}{\partial \epsilon}\right) d\epsilon$$

$$= \frac{-ne^2}{ms(\epsilon_F)} \cdot \int_0^\infty \frac{\lambda \epsilon}{\epsilon_F} \left(\frac{\partial f(\epsilon)}{\partial \epsilon}\right) d\epsilon \quad (3\text{-}74)$$

The second step in Equation 3-74 is designed to bring the reader up with a start. This transformation has been effected by employing Equation 3-49 and by writing $s(\epsilon_F) = (2\epsilon_F/m)^{1/2}$ as the speed of an electron with the Fermi energy.

Contributions to the integral in Equation 3-74 can be made only in the energy range within a few $k_0 T$ of the Fermi energy ϵ_F, since $(\partial f/\partial \epsilon)$ vanishes for both higher and lower energies. Figure 3-13 illustrates $f(\epsilon)$ and its derivative for a finite temperature; the maximum value of the derivative is $(-1/4k_0 T)$ at $\epsilon = \epsilon_F$. Further study of this interesting derivative is suggested in Problem 3.15. In view of a delta function-like behavior for $(\partial f/\partial \epsilon)$, we can say of *any* slowly varying function $\beta(\epsilon)$ of energy

[19] A. Sommerfeld, Z. Physik **47**, 1 (1928). See also A. Sommerfeld and N. Frank, Rev. Mod. Phys. **3**, 1 (1931).

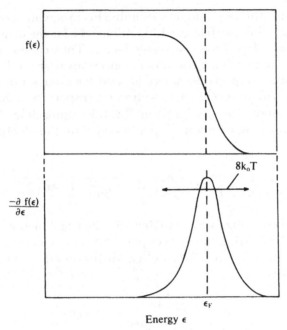

Figure 3-13 The Fermi-Dirac occupancy factor $f(\epsilon) = \left[1 + \exp\left(\dfrac{\epsilon - \epsilon_F}{k_0 T}\right)\right]^{-1}$ and its derivative, as functions of energy. $(-\partial f/\partial \epsilon)$ is symmetrical in energy about ϵ_F and has a maximum value of $(1/4 k_0 T)$.

that

$$-\int_0^\infty \beta(\epsilon) \left(\frac{\partial f(\epsilon)}{\partial \epsilon}\right) d\epsilon = \beta(\epsilon_F) \qquad (3\text{-}75)$$

Thus

$$-\int_0^\infty \frac{\epsilon \lambda}{\epsilon_F} \left(\frac{\partial f(\epsilon)}{\partial \epsilon}\right) d\epsilon = \lambda(\epsilon_F), \qquad (3\text{-}76)$$

the mean free path for an electron of just the Fermi energy. Equation 3-76 can be substituted into Equation 3-74 to give the electrical conductivity in terms only of the total electron density and of properties of Fermi energy electrons:

$$\sigma = \frac{ne^2 \lambda(\epsilon_F)}{ms(\epsilon_F)} \qquad (3\text{-}77)$$

As a pictorial comment on Equations 3-72 through 3-77, we may compare the displacement of the diffuse Maxwellian distribution in Figure 3-5 by an electric field with the displacement of a rather sharply defined Fermi sphere in velocity space or k-space in Figure 3-14. In the latter case, application of an electric field totally depopulates a thin curved section of v-space on one side of the Fermi sphere and totally

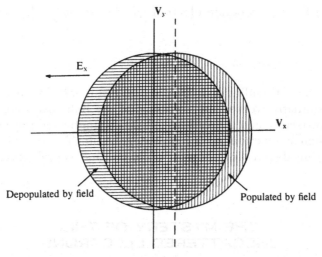

Figure 3-14 Displacement of the occupied region in velocity space by an electric field, for a highly degenerate electron gas. The situation is drawn here for T = 0, but would not look much different for a finite temperature. With no field, the occupied region is a sphere centered on the origin of v-space. The outer surface of this volume is called the *Fermi sphere*, particularly when the coordinates are changed to make this a representation in k-space. An electric field shifts the Fermi sphere bodily, to an extent which depends on $\lambda(\epsilon_F)$. We shall find in the band theory of solids that the *Fermi surface* may be warped from a spherical shape, but that displacement of the Fermi distribution by applied fields will occur more or less as illustrated above.

populates the mirror image section on the opposite side of the occupied region. All of the states with altered occupancy correspond to electron energies close to the Fermi energy. Thus the conductivity of Equation 3-77 depends only on $s(\epsilon_F)$ and $\lambda(\epsilon_F)$.

We may wish to express conductivity in terms of the mean free time between scattering collisions for conduction electrons, which is

$$\tau_m = [\lambda(\epsilon_F)/s(\epsilon_F)] \qquad (3\text{-}78)$$

for electrons of the Fermi speed. Accordingly, the conductivity is

$$\sigma = (ne^2\tau_m/m) \qquad (3\text{-}79)$$

which we may note has exactly the same *form* as that in Drude's earliest model. When we were busy developing Equation 3-16 of the Drude model, we remarked that later theories of conduction would differ only in how quantities such as τ_m would have to be defined. We can understand now that for a degenerate electron gas, the mean free time for only a small minority of the total electron population is significant. It has been remarked that Sommerfeld drew an important distinction between the large number of "free electrons" and the much smaller number of "conduction electrons."

The Sommerfeld approach similarly yields an expression for the electronic thermal conductivity, and a value $(\pi^2 k_0^2/3e^2)$ for the Lorenz number which is in good agreement with experiments at ordinary temperatures. For a degenerate electron gas the result for the Hall coefficient is

$$R_H = -1/ne \tag{3-80}$$

some 15 per cent smaller than the classical result of Equation 3-38. Sommerfeld similarly worked out solutions of the Boltzmann equation for magnetoresistance and the various galvanomagnetic and thermoelectric effects when an electron gas is highly degenerate. These can be applied with varying degrees of success to simple and complicated metallic solids.

THE MYSTERY OF THE UNSCATTERED ELECTRONS

In the last subsection, we noted that Sommerfeld used Fermi-Dirac statistics, but retained the other assumptions of Lorentz in solving the transport equation. These included the assumption that electrons would suffer elastic collisions at ion cores. As a consequence, his result for the electrical conductivity, Equation 3-77, was found to disagree with the known magnitude and temperature dependence of conductivity in metals.

Since the Sommerfeld result contains the temperature-independent speed

$$s(\epsilon_F) = (2\epsilon_F/m)^{1/2} \sim 10^6 \text{ m/s} \tag{3-81}$$

in the denominator, it appears that the only quantity in Equation 3-77 which *can* be temperature-dependent is the mean free path. Experimental magnitudes of conductivities (such as those in Tables 3-1 and 3-2) imply that $\lambda(\epsilon_F)$ must be several hundreds of Angstrom units at room temperature in good conductors, and much greater at low temperatures,[20] a far cry from the $\sim 10\text{Å}$ value one expects at all temperatures from Equation 3-7. Thus the following two questions must somehow be answered:

(1) Why do the expected elastic collisions with ion cores not occur?

(2) What scattering processes *do* control the mean free path?

We shall find it simpler to start with the second of these questions. Electrons certainly should be scattered both by phonons and by permanent imperfections in the lattice. The effects of permanent imperfections include the scattering by foreign atoms, vacancies, interstitials, disloca-

---

[20] For a given electrical conductivity, the Sommerfeld model (Equation 3-77) dictates a considerably larger mean free path than a classical model (Equation 3-19 or 3-34) because the speed of conduction electrons in a degenerate gas is much larger than the mean speed of a Maxwellian distribution. Thus, mean free paths such as those calculated for Table 3-2 must be increased by nearly two orders of magnitude when calculated in quantum terms. This difference is investigated in Problems 3.3 and 3.16.

tions, grain boundaries, and external surfaces.[21] The mean free path λ_{imp} set by imperfection scattering alone should not depend on temperature, whereas the mean free path λ_{phon} for phonon scattering will decrease with rising temperature. For the combination of the two processes (and dismissing elastic scattering as something which just does *not* happen), the mean free path should take the form

$$(1/\lambda) = (1/\lambda_{imp}) + (1/\lambda_{phon}) \qquad (3\text{-}82)$$

and the electrical resistivity should show the same combination

$$(1/\sigma) = (1/\sigma_{imp}) + (1/\sigma_{phon}) \qquad (3\text{-}83)$$

as shown in Figure 3-15. This combination of a temperature-dependent resistivity which is independent of the purity, and a temperature-independent component which depends on the purity and crystal perfection[21] is well known in experimental work and is known as *Matthiessen's rule* (1860).

---

[21] If imperfections such as vacancies, interstitials, and so forth are indeed "permanent," then their contribution to the electrical resistivity can be independent of temperature, as assumed in the application of Matthiessen's rule. This is usually a reasonable assumption for temperatures very much smaller than the melting point, when we can expect to find λ_{imp} comparable with or smaller than λ_{phon}. Of course at temperatures approaching the melting point of a solid, vacancies and interstitials are thermally generated (as exemplified by Figure 1-54), and Matthiessen's rule obviously does not work under these conditions.

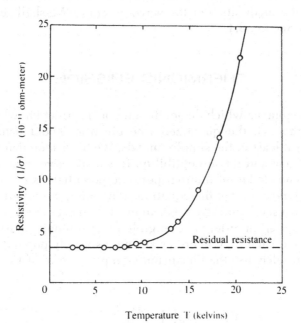

Figure 3-15 The electrical resistivity of a metal, showing the low temperature limit set by imperfection scattering (the residual resistance), and the superimposed resistance caused by phonon scattering. According to Matthiessen's Rule, the two resistivities add simply. The data here are for sodium, after D. K. C. MacDonald and K. Mendelssohn, Proc. Roy. Soc. **A202**, 103 (1950).

The mean free path controlled by phonon scattering might well be expected to vary as T^{-1} for temperatures well above the Debye temperature, since for such large temperatures the phonon population of all frequencies is proportional to T. At the opposite end of the temperature scale, we expect the phonon density to vary as T^3 when $T \ll \theta_D$. Moreover, the phonons excited at low temperatures are primarily long wave phonons (see Problem 2.16) which can scatter electrons only through small angles. The combined effect of the number and spectrum of phonons should be that the phonon-controlled mean free path would vary as T^{-5} for low temperatures. In an attempt to interpolate between the behavior in the two asymptotic ranges, Gruneisen (1933) suggested the semiempirical expression

$$\frac{1}{\lambda_{phon}} = CT^5 \int_0^{(\theta_D/T)} \frac{x^5 \cdot dx}{(e^x - 1)(1 - e^{-x})} \qquad (3\text{-}84)$$

which has proved satisfactory in fitting conductivity-temperature data for a variety of metals. The curve for copper in Figure 3-1 is in fairly good accord with the Gruneisen expression until imperfection scattering takes over below 20K.

Thus progress can be made in accounting for the magnitude and temperature dependence of mean free path and conductivities *if* the idea of elastic scattering at the array of ion cores can legitimately be abandoned. That such a concept *should* be discarded is entirely natural when we consider an electron in a perfect crystal as a wave which already includes the effects of the periodic electric potential via a spatially periodic modulation of the wave-function. We shall see how this happens in Section 3.4.

THERMIONIC EMISSION

For phenomena which do not depend on a correct knowledge of the scattering process, the quantized free electron model usually works rather well, at least in the simpler metals. We have seen that this is true for specific heat and for susceptibility. It is instructive to note now that no special knowledge of internal periodic potentials is necessary for a general understanding of the equations governing thermionic emission.

As we can see from Figure 3-3, an electron must have kinetic energy of at least $(\phi + \epsilon_F)$ in order to be eligible for thermionic emission from a solid. States at such high energies are very sparsely populated, and for them we can safely use the Boltzmann asymptotic form of Equation 3-45, namely,

$$f(\epsilon) \rightarrow \exp\left(\frac{\epsilon_F - \epsilon}{k_0 T}\right), \qquad \epsilon \gg \epsilon_F \qquad (3\text{-}85)$$

In other words, for such high energies, the density of electrons (counting both spin options) in velocity space is governed by an asymptotic form of Equation 3-73:

$$f_0 dv_x dv_y dv_z = \frac{1}{4} \left(\frac{m}{\pi \hbar}\right)^3 \exp\left(\frac{\epsilon_F}{k_0 T}\right) \exp\left[\frac{-m(v_x^2 + v_y^2 + v_z^2)}{2k_0 T}\right] dv_x dv_y dv_z,$$
$$\epsilon \gg \epsilon_F \qquad (3\text{-}86)$$

As a further restriction on the density of electrons which are capable of passing out of a crystal surface, note that it is necessary that electrons approach the surface with an x-component of velocity

$$v_x \geq [2(\phi + \epsilon_F)/m]^{1/2} \qquad (3\text{-}87)$$

in order to penetrate the surface in the x-direction, regardless of the velocity components in the yz plane. Of course, once an electron has escaped from the solid, it has traded most of its kinetic energy for potential energy and is moving in the x-direction at the much slower rate

$$u_x = [v_x^2 - 2(\phi + \epsilon_F)/m]^{1/2} \qquad (3\text{-}88)$$

The speed and energy distributions of the escaping electrons are suggested as a topic for study in Problem 3.18.

From Equation 3-87,

$$n(v_x)dv_x = \frac{1}{4} \left(\frac{m}{\pi \hbar}\right)^3 \exp\left(\frac{\epsilon_F}{k_0 T}\right) \exp(-mv_x^2/2k_0 T) \mathscr{I}^2 dv_x \quad (3\text{-}89)$$

where $\mathscr{I}$ denotes the integral

$$\mathscr{I} = \int_{-\infty}^{\infty} \exp(-mv_y^2/2k_0 T)dv_y = (2\pi k_0 T/m)^{1/2} \qquad (3\text{-}90)$$

or the corresponding integral of infinite range with respect to v_z. Incorporating Equation 3-90 into Equation 3-89, we see that the number of qualified electrons within an infinitesimal range of v_x is

$$n(v_x)dv_x = (m^2 k_0 T/2\pi^2 \hbar^3) \exp(\epsilon_F/k_0 T) \exp(-mv_x^2/2k_0 T)dv_x \quad (3\text{-}91)$$

The current density comprising the outward flow of energetic electrons in the positive x-direction from the metal into a surrounding vacuum will be

$$J = \int_{[2(\phi + \epsilon_F)/m]^{1/2}}^{\infty} ev_x[1 - r(v_x)]n(v_x)dv_x \qquad (3\text{-}92)$$

Here $r(v_x)$ is a reflection coefficient which is required by quantum-mechanical considerations even though the electron has sufficient energy to clear the barrier. The fraction of electrons which is reflected depends on just how the surface potential barrier varies with distance on an atomic scale, something we shall comment on further in the next subsection.

It is usually assumed that $r(v_x)$ does not vary appreciably over the small velocity range in which the major contributions are made to the integral of Equation 3-92. This assumption permits us to take a factor $(1 - r)$ outside the integral sign. Then substituting for $n(v_x)$ from Equation 3-91 the emerging current density is

$$J = (em^2 k_0 T/2\pi^2 \hbar^3)(1 - r) \exp(\epsilon_F/k_0 T) \int_{[2(\phi+\epsilon_F)/m]^{1/2}}^{\infty} v_x \exp(-mv_x^2/2k_0 T) dv_x$$

(3-93)

The integral in Equation 3-93 can be evaluated simply by changing the variable to $y = (mv_x^2/2k_0 T)$, for then

$$\int_{[2(\phi+\epsilon_F)/m]^{1/2}}^{\infty} v_x \exp(-mv_x^2/2k_0 T) dv_x = \int_{[(\phi+\epsilon_F)/k_0 T]}^{\infty} e^{-y} dy$$

$$= \exp\left[\frac{-(\phi + \epsilon_F)}{k_0 T}\right] \quad (3\text{-}94)$$

This result may be substituted into Equation 3-93 to yield a thermionic emission current density

$$J = (emk_0^2 T^2/2\pi^2 \hbar^3)(1 - r) \exp(-\phi/k_0 T) \quad (3\text{-}95)$$

which is controlled by the ratio of the work function to the thermal energy $k_0 T$. Equation 3-95 is the *Richardson-Dushman* relation for thermionic emission,[22] usually expressed in the form

$$J = AT^2(1 - r) \exp(-\phi/k_0 T) \quad (3\text{-}96)$$

where

$$A = (emk_0^2/2\pi^2 \hbar^3) = 1.2 \times 10^6 \text{ amp/meter}^2 \text{ kelvin}^2 \quad (3\text{-}97)$$

The values of $A(1 - r)$ and of ϕ can be determined from the intercept and slope of a *Richardson plot* of $\log_{10}(J/T^2)$ against $1/T$, as exemplified in Figure 3-16.

The numbers reproduced in Table 3-3 show that values determined for $A(1 - r)$ from a Richardson plot with uncoated elements are usually of the order of magnitude of the calculated value for A. Extensive studies of thermionic emission[23] show, however, that the work function ϕ is a rather variable "parameter." Attempts to measure ϕ are clouded by several effects, including

[22] S. Dushman, Rev. Mod. Phys. **2**, 381 (1930).
[23] See, for example, C. Herring and M. H. Nichols, Rev. Mod. Phys. **21**, 185 (1949). Also W. B. Nottingham, *Handbuch der Physik*, Vol. 21, p. 1 (Springer, 1956).

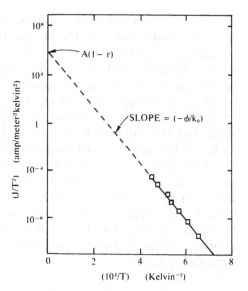

Figure 3–16 A Richardson plot for tungsten, from which (in principle) the work function and the Richardson constant can be determined. After G. Herrmann and S. Wagener, *The Oxide Coated Cathode* (Chapman and Hall, 1951).

1. A temperature dependence of ϕ due to thermal expansion.

2. A small difference in the work function between one crystallographic plane and another. This can be understood in the light of band theory, but is not explainable for completely free electrons. In a polycrystalline metallic sample, the measured ϕ is a composite, but is dominated by any crystal planes in the surface for which ϕ is especially small.

3. Modification of the emission probability by surface contamination.

Item 3 can be either a nuisance or a blessing, depending on one's point of view. The effective work function for a metal in which the Fermi energy is far below the vacuum level is quite sensitive to the presence of

TABLE 3-3 THERMIONIC WORK FUNCTION AND RICHARDSON CONSTANT, AS DETERMINED FROM TEMPERATURE-DEPENDENT THERMIONIC EMISSION[*]

Metal	Work Function ϕ eV/electron	Richardson Constant $A(1 - r)$ amp/meter² kelvin²
Platinum	5.3	0.32×10^6
Tungsten	4.5	0.72×10^6
Molybdenum	4.4	1.15×10^6
Tantalum	4.1	0.37×10^6
Calcium	3.2	0.60×10^6
Thoriated Tungsten	2.6	0.03×10^6
Barium	2.5	0.25×10^6
Cesium	1.8	1.60×10^6
Cesiated Tungsten	1.4	0.03×10^6

[*] Data from C. Herring and M. H. Nichols, Rev. Mod. Phys., **21**, 185 (1949).

adsorbed gases, to impurities and coatings, and to the valence condition of surface states caused by the interruption of lattice periodicity. A small change in ϕ can be quite noticeable; thus a 1 per cent reduction in a work function of 5 eV per electron will cause the emission current to increase by 80 per cent at a temperature of 1000K. An enormous increase in emission current occurs when atoms of a low work function material are distributed over the surface of a metal with large ϕ. (Problem 3.19 deals with this subject.) A layer equivalent to one atomic layer or less is sufficient to mask the work function of the host solid.

The reduction of work function by surface coating is of great technical importance in permitting vacuum tube operation without "white hot" filaments, and there is extensive literature on the applied science of cathodes for vacuum tubes.[24] Despite the present-day extensive use of transistors, there are still many uses for vacuum tubes with thermionic emitters. These include high-power tubes for radio transmitters, and cathode ray tubes.

FIELD-AIDED EMISSION
AND STRONG-FIELD EMISSION

In discussing thermionic emission, we have assumed that the potential energy of an electron jumps from zero inside a metal to $(\phi + \epsilon_F)$ immediately outside the surface, as illustrated in Figure 3-3, or in line (a) of Figure 3-17. The quantum-mechanical reflection factor is considerable for an abrupt barrier, even for electrons with energy larger than the barrier height.

The potential barrier encountered by a departing electron should be more gradual, as was first noted by Schottky (1914). We might expect $\mathscr{V}(x)$ to increase linearly with x at first; but when the electron is more than a few Å beyond the surface it should experience the attractive image force of a charge $-e$ with respect to a homogeneous conducting plane, to give a potential energy

$$\mathscr{V}(x) = (\phi + \epsilon_F) - (e^2/16\pi\epsilon_0 x), \quad \text{(large x)} \quad (3\text{-}98)$$

The correct asymptotic behavior for very small and very large x is given in the form

$$\left. \begin{array}{ll} \mathscr{V}(x) = & 0, \quad x < 0 \\ \mathscr{V}(x) = \dfrac{(\phi + \epsilon_F)^2}{(\phi + \epsilon_F) + (e^2/16\pi\epsilon_0 x)}, & x > 0 \end{array} \right\} \quad (3\text{-}99)$$

which is illustrated by curve (b) of Figure 3-17. For an electron with an initial kinetic energy slightly larger than $(\phi + \epsilon_F)$, the quantum-mechanical probability of reflection is considerably smaller for a barrier represented by Equation 3-99 than for a step function barrier.

---

[24] See, for example, G. Herrmann and S. Wagener, *The Oxide Coated Cathode* (Chapman and Hall, 1951).

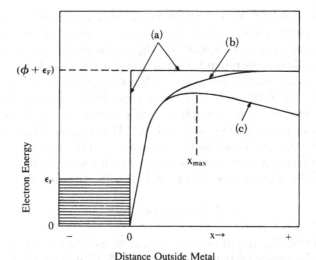

Figure 3–17   Electron energy inside a metal and in a vacuum surrounding the metal. The origin of energy is chosen as a state of rest inside the crystal. (a), The completely abrupt potential energy barrier against thermionic emission when image force is ignored. (b), The form of the barrier expected because of image force between an emitted electron and the metallic surface. (c), Modification of this barrier by an external electric field, to produce a slight lowering of the barrier height.

Suppose that an electric field in the x-direction is created in the vacuum outside a heated metal crystal. This modifies the potential energy function to

$$\mathscr{V}(x) = (\phi + \epsilon_F) - (e^2/16\pi\epsilon_0 x) - exE_x \qquad (3\text{-}100)$$

for locations more than a few Å outside the metal. A curve for the field-modified potential energy is shown as (c) in Figure 3-17. Differentiating $\mathscr{V}(x)$ with respect to x, we find that there is a maximum in the height of the potential barrier at

$$\left.\begin{aligned} x_{max} &= (e/16\pi\epsilon_0 E_x)^{1/2} \\ \mathscr{V}_{max} &= (\phi + \epsilon_F) - (e^3 E_x/4\pi\epsilon_0)^{1/2} \end{aligned}\right\} \qquad (3\text{-}101)$$

Thus the presence of an external electric field produces a slight lowering of the effective work function. This is the phenomenon of field-aided emission, or Schottky emission (Schottky, 1914, 1923). As verified in Problem 3.20, the lowering of the work function is small for fields of a few thousand volt/m, and the potential maximum is then many angstrom units out from the crystal surface. Even a small reduction of work function makes thermionic emission possible for many electrons which are insufficiently energetic at zero field. Schottky emission has been studied extensively.[23]

A different phenomenon can occur when the electric field reaches a strength of some 10^8 V/m. This is *strong field emission*[25] or *cold emission*. The latter name is sometimes used because the process occurs equally readily at any temperature; thermal activation *over* a potential barrier is no longer dominant. Instead, as illustrated in Figure 3-18, the surface potential barrier becomes so thin that quantum-mechanical tunneling can occur through it. At a critical field strength, the barrier thickness becomes thin enough so that electrons of the Fermi energy have a finite transmission probability. [This occurs when (ϕ/eE_x) is about 10 Å.] For any larger field, the thinness of the barrier permits tunneling by electrons of even less energy. Fowler and Nordheim[25] suggest that the field-dependence of current density should be of the form

$$J = \alpha E^2 \exp(-\beta\phi/E) \tag{3-102}$$

by considering the quantum-mechanical tunneling probability through the triangular potential barrier of Figure 3-18. Comparing Equation 3-96 with Equation 3-102, we can see that for field emission the electric field replaces temperature as the operating variable both inside and outside the exponential.

The probability of tunneling by electrons of the Fermi energy should be very slight unless the barrier presented to them is less than 10 Å thick. Thus it might be expected (since ϕ is typically some 3 eV/electron) that a minimum field strength for cold emission ought to be about 3×10^9 V/m. Instead, such emission is seen, as already noted, when the applied voltage is sufficient for a macroscopic electric field about 30 times smaller than that. It is probable that local surface irregularities permit extremely large electric fields on a highly local scale, and that almost all strong-field emission comes from these points. Experimental ob-

[25] R. H. Fowler and L. W. Nordheim, Proc. Roy. Soc. **A119**, 173 (1928). For a more recent tutorial account of field emission see R. H. Good and E. W. Müller, *Handbuch der Physik*, Vol. 21, p. 176 (Springer, 1956); also, E. W. Müller and T. T. Tsong, *Field Ion Microscopy: Principles and Applications* (Elsevier, 1969). These two references discuss field emission with special emphasis on the microscopy of metallic lattices on an atomic scale. C. B. Duke, *Tunneling in Solids* (Academic Press, 1969), discusses field emission in addition to the tunneling that can take place from one solid to another across a very small gap of vacuum or insulator.

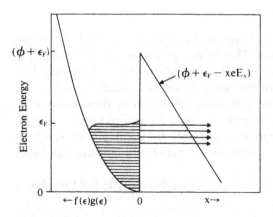

Figure 3–18 Potential barrier at the surface of a metal when a very large electric field is applied. If the external energy falls below the Fermi level for distances within a few Å of the surface, strong-field emission can take place.

servations of cold emission[25] are in reasonable accord with the Fowler-Nordheim formula, Equation 3-102, which admittedly contains two adjustable parameters.

PHOTOEMISSION

We have just been concerned with electrons emitted from a metal under the influence of heat or of a very strong electric field. Two other effects which result in electron emission are mentioned briefly here and in the next sub-section.

Photoemission, otherwise known as photoelectric emission or as the external photoelectric effect, is the emission of electrons when a metal absorbs photons of energy exceeding ϕ, a process correctly described in one of Einstein's contributions to the physics literature of 1905. A photon of energy $\hbar\omega$ elevates an electron of energy ϵ to a state of energy $(\epsilon + \hbar\omega)$, and this can be emitted if $(\epsilon + \hbar\omega) > (\epsilon_F + \phi)$, and if the motion of the "hot" electron is toward the nearest surface.

A monochromatic stream of photons results in the emission of photoelectrons of kinetic energy ranging from $(\hbar\omega - \phi)$ downward to zero, for the initial electronic energies will range from ϵ_F downward, and a fraction of the photoelectrons will have lost some kinetic energy in one or more collisions before leaving the metal. The spectrum of photoelectron energies can be studied by applying a retarding electric field in the vacuum outside the metal.

Photoemission is a three step process, (1) creation of the "hot" electron, (2) motion to the surface, and (3) escape over the surface potential barrier. Process (1) is limited to a layer a few hundred Å thick at the surface, since a metal has an optical absorption coefficient of from 10^7 to 10^8 m^{-1} for photons with an energy of a few eV. In practice, a "hot" electron energized only within a few tens of Å of the surface is likely to reach the surface without having lost energy in collisions with other electrons. Thus the character of the metal within a few atomic layers of the surface is crucial for photoemission.

Photoemission is studied both for what it can reveal about the physics of solids, and by virtue of the practical importance of photoemissive cathodes.[26] Such cathodes are used in a wide variety of practical sensing devices, including vacuum and gas-filled photodiodes, photomultipliers (of which more in the next sub-section), image converters and intensifiers, and television camera tubes. If a response is desired for small photon energies in the red or infrared, one of several alternatives is the use of cesium or some other low work function metal as a thin surface coating of the cathode.[26]

Photoemission is also a widely used technique in the study of the spectrum of electron states in a solid.[27] The need for such complex

---

[26] A. H. Sommer, *Photoemissive Materials* (Wiley, 1968).

[27] The experimental techniques for use of photoemission to deduce the density of states are described well by W. E. Spicer, Phys. Rev. **112**, 114 (1958). A comprehensive account of this technique applied to copper and silver is given by W. E. Spicer and C. N. Berglund, Phys. Rev. **136**, A1044 (1964).

studies will become much more apparent once we get to the band theory of solids, for then it can be seen that the "hot" electron of photoemission is the result of excitation from one band state below the Fermi level to a state [above the vacuum level ($\epsilon_F + \phi$)] in *another* band. Thus the probability of photoexcitation depends on the densities of states in *two* bands, and upon the inter-band matrix element. A considerable quantity of experimental data suggests that such excitation often involves abandonment of the principle of conservation of crystal momentum.[28] In this connection, it is worth noting that the wave-vector cannot be conserved when an electron in a quantized free electron gas is photoexcited and emitted. For if energy and wave-vector are related by Equation 3-40, then an electron of energy ($\epsilon + \hbar\omega$) *must* have a wave-vector which differs from that of an electron with energy ϵ. The apparent change in wave-vector is many times larger than the wave-vector (10^7 m^{-1} or so) of the participating photon. When photoemission is analyzed for a solid in terms of the band theory of solids, some photoexcitation events are found to satisfy k-conservation, while other events do not.

SECONDARY ELECTRON EMISSION

Electrons can be emitted by the surface of a solid when it is exposed to bombardment by energetic charged particles. This was first observed by Austin and Starke in 1902, using moderately fast electrons. They showed that this was not just a matter of part of the incoming electrons being reflected at the metal surface, since sometimes more electrons were emitted than were incident. The process is called *secondary electron emission*,[29] and is of particular interest when the secondary yield δ (ratio of emitted electrons to primary electrons) exceeds unity.

The emission flux actually consists of three families of electrons:

(a) primary electrons which have been elastically reflected;

(b) primaries which have been inelastically reflected, perhaps after they have been responsible for one or more secondaries; and

(c) true secondary electrons, which may or may not have lost part of their energy in collisions with *further* electrons. Such further collisions limit secondary emission to a layer at the surface a few tens of atomic layers thick.

Figure 3-19 illustrates the energy spectrum of electrons emitted by silver when 155 eV primary electrons are incident, with an identification of groups (a) through (c). The relative importance of the three groups depends on the angle of incidence and the incident energy, as well as on the characteristics of the material being bombarded.

Secondary electrons result from Coulomb interaction of an incoming electron (or other charged particle) with electrons lying within

---

[28] W. E. Spicer, "Photoelectric Emission" in *Optical Properties of Solids* (North-Holland, 1972), p. 755.

[29] A good review is given by A. J. Dekker, in *Solid State Physics, Vol. 6* (edited by F. Seitz and D. Turnbull) (Academic Press, 1958), p. 251.

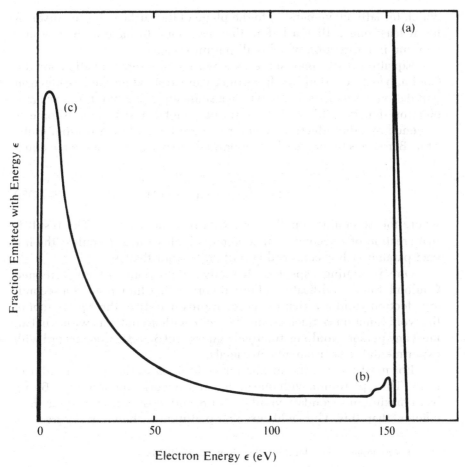

Figure 3-19 The energy spectrum of secondary electrons emitted by silver, when 155 eV primaries are incident. From E. Rudberg, Proc. Roy. Soc. (London) A127, 111 (1930). The electron groups marked (a), (b) and (c) are discussed in the text.

the Fermi sphere in **k**-space. Baroody[30] has considered this mechanism for the unscreened Coulomb force

$$\mathbf{F} = (e^2\mathbf{r}/4\pi\varepsilon_0 r^3) \tag{3-103}$$

exerted by a primary electron whose trajectory at velocity **v** results in an impact parameter (minimum distance of separation) of b with respect to an electron in a state with the Fermi energy or less. He shows (as suggested for Problem 3.21) that the impulse of the passing fast electron results in momentum transferred to the conduction electron of

$$\Delta\mathbf{p} = (e^2 b/2\pi\varepsilon_0 b^2 v) \tag{3-104}$$

which it will be noted is in a direction perpendicular to **v**. Thus a primary incident to the solid at normal incidence creates secondaries

---

[30] E. M. Baroody, Phys. Rev. **78**, 780 (1950).

which initially move parallel to the plane of the surface; these must suffer at least one collision before they can hope to have their new momentum in a direction which will permit escape.

Equation 3-103 may serve as a reasonable approximation for the Coulomb force exerted by a fast primary electron when the free electron population is small, as is the case in a semiconductor. For the very large electron density (10^{28} m^{-3} or more) of a metal, the Coulomb force is screened by other electrons when r is more than a few Ångstrom units. Thus Baroody's theory has been modified[31] to use a screened force equation

$$\mathbf{F} = (e^2\mathbf{r}/4\pi\epsilon_0 r^3)\exp(-r/\lambda) \qquad (3\text{-}105)$$

where the screening length λ is typically some 10^{-10} m. The result is that creation of a secondary in a metal is highly unlikely unless the impact parameter b is comparable with or smaller than λ.

One intriguing aspect of Baroody's theory and of the screened Coulomb force modification of this theory is that they predict a secondary electron yield δ which *increases* approximately as the *square root* of the work function ϕ. Such an increase of δ with ϕ contrasts with what almost any person would instinctively guess, but $\delta \sim \phi^{1/2}$ does accord with experimental measurements on metals.

For most metals, the maximum yield δ_m is of the general order of unity. Thus, platinum, with one of the largest work functions ($\phi \approx 6$ eV), has a maximum secondary yield $\delta_m \approx 1.8$, and most other metals are less efficient than this. The yield curve for platinum is shown in Figure 3-20,

---

[31] For example, A. van der Ziel, Phys. Rev. **92**, 35 (1953).

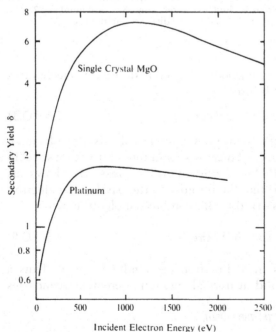

Figure 3–20 Secondary electron yield δ as a function of primary electron energy for one of the most efficient metallic emitters, and a very efficient insulator. Data for platinum from P. L. Copeland, reported by Dekker in *Solid State Physics* (Prentice-Hall, 1957). Data for MgO by J. B. Johnson and K. G. McKay, Phys. Rev. **91**, 582 (1953).

which also shows the much larger secondary yield attainable with insulating MgO. A number of insulators and semiconductors have been found to be very efficient secondary emitters, and the practical applications of secondary emission are almost all centered on either the use of single crystal nonmetals or the creation of a thin layer of a suitable nonmetallic compound on the surface of a metal host. Thus oxidized alloys of Mg-Ag or Be-Cu, intermetallic compounds such as Cs_3Sb, and crystals of gallium phosphide or gallium arsenide have all been used with success as *dynode electrodes* in *photomultipliers*.[32] A photomultiplier comprises a photoemissive cathode within an evacuated envelope, with opportunities for photoelectrons to be accelerated electrically and to cause current amplification at a series of secondary emitting surfaces (dynodes). Clever design can result in the production and detection of 10^9 or more electrons through the absorption of a single photon at the photoemissive cathode.[33]

PLASMA OSCILLATIONS OF AN ELECTRON GAS

A screened Coulomb force equation (Equation 3-105) was used in the discussion of the mechanism for secondary electron creation. The factor $\exp(-\lambda/r)$ which modifies this from the conventional Coulomb force (Equation 3-103) arises from the collective motion of a free electron gas, which behaves as a plasma[34] of mobile charged particles.

The name "plasma" was suggested by Irving Langmuir in 1929 to describe the collective electrical properties which he noted in an ionized gas, and the field of gaseous plasma physics is well established. Many of the phenomena observed for a gaseous plasma can be reproduced in the electron gas of a metal or a semiconductor, with some interesting differences and complexities.

Suppose an electron gas (classical or quantized) in a solid, superimposed on the fixed opposite charge density of the ion cores. Any displacement of the electron gas with respect to the core assembly will set up a restoring electric field; freed from restraint, the electron gas would oscillate about its equilibrium position at a natural *plasma frequency* ω_p. For a gas with n electrons per cubic meter, all displaced by distance x, the restoring field is $E_x = (nex/\varepsilon_0)$. Thus the force equation for each electron in the gas is

$$m(d^2x/dt^2) = -eE_x = -(ne^2x/\varepsilon_0) \qquad (3\text{-}106)$$

[32] V. K. Zworykin and E. G. Ramberg, *Photoelectricity and its Applications* (Wiley, 1949).

[33] S. K. Poultney, *Advances in Electronics and Electron Physics, Vol. 31* (edited by L. Marton) (Academic Press, 1972), p. 39.

[34] An elementary discussion of plasmas in solids is given by M. F. Hoyaux, *Solid State Plasmas* (Pion, 1970). A more advanced treatment is given by M. Glicksman in *Solid State Physics, Vol. 26* (Academic Press, 1971), p. 275.

in the absence of damping. Equation 3-106 describes harmonic motion at the natural angular frequency

$$\omega_p = 2\pi \nu_p = (2\pi c/\lambda_p) = (ne^2/m\varepsilon_0)^{1/2} \qquad (3\text{-}107)$$

This frequency corresponds with a wavelength which lies in the microwave region of the electromagnetic spectrum for gaseous plasmas ($\lambda_p = 0.1$ m for $n = 10^{17}$ m^{-3}), but is typically in the infrared for semiconductor plasmas, and in the ultraviolet for plasmas in metals ($\lambda_p = 0.3$ μm for $n = 10^{28}$ m^{-3}).

A reason for referring to the plasma resonance condition in terms of portions of the electromagnetic spectrum is that a plasma is opaque and totally reflecting for electromagnetic radiation of frequency less than ω_p. An electromagnetic wave with electric vector $E_\omega = E_m \exp{(i\omega t)}$ produces an electric displacement $D_\omega = D_m \exp{(i\omega t)}$ such that the relative dielectric constant is

$$\kappa_\omega = (D_\omega/\varepsilon_0 E_\omega) = [1 - (\omega_p^2/\omega^2)] \qquad (3\text{-}108)$$

Thus the dielectric constant is negative (forbidding transmission) for electromagnetic radiation with $\omega < \omega_p$, and becomes positive (permitting transmission) for $\omega > \omega_p$. Simple metals (such as the alkali elements) are indeed opaque and totally reflecting in the visible, yet transparent in the ultraviolet below a wavelength $\lambda_p = (2\pi c/\omega_p)$, which corresponds quite well with the spectral position expected from Equation 3-108. (See Problem 3.22 for an example of this correspondence.)

The large plasma resonance frequency of a metallic electron gas means that such a gas responds very rapidly to any change in charge distribution. Redistribution of the charge of the gas serves to screen out the influence of any newly introduced charge outside a radius $\lambda = \bar{s}/\omega_p$. For a cold Fermi-Dirac electron gas, the appropriate average electron speed is $\bar{s} = (\epsilon_F/m)^{1/2}$; thus the *Thomas-Fermi screening radius* is

$$\lambda_{TF} = (\bar{s}/\omega_p) = (\varepsilon_0 \epsilon_F/ne^2)^{1/2} = \left(\frac{3\pi\hbar^2\varepsilon_0}{2me^2}\right)^{1/2} \left(\frac{\pi}{3n}\right)^{1/6} \qquad (3\text{-}109)$$

using Equation 3-107 for the plasma frequency, and Equation 3-49 for the relation between electron density and Fermi energy. The screening length of Equation 3-109 is very close to 1 Å for an electron density of 10^{28} m^{-3}, as is typical for many metals. Thus the effect of a moving charge at a distance of more than an Ångstrom unit or so from an atom is effectively screened in the manner indicated by Equation 3-105.

For an electron gas in a semiconductor (which we shall consider at length in Chapter 4), the velocity distribution may well be Maxwellian rather than Fermi-Dirac. In such a *non-degenerate* electron gas, the average electron speed is $\bar{s} = (k_0 T/m)^{1/2}$, and the *classical screening length* is

$$\lambda_c = (\bar{s}/\omega_p) = (\varepsilon_0 k_0 T/ne^2)^{1/2} \qquad (3\text{-}110)$$

For the relatively small electron densities which are possible in a semi-conductor, λ_c can be very large compared with an interatomic distance, though electron screening is still important for many electronic processes in a semiconductor.

The electron gas in a metal can be encouraged to oscillate longitudinally at the plasma frequency by firing fast electrons through a thin metal foil or by causing an electron or a photon to suffer inelastic reflection at the surface of a metal film. Energy is interchanged as a multiple of the quantized *plasmon energy* $\hbar\omega_p$. The energy loss of the electron or photon stimulating plasma oscillations can be measured (typically 5 to 15 eV), as a cross-check on the value for ω_p deduced from electron density. The two values for ω_p do not necessarily agree to better than a few per cent, because of inadequacies in the free electron model.

Despite the very large value of ω_p for a metallic electron gas, waves of a different character can propagate with low speed and low frequency when a magnetic field is also present. Since these have circular polarization, these have been named *helicon waves* by Aigrain,[35] and have been extensively studied in both metals and semiconductors. For a semiconductor or semimetal in which the densities of electrons and holes are equal, the helicon wave is replaced by two modes of propagation in a magnetic field: the fast and slow *Alfven waves*.[36]

---

[35] P. Aigrain, *Proceedings 5th International Semiconductor Conference* (Czechoslovak Academy of Sciences, 1961), p. 224.

[36] Helicon and Alfven waves are discussed by Glicksman (Ref. 34). See also P. M. Platzman and P. A. Wolff, *Waves and Interactions in Solid State Plasmas* (Academic Press, 1973); or M. C. Steele and B. Vural, *Wave Interactions in Solid State Plasmas* (McGraw-Hill, 1969).

The Band Theory of Solids

ELECTRON MOTION IN A SELF-CONSISTENT FIELD

As we have remarked before, quantized free electron theory is rather successful in explaining many metallic properties. The difficulties arise when it is necessary to make assumptions about the scattering process. In the traditional free electron theory, the mean free path λ or time τ_m was used in the calculation of conductivity upon the assumption that every ion core would act as a site for elastic scattering. Yet the actual *magnitude* of the conductivity in a metal is incompatible with the existence of this process (quite apart from the incorrect temperature dependence predicted with elastic scattering). Although we can superpose inelastic scattering events (by phonons or by defects) onto the Sommerfeld model, we ought first to explain how it can be that electrons in certain states can have an infinite free path in the absence of these inelastic processes. We should also like to be able to explain why some materials are metals, why others are good insulators, and why others are of intermediate conductivity (semiconductors). Figure 3-21 demonstrates the enormous range of room temperature conductivity values for the three types of solid. We know that lack of conductivity is not necessarily due to lack of electron mobility, since many insulators are sensitive photoconductors.

The band theory of solids deals with these problems by prescribing a non-zero periodic potential energy $\mathscr{V}(\mathbf{r})$ in the Schrödinger equation for an electron in a crystal, using the following set of assumptions:

1. The wave-function ψ for an eigenstate compatible with a certain eigenenergy ϵ is calculated for a *perfect* periodic lattice. Scattering is introduced afterwards as a perturbation. Certainly there are some situations for which the neglect of thermal lattice vibrations until the end seems to be a rather unhealthy procedure, but from a pragmatic point of view we must observe that the technique usually works rather well, and that it does simplify the calculation.

2. Band theory is developed as the theory of a single electron. It is supposed that for any electron, *everything* else in the crystal can be represented by an *effective potential energy* $\mathscr{V}(\mathbf{r})$. The value required for $\mathscr{V}(\mathbf{r})$ may be different for different electron energies.

3. Adoption of a one-electron model permits the use of the one-electron Schrödinger equation

$$[(-\hbar^2/2m)\nabla^2 + \mathscr{V}(\mathbf{r})]\psi = \epsilon\psi \qquad (3\text{-}111)$$

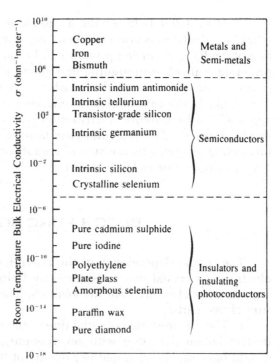

Figure 3–21 The range of electrical conductivity encountered in measurements at room temperature, with some typical members of the three classes of conductor.

and requires that the eigenstates discovered from the solution of this equation be filled in accordance with the Fermi-Dirac occupancy law, Equation 3-45.

The one-electron approach is compatible with the use of self-consistent field techniques, as used by Hartree (1928) and by Fock (1930), for the electron distributions in multi-electron atoms. Hartree or Hartree-Fock approximations replace the dynamic interaction of an electron with the other electrons in the crystal by averages of the interaction over the occupied electron states. In applying self-consistent methods to the calculation of electronic spectra in a solid, we could start (in principle) knowing only the crystal symmetry of a lattice and guessing a trial $\mathscr{V}(r)$ which was compatible with this symmetry. Solutions of the Schrödinger equation, iterated several times, should improve $\mathscr{V}(r)$ to the proper function and should create sets of values for ψ and for ϵ which could be regarded as optimized self-consistent values from a first-principles calculation. Some calculations of electronic bands in solids are made in this way,[37] though there are other theoretical schools of thought which argue that it simplifies and assists a calculation when some numbers based on experimental observations of the solid are inserted at the beginning and are used as constraints for the theory.

---

[37] An eloquent argument in favor of "first-principles" energy band calculations is made by F. Herman, *Proceedings of 7th International Semiconductor Conference* (Academic Press, 1964). For the same author's more recent views on the blend of experimental data with calculated band parameters, see F. Herman et al., *Proceedings of the Symposium on Energy Bands in Metals and Alloys, Los Angeles, 1967* (Gordon and Breach, 1968) p. 19.

The correlation between the motion of an electron and the motion of other free electrons is crudely taken care of in band theory by the operation of the Pauli exclusion principle in Fermi-Dirac statistics. The Pauli principle is invoked in arguing that electron-electron collisions should never occur. As we remarked earlier, many-body techniques which would take better care of electron-electron correlation are complicated, and void the Fermi-Dirac distribution law. It is possible to get around this by the *method of elementary excitations,* which replaces a system of interacting electrons by the sum of energies of a set of "quasi-particles" which *do* obey Fermi-Dirac statistics.[38]

BLOCH FUNCTIONS

The spatial dependence of the potential experienced by an outer electron in a crystal was considered by Felix Bloch,[39] who drew some important conclusions from his analysis. The total potential V(r) is the sum of two parts:

1. The electrostatic potential due to the array of atomic cores. For a perfect lattice (i.e., one with no phonons), this contribution to V(r) should have the translational periodicity of the lattice.

2. The potential due to all other outer electrons. Bloch assumed that the charge density from this source would have the same long-term average value in every unit cell of the crystal, and would thus also be periodic. Such an assumption certainly satisfies the requirements of electrical neutrality and crudely takes account of electron-electron repulsion.

Thus, Bloch argued, the total $\mathscr{V}(r) = -eV(r)$ to be substituted into the Schrödinger equation, Equation 3-111, has the periodicity of the lattice. He concluded that the wave-functions which satisfy Equation 3-111 subject to such a potential must be of the form

$$\psi_k(\mathbf{r}) = U_k(\mathbf{r}) \exp\,(i\mathbf{k} \cdot \mathbf{r}) \qquad (3\text{-}112)$$

where $U_k(\mathbf{r})$ is some function (depending on the value of the wave-vector $\mathbf{k}$) which *also* has the complete periodicity of the lattice.

Equation 3-111 belongs to the family of differential equations known in mathematics as Hill's equation; that periodicity of $\mathscr{V}(r)$ requires the existence of solutions of the type exhibited in Equation 3-112 has been known to mathematicians as Floquet's theorem. However, when electrons in periodic lattices are being discussed, the theorem leading to Equation 3-112 is usually described as the Bloch theorem, and the functions themselves are called Bloch functions. These functions are very useful.

---

[38] The interested (and well-prepared) reader will find references to a number of many-body techniques in *Concepts in Solids* by P. W. Anderson (Benjamin, 1964).

[39] F. Bloch, Z. Physik, **52**, 555 (1928).

In order to discuss the Bloch theorem for a very simplified geometry, let us imagine a linear array of atoms, all identical, with interatomic
spacing a. The atoms have liberated electrons to form a degenerate one-
dimensional electron gas, and the positive ion cores make the electric
potential go through a periodic series of maxima, as in Figure 3-22.
[Thus the electronic potential *energy* $\mathscr{V}(x) = -eV(x)$ goes through a periodic series of minima.] Since the potential has the periodicity of the lattice, then

$$V(x) = V(x + a) \tag{3-113}$$

for any value of x.

Solutions of the one-dimensional simplification of Equation 3-111
exist for the potential energy $\mathscr{V}(x)$. We now impose the Born–von Kármán
type of cyclic boundary condition, that the wave-function solutions
repeat after N unit cell lengths, where N is a large and rather arbitrary
number. Then

$$\psi(x) = \psi(x + Na) \tag{3-114}$$

for any value of x. Now in view of the translational symmetry in our
model, it seems plausible that the wave-function in one unit cell should
be related to that in the next by an expression of the kind

$$\psi(x + a) = J\psi(x) \tag{3-115}$$

where J is a function to be determined. This means that for locations separated by several unit cells,

$$\psi(x + ma) = J^m\psi(x) \tag{3-116}$$

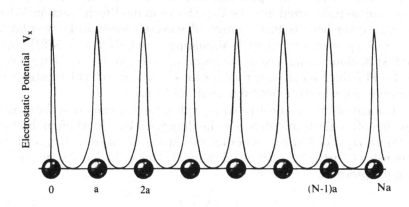

Figure 3–22 Periodic potential for a linear monatomic lattice. After some arbitrary
number N of unit cells, we may suppose that the wave-functions of *all* the eigen-states
must repeat.

and in particular that

$$\psi(x + Na) = J^N\psi(x) \qquad (3\text{-}117)$$

Comparison of Equations 3-114 and 3-117 shows that J must be one of the N roots of unity:

$$J = \exp(2\pi iM/N) \qquad (3\text{-}118)$$

where M is an integer.

Our requirements for the wave-function can evidently all be met if we write $\psi(x)$ as a product of two functions,

$$\psi(x) = U_M(x) \exp(2\pi iMx/Na) \qquad (3\text{-}119)$$

where the first factor, $U_M(x)$, is a function which, like $V(x)$, has the translational periodicity of the lattice.

Now a particular value for M can be associated with a particular value for the wave-vector **k** which controls the periodicity of the wave-function. Since the second factor in Equation 3-119 goes through M complete cycles in the distance Na, we can relate k and M by

$$k = (2\pi M/Na) \qquad (3\text{-}120)$$

Insertion of this identification into Equation 3-119 gives us a wave-function

$$\psi(x) = U_k(x) \exp(ikx) \qquad (3\text{-}121)$$

which is the Bloch result in one dimension. The result can be validated in two and three dimensions as well.

In Equation 3-119 it is not necessary to use a value for M which is larger than $\pm(N/2)$, corresponding with $k = \pm(\pi/a)$. Any excess periodicity can be transferred into the $U_k(x)$ factor of the Bloch function. What we are saying here is that a range of wave-vector from $k = (-\pi/a)$ to $k = (+\pi/a)$ [corresponding with M ranging from $(-N/2)$ to $(+N/2)$] forms the first Brillouin zone, just as we found in Chapters 1 and 2. Larger values for M (which mean larger values for k) correspond with locations in Brillouin zones remote from the origin of k-space.

Choice of the smallest possible value for k means that the wave-function of an electronic state is to be described by a location in the first Brillouin zone of k-space, whether this space be of one, two, or three dimensions. Just as for a lattice vibrational state, a change of wave-vector by a reciprocal lattice vector,

$$k' = k \pm G \qquad (3\text{-}122)$$

results in a state equivalent to **k** in terms of the observable electron motion. The transformation of Equation 3-122 applied to Equation 3-112 produces

$$\psi_{k'}(\mathbf{r}) = [U_k(\mathbf{r}) \exp(i\mathbf{G}\cdot\mathbf{r})] \exp(i\mathbf{k'}\cdot\mathbf{r}) \qquad (3\text{-}123)$$

Let us consider this factor $\exp(i\mathbf{G}\cdot\mathbf{r})$ which has now appeared in the Bloch function, and demonstrate that this has the periodicity of the lattice. For a translation in real space to

$$\mathbf{r'} = \mathbf{r} + \mathbf{T} \qquad (3\text{-}124)$$

then

$$\exp(i\mathbf{G}\cdot\mathbf{r'}) = \exp(i\mathbf{G}\cdot\mathbf{r}) \exp(i\mathbf{G}\cdot\mathbf{T}) \qquad (3\text{-}125)$$

and $\exp(i\mathbf{G}\cdot\mathbf{T})$ must be unity if $\mathbf{G}$ and $\mathbf{T}$ are given by Equations 1-43 and 1-42. Thus $\exp(i\mathbf{G}\cdot\mathbf{r})$, and the entire quantity in square brackets in Equation 3-123, have the periodicity of the lattice, just like $U_k(\mathbf{r})$ itself. Hence, $\mathbf{k}$ can be confined to the first Brillouin zone if so desired.

For the lattice vibrations of a monatomic solid, it was found that there was only one solution (one frequency, one phonon energy) for a given wave-vector. However, there were two possible phonon energies for a given wave-vector in a solid with a multi-atomic basis—the acoustic and optical mode branches. The situation is more complicated still for electron states according to the band theory; Equation 3-111 has a sequence of Bloch function solutions for a given value of $\mathbf{k}$, at successively larger energies.

Band theory solutions are said to be expressed in the *reduced zone* representation when electron energy is shown as a multi-valued function of $\mathbf{k}$ within the wave-vector range of the first Brillouin zone. This representation is often used. However, it is sometimes useful to employ the *extended zone* representation, with energy as a single-valued function of wave-vector over two or more zone-widths. This latter representation is useful when we wish to study energy as a function of $\mathbf{k}$ in passing from one Brillouin zone to the next, and in observing the energy discontinuity at the zone boundary created by a finite periodic potential.

ALLOWED ENERGY BANDS
AND FORBIDDEN GAPS

In the free electron model, we were able to discuss electron energy and wave-vector without regard for the crystallography of the solid. The various Brillouin zone surfaces in $\mathbf{k}$-space become important as soon as periodic potentials have to be taken into account. They become particularly important if the Fermi surface in $\mathbf{k}$-space which we should calculate for a free electron model is intersected by Brillouin zone boundaries of the lattice.

The relationship between energy and momentum for electrons is very simple in free electron theory. Since the potential energy is assumed to be zero everywhere, we can write

$$\epsilon = (p^2/2m) \qquad (3\text{-}126)$$

The de Broglie wave describing this electron has a wave-vector which is related to the energy by the dispersion relationship

$$\epsilon = (\hbar^2 k^2/2m) \qquad (3\text{-}127)$$

previously quoted as Equation 3-40.

Since Equation 3-111 has been introduced as the foundation of band theory, with solutions sought in the form of Equation 3-112, it is worthwhile to establish the asymptotic connection between band theory and free electron theory. For the free electron asymptotic limit of vanishingly small periodic potential energy in Equation 3-111, the Bloch solutions are plane waves [$\psi \sim \exp(i\mathbf{k} \cdot \mathbf{r})$] with wave-vector and energy related simply by Equation 3-127. For this free electron limit, any energy is possible for some real value of $\mathbf{k}$.

Figure 3-23 plots the ϵ–$\mathbf{k}$ relationship of Equation 3-127 for a one-dimensional solid of lattice constant "a" and a vanishingly small peri-

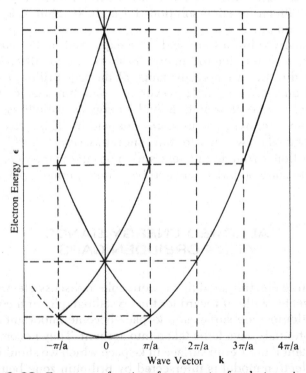

Figure 3–23 Energy as a function of wave-vector for electrons in a one-dimensional crystal of lattice constant "a," when the amplitude of periodic potential is zero. Energy is then a continuous function of wave-vector. Energy is shown as a multi-valued function of k in the shaded first Brillouin zone, the reduced zone representation.

odic potential. This figure shows both the *extended zone* and *reduced zone* representations of the solution. For the latter type of representation, portions of the ϵ–k solution have been moved horizontally through a reciprocal lattice vector in order to bring the range of k used within the first Brillouin zone. [For a one-dimensional solid of lattice constant "a", any reciprocal lattice vector is an integral multiple of $(2\pi/a)$.]

This delightful simplicity is lost once we consider the eigenfunctions of Equation 3-111, their wave-vectors and their permitted eigenvalues (energies) for a finite periodic potential energy $\mathscr{V}(\mathbf{r})$. It turns out that whatever the form of this periodic $\mathscr{V}(\mathbf{r})$, some energies are compatible with real values for k, while other energies correspond only with imaginary values for k.

A range of energy for which a Bloch function (Equation 3-112) solution of Equation 3-111 enjoys real values for k is known as an *allowed energy band* of the solid. There will be several ranges of energy for which different Bloch solutions exist with real k. The range of energy lying above one allowed band and below the next is called a *forbidden gap* of the material. Figure 3-24 sketches the anticipated alternation of

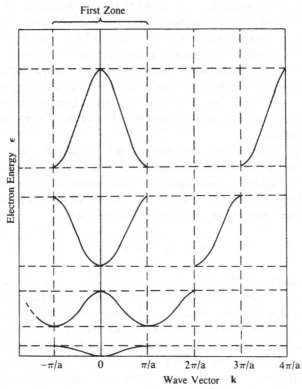

Figure 3-24 The extended zone and reduced zone representations of energy versus wave-vector for electrons in a one-dimensional solid of lattice constant "a," when the periodic potential is of finite amplitude. Well-behaved nonlocalized wave-functions can be constructed (i.e., wave-functions for real k) only within the shaded regions of eigenenergy. These are the allowed electron bands, separated from each other by forbidden energy gaps.

bands and gaps for a hypothetical one-dimensional solid with lattice constant "a". The distinction between Figures 3-23 and 3-24 is that the periodic potential is assumed to be infinitesimal in the former case and finite in the latter.

When **k** is real (as it is for the energy ranges we have called the allowed bands), the Bloch function solution of Equation 3-112 is a running wave, modulated with the periodicity of the lattice. A Bloch function wave extends without attenuation through a crystal, and the long-term electric charge density $-e|\psi|^2$ has the same value in every unit cell of the crystal.

A "band" electron in a state describable by a Bloch function with real **k** will travel forever without any attenuation of its wave-function. Thus, as long as things remain perfectly periodic, a band electron has an infinite mean free path. In this way quantum mechanics is able to explain the total absence of elastic scattering by the array of positive ion cores. The free path of an electron terminates when it encounters an interruption of the perfect periodicity. Any interruption makes a Bloch function inappropriate and encourages scattering, usually to some other direction. Crystal surfaces, static point defects, and phonons provide such interruptions and result in a finite mean free path.

As we can see from Equation 3-112, Bloch functions are not well behaved when **k** is imaginary. Substitution of an imaginary **k** will make the wave-function diverge for one direction in real space. Thus it is not possible for an electron whose energy satisfies Equation 3-111 for imaginary **k** to move freely throughout a crystal. The ranges of electron energy for which **k** must be imaginary (for the correct potential) are accordingly forbidden ranges for mobile electrons in the solid. An energy gap in the electronic energy spectrum of a solid corresponds in its nature with the gap we found between the acoustic branch and the optical branches in lattice vibrations.

The only kind of wave-function which is valid for one of the gap ranges of energy is a wave-function localized around a defect or at the surface of the crystal. A foreign atom, lattice vacancy, or other interruption of perfect periodicity can lead to the existence of a *bound state* wave-function centered on the defect location r_0:

$$\psi = A \exp[-ik|r - r_0|] \qquad (3\text{-}128)$$

Since k is imaginary, the wave function of Equation 3-128 decays steadily as we move away from r_0 in any direction, which implies a localization of the charge density $-e|\psi|^2$. An electron can be trapped by the local electric field of the defect, and has an energy lying within one of the energy gaps; an energy which is forbidden for freely moving electrons. Such localized states, which depend on impurities and disorder, are much less numerous than the band states which arise from the regular atomic array. As a result, the electronic consequences of localized defect states in a metal are often negligible. However, the localized defect states are usually of great importance in an insulating or semiconducting solid.

It was first pointed out by Tamm[40] that the interruption of a periodic lattice at the crystal surface will cause localized surface states, consistent with Bloch functions for an imaginary value of **k**, and with an energy lying within one of the "forbidden gaps." Extensive experimental work on surfaces has verified that Tamm surface states do exist, but that contamination of a crystal surface (such as with adsorbed oxygen) is likely to create many more surface states. Much of the literature concerns semiconductor surface states[41] rather than metals, and we shall not go into further details here.

THE KRONIG-PENNEY MODEL

Before we turn to the reality of energy bands and forbidden gaps in real three-dimensional solids (for which exact solutions of Equation 3-111 are often not attainable), there is much we can still learn from one-dimensional models. One such model was suggested by Kronig and Penney,[42] using the simple square potential barriers of Figure 3-25. The form of the variation of energy with wave-vector then depends on the width b and height $\mathscr{V}_0$ of these barriers.

At the extreme of vanishing small product ($b\mathscr{V}_0$), electrons of any energy can move at will wherever they please, and the dispersion relation will be the free electron one of Equation 3-127. At the opposite extreme of barriers so thick that electrons of energy less than $\mathscr{V}_0$ have zero probability of tunneling from one atomic space to the next, the only possible energies are the series

$$\epsilon_j = (\pi^2 \hbar^2 j^2 / 2ma^2), \quad j = 1, 2, 3. \ldots \qquad (3\text{-}129)$$

[40] I. Tamm, Z. Physik **76,** 849 (1932). See S. G. Davison and J. D. Levine in *Solid State Physics, Vol. 25* (Academic Press, 1970), p. 1.

[41] D. R. Frankl, *Electrical Properties of Semiconductor Surfaces* (Pergamon, 1967).

[42] R. de L. Kronig and W. G. Penney, Proc. Roy. Soc. (Lond.) **A.130,** 499 (1930).

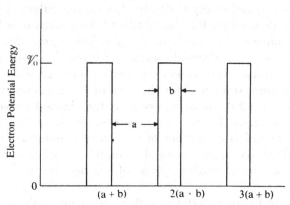

Figure 3-25 One-dimensional periodic electron potential energy supposed in the Kronig-Penney model.

until ϵ_j exceeds $\mathscr{V}_0$. Equation 3-129 is the conventional quantum-mechanical result for the energies of a particle in a one-dimensional box of length "a".

What now of the situation of large $\mathscr{V}_0$, yet small enough b so that there is a finite possibility for an electron with energy less than $\mathscr{V}_0$ to tunnel from one cell to the next? Kronig and Penney analyzed this intermediate situation, requiring that the wave-functions in the shaded and unshaded portions of Figure 3-25 match at the interfaces. The solution is worked in full in many elementary texts on quantum mechanics: it assumes a particularly simple form when the limit is taken of barriers which are very tall but very narrow in such a way that the product ($b\mathscr{V}_0$) remains finite. Then wave-vector k and energy ϵ for an electron are related by

$$\cos(ka) = \cos(\alpha a) + P\left[\frac{\sin(\alpha a)}{\alpha a}\right] \qquad (3\text{-}130)$$

where α expresses the energy through

$$\alpha = [2m\epsilon/\hbar^2]^{1/2} \qquad (3\text{-}131)$$

and the quantity P is a measure of the strength of the barriers:

$$P = \lim_{\substack{b \to 0 \\ \mathscr{V}_0 \to \infty}} [(b\mathscr{V}_0)ma/\hbar^2] \qquad (3\text{-}132)$$

The limit of P $\to$ 0 results in $\alpha = k$, or the free electron dispersion relation of Equation 3-127. The limit of P $\to \infty$ results in the set of energies given by Equation 3-129. However, the solution of Equation 3-130 for a finite value of P results in an alternation of energy regions for real and imaginary k. For k must be imaginary whenever the right side of Equation 3-130 has an absolute magnitude larger than unity, and will be real whenever the absolute magnitude of the R.H.S. is less than unity. Figure 3-26 plots the right side of Equation 3-130 versus α, for a finite value of the parameter P. Note that the ordinate is $(1 + P)$ for $\alpha = 0$, and is ± 1 when α is equal to any multiple of (π/a), regardless of the value of P. What *does* depend on P is the width of the ranges of α for which k must be imaginary, the width of the *forbidden gaps* in this model.

The alternation of allowed bands and forbidden gaps is shown in Figure 3-27, which displays energy versus *reduced* wave-vector (the reduced zone representation) for the case of P = 2 (as was used for the curve of Figure 3-26), superimposed on the dashed curves for completely free electrons (P = 0). The upper energy limit for each allowed band in this model is a member of the set of Equation 3-129, but the lower limit depends on the assumed value for P. As P is increased, each band pulls up into a smaller range of energy, leaving a larger gap beneath it.

The Kronig-Penney model is highly idealized, both in using a one-dimensional analog of a solid, and in using an artificially simple poten-

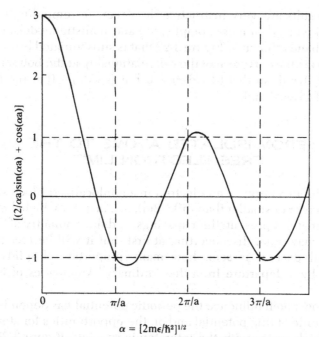

$$\alpha = [2m\epsilon/\hbar^2]^{1/2}$$

Figure 3–26 A plot of the right side of Equation 3-130 (for the parameter P = 2) against the quantity α, which depends on the electronic energy in the Kronig-Penney model.

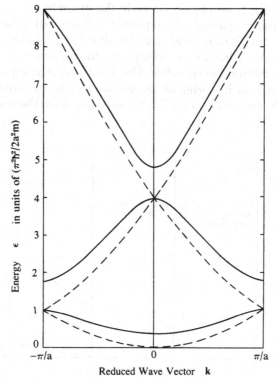

Figure 3–27 A reduced-zone representation of energy versus wave-vector for the Kronig-Penney model when P = 2 (as was chosen in Figure 3-26). The corresponding curves for P = 0 (free electrons) are shown as dashed curves.

tial. The results are very instructive, however, in reminding us of the *principles* involved in more complicated and realistic models. One feature of the bands shown in Figure 3-27 that is substantiated by more realistic models is the curvature of the ϵ–k relationship at the bottom and the top of each band, so that $(d^2\epsilon/dk^2) \neq 0$ but $(d\epsilon/dk) = 0$ at the terminal energies of every band.

FROM ISOLATED ATOMS TO THE FREE ELECTRON LIMIT

The kinetic energy of an electron in a nonlocalized Bloch state may be either larger or smaller than $(\hbar^2k^2/2m)$, in a manner which depends on the types of atoms present, their spacings, and the symmetry of the structure. This may seem disconcerting at first, but it will be recognized as very useful once it is seen how many properties of solids can be explained by a departure from the "ordinary" kinematics of Equation 3-126.

Just how much influence the periodic potential has depends both on the magnitude of this potential and on the opportunities for atoms to interact—which varies with the interatomic spacing. Figure 3-28 visualizes the various stages of interatomic interaction, from zero on the left to total at the right. When atoms are sufficiently far apart, there is no interaction, and electrons are restricted to the discrete set of atomic states. For an interatomic spacing which permits some overlap between atoms (but not very much), the bands can be simulated by the *tight binding* model, of which we have more to say later in connection with the bands of a real three-dimensional solid. The tight binding approach was assumed by Bloch in his original discussion of energy bands in a solid. This approach is also called the LCAO method, since the wave-functions

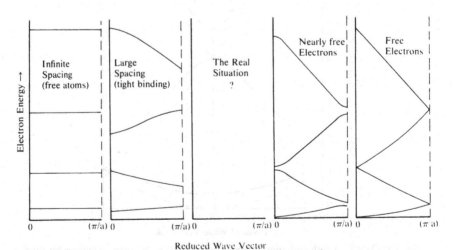

Figure 3-28 A one-dimensional visualization of the dependence of energy on wave-vector for electron states as the coupling between adjacent atoms varies from zero (atoms infinitely separated) to the total abolition of interatomic barriers.

for electrons in the relatively narrow bands of a tight-binding model can be approximated by a *Linear Combination of Atomic Orbitals.*

The *nearly free electron* model for electron states in a solid is appropriate when the periodic potential is rather small (and/or the interatomic separation is small enough to enforce considerable overlap between neighboring atoms). Departures from the free electron dispersion relation of Equation 3-127 are then apparent only in the regions of **k**-space near the Brillouin zone boundaries. In comparing Figures 3-28 and 3-24, it will be clear that the latter was drawn with "nearly free" electrons in mind; the ϵ–**k** relationship is essentially unperturbed for the regions well within the bands.

In a later sub-section, we shall comment on the applicability of the nearly free electron model for a metal such as aluminum. However, for most solids the effective interatomic potential constraining the band shapes lies intermediate between the tight-binding and nearly free electron situations. Therein lies the complexity of band calculations for real solids, as symbolized by the question mark in the central section of Figure 3-28.

Nonetheless, whether the periodic potential is large or small, we can visualize the origin of bands and gaps from the broadening of discrete atomic levels, as sketched in Figure 3-29. A system of widely separated atoms has many states, all at the same energy, and if the interatomic spacing is steadily decreased to the equilibrium spacing in the solid, interatomic coupling causes each atomic level to spread out over a range of energy. Each "band" accommodates as many electron states as there were states available in the original atomic level. As we shall discuss further (and explore in Problems 3.23 and 3.24), an elemental solid comprising N atoms has a volume of **k**-space available in each Brillouin zone adequate for the accommodation of 2N electrons.

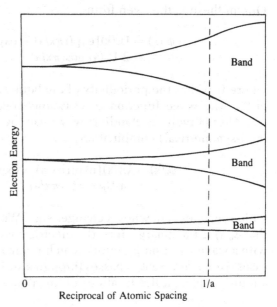

Figure 3-29 The broadening of each atomic energy level into a band of finite width as separated atoms move together to form a solid. The broadening is usually less prominent for the lower-lying levels, which correspond with the inner electronic shells.

BRAGG REFLECTION AND
THE ENERGY GAP

Let us consider an electron state for which k is on a zone boundary, so that the energy is either just above or just below one of the gaps. In the one-dimensional world of Figures 3-24, 3-27, and 3-28, this means a state with $k = (\pm\pi/a)$. We should recall that for lattice vibrational modes at the very top of the acoustic branch and at the bottom of the optical branch, the group velocity of waves becomes zero; such modes are described by standing waves. In a diatomic lattice, one kind of atom carries the vibrations for the mode just below the gap, and the other carries those for the lowest energy optical mode.

An analogous situation exists for electron states at the edge of a gap. The Bloch function for a band-edge state is a standing wave rather than a running wave, because an electron with this wave-vector finds that it satisfies the Bragg diffraction condition

$$2\mathbf{k} \cdot \mathbf{G} + \mathbf{G}^2 = 0 \qquad (3\text{-}133)$$

(which was previously reported as Equation 1-54) when it tries to move with a momentum component (in real space) perpendicular to the zone boundary (in k-space). This condition sounds complex when quoted with reference to multiple dimensions, but we can easily see what it means if we take the one-dimensional simplification

$$k = \pm G/2 = \pm\pi/a \qquad (3\text{-}134)$$

With the wave-vector at this value, there is a phase change of π between the waves reflected from two successive atoms. Then the solution for this value of k comprises equal components of waves traveling to the left $[k = (-\pi/a)]$ and to the right $[k = (+\pi/a)]$. This is a standing wave.

The waves traveling in opposite directions can add up in two ways. One of these is the even form

$$\begin{aligned}\psi_e(x) &= U_e(x)[\exp(i\pi x/a) + \exp(-i\pi x/a)] \\ &= 2\,U_e(x)\,\cos(\pi x/a)\end{aligned} \qquad (3\text{-}135)$$

where $U_e(x)$ has the periodicity of the lattice and the cosine has a period of 2a. The wave function $\psi_e(x)$ is unchanged if we replace x by $-x$.

Alternatively, a standing wave can be constructed from the odd (anti-symmetrical) combination

$$\begin{aligned}\psi_0(x) &= U_0(x)[\exp(i\pi x/a) - \exp(-i\pi x/a)] \\ &= 2i\,U_0(x)\,\sin(\pi x/a)\end{aligned} \qquad (3\text{-}136)$$

which changes sign when x changes sign. We need not be alarmed to see that $\psi_0(x)$ is imaginary, since the electric charge density to be associated with a wave function is $-e|\psi|^2$, which is a real, negative quantity in both cases. However, $-e|\psi_e|^2$ passes through a series of maxima whenever x is a multiple of a, at the locations of the atomic cores. The charge density associated with $\psi_0(x)$ is zero at each of the atomic sites, since the sine function vanishes for these values of x.

Thus we can represent an electron with $k = (\pm\pi/a)$ *either* by a wave-function which requires the particle to spend most of its time near to the atomic cores *or* by a wave-function which tells the electron to stay in the spaces between the cores. Since the electrostatic potential is large and positive at each atomic site, and since an electron has a negative charge, the energy required for the wave-function $\psi_e(x)$ is lower than that for $\psi_0(x)$. The periodic potential creates an energy gap in the one-electron spectrum which did not exist in a free electron theory.

In one dimension, the nth free electron band in the reduced zone comes from taking part of the free electron parabola centered on the origin (Equation 3-127), and moving this through a reciprocal lattice vector large enough to execute the transition from the nth zone to the first zone. In general, each time a free electron k value crosses a Brillouin zone boundary, a gap appears for finite interatomic potential.

We must now consider the application of this principle to multi-dimensional solids.

BRILLOUIN ZONES FOR MULTI-DIMENSIONAL SOLIDS

It is now necessary to consider Brillouin zones once more for structures in two and three dimensions. A two-dimensional Brillouin zone for an oblique lattice was introduced as Figure 1-52, defined as centered on one reciprocal lattice point and bounded by the set of bisectors of reciprocal lattice vectors. This zone had the same area as a cell defined by the basic reciprocal lattice vectors. More generally, for a solid of m dimensions, the volume of the unit cell in reciprocal space and of each Brillouin zone is $(2\pi)^m$ divided by the volume of the unit cell in real space. Since the allowed density of electron states in reciprocal space is determined by the considerations of Figure 3-7, this accounts for the statement earlier in this section that an elemental solid totalling N atoms can accommodate 2N electron states within a Brillouin zone. (The factor of two arises from spin degeneracy.) It must be emphasized that not only does the *first* Brillouin zone have space for 2N states, but that every higher zone of the extended representation must have the same amount of (reciprocal) space. This must be so since any higher zone can be converted into the reduced zone by translations of well chosen reciprocal lattice vectors.

It is easy to see that all zones of a one-dimensional lattice have the same length in k_x. It will be seen in Figure 3-24 that the mth zone extends from $k_x = [-m\pi/a]$ to $[-(m-1)\pi/a]$, and again from $k_x = [(m-1)\pi/a]$ to $[m\pi/a]$. Each zone thus has a total volume (i.e., length) of $[2\pi/a]$. Before we can be equally convinced that all zones in two or three dimensions have the same volume, it is necessary to find out how a zone beyond the first is constructed. This was formulated by Brillouin.[43]

---

[43] L. Brillouin, J. Phys. Radium **1**, 377 (1930). See L. Brillouin, *Wave Propagation in Periodic Structures* (Dover, 2nd edition, 1953).

If the de Broglie wave associated with an electron (or any other wave, for that matter) propagates through a periodic lattice in such a manner that its wave-vector measured from the origin of k-space terminates on a perpendicular bisector of a reciprocal lattice vector, then a discontinuity occurs in the ϵ–k relationship. (Such a value for k satisfies the Bragg diffraction condition, Equation 1-54 or 3-133.) Brillouin remarks that the purpose of introducing zones is to eliminate discontinuities in the ϵ–k relationship *except* at the zone boundaries.

Every zone must be bounded by perpendicular bisectors of reciprocal lattice vectors initiated from a particular reciprocal lattice point which is designated as the origin of k-space. Since reciprocal space is infinite for a perfect lattice, *any* point of the reciprocal lattice will serve equally well. For a two-dimensional lattice, the perpendicular bisectors are lines, as in the prior example of Figure 1-52. For a three-dimensional lattice, each bisector of a reciprocal lattice vector is a plane. No bisector may pass through the interior of a zone. Then the volume to be associated with the nth Brillouin zone is the volume enclosed by the nth such surface *minus* the volume enclosed by the (n − 1)th surface.

Zone boundaries for a hypothetical two-dimensional square lattice of lattice constant "a" are set by solutions of

$$(mk_x + nk_y) = (\pi/a)(m^2 + n^2) \qquad (3\text{-}137)$$

where m and n are integers. The lines satisfying this equation border the first ten zones for a square lattice, shown in Figure 3-30. Note the frag-

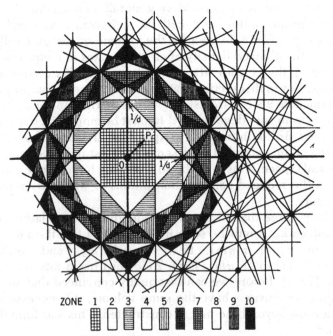

Figure 3-30 The first ten zones for a two-dimensional square lattice. From L. Brillouin, *Wave Propagation in Periodic Structures* (Dover, 2nd edition, 1953).

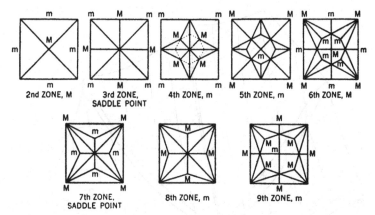

Figure 3–31 Reduction into the first zone for the second through ninth zones of the two-dimensional square lattice. From L. Brillouin, *Wave Propagation in Periodic Structures* (Dover, 2nd edition, 1953).

mentation of k-space for high order zones. Yet each high order zone can be converted into the space and shape of the first (reduced) zone when translations of suitable **G** are made. Figure 3-31 shows the results of this "jigsaw puzzle" construction for zones 2 through 9 of the square lattice. This shows that each zone does indeed have an area of $(2\pi/a)^2$.

The reader is very possibly wondering at this point when the reduced zone representation will be the best way to view the energy bands of a solid, and when the extended zone representation helps to demonstrate what is going on. Figure 3-32 gives a good perspective of the variation of energy with wave-vector for a hypothetical square lattice. This perspective diagram shows energy vertically and the vectors k_x and k_y in the horizontal plane, for the first zone and for appropriate parts of the second through fourth zones. The discontinuities in energy as one requires **k** to increase through a zone boundary appear as vertical cliffs. The corresponding figure for a free electron model of negligible periodic potential would have the same general increase of ϵ with **k**, but without the discontinuities.

Thus the extended zone method is useful for conveying the information of Figure 3-32. The reduced zone representation is more useful when one wishes to see the consequences of changes in wave-vector for an electron in a band above the first. Problem 3.23 examines the appearance of a trajectory in k-space for the fourth zone of a square lattice.

The geometric principles involved in creating zones beyond the first for the square lattice work for other two- and three-dimensional lattices. Figure 3-33 shows the first five zones for the two-dimensional hexagonal lattice, and Problem 3.24 suggests a reduction of higher zones into the first. Problem 3.25 is concerned with rectangular lattices as far as the fourth zone.

Geometric complications of zone surface topology are inevitable in three dimensions when we consider any structure more complex than simple cubic (for which things work out exactly as in one and two dimen-

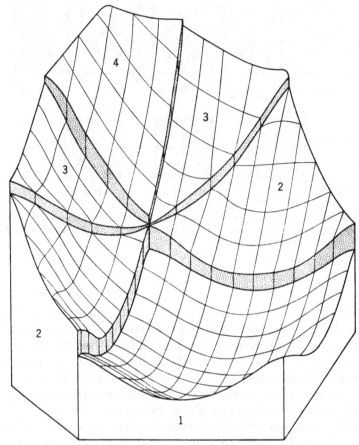

Figure 3–32 A perspective drawing of the $\epsilon - \mathbf{k}$ relationship for a hypothetical two-dimensional square lattice, using the extended zone representation. Energy is the ordinate, with the k_x and k_y axes in the horizontal plane. Portions of zones 1, 2, 3, and 4 are displayed, with a finite periodic potential supposed to create finite energy discontinuities at the zone boundaries. From J. C. Slater, *Quantum Theory of Molecules and Solids, Vol. 2* (McGraw-Hill, 1965).

sions). Figure 3-34 shows the first four zones for simple cubic, while Figures 3-35 and 3-36 illustrate the first few zone boundaries for B.C.C. and F.C.C. space lattices respectively. The B.C.C. space lattice has a simple first zone, but the outer surface of the second zone shows the fragmentation of **k**-space required in order that regions of successive zones should be separated by reciprocal lattice vectors.

For an F.C.C. crystal with cube edge a, the reciprocal lattice has B.C.C. symmetry with a cube edge of $(4\pi/a)$. Bisection of reciprocal lattice vectors results in a first zone which is a truncated octahedron with a volume of $4(2\pi/a)^3$, four times as large as for an S.C. crystal of the same a. The point here is that the cube edge in an F.C.C. crystal is larger than the primitive translation distance of the rhombohedral primitive cell. The F.C.C. cube accommodates four atoms, and once again the Brillouin zone has a volume suitable for two opposing spin states for each atom. This is a general rule for an elemental solid, regardless of the structure.

(See Problem 3.26 in connection with the H.C.P. lattice and its zone.) In a compound there is a set of zones for each component.

Since there is a correspondence between the discrete states of an isolated atom and the Brillouin zones in a solid, we may expect to find

1 zone traceable to 1s states

1 zone traceable to 2s states

3 zones traceable to 2p states

1 zone traceable to 3s states

3 zones traceable to 3p states

5 zones traceable to 3d states, and so on.

In the crystalline field, a p or d atomic state of 6 or 10 electrons per atom may stay together or at least remain degenerate at one value of **k**, as suggested in Figure 3-37. This gives a band of *energy* comprising more than a single zone and having more than 2N electron states. Alternatively, the various components of a p or d atomic state may separate into non-overlapping ranges of energy. Thus actual bands may have 2N, 4N, 6N, . . . states altogether.

In a real (three-dimensional) solid, overlap in energy can occur for states which might have arisen from quite different types of atomic level. This possibility is illustrated in Figure 3-38, which shows the dependence of ϵ on **k** for three different directions in **k**-space. Since a non-vanishing periodic potential is supposed, there must be an increase of energy involved as **k** progresses across the zone boundary in any of

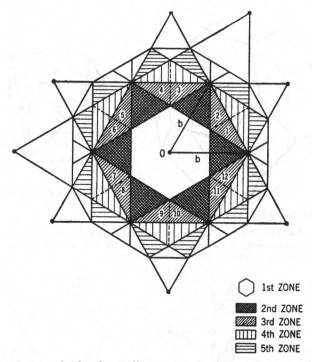

Figure 3–33 The first five Brillouin zones for the two-dimensional hexagonal lattice. From L. Brillouin, *Wave Propagation in Periodic Structures* (Dover, 2nd edition, 1953).

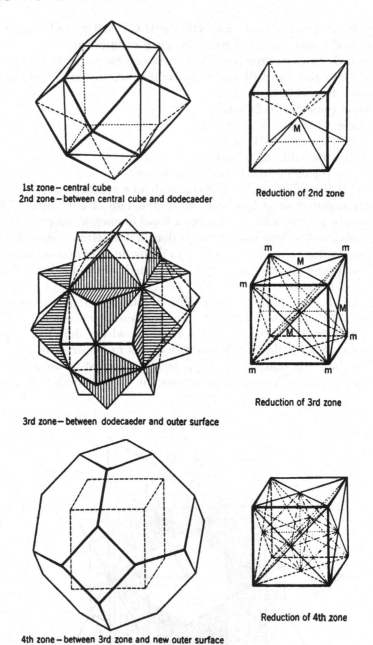

1st zone – central cube
2nd zone – between central cube and dodecaeder

Reduction of 2nd zone

3rd zone – between dodecaeder and outer surface

Reduction of 3rd zone

4th zone – between 3rd zone and new outer surface

Reduction of 4th zone

Figure 3–34 The first four Brillouin zones and the reduction of these into the size and shape of the first zone for the simple cubic lattice. From L. Brillouin, *Wave Propagation in Periodic Structures* (Dover, 2nd edition, 1953).

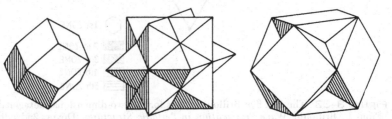

Figure 3–35 The first three Brillouin zones for a solid which crystallizes in the B.C.C. structure (thus endowing reciprocal space with F.C.C. symmetry). The faces of the first zone are parallel to (110) planes. From J. C. Slater, Rev. Mod. Physics **6**, 212 (1934).

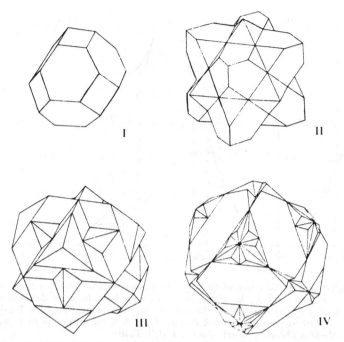

Figure 3–36 The first four Brillouin zones for a solid which crystallizes in the F.C.C. structure. Reciprocal space then has B.C.C. symmetry. Brillouin zones of the same shape are found for structures derived from F.C.C., including the diamond and zincblende lattices. From W. A. Harrison, *Solid State Theory* (McGraw-Hill, 1970).

directions k_a, k_b, or k_c. For the situation in the right half of the figure, *any* energy in the second band is higher than *any* energy in the first band. However, for the situation on the left of the figure, the highest Band 1 energy in direction k_c is larger than the lowest possible energy for Band 2 in other directions of **k**-space. Thus it is possible for an electron in

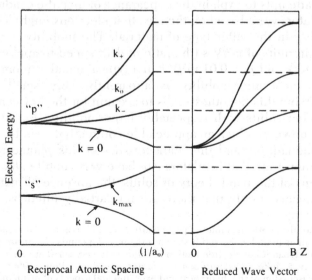

Figure 3–37 A hypothetical example of what *could* happen to the components of an atomic "p" state when atoms are brought together to make a solid. In this case, the "p" states form a band of 6N states in total, with the largest density of states in the lower portion.

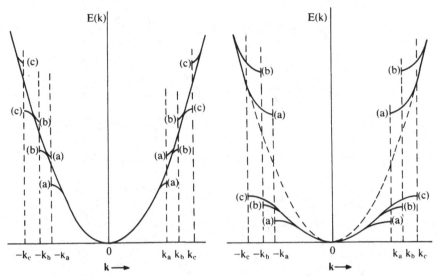

Figure 3–38 Variation of energy with wave-vector for two successive bands in a three-dimensional solid. k_a, k_b, and k_c are at the zone boundaries in different directions. Overlap occurs in the situation on the left, because the highest possible Band 1 energy exceeds the lowest possible Band 2 energy (for a different direction in **k**-space). After F. Seitz, *Modern Theory of Solids* (McGraw-Hill, 1940).

Band 1 to be scattered into Band 2 without change of energy, and the bands then form a composite band of 4N states. Overlap seems likely between the bands shown in the perspective drawing of Figure 3-32.

METALS, INSULATORS, AND SEMICONDUCTORS

Early attempts to explain the appearance of metallic conductors and good insulators were based on the idea that electrons might have a very low *mobility* in the latter type of material. The mobility $\mu = (\sigma/ne)$ is expressed in units of m²/V s (the ratio of drift speed to applied electric field strength), and $\mu \simeq 0.01$ m²/V s in a typical metal at room temperature. The electronic mobility is apparently very small in *some* insulators,[44] but this hypothesis fails to account for the low conductivity of many other solids with respectably large mobility. For many insulators are known in which an appreciable conductivity can be produced under illumination,[45] and the characteristics of this photoconductivity clearly demonstrate that insulators *can* have very mobile electrons.

In terms of the band theory of solids, the absence of metallic conductivity means simply that there are no partially filled bands. In an

---

[44] Low mobilities are encountered particularly in thin films and in amorphous nonconductors, as well as in molecular solids such as organic compounds. There are problems in describing the dynamics of electron motion in solids with very small apparent mobility, as discussed by A. F. Ioffe, *Physics of Semiconductors* (Academic Press, 1960).

[45] A full discussion of the theoretical and experimental aspects of insulating and semiconducting photoconductors is given by R. H. Bube, *Photoconductivity of Solids* (Wiley, 1960).

insulator, every band is either completely full or completely empty. For example, sodium has 11 electrons per atom. In an isolated atom these electrons have the configuration

$$Na = [1s^2\, 2s^2\, 2p^6]\, 3s^1 \tag{3-138}$$

The 10 inner electrons form closed shells in the isolated atom, and we can expect them to form very narrow bands in the solid, using up the first five zones in extended k-space. The single outer electron per atom will half-fill the next Brillouin zone, and sodium has to be a metal. Figure 3-39 shows the sodium 2p and 3s states as visualized by Slater.

The next element in the periodic sequence is

$$Mg = [1s^2\, 2s^2\, 2p^6]\, 3s^2 \tag{3-139}$$

which apparently could be an insulator *if* there was no overlap in energy with the states of the 3p bands. Since we know in practice that Mg is metallic, some quirk of the periodic potential must produce the overlap sketched in the left portion of Figure 3-40. The Fermi energy is shown to occur for magnesium at an energy which fills the 3s band some 90%, with just a few per cent occupancy for the overlapping 3p band.

The density of states curves of Figure 3-40 are not intended to be exact, but illustrative rather of the features which make magnesium a metal and silicon an insulator. Section 3.5 takes up details of the evaluation of the scalar $g(\epsilon)$ from the ϵ–k relation. One aspect of Figure 3-40 which deserves comment before going on concerns the *Van Hove singu-*

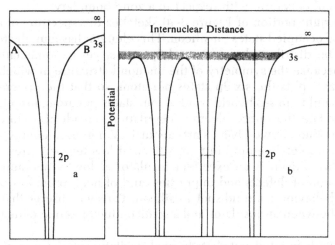

Figure 3–39 A self-consistent model for a sodium atom (at left) and metallic sodium (at right), from J. C. Slater, Physics Today **21** (No. 1), 43 (1968). The 3s state of an isolated atom is broadened into a band, which is half-filled. For this particular case, the half-filled band which locates the Fermi level happens to coincide with the maxima of the periodic potential energy, according to the calculation on which this figure is based. Note that the 3s band is lower in energy than the isolated 3s state. It is this which binds the solid together; the lowering is the cohesive energy.

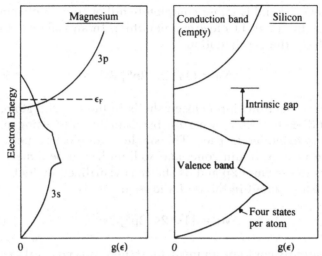

Figure 3–40 A contrast between the energy distribution of states in the solid derived from the 3s and 3p states in metallic magnesium as against insulating silicon. (The bands are not drawn to scale, and the shapes are not intended to be accurate.) Magnesium is saved from nonmetallic status by the overlap of the 3s and 3p bands. For silicon the valence band is an s-p hybrid which just has room to accommodate the four bonding electrons per atom, leaving a separate and empty band above. In each imagined density of states curve, abrupt changes in slope are drawn for the energies of van Hove singularities.

larities shown at certain energies. These were mentioned before in Chapter 2, since similar singularities or *critical points* occur in the spectrum of lattice vibrations (see Figure 2-7). A Van Hove singularity[46] occurs at an energy for which the constant energy contour in k-space either touches a zone boundary for an important direction, or changes its direction of curvature with respect to a zone boundary.

The right portion of Figure 3-40 sketches the spectrum of allowed states and occupied states for silicon. This element has four electrons per atom to be distributed over 3s and 3p states, yet is nonmetallic. This occurs because the symmetry of the diamond structure in which silicon crystallizes splits the six 3p states per atom, so that the uppermost occupied band is an s-p hybrid band of four states per atom, just enough to accommodate the entire electron population outside the closed shell cores. Another allowed band starts about 1 eV higher in energy, but this band is completely empty in pure silicon at low temperatures. Accordingly, pure silicon is an excellent insulator at low temperatures. The combination of defects and impurities, and of heat, promotes *semiconducting* behavior in a solid such as silicon. Thus we observe that the difference between an insulator and a semiconductor is one of degree, not of type.

Wilson[47] pointed out that thermal excitation could empty a small fraction of the states in the uppermost occupied band of an insulator (the valence band), simultaneously populating a few of the lower states in the

[46] L. Van Hove, Phys. Rev. **89**, 1189 (1953).
[47] A. H. Wilson, Proc. Roy. Soc. **A133**, 458 (1931).

next band (the conduction band). Conduction by electrons in the conduction band and by the same number of *positive holes* in the valence band is the situation in an *intrinsic semiconductor,* and the energy separation between valence and conduction bands is the *intrinsic gap.* Wilson[48] further suggested that at temperatures too low for appreciable intrinsic electron-hole generation, localized electron states within the forbidden gap associated with defects and impurities could still be thermally ionized to generate *either* free electrons *or* free holes. This second form of semiconducting behavior is termed *extrinsic,* since it depends on non-intrinsic properties of the conducting medium.

We shall have more to say about insulators and semiconductors in Chapter 4. For the present we may observe that nonmetallic solids in which the intrinsic gap is less than about 2 eV, or for which large densities of localized defect states can be created near to the edge of either the valence band or conduction band, are likely to be recognized as semiconductors. Solids in which the intrinsic gap is wide and the defect levels are far from either band are insulators at ordinary temperatures.

In discussing silicon, we noted that the distribution of 3s and 3p states which makes silicon nonmetallic is a consequence of the symmetry of the diamond lattice. Similar considerations are found in other elemental solids and in a variety of compounds. A material which exists in two or more allotropic forms may have a metallic overlap of Brillouin zone energy ranges for one possible crystal structure but not for another.

CALCULATION OF ENERGY BAND SHAPES

A full discussion of the techniques involved in calculations of energy bands and gaps in three-dimensional solids would be too long and complex to include in this book. It is sufficient that we mention some of the principal theoretical procedures, noting that the bibliography at the end of this chapter lists several extended accounts.

One of the earlier theoretical methods was the *cellular method* of Wigner and Seitz (1933). This method proceeds from a wave-function which is a solution of the Schrödinger equation for the chosen electrostatic potential, and which must then be made continuous and periodic from one atomic site to the next. Wigner and Seitz envisaged a polyhedron constructed in real space centered on each atom of a crystal, such that the set of polyhedra, or *cells,* filled all the space. They then conjured up wave-functions for electrons which could be continuous in magnitude and derivative at important places on the boundary separating one cell from the next. In order to solve a problem by the Wigner-Seitz method, it was often found convenient to replace a polyhedral cell by a spherical cell of the same volume.

The cellular method has proved quite successful with the simple alkali metals. Figure 3-39 exemplifies the use of this method in calculating energy bands for sodium. Another early success in describing

---

[48] A. H. Wilson, Proc. Roy. Soc. **A134**, 277 (1931).

energy bands using the Wigner-Seitz technique was Kimball's calculation of the bands derived from the 2s and 2p states of carbon in the diamond form. Kimball carried out these calculations as a function of interatomic spacing. His result is illustrated in Figure 3-41 and correctly forecasts that two lower and six upper states per atom (for isolated atoms) are replaced by four lower and four upper states per atom below a critical interatomic spacing r_x. This transformation explains the nonmetallic nature of diamond and of the homologous series of solids (silicon, germanium, SiC, GaAs, and so forth).

Despite such successes, calculation techniques other than the cellular technique have been dominant in nonmetallic solids in recent years and have permitted improved quantitative accuracy. In most such band calculation procedures, a wave-function which is continuous and periodic is required from the outset, even though this may mean starting with a wave-function which does not satisfy the Schrödinger equation until the potential energy term has been adjusted in some manner.

At one extreme of these procedures is the *tight binding method*, which was introduced by Bloch (1928) in his initial discussion of energy bands.[39] Tight binding was visualized in Figure 3-28 as the weakest type of interaction between neighboring atoms. In this approach, atomic orbitals (each localized on a particular atom) are combined linearly in a fashion consistent with the Bloch-Floquet theorem in order to represent a state running throughout the crystal. The structure of the resulting band then depends on overlap integrals which are quite sensitive to the details of the orbitals outside the cores, and to the lattice spacing, since an electron is little affected in the tight binding approach by atoms more than a single atomic spacing away.

The simplest approach to a tight binding model[49] suggests that for a

---

[49] The interested reader with some background in quantum mechanics can find this model discussed at a level which is not too complex in Section 3.4 of J. Ziman, *Principles of the Theory of Solids* (Cambridge, 2nd edition, 1972) or in Section 10-8 of A. J. Dekker, *Solid State Physics* (Prentice-Hall, 1957).

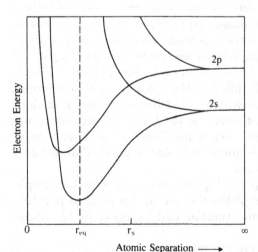

Figure 3–41 Electron energy as a function of interatomic spacing for diamond, as calculated by the cellular method by G. E. Kimball, J. Chem. Phys. 3, 560 (1935). For spacings larger than r_x, the bands are broadened forms of the atomic 2s and 2p states, with capacities of two and six electrons per atom respectively. (The 1s core states are at a much lower energy and are omitted.) For spacings less than r_x, the bands are s-p hybrids; the lower one (four electrons per atom) is full and the upper one is empty. At the equilibrium spacing r_{eq}, Kimball deduced an intrinsic energy gap of 7 eV.

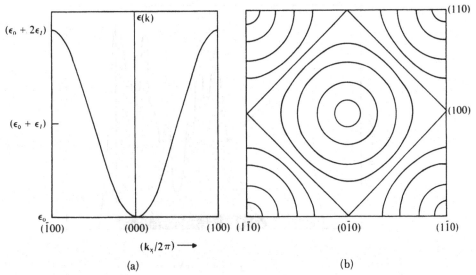

Figure 3–42 (a), Variation of electron energy with wave-vector in the k_x direction for a simple cubic lattice according to a tight binding model (Equation 3-140). (b), Contours of constant energy in the $k_z = 0$ plane of the Brillouin zone for this model.

simple cubic lattice the dependence of energy on wave-vector will take the form

$$\epsilon(\mathbf{k}) = \epsilon_0 - \epsilon_1[\cos(ak_x) + \cos(ak_y) + \cos(ak_z) - 3] \quad (3\text{-}140)$$

which extends over the energy range ϵ_0 to $(\epsilon_0 + 6\epsilon_1)$. Figure 3-42 shows the ϵ–$\mathbf{k}$ variation in the direction of k_x, and the set of energy contours in $\mathbf{k}$-space for the k_xk_y plane. Only for the lowest energies (near the center of the zone) does this qualitatively resemble a free electron spectrum. This description of a band is most likely to be correct when the interatomic interaction is small and the band is a slightly broadened form of an atomic level.

When there is an appreciable interatomic interaction, the tight binding approach must use a *linear combination of atomic orbitals* (the LCAO method), in which a quantum-mechanical variational procedure is used to find the combination of s, p, and d orbitals that corresponds to the lowest energy in the system. In one modification of this method (Löwdin, 1950), orbitals are chosen in such a way that functions centered on the different atoms are automatically orthogonal.

The LCAO method has been used in many band calculations, both for first principles calculations and for interpolation when some information about the bands in a solid is known from other sources. Tight binding methods are, not surprisingly, most successful when the effect of the periodic core potential is quite large. This is true, for example, in ionic solids such as NaCl. It is also true for the band derived from 3d states in the first series of transition elements. The reader will remember that isolated atoms of elements in the first transition series have one or two electrons in the 4s sub-shell and a partially filled 3d shell. When such atoms are brought together to make a solid, overlap of 4s states is very

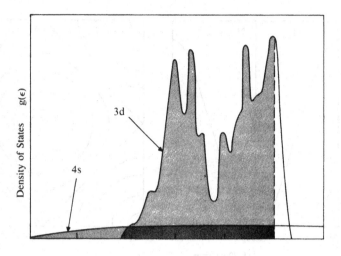

Density of States $g(\epsilon)$

3d

4s

$(\epsilon - \epsilon_F)$ (eV)

Figure 3–43 Density of states in F.C.C. nickel for the bands close to the Fermi energy, as calculated by the tight binding method by G. F. Koster, Phys. Rev. **98**, 901 (1955). The number of electrons per atom in nickel would be just sufficient to fill the 3d band if the 4s band were entirely at higher energy. However, the 4s band has a very low density of states and spreads over a wide energy range, from below the bottom of the 3d band to 5 eV above the top of it. The Fermi energy which results makes the 3d band 94 per cent full, so that the 4s band contains 0.6 electrons per atom.

strong but overlap of states from the unfilled inner level is rather weak. The result for the density of states as a function of energy is typified by the curves of Figure 3-43 calculated for nickel. The 3d band extends over only a moderate range of energy, and the density of states within this range must be very large. Overlap in energy between the 3d and 4s bands for transition elements results in a fractional electron per atom occupancy of the 4s band. The few highly mobile electrons in the 4s band do most of the work of conduction in such a metal, but it is the large 3d density of states at the Fermi energy which controls the electronic specific heat and the magnetic properties.

At the opposite extreme of techniques from the tight binding method is the *nearly free electron method* (NFE method), which is suitable for situations of strong overlap. The NFE method supposes that the periodic potential produces only a small perturbation of the free electron spectrum of states. As a justification of this latter approach for the electron gas in a metal, remember that the atomic cores are regions of negative electronic potential energy compared with the inter-core regions. From the conservation of total energy, an electron must move more rapidly (having more kinetic energy) when it *does* pass through a core region. In combination with the preponderance of space outside to space inside cores (see Problem 3.1), we can see that an electron spends almost all its time in the relatively field-free regions outside the cores, and its condition is only slightly affected by the exact form of the core potential. Thus a tolerable first approximation to the electronic wave-function for conditions of strong overlap should be a *free* electron wave-function (a

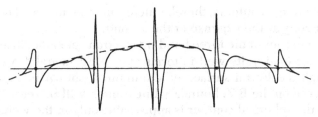

Figure 3–44 A wave-function for an electron in the 3s band of metallic sodium as a function of position, for the eigen-energy which corresponds with $k = \pi/4a$. After J. C. Slater, Rev. Mod. Phys. **6**, 209 (1934). The dashed curve shows the simple plane wave which approximates the wave-function fairly well in the regions of space outside the atomic cores.

plane wave). This approximation is illustrated as the dashed curve of Figure 3-44 for a bonding electron in sodium.

We recall that a surface of constant energy in **k**-space for *perfectly free* electrons is a sphere centered on **k** = 0. The surface for $\epsilon = \epsilon_F$ (the Fermi surface) is a perfect sphere for such electrons (Figure 3-14). Then it is argued in the NFE method that a *small* periodic potential will not warp constant-energy surfaces such as the Fermi surface far from a spherical shape. This argument serves as a guide as to which portions of **k**-space will be occupied and which will remain empty.

The NFE approach is exemplified in Figure 3-45 for a hypothetical square lattice. All electron states within a certain radial distance from the origin of **k**-space are shown as being occupied in the extended zone scheme. The right portion of Figure 3-45 views the electron distribution in reduced zone terms, and it can be seen that there are "holes" in the corners of the first zone and pockets of electrons in the second zone. An increase in the amplitude of the periodic potential will warp the shapes

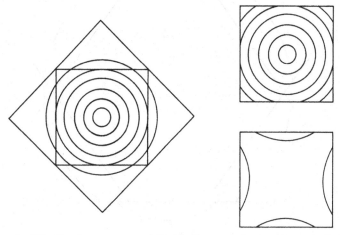

Figure 3–45 The first two zones of a simple square lattice, shown in both the extended zone and reduced zone representations. Enough electrons are added to approximately fill one zone, but if the periodic potential is very small (nearly-free electron model), the highest energy corners of the first zone will remain empty and there will be electron pockets in the second zone. An increase in the periodic potential will warp the shape of equal-energy contours in **k**-space, and will probably force more electrons to remain in the first zone.

of constant-energy contours, thereby increasing the degree of occupancy of the first zone at the expense of the second.

The properties of the three bonding electrons per atom in aluminum can be understood quite well in terms of an NFE model.[50] A cross-section through the Fermi surface of aluminum is shown in Figure 3-46, superimposed on the B.Z. boundary lines, and it will be seen that modification of the spherical contour is appreciable only in the vicinity of the actual boundaries. The first zone is completely full, and there is partial occupancy of each of the next three zones. Harrison[50] pictures the occupied regions of the first four zones as in the perspective sketches of Figure 3-47, and some aspects of this picture have been confirmed by experiment.[51]

The Kronig-Penney model can be regarded as a one-dimensional form of the NFE approach when the parameter P of Equation 3-132 is small. For a larger value of P, the Kronig-Penney model becomes a one-dimensional form of the APW method we shall discuss in a moment.

There are several ways in which the plane wave solutions of the NFE method can be improved. We should reflect that a single plane wave describes the wave-function in Figure 3-44 quite well *outside* the

---

[50] W. Harrison, Phys. Rev. **118**, 1183–1190 (1960). For a fascinating discussion of theoretical approaches (such as the NFE method and various others) coupled with experimental results on the shapes of bands and shapes of Fermi surfaces in metals, see *The Fermi Surface*, edited by W. A. Harrison and M. B. Webb (Wiley, 1960).

[51] G. N. Kamm and H. V. Bohm, Phys. Rev. **131**, 111 (1963); F. W. Spong and A. F. Kip, Phys. Rev. **137**, A431 (1965).

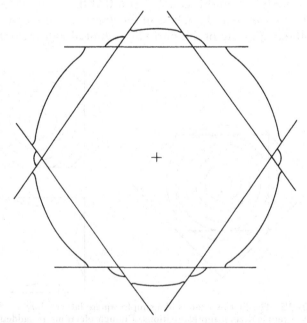

Figure 3–46 A cross-section through the (110) plane in reciprocal space for a solid which crystallizes in F.C.C., showing the first zone and portions of the second, third, and fourth zones. Superimposed is the contour of the Fermi energy for aluminum (three outer electrons per atom), assuming that the electrons are nearly free. After W. Harrison, Phys. Rev. **118**, 1190 (1960).

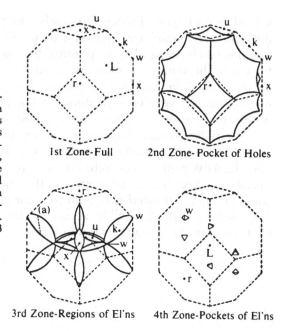

Figure 3–47 A perspective view of the first four zones in k-space for aluminum (which has an F.C.C. structure). This shows the occupied portions in the second, third, and fourth zones, which together make up the Fermi surface. The six small pockets of electrons in the fourth zone disappear when the core potential is increased. From W. A. Harrison, Phys. Rev. **118**, 1183 (1960).

1st Zone-Full

2nd Zone- Pocket of Holes

3rd Zone-Regions of El'ns

4th Zone-Pockets of El'ns

core regions, but that a combination of an enormous number of plane waves would be necessary to describe the complete function, since it must vary rapidly in the vicinity of the cores. However, the difference between a single plane wave (the dashed curve) and the complete wave-function can be described rather simply in terms of *atomic orbitals* at each atomic site. This reasoning led Herring (1940) to suggest the method of *orthogonalized plane waves* (OPW's), functions which are linear combinations of plane waves and mixtures of atomic wave-functions from the occupied states of the ion cores. Thereby the properties of an electron can be described more reliably both inside and outside the core regions. The OPW method has been used in numerous calculations of band shapes in both metallic[52] and nonmetallic[53] solids.

An alternative approach to improving upon the NFE method has developed from an examination of the periodic potential itself. Slater[54] argued that a reasonable approximation to the potential should be a *constant* potential outside the cores, and an ordinary atomic potential inside a sphere surrounding each ion core. The Schrödinger equation is now

[52] The method was demonstrated first for beryllium by C. Herring and A. G. Hill. Phys. Rev. **58**, 132 (1940). Among numerous subsequent OPW calculations for metals we may mention that for aluminum by V. Heine, Proc. Roy. Soc. **A240**, 340 (1957), which provided the justification for Harrison's approach (see footnote 50).

[53] F. Herman [Phys. Rev. **93**, 1214 (1954)] demonstrated the results of the OPW method for diamond and germanium. His demonstration of germanium as a semiconductor in which the lowest conduction band energy and highest valence band energy occur in different parts of the B.Z. opened a new era in the understanding of semiconductors. See footnote 37 of this chapter in connection with more refined OPW calculations of germanium bands by Herman.

[54] J. C. Slater, Phys. Rev. **51**, 846 (1937). The difficulties of using the APW method are not as formidable as they might first appear, as pointed out by the same author in Phys. Rev. **92**, 603 (1953). The method has been used successfully both for metallic and semiconducting solids.

solved for each type of space separately, and the solutions matched over the spherical boundaries between them. Within each core the wavefunction can be expanded in spherical harmonics. For the space outside the cores, the solution is of course a superposition of plane waves. The combination is known as an *augmented plane wave* (APW). The APW method sounds very complicated, but has (with the aid of large computers) been made to work for both metallic and semiconducting solids.

A variety of other techniques, too numerous to list here, have been devised for energy band calculations. A feature of many of them is that (like the APW method) the Schrödinger equation is solved for a potential distribution which is a modification of the true periodic potential. Use of such a *pseudo-potential* may lead to incorrect wave-functions, but can often tell us with acceptable accuracy how energy varies with wavevector within the various Brillouin zones.

Dynamics of Electron Motion

3.5

PHASE AND GROUP VELOCITIES

A rather disturbing thing about wave-functions which describe electrons in real crystals is that the phase velocity of a wave may not coincide either in magnitude or in direction with the actual electronic motion. As an illustration of this point, consider the point $\mathbf{K}$ in the k-space of Figure 3-48. Waves with a wave-vector near $\mathbf{K}$ move in the direction $\mathbf{OK}$ with a phase velocity $v_p = (\omega/k)$. This would appear to make the total value of k important, something which we have swept aside in embracing the idea of the reduced zone.

However, the phase velocity is not really of physical importance. The *wave-packet* around $\mathbf{K}$ which comprises the electron moves in real space through the lattice at the group velocity $v_g = (\partial\omega/\partial\mathbf{k})$. Now if an electron state is represented by a wave-vector $\mathbf{K}$ and a corresponding energy ϵ_K, then its angular frequency (from the time-dependent Schrödinger equation) is $\omega = (\epsilon_K/\hbar)$. Thus the group velocity is

$$v_g = \left(\frac{\partial\omega}{\partial\mathbf{k}}\right) = \frac{\partial}{\partial\mathbf{k}}\left(\frac{\epsilon_k}{\hbar}\right) = \frac{1}{\hbar}\nabla_k\epsilon \qquad (3\text{-}141)$$

This requires that we know the value of k only with respect to the zone boundary or with respect to the center of the reduced zone.

In order to understand this statement, we should remember that the probability wave

$$\psi = U_k(\mathbf{r})\exp[i(\mathbf{k}\cdot\mathbf{r} - \omega t)] \qquad (3\text{-}142)$$

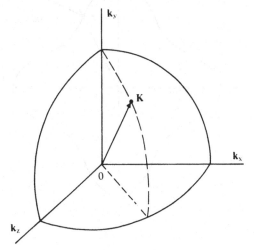

Figure 3–48 Description of the state of motion of an electron by designation of the appropriate point in k-space.

moves through the solid in such a way that the charge density $-e|\psi|^2$ has the same long-term value in every unit cell. The phase velocities of individual waves are mathematical fictions, and it is only the motion of the center of charge of the wave packet which traces out the actual path of an electron. Thus the value of **k** relative to the zone edge determines the speed, direction of motion, and energy of an electron.

Even for a perfectly free electron gas, the *magnitudes* of the phase and group velocities are different (see Problem 3.28), though the group velocity *is* in the direction **OK** for an electron with wave-vector **K**. However, for the distribution of electron states which results from a finite periodic potential, surfaces of constant energy are not necessarily spherical. Figure 3-49 shows a cross-section through such a warped surface. A packet of waves (each individual wave with a phase velocity parallel to **OK**) makes up the electron state at **K**, but any electron which occupies this state will move through real space at the velocity $v_g = (1/\hbar)\nabla_k\epsilon$. This movement is in the direction normal to the constant energy contour. As demonstrated in Figure 3-49, v_g is likely to be in a direction very different from that of **OK**.

The dynamics of electrons are strongly affected by the shapes of constant energy surfaces. When an electron is left alone, it remains in a state of uniform motion in a straight line in real space, with velocity v_g. The electron can then be represented by the same location in k-space, independent of time. Three kinds of influence can upset this situation:

1. The electron can be scattered. This usually involves some change in both energy and wave-vector. Thus the electron disappears from one location in k-space and appears at a different location.

2. The electron can experience an electric field. This requires that the electron energy change. The progression of the electron through k-space and real space is determined by the rate of change of energy.

3. The electron can experience a magnetic field. Such a field alone

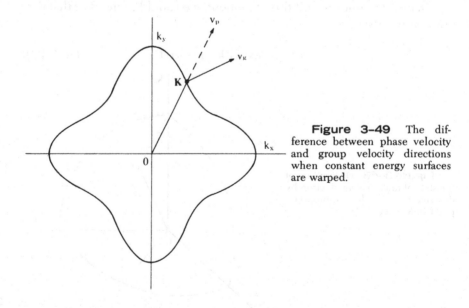

Figure 3-49 The difference between phase velocity and group velocity directions when constant energy surfaces are warped.

requires that the direction of motion (and the occupied location in k-space) change without any modification of electron energy. Thus the electron moves over a constant energy surface in k-space in response to a magnetic field.

We shall now look into the effect of electric and magnetic fields in more detail. In so doing, we must consider the *effective mass* characterizing the inertia of an electron which experiences a periodic potential. Though semiconductors are the official topic of Chapter 4, we shall find it convenient to examine electron dynamics in this section as it relates to both metals and semiconductors.

THE EFFECTIVE MASS

Consider an electron in a periodic solid which is subject to an external electric field **E**. The force on the electron is $-e\mathbf{E}$. Thus the rate of change of electron energy is

$$\frac{d\epsilon}{dt} = -e\mathbf{E} \cdot \mathbf{v} \tag{3-143}$$

(where **v** is the velocity of the electron itself; that is, $\mathbf{v}_g$ rather than $\mathbf{v}_p$ of the previous subsection). Now an infinitesimal change of energy can be written as $d\epsilon = \nabla_k \epsilon \cdot d\mathbf{k}$. Also, we can substitute the result of Equation 3-141 for velocity into Equation 3-143. As a result of these two substitutions,

$$\hbar(d\mathbf{k}/dt) = -e\mathbf{E} \tag{3-144}$$

and, remembering that force is rate of change of momentum, we can see that $\hbar\mathbf{k}$ has the proper behavior for electron momentum as far as external forces are concerned.

The acceleration of an electron in an externally applied electric field is:

$$\mathbf{a} \equiv \left(\frac{d\mathbf{v}}{dt}\right) = \frac{1}{\hbar} \nabla_k \left(\frac{d\epsilon}{dt}\right)$$

$$= \frac{1}{\hbar} \nabla_k(-e\mathbf{E} \cdot \mathbf{v})$$

$$= \frac{1}{\hbar} \nabla_k \left[\frac{(-e\mathbf{E})}{\hbar} \cdot \nabla_k \epsilon\right]$$

$$= \frac{-1}{\hbar^2} \nabla_k \nabla_k \epsilon \cdot (e\mathbf{E}) \tag{3-145}$$

Thus the acceleration has components

$$a_i = \frac{-1}{\hbar^2} \sum_j \frac{\partial^2 \epsilon}{\partial k_i \partial k_j} (eE_j) \tag{3-146}$$

and so forth. A comparison of Equations 3-145 and 3-146 with Newton's laws of motion shows that the tensor quantity $\left(\frac{1}{\hbar^2}\,\nabla_k\nabla_k\epsilon\right)$ has dimensions of (mass)$^{-1}$. Thus

$$[m_{ij}] = \hbar^2\,[\partial^2\epsilon/\partial k_i\partial k_j]^{-1} \qquad (3\text{-}147)$$

is known as the *tensor of effective mass* for an electron subject to a periodic potential.

In some rather complex solids, the off-diagonal components of this tensor are large, and an electric field in one direction produces acceleration in a different direction. This will happen for surfaces of constant energy in k-space which are markedly non-spherical, as exemplified by the curve in Figure 3-49. For an ideally simple and isotropic solid, all off-diagonal components of the effective mass tensor vanish, and the three diagonal ones are permitted by the form of the periodic potential to be the same. In such an ideal solid we have a *scalar effective mass*

$$m^* = \left(\frac{\hbar^2}{d^2\epsilon/dk^2}\right) \qquad (3\text{-}148)$$

for at least part of the electron energy range (that is, in at least some parts of the Brillouin zone).

When the dependence of energy upon wave-vector is quadratic, we can say that electrons experience an "energy independent effective mass," since

$$\epsilon = \frac{\hbar^2 k^2}{2m^*} \qquad (3\text{-}149)$$

Over a large energy range, however, the dependence of ϵ on k inevitably departs from the simplicity of Equation 3-149. Such a departure is often described as an energy dependence of the effective mass, though the manner in which the ϵ–k dependence deviates from quadratic could be described equally well in other ways. The "effective mass" is regarded merely as a convenient vehicle for describing many of the complications in a solid.

As an example of these complications, we should remark that in a solid with a highly anisotropic crystal structure, it is quite likely that the effective mass tensor will have positive components for some directions and negative components for other directions.

For any periodic potential, there will be some locations in the Brillouin zone for which the Schrödinger equation gives the lowest possible value for electron energy. This may be a single location (at the center of the zone), several locations (equivalent to each other by symmetry) on the surface of the zone, or a number of locations (again equivalent by symmetry) at some interior points of the zone. Figure 3-50 is an example of the latter situation, whereby the minimum energy occurs at six equivalent points at intermediate locations along the (100) axes of the

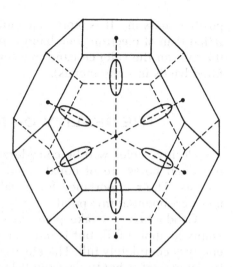

Figure 3–50 An example of a band for which the minimum electron energy occurs neither at the center of the zone nor on the outer surface of the B.Z. Instead, the minimum energy occurs for a wave-vector of (a00) and the five other locations equivalent to this by symmetry. For a slightly larger energy, the condition of constant energy is satisfied over the surface of six ellipsoidal surfaces. The lowest portion of the conduction band in the semiconducting element silicon is of this form, as is the second lowest conduction band in gallium arsenide.

B.Z., and the surface of constant energy for a slightly larger energy must be six prolate ellipsoids centered on these locations.

At any rate, let us say that the electron energy is a minimum when $\mathbf{k} = \mathbf{k}_1$ (and any appropriate number of locations equivalent to $\mathbf{k}_1$). Then the three principal components of m_{ij} are *usually* all positive in the vicinity of $\mathbf{k}_1$ (and in any equivalent regions of $\mathbf{k}$-space). Moreover, the effective mass is *usually* fairly energy-independent for energies within a fraction of an electron volt of ϵ_1:

$$\epsilon \approx \epsilon_1 + A_1 (\mathbf{k} - \mathbf{k}_1)^2 \qquad (3\text{-}150)$$

where, however, A_1 will *usually* depend on the direction of $(\mathbf{k} - \mathbf{k}_1)$. Thus for electrons at the bottom of a permitted energy band we can say that

$$\epsilon = \epsilon_1 + \frac{\hbar^2}{2m_1} (\mathbf{k} - \mathbf{k}_1)^2 \qquad (3\text{-}151)$$

where m_1 is the effective mass *for this direction* in $\mathbf{k}$-space away from the band minimum.

In the parts of the B.Z. for which the required electron energy is highest, all three components of $(\partial^2\epsilon/\partial k_j{}^2)$ are likely to be *negative*, so that we should write

$$\epsilon = \epsilon_2 - \frac{\hbar^2}{2m_2} (\mathbf{k} - \mathbf{k}_2)^2 \qquad (3\text{-}152)$$

An electron in such a group of energy states responds to a field as though it had a negative mass. Thus the location in real space of an *unoccupied* state near to the top of a band is likely to move through the crystal with the behavior of a particle commonly known as a *positive hole*. A hole has positive charge (compared with the charge of occupied states) and positive mass m_2. It is more convenient to discuss the motion of holes rather than of electrons in a band which is filled almost to the top (as is the case for the valence band in a semiconductor, and for incompletely filled bands in some metals).

THE REALITY OF POSITIVE HOLES

Positive holes will be considered further in Chapter 4, since electrons and holes are of equal significance in conduction by semiconductors. Their importance for metals and semimetals is sufficient to merit comment at this point.

Conduction can take place in a band which ranges from almost empty to almost full, but is impossible for a band which is completely empty or completely full. The electrons in a completely full band can interchange states, but this accomplishes nothing since electrons are indistinguishable. Conduction in an almost empty band is performed by electrons, of positive mass and negative charge. Conductive effects (such as the Hall effect) which depend on the first power of the charge for a mobile carrier accordingly have a negative sign.

For a band which is almost full, it is said that conduction occurs by movement of the small number of unoccupied states—the positive holes, each with a positive effective mass and a positive charge. Shockley[55] has looked in detail at the wave-function for a band which is full except for a single state, and shows that electric and magnetic fields cause a time-dependence of momentum

$$(dp/dt) = +e[\mathbf{E} + (\mathbf{v} \times \mathbf{B})] \qquad (3\text{-}153)$$

(compare with the sign in Equation 3-35) which is appropriate for conduction by a hole.

The concept of hole conduction explains the positive Hall coefficients reported in Table 3-2 for cadmium and iron. The bands in cadmium are similar to those illustrated in Figure 3-40 for magnesium, with a small degree of overlap between a band derived from 5s states (almost filled) and the almost empty 5p band. The Hall coefficient is encouraged to be positive by holes in the 5s band, and to be negative by electrons in the 5p band—and this particular struggle is won by the holes. For iron, the positive Hall coefficient is a consequence of 80% occupancy of the 3d band.

If the reader still has qualms about the validity of the *absence* of a particle as a mobile charge carrier, some insight may be drawn from the following very crude analogy. Consider four glass tubes, each sealed at both ends. Let one be empty, another totally filled with water, another one almost completely filled with water, and the fourth containing only a small amount of water. If one end of each tube is raised and lowered, there will obviously be no transport in the empty tube. Nor will there

[55] W. Shockley, *Electrons and Holes in Semiconductors* (Van Nostrand, 1950), Section 15.8.

be transport in the full tube, for water molecules (like electrons) can then only interchange positions. However, water can be seen to run from one end to the other in the tube containing a little liquid, while *air bubbles* move in the opposite direction in the almost full tube. Air bubbles in a tube otherwise filled with water react in the same way as holes in an otherwise filled band. Holes rise to the highest attainable states of electronic energy, and move toward the negative terminal in an electric field.

COMPLICATIONS OF THE EFFECTIVE MASS

It is now necessary to return to the properties of the effective mass tensor over the entire energy range of a band. We must recognize that the quantities m_1 and m_2 of Equations 3-151 and 3-152 are not equal to each other (except by accident), nor is either of them in general equal to the ordinary electronic mass. The quantities m_1 and m_2 represent the ability of an electron at the bottom of a band or a hole at the top of a band to respond to an external field *combined with* the interatomic periodic potential. Depending on the form of that potential, an effective mass may be either larger than or smaller than the ordinary mass of an isolated electron.

For intermediate parts of the energy range within a band, those well above ϵ_1 but well below ϵ_2, more complex forms for the effective mass tensor are inevitable. The character of these complications is dictated by the crystal symmetry and the magnitude of the periodic potential. Whether this potential is small (as was supposed in drawing the energy contours in Figure 3-45) or large (as was supposed for Figure 3-42), energy contours for the intermediate energies must depart from a spherical shape, and must depart from the quadratic spacing of Equations 3-151 and 3-152. As an example of the shape of a contour of intermediate energy, Figure 3-51 gives a perspective view of the surface in k-space for

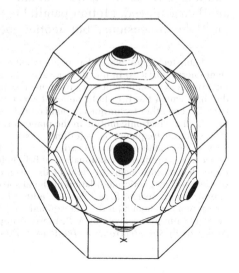

Figure 3-51 The Fermi surface for copper. The Fermi surface in this metal is formed by electrons in the half-filled 4s band. This shape was deduced by Pippard (1957) from measurements of the anisotropy of the anomalous skin effect in copper. From G. E. Smith, in *The Fermi Surface,* edited by W. A. Harrison and M. B. Webb (Wiley, 1960).

the Fermi energy in copper. This surface can be derived from a sphere by pulling outward along the (111) directions (so that contact is made with the B.Z. boundary along those directions) and pushing inward in the various (110) directions. As a consequence, the "effective mass" to be associated with a state having the Fermi energy depends to a large extent on the direction of the corresponding wave-vector.

The Fermi surface shown in Figure 3-51 has been deduced from the results of a variety of sophisticated experiments[56] using techniques developed since the mid 1950's. We shall have more to say about some of these experimental tools later in this section. Through the study of electron dynamics it is possible to observe the motion of an electron which has some positive and some negative components of effective mass, as is the case for electrons on the Fermi surface of copper near to the areas of contact with the zone boundary.

The problems experienced by an electron in these regions of k-space can be seen more clearly in the sketch of Figure 3-52. This figure uses the *repeated zone representation* whereby the situation on one side of a zone boundary is repeated in mirror image form. (Such a method of presentation is valid because states separated by a reciprocal lattice vector are the same state. Then the region in the right portion of Figure 3-52 portrays the states which could be drawn equally well one zone-width to the left.) The hyperbolic energy surface of Figure 3-52 requires that the effective mass be negative in the direction k_a perpendicular to the B.Z. boundary plane and positive in the directions orthogonal to k_a. The dynamics of an electron on this hyperbolic energy surface reflect the positive and negative mass components, which can be identified experimentally.

Before leaving Figure 3-52 it should be noted that the energy surface intersects the zone boundary at right angles. This happens because an electron at the zone boundary has no velocity component (in real space) perpendicular to the B.Z. boundary plane (in $k =$ space). It will be remembered from our one-dimensional discussions of zone boundaries that we concluded that an electron state is a *standing* rather than a *running* wave at the zone boundary, since Bragg reflection conditions are then achieved. Motion parallel to the zone boundary is permitted in multiple dimensions, but motion perpendicular to the boundary is

---

[56] One technique which has been valuable in the exploration of Fermi surfaces (constant energy surfaces) in metals is measurement of the anomalous skin effect; that is, the high frequency conductance of a pure metal sample for high frequencies such that the skin depth to which the R.F. field penetrates is much smaller than the electronic mean free path. Several other techniques depend on the quantization of electron orbits in a magnetic field, so that electronic motion in the plane perpendicular to a magnetic field is periodic at the cyclotron frequency $\omega_c = (eB/m^*)$. Associated with this quantization (which is discussed later in the section) is a variety of observable phenomena, including cyclotron resonance at an appropriate combination of R.F. field and steady magnetic field, an oscillatory dependence upon magnetic field of the magnefic susceptibility and electronic conductivities, and magnetic modulation of the absorption coefficient for ultrasonic waves. The various papers in *The Fermi Surface*, edited by W. A. Harrison and M. B. Webb (Wiley, 1960), report data obtained by these methods. A comprehensive review of useful experimental techniques is given by A. B. Pippard, Reports on Progress in Physics **23**, 176 (1960). Alternatively see *Electrons in Metals* by J. Ziman (Taylor and Francis, 1963).

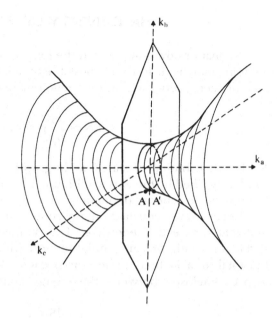

Figure 3-52 A detail of the Fermi surface for copper, shown in the *repeated zone* representation. For the energy states at the zone boundary, the effective mass is positive in the directions of k_b and k_c, but negative in the direction k_a perpendicular to the boundary plane. This type of situation is known in the literature as a "neck." In a magnetic field, an electron can be made to precess about a "neck orbit" of constant energy.

inhibited. For this reason, the states A and A' in Figure 3-52 represent identical conditions of motion.

It was remarked in connection with Figure 3-27 of the one-dimensional Kronig-Penney model that this correctly shows $(d\epsilon/dk)$ reaching zero at the top and bottom of each band, while $(d^2\epsilon/dk^2)$ remains finite at those locations. The finite second derivatives determine the values for hole and electron effective masses, while the vanishing first derivatives affirm (see Equation 3-141) that the Bragg condition of a standing wave with zero group velocity has been achieved. The same curvature of the ϵ–k characteristic was deliberately drawn in Figures 3-24, 3-28, 3-37, and 3-38 for the same reasons.

Has this policy been adhered to in drawing contours of constant energy in a multi-dimensional k-space? It has just been noted in connection with Figure 3-52 that the Bragg condition (Equation 3-133) requires that an electron at the zone boundary have zero component of motion in real space perpendicular to the B.Z. boundary plane. Thus a surface of constant energy in k-space must intersect the zone boundary plane at right angles. On looking back at previous figures, we see that Figure 3-42 (b) shows the contour lines around the zone corners reaching the boundary at right angles. The straight lines for the middle energy in the band should show a tiny curvature to normal incidence at the last moment.

Normal incidence is *not* drawn for energy contours cutting the zone boundary in Figure 3-45, and we now recognize that this must be done if the periodic potential is finite. Appropriate curving of portions of the Fermi energy contour has been done for the NFE treatment of aluminum in Figure 3-46.

THE DENSITY OF STATES

No matter how complicated the form and variety of constant energy contours may be, it is always possible to define a scalar density of states $g(\epsilon)$ for each energy within the range of a band, so that

$$2N = \int_{\epsilon_1}^{\epsilon_2} g(\epsilon)\, d\epsilon \qquad (3\text{-}154)$$

Such a distribution of states, adding up to two states per atom in the solid, is typified by the curve of Figure 3-53. This curve can be described in terms of an (energy-dependent) density-of-states effective mass.

Let us draw a constant energy contour in the reduced k-space of the first zone for electron energy ϵ. (This will have to be a set of equivalent surfaces if conditions are similar to those illustrated in Figure 3-50). Let the total area of this surface (or surfaces) be S. Then the volume of k-space enclosed between the energy contours for ϵ and $(\epsilon + d\epsilon)$ is

$$\int \left(\frac{dS}{\nabla_k \epsilon}\right) d\epsilon \qquad (3\text{-}155)$$

Since the volume of k-space per electron state is $(4\pi^3/V)$ for a crystal of volume V, we can define the density of states (for unit volume) by the equation

$$g(\epsilon)d\epsilon = \left[\int \frac{dS}{4\pi^3 \nabla_k \epsilon}\right] d\epsilon \qquad (3\text{-}156)$$

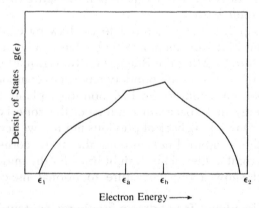

Figure 3-53 Electron Energy ——▶

Figure 3–53 A typical curve for the energy dependence of the density of states within a band. The total area under the curve is the total number of states in the band, which is just twice the number of atoms in the crystal for a simple band. For energies not far above ϵ_1, the curvature is controlled by the mass parameter m_1. The "hole" mass m_2 controls the curvature for the uppermost part of the band. For intermediate energies, $g(\epsilon)$ varies neither as $(\epsilon - \epsilon_1)^{1/2}$ nor as $(\epsilon_2 - \epsilon)^{1/2}$. Van Hove discontinuities in the slope of the density of states curve are supposed for energies ϵ_a and ϵ_b. Energy ϵ_a is the lowest energy for which the constant energy surface comes into contact with the zone boundary.

The simplest situation is the one for which

$$\epsilon = \epsilon_1 + \frac{\hbar^2}{2m_1} (k - k_1)^2 \qquad (3\text{-}157)$$

with no dependence of m_1 on the direction of $(k - k_1)$. In this happy situation, $(d\epsilon/dk) = (\hbar^2 k / m_1)$, while $\int dS = 4\pi k^2$, so that

$$g(\epsilon) = \frac{1}{2\pi^2} \left(\frac{2m_1}{\hbar^2} \right)^{3/2} (\epsilon - \epsilon_1)^{1/2} \qquad (3\text{-}158)$$

This differs from Equation 3-44 of the Sommerfeld free electron model only in the use of m_1 instead of the ordinary electronic mass.[57]

No matter how complicated the constant energy surfaces may be in k-space (multiple surfaces, warped surfaces, re-entrant surfaces, and so forth), whenever $g(\epsilon)$ is such that

$$\frac{d}{d\epsilon} [g(\epsilon)]^2 \to \text{constant} \qquad (3\text{-}159)$$

then the density of states can be said to be characterized by a *density of states effective mass* m^*. For if $g(\epsilon)$ is finite and satisfies Equation 3-159, we can force it into the form

$$g(\epsilon) = \frac{1}{2\pi^2} \left(\frac{2m^*}{\hbar^2} \right)^{3/2} (\epsilon - \text{constant})^{1/2} \qquad (3\text{-}160)$$

---

[57] When energy increases quadratically with wave-vector (that is, when the effective mass m^* is energy-independent), then the solutions of the Schrödinger equation

$$[(-\hbar^2/2m)\nabla^2 + \mathscr{V}(r)]\psi = \epsilon\psi$$

are just those of

$$(-\hbar^2/2m^*)\nabla^2\psi = \epsilon\psi$$

over the appropriate energy range. In other words, the freedom of the Sommerfeld model is recreated by a mass renormalization, which "takes care of" the effects of the periodic potential. (An analogy may be drawn with mass renormalization in quantum electrodynamics.)

Suppose the states so obtained are the ones near the bottom of a band. Then the question is often asked as to whether the same value of m^* can be useful in describing states *below* the bottom of the band, that is, bound states associated with a localized defect. The Schrödinger equation for an electron bound by a fixed positive charge is

$$[(-\hbar^2/2m^*)\nabla^2 - (e^2/4\pi\kappa\varepsilon_0 r)]\psi = \epsilon\psi$$

where κ is hopefully the macroscopic dielectric constant. This equation generates states of the form of Equation 3-128, with an energy spectrum scaled by the factor $(m^*/\kappa^2 m)$ from the hydrogenic spectrum. Such a "hydrogenic model" or "effective mass model" can be fairly successful in practice, in describing bound states of small binding energy (states which extend over a region of real space much larger than a unit cell). The actual nature of the defect responsible for bound states plays a more important role for states of large binding energy and small Bohr radius, since the dielectric constant concept is then of dubious validity. These defect state considerations are of course more important in insulators and extrinsic semiconductors than in metals.

Situations for which Equation 3-159 does not hold are said to need an "energy-dependent" density-of-states mass. Part of the complexity in understanding electronic phenomena in metals as contrasted with semi-conductors lies in the fact that the electrons studied in a semiconductor are in states close to the edge of a band (where the chances are good that Equation 3-160 can be a reasonable approximation), whereas electrons with the Fermi energy in a metal occupy those states in the intermediate energy range for which Equation 3-159 is usually a poor approximation.[58]

In order to bring together the various things which are associated with the terms "density of states" and "effective mass," we may observe that if a band covers a small total range of energy, then:

1. The energy contours must be far apart in k-space, i.e., $\Gamma_k \epsilon$ is small.

2. The density of states $g(\epsilon)$ then has a large average value over the energy range of the band.

3. This is a band for which electrons have a large effective mass m^*.

4. For these "heavy" electrons, the acceleration in an applied electric field is small.

5. The electron mobility will thus be rather small in this narrow band (mobility is likely to vary inversely as the first or second power of the effective mass, the exponent depending on the principal scattering mechanism).

6. Electrons in this band will have a large electronic specific heat ($C_v \propto m^*$).

7. Electrons in this band will also have a large Pauli paramagnetic susceptibility ($\chi_m \propto m^*$).

By the same token, a band which extends over a broad energy range connotes electron states of small effective mass, and electrons in these states will have a large mobility, and so forth.

VELOCITY IN AN ELECTRIC FIELD

If every state in an energy band is full, then the Pauli principle does not permit any response to an electric field. For a band which is half full,

---

[58] Properties of metals which depend upon conduction electrons necessarily tell us about the situation for electrons of the Fermi energy but are uninformative about states of either higher or lower energy. However, an electron can be excited from part of a band lying below the Fermi energy to a higher unoccupied state by absorption of a photon. The strength of such an optical transition depends on the nature of the wave-functions which describe the upper and lower states, and on the densities of states in the two bands. The important state density is the *joint density of states* connecting upper and lower energy states for a given energy difference (that is, photon energy). Thus the interpretation of experimental data is far from straightforward. A singularity in the optical reflectivity is expected when the photon energy scans through a value for which there is a van Hove singularity (see footnote 46) in the density of states for either band, or in the joint density of states. As an alternative to reflectivity, the optical coupling of states below ϵ_F with other states above ($\epsilon_F + \phi$) can also be studied by photoemission (discussed in Section 3.3). Photoemission and reflectivity (including variants such as piezo-reflectance and electric field-modulated reflectance) are useful diagnostic tools for the study of metals in that they can give us some (admittedly indirect) information about bands *below* and *above* the Fermi energy. M. Cardona, *Modulation Spectroscopy* (Academic Press, 1969), discusses what extra information can be obtained about electron states in a solid when a property such as reflectance is measured while a variable such as temperature or electric field is forced to be a periodic function of time.

the response to an external influence such as an electric field is controlled by the states near in energy to the Fermi energy. In order to discuss what an electron *ought* to do in an electric field, let us imagine a band which is empty except for one electron (or full except for one hole). We shall imagine that there is no scattering for this electron, so that it can progress smoothly from one state to another state of different energy when the field **E** is turned on.

Let us suppose first a one-dimensional lattice, with a periodic potential of such a form that the electron energy is a minimum (say $\epsilon = \epsilon_1$) at the center of the zone, and a maximum (say $\epsilon = \epsilon_2$) at the zone boundaries, $k_x = \pm(\pi/a)$. The ϵ–**k** curve for intermediate energies is sketched in Figure 3-54(a). At the bottom of this band,

$$\epsilon = \epsilon_1 + (\hbar^2 k_x^2/2m_1) \qquad (3\text{-}161)$$

and the corresponding electron velocity is

$$v_x = \frac{1}{\hbar}\left(\frac{d\epsilon}{dk_x}\right) = \frac{\hbar k_x}{m_1} \qquad (3\text{-}162)$$

which varies linearly with k_x. However, v_x reaches a maximum (positive or negative) for some energy near the middle of the band. For higher energies (values of k_x near to the left or right zone boundaries), the electron motion is less rapid. Since

$$\epsilon = \epsilon_2 - \frac{\hbar^2}{2m_2}\left[k_x \pm \frac{\pi}{a}\right]^2 \qquad (3\text{-}163)$$

near the top of the zone, then

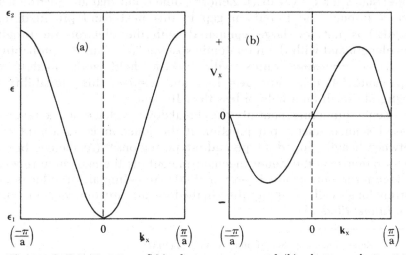

Figure 3-54 Variation of (a), electron energy, and (b), electron velocity, with reduced wave vector for states in a band of a one-dimensional crystal. The slope of v_x with respect to k_x is controlled by the effective mass m_1 near the center of the zone, and by the "hole" effective mass m_2 for the regions near the zone edges.

$$v_x = \frac{-\hbar}{m_2} \left[k_x \pm \frac{\pi}{a} \right] \tag{3-164}$$

describes the velocity behavior near the zone boundary, and an electron in a state *at* the zone edge is not moving at all, in conformity with the previous discussions of Bragg diffraction and standing waves for the zone-edge states. The complete velocity curve for this one-dimensional crystal is as indicated in Figure 3-54(b).

When there is no scattering and no applied electric field, an electron is described by *one location* in the B.Z., and it moves forever in a straight line with constant speed. With a finite applied field E_x (and still no scattering), the one-dimensional simplification of Equation 3-144 suggests that k_x should change with time in accordance with

$$(dk_x/dt) = -eE_x/\hbar \tag{3-165}$$

Suppose now that a constant E_x is applied for a long time. Equation 3-165 says that the location in the zone describing our electron will move to the left at a constant rate until $k_x = -\pi/a$ (and $v_x = 0$). What will happen with continued application of the field? It depends on how large the field is.

ZENER AND OTHER TUNNELING

If the field is very large, there is a finite possibility that the electron may make a transition across the forbidden gap to the state of $k_x = -\pi/a$ in the next higher band. Suppose that the forbidden gap between the top of the band under consideration and the next higher band is $\Delta\epsilon$. Then an electric field E_x brings the lower states for x = 0 on a level with the upper states for x = $(\Delta\epsilon/eE_x)$. Zener[59] pointed out that an electron may *tunnel* through the forbidden gap to this next band provided that $(\Delta\epsilon/eE_x)$ is not too large compared with the electron de Broglie wavelength and with the interatomic spacing "a". Thus the probability of *Zener tunneling* varies with electric field approximately as $\exp(-am\Delta\epsilon^2/4\hbar^2eE_x)$. For $\Delta\epsilon = 1$ eV and a = 1 Å, this probability is negligibly small for a field of less than 10^7 V/m.

Zener tunneling controls the breakdown voltage for a reverse-biased semiconductor p-n junction if the junction is an abrupt one between heavily doped n-type and p-type regions.[60] (Avalanche breakdown rather than the Zener mechanism controls the maximum reverse voltage for a more gradual p-n junction). Tunneling also produces the current for a small voltage applied in the forward direction for a p-n junction *tunnel diode*.[61]

---

[59] C. Zener, Proc. Roy. Soc. (Lond.) **A145**, 523 (1934).

[60] See any of the books listed under *Solid State Electronics* in the general bibliography.

[61] L. Esaki, Phys. Rev. **109**, 603 (1957).

A very large electric field to provoke Zener tunneling cannot be produced *within* a metal, but it can be produced in vacuum or an insulator at the *surface* of a metal. In Section 3.3 we discussed field emission at the surface of a metal, which depends on tunneling. Variants on this are tunneling through a very thin (100 Å or less) layer of vacuum or insulator from one metal to another, with particular emphasis on situations in which the metals on one or both sides of the insulator are in the superconducting state. In Section 3.6 we shall comment briefly on the BCS theory of superconductivity and its successful prediction of a small energy gap at the Fermi energy. The smallness of this energy gap means that the voltage required to initiate tunneling between superconductors can be as small as 100 μV.

Tunneling involving states above and below the energy gap of a superconductor in a metal-insulator-metal sandwich was first detected experimentally by Giaever.[62] Subsequently, Josephson[63] pointed out that if the insulating layer in such a superconducting sandwich is thin enough, then *supercurrent* in the form of *electron pairs* can flow across the interface. Such tunneling forms the basis for a number of applications,[64] to which we shall return at the end of Section 3.6.

BRAGG REFLECTION CAUSED BY AN ELECTRIC FIELD

Let us return to consideration of a band with one electron in it, conditions of no scattering, and with an electric field small enough to preclude Zener tunneling. What happens now when an electric field E_x is applied for a long enough time so that k_x (in accordance with Equation 3-165) reaches the zone boundary at $k_x = -\pi/a$? A model developed by Houston[65] predicts that this electron will then suffer Bragg reflection, and will thereafter instantaneously be at the state within the same band for $k_x = +\pi/a$. Continued application of the steady electric field would cause k_x for the electron to move toward the left of the zone again.

This model of Houston's predicts that application of a constant electric field will cause the position of the electron in real space to move back and forth, coming to rest whenever $k_x = 0$ and whenever $k_x = \pm(\pi/a)$. Thus a single unscattered electron in an empty band would have a *zero* contribution to the conductivity when averaged over a long time period. How long a complete oscillation takes through k-space and real space is queried in Problem 3.29, and the next problem looks at the response of an electron in such an empty band to a time-varying field.

One way of describing this curious apparent result for the zero long term conduction of an electron in an empty band is to say that the

[62] I. Giaever, Phys. Rev. Letters **5**, 147 (1960).

[63] B. D. Josephson, Phys. Letters **1**, 251 (1962); Rev. Mod. Phys. **36**, 216 (1964). The 1973 Nobel prize in physics was shared among Josephson, Esaki, and Giaever for their respective contributions to our understanding of tunneling phenomena.

[64] See, for example, L. Solymar, *Superconductive Tunneling & Applications* (Chapman and Hall, 1972).

[65] W. V. Houston. Phys. Rev. **57**, 184 (1940).

response of the electron assists conduction while its mass is positive, but works against conduction when it lies in the portion of the band for which *hole mass* is a positive quantity.

It has been argued by Rabinovitch and Zak[66] that Houston's picture of a Bloch electron oscillating in an empty band is incorrect. These authors point out that the band states are perturbed by the presence of an applied electric field (i.e., the Stark effect), and that this perturbation is comparable with interband corrections to the Bloch one-electron description of the available states. Of course, the Houston model would be hard to duplicate in practice anyway, since any real band contains more than one electron, and any electron is subject to from 10^9 to 10^{14} scattering events per second.

Despite the cloud of uncertainty raised by the last paragraph, we shall find it convenient to continue to use Equation 3-144 for the progression of an electron through k-space in a multi-dimensional solid when an electric field is applied. We shall also assume Bragg reflection when k reaches a boundary.

As noted previously, the electron velocity in a multi-dimensional crystal does not necessarily vanish when the state occupied is on the zone boundary, but the component of velocity perpendicular to the boundary vanishes. Figure 3-55 illustrates the locations in k-space successively occupied by an electron in a two-dimensional lattice as a result of an electric field E and no scattering. The field is turned on when the electron state occupied is at the location P, and the velocity of the electron in real space at that time is indicated by the arrow extending from P. The length and direction of that arrow are determined by the value of $\nabla_k \epsilon$ in accordance with Equation 3-141.

Application of the field **E** causes the wave vector of the electron to change first to Q and then to R, in accordance with Equation 3-144. If we

---

[66] A. Rabinovitch and J. Zak, Physics Letters **40A**, 189 (1972).

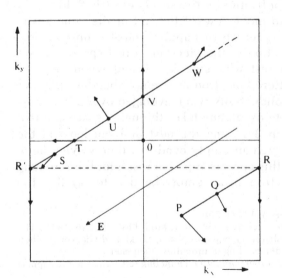

Figure 3-55 The effect of an electric field E upon the location of an electron in an empty band as a function of time (assuming no scattering). For each electron state, from P onward, the length and direction of the arrow show the velocity which might occur in real space for a plausible set of energy contours. Bragg reflection occurs when k_x reaches the zone boundary, and $v_x = 0$ for this condition. In the absence of scattering, further Bragg reflections will occur.

adopt Houston's view of Bragg reflection at the zone boundary, then reflection of the electron at R causes the location of the electron in k-space to follow the sequence indicated from R' to W, and so on. The motion in real space is dictated by the varying values of $\nabla_k \epsilon$ in the regions of k-space being traversed. Thus at times the electron is acquiring energy from the external field, and at other times it is delivering energy to that field.

The motion of an electron in a real three-dimensional solid occurring in response to an electric field departs from the picture we have just painted because of the Pauli principle and because of scattering. The Pauli principle prevents an electron from progressing to a state below the Fermi energy in response to a field, and scattering by defects and phonons in both metallic and nonmetallic solids usually interrupts the field-induced progression of k and associated properties after a small part of the entire zone-width has been traversed. The smaller the phonon density and the more perfect the crystal, the better the chance of observing electron progression over a finite fraction of the zone. Accordingly, studies of band shapes and Fermi surface shapes are usually pursued at low temperatures with carefully oriented pure single crystals.

CYCLOTRON RESONANCE

According to Maxwell's equations, a magnetic field applied to an electron tends to change the electron's direction of motion without changing its energy. This is implicit in the Lorentz force equation (Equation 3-35). Thus a magnetic induction B_z affects motion in the xy plane without affecting motion in the z-direction. If the electron is not scattered, it executes an orbit in the xy plane, superimposed on any trajectory it may have in the z-direction.

For a quasi-free electron of scalar mass m^*, the orbit executed is a circular one, of radius r and angular frequency ω_c. These quantities are related by the requirement that the centrifugal force $(m^* \omega_c^2 r)$ must balance the Lorentz force $(r \omega_c e B_z)$. Thus the angular *cyclotron frequency* is

$$\omega_c = (e B_z / m^*) \tag{3-166}$$

regardless of the electronic kinetic energy. [Energy does control the size of the orbit in real space, since $\epsilon = (m^* \omega_c^2 r^2 / 2)$.] The cyclotron frequency lies in the radio or microwave parts of the electromagnetic spectrum for normally attainable magnetic fields, since

$$\nu_c \equiv (\omega_c / 2\pi) = 28.0 \, (B_z m / m^*) \qquad \text{GHz} \tag{3-167}$$

for a magnetic induction expressed in teslas.

The motion of the electron in real space under the influence of a magnetic field is accompanied by a precession through k-space on a path *of constant energy* in the Brillouin zone. Of course, for the highly degenerate electron population in a metal, this motion is seen only for electrons at the Fermi energy, electrons which perform orbits in k-space around the Fermi surface. Since there is inevitably *some* scattering of electrons by phonons and defects even in a nearly perfect crystal at low temperatures, a well-defined cyclotron motion can be resolved only if $(\omega_c \tau_m) > 1$, so that an electron can move through a significant part of a magnetic orbit before it is scattered.

Of the electrons at the Fermi energy, most have a non-vanishing component of momentum parallel to B_z. These electrons execute a circular path in k-space which has a radius smaller than the radius of the Fermi sphere; and their path in real space is the sum of a circular motion in the xy plane and linear motion in the z-direction. However, some of the electrons at the Fermi energy have a zero component of momentum in the z-direction. These latter electrons are required by the field B_z to move in an equatorial "great circle" path around the Fermi sphere, and their motion in real space is similarly a circular one without any super-imposed linear motion. This equatorial orbit around a Fermi sphere is the simplest kind of *extremal orbit*, a class of orbit which is of great importance in *cyclotron resonance* experiments. Even when the Fermi surface is far from spherical, there are still certain extremal paths which can be identified and used in characterizing the surface topology.

It must be apparent by now that only under fortuitous circumstances will a spherical Fermi surface be found for a metal. Far more commonly, the application of a magnetic field B_z causes electrons of the Fermi energy to move around the Fermi surface in k-space along a path in which the effective mass changes constantly. Then the rate at which wave-vector changes with time is non-uniform, something we can see by noting that the magnetic force on the electron is equal to $\hbar(dk/dt)$ and is also equal to $-e(v \times B)$. In consequence, the orbit performed in real space is executed at a non-uniform rate, as examined for a simple case by Problem 3.32.

Experiments of *cyclotron resonance* are based on the absorption of radio-frequency energy at a frequency ω when a steady magnetic induction B is adjusted to make ω coincide with ω_c. Measured combinations of ω and B then allow us (in principle) to deduce information about the tensor of effective mass at the Fermi energy. The theory of cyclotron resonance is in practice rather more complicated, both for semiconductors and for metals.

For a semiconducting material in which the free electron density is small, cyclotron resonance experiments can be carried out with e.m. waves penetrating throughout the solid. The complications which arise are associated with the topology of constant-energy surfaces, and with hybrid plasma resonances[67] if the free electron density is not very small.

The frequencies used for cyclotron resonance studies in a metal are invariably much smaller than the plasma frequency (because the elec-

---

[67] Band theory changes the plasma frequency from the value of Equation 3-107 to a value characterized by the effective mass: $\omega_p = (ne^2/m^\circ \varepsilon_0)^{1/2}$. Then unless the electron density in a semiconductor is so small that $\omega_p \ll \omega_c$, the maximum coupling of an e.m. field to modes of electronic motion will occur at the upper hybrid frequency $(\omega_c{}^2 + \omega_p{}^2)^{1/2}$ rather than at ω_c itself. The interference of plasma resonance with the measurement of cyclotron resonance is discussed by G. Dresselhaus, A. F. Kip, and C. Kittel, Phys. Rev. **100**, 618 (1955).

tron density in a metal is so large, which makes ω_p in turn very large). As discussed in Section 3.3, for $\omega < \omega_p$, the real part of the dielectric constant is negative. Accordingly, a metal is opaque for such frequencies, with a penetration depth (skin depth) δ much smaller than the probable thickness of a sample. The magnitude of the electronic mean free path λ then determines whether the surface electrical characteristics for R.F. waves are controlled by the *normal skin effect* or by the *anomalous skin effect*. The former situation holds if $\lambda < \delta$, and the latter when λ is large compared with the skin depth.

Under this latter set of conditions, it is possible to stimulate cyclotron motion in a combination of constant magnetic induction (say B_z) and an R.F. electromagnetic field, using the geometric arrangement suggested by Azbel and Kaner,[68] which is illustrated in Figure 3-56. Azbel and Kaner noted that if the static induction B_z lies in a direction in the plane of the surface, then cyclotron motion must take place in a plane which intersects the surface. Some cyclotron orbits will extend into the R.F. skin depth layer, as shown in the figure. Electrons performing these orbits which reach into the surface can experience the radio frequency field at angular frequency ω, and can gain or lose energy from that field in a manner which depends on the relationship between ω and the imposed cyclotron frequency ω_c. Thus the surface impedance of the crystal with respect to the R.F. radiation is a function of the strength of the magnetic induction.

For successful observation of resonance phenomena, a pure, perfect, single crystal must be used at low temperatures, so that the electronic mean free path is large compared with the size of the cyclotron orbit.

---

[68] M. Ya. Azbel and E. A. Kaner, Soviet Physics J.E.T.P. 3, 772 (1956); also M. Ya. Azbel and E. A. Kaner, J. Phys. Chem. Solids 6, 113 (1958).

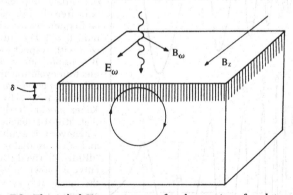

Figure 3-56 The Azbel-Kaner geometry for observation of cyclotron resonance in a metallic crystal. The shaded layer is the skin depth δ for r.f. radiation of angular frequency ω. A possible electron orbit which extends through this surface layer is shown. This could be the cyclotron motion attendant upon a static magnetic induction B_z which is applied in a direction in the plane of the surface. For Azbel-Kaner resonance, a metallic single crystal of high purity and perfection has one face [such as a (100) or (111) plane] prepared with great care, so that at low temperatures the mean free time (and hence the mean free path) can be large within both the body of the crystal and the skin depth. Energy from the r.f. field can be coupled to the circulatory motion of the electrons, provided that $\lambda > \delta$. Provided also that $(\omega_c\tau_m) \gg 1$, a sharp cyclotron resonance can be detected when $\omega = \omega_c$ or a multiple of ω_c.

The surface which is to receive the R.F. field must be of good quality so that λ is as large in the surface layer as it is in the bulk. Under these conditions λ will also be large compared with the skin depth δ, and a circulating electron will interact with the R.F. field during only a small fraction of its orbital period. Azbel and Kaner pointed out that when $\lambda > \delta$ and $(\omega_c \tau_m) \gg 1$, it is possible to secure an interaction between the R.F. field and the cyclotron motion when ω is any reasonably small multiple of ω_c, as well as when $\omega = \omega_c$. Let B_c be the magnetic induction required to make $\omega = \omega_c$. For a magnetic induction which is a submultiple of B_c, the R.F. field must go through several cycles between one trip of a given electron through the surface layer and the next. However, if B_z *is* a submultiple of B_c, then the R.F. field is once again in a position to repeat its previous message when the electron *does* pay a return trip to the surface. Azbel and Kaner find that the complex surface impedance varies with magnetic induction according to

$$Z(B) = Z_0 [1 - \exp(-2\pi/\omega\tau_m)\,\exp(-2\pi i \omega_c/\omega)]^{1/3} \qquad (3\text{-}168)$$

where the magnetic induction is represented through the value of ω_c. The oscillatory behavior of the real part of this impedance (the surface resistance) is shown in Figure 3-57. This figure also indicates the course of (dR/dB), a quantity which an experiment can be made to measure directly.

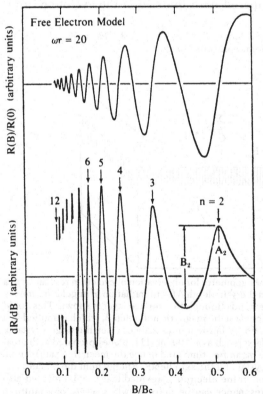

Figure 3–57 The upper curve shows the variation of the surface resistance (the real part of the surface impedance) for a metallic free electron gas at a r.f. angular frequency ω, as a function of the induction **B**. The ordinate is normalized with respect to the induction $B_c = (\omega m/e)$ for which ω and the electron cyclotron frequency coincide. This curve can be calculated from Equation 3-168 of the Azbel-Kaner model. It should be noted that (dR/dB) reaches a maximum whenever B is a sub-multiple of B_c, and the oscillatory behavior of (dR/dB) derived from the upper curve is shown in the lower half of the figure. It proves convenient to arrange a cyclotron resonance experiment in such a way that (dR/dB) is measured and plotted directly as a function of magnetic induction for a fixed r.f. frequency. From P. Häussler and S. J. Welles, Phys. Rev. **152**, 675 (1966).

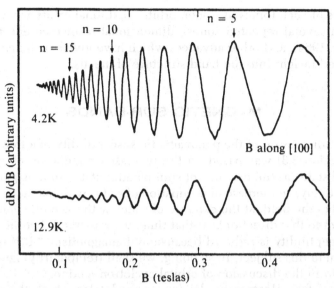

Figure 3-58 Experimental Azbel-Kaner resonance for a pure copper crystal at two temperatures. The curve for the higher temperature is impaired by increased thermal scattering of the circulating electrons. The crystal surface is a (110) plane, with the magnetic field along a [100] direction lying in this surface. Resonance occurs for electrons in the extremal "belly orbit," which encompasses the main body of the Fermi surface (see Figure 3-51). From P. Häussler and S. J. Welles, Phys. Rev. **152**, 675 (1966).

Figure 3-58 displays an example of Azbel-Kaner experimental data for a very pure copper sample at low temperatures. The contrast between the two curves shows how important it is that electron scattering be minimized. For the curve taken at 4.2K, a direct comparison with the provisions of the Azbel-Kaner theory is possible and permits a determination of the size of the orbit for Fermi electrons in copper. For the orientation of fields when the data of Figure 3-58 were taken, the important electron orbit was the "belly orbit," in which an electron progresses in an almost circular path in k-space around the main girth of the Fermi surface illustrated in Figure 3-51. The belly orbit is an *extremal* orbit which *maximizes* the cyclotron period, just as the "neck orbit" around the neck shown at the zone boundary in Figures 3-51 and 3-52 is *extremal* in the sense of *minimizing* the cyclotron period with respect to neighboring orbits.

The importance of extremal orbits lies in the fact that electrons precessing in orbits on a non-spherical Fermi surface have a variety of periods for a given magnetic field. Yet the contributions of electrons in non-extremal orbits tend to cancel each other because of the variety of phases. The dominant contribution comes from an extremal region in which the first derivative of the period with respect to the component of k along the magnetic axis disappears. Such a region accounts for an appreciable signal all in phase.

The subject of electron orbits in a metal when the strength of magnetic induction is appreciable has flourished vigorously since the mid 1950's, and there is much more to be said. The interested reader is referred to the bibliography at the end of this chapter for extensive discussions of such topics as "open orbits" (extremal orbits which extend through several repeated zones), dimensional resonance effects (Gantmakher, 1964), and other advances which have given such dramatically improved insight into the band structure of metals.

MAGNETIC SUB-BANDS

Our discussion of the paramagnetic susceptibility of a free electron gas (Section 3.3) was based on the increased population of spin anti-parallel states at the expense of spin-parallel states in a magnetic field (that is to say, an increase of occupation of states with magnetic moment parallel to the field, at the expense of states with magnetic moment antiparallel to that direction). At that time, we remarked that the electron gas susceptibility is reduced because of diamagnetism.[69] It is now time to return to this subject. In so doing, we shall (temporarily) ignore electron spin in the discussion of orbital cyclotron motion.

For a free electron model (i.e., a model for which the potential energy is constant through space), the Schrödinger equation

$$(-\hbar^2/2m)\nabla^2\psi = \epsilon\psi \qquad (3\text{-}169)$$

is appropriate in the absence of a magnetic field. We know that the solutions of this equation are plane waves, with wave-vectors and eigenenergies related by

$$\epsilon = (\hbar^2 k^2/2m) \qquad (3\text{-}170)$$

However, we know that the application of a magnetic induction B_z will affect the motion of each electron in the xy plane, while leaving motion along the z-direction undisturbed. The Lorentz force converts the electron motion into the sum of linear motion along the z-direction and cyclotron motion in the xy plane. Thus an electron follows a helical path.

Landau described this by showing that the Hamiltonian operator of the Schrödinger equation has two additional terms when a finite field is present. With magnetic induction B_z, Landau generalized Equation 3-169 to

$$\left\{(-\hbar^2/2m)\nabla^2 + (m\omega_c^2/8)(x^2 + y^2) - (i\hbar\omega_c)\left[x\left(\frac{\partial}{\partial y}\right) + y\left(\frac{\partial}{\partial x}\right)\right]\right\}\psi = \epsilon\psi \qquad (3\text{-}171)$$

where the intensity of magnetic induction is expressed in terms of the free electron cyclotron frequency $\omega_c = (eB_z/m)$. Equation 3-171 imposes

---

[69] The diamagnetic organization of electron states was first described by L. Landau [Z. Physik **64**, 629 (1930)]. Landau assumed a "free electron gas" with no periodic potential. R. E. Peierls [Z. Physik **80**, 763 (1933)] showed that electrons in a band derived from a tight binding model have similar properties in a large magnetic field. J. M. Luttinger [Phys. Rev. **84**, 814 (1951)] then pointed out that Landau's result can be used for any band in which energy varies as the square of wave-vector, just by substituting an effective mass m° for the ordinary electronic mass.

a constraint on motion in the xy plane, which can be described in the form

$$x = x_0 + [(2p + 1)\hbar/m\omega_c]^{1/2} \cos(\omega_c t)$$
$$y = y_0 + [(2p + 1)\hbar/m\omega_c]^{1/2} \sin(\omega_c t) \qquad (3\text{-}172)$$

where p can be any integer (including zero). This set of possible modes of motion in the xy plane is the set of states of a two-dimensional quantum harmonic oscillator.

For a given value of the quantum number p, the cyclotron motion has a radius (in real space) of

$$r_p = [(2p + 1)\hbar/m\omega_c]^{1/2} \qquad (3\text{-}173)$$

Identifying k_x with $(m/\hbar)(dx/dt)$ and k_y with $(m/\hbar)(dy/dt)$, we can see that the electronic precession through k-space is required to satisfy

$$(k_x{}^2 + k_y{}^2) = [(2p + 1)m\omega_c/\hbar] \qquad (3\text{-}174)$$

so that all states corresponding with a given value of the quantum number p fall on a cylindrical surface in k-space of radius

$$k_p = [(2p + 1)m\omega_c/\hbar]^{1/2} \qquad (3\text{-}175)$$

This situation, which is illustrated in Figure 3-59, is startlingly different from the uniform distribution of electron states in k-space which we have employed up to the present. The assembly of electron states which corresponds with a given value for the quantum number p is known as a *magnetic sub-band*, or as a *Landau level*.

The eigenenergies for Landau's modification of the Schrödinger equation (Equation 3-171) are:

$$\epsilon = (\hbar^2 k_z{}^2/2m) + (p + \tfrac{1}{2})\hbar\omega_c \qquad (3\text{-}176)$$

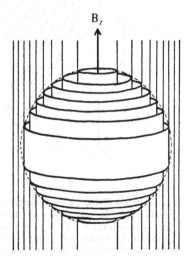

Figure 3–59 The effect of magnetic induction B_z on the regions of k-space which can be occupied by electrons, according to Landau's model. The states allowed by Equation 3-174 constitute a series of infinite cylinders, all coaxial and parallel to B_z. The limit of occupancy set by the Fermi energy resembles the field-free Fermi sphere in coarse detail. The sharpness of these cylinders will be blurred in practice by electron scattering. [After R. G. Chambers, Canad. J. Phys. **34**, 1395 (1956).]

Of this total energy, the first term on the right is the free particle eigenenergy in the direction of the applied field (which is independent of the magnitude of that field), while the second is the energy of the cyclotron motion depicted by Equations 3-172 through 3-175 for the xy plane.

In a moment we must discuss the effect of the magnetic field upon the density of states function $g(\epsilon)$. However, whatever this function may become, we know that a Fermi energy can be defined by the condition

$$n = \int_0^\infty g(\epsilon) \left[1 + \exp\left(\frac{\epsilon - \epsilon_F}{k_0 T}\right) \right]^{-1} d\epsilon \qquad (3\text{-}177)$$

We also know that states below ϵ_F will be full and states above ϵ_F will be empty. Figure 3-60 shows how the energy of Equation 3-176 varies with k_z for the first few sub-bands, when the magnetic field strength is so large that all other Landau levels lie entirely above the Fermi energy. Remember that for each Landau level the permitted values of k_x and k_y are only those lying on a cylindrical surface (Equation 3-174) coaxial with the k_z axis; it will then be evident that the ϵ–k_z curves of Figure 3-60 require the occupied regions of k-space to be as those sketched in Figure 3-59.

Magnetic reorganization of the density of states can be ignored in a weak magnetic field (or with a small electron mobility), since the emergence of distinguishable Landau levels will not be apparent unless $(\omega_c \tau_m) \gg 1$. When B_z is large enough to produce a magnetic redistribution of $g(\epsilon)$, it should not be assumed that a band has states for fewer electrons in the magnetic environment than it has in the absence of that

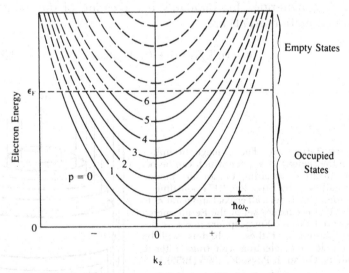

Figure 3–60 Electron energy as a function of k_z for the first seven magnetic sub-bands when the magnetic induction is B_z. It can be seen from Equation 3-176 that the minimum energy for each sub-band is $(p + \frac{1}{2})\hbar\omega_c = (p + \frac{1}{2})(e\hbar B_z/m)$. A Fermi energy is supposed for which sub-bands beyond the seventh are completely unoccupied. The distribution of occupied states in k-space is then as shown in Figure 3-59.

field. Each Landau level accommodates as many states as does an area $(2\pi m\omega_c/\hbar)$ in the $k_x k_y$ plane of the field-free metal. What a magnetic field does is collect a bundle of $(m\omega_c/\hbar)$ states (per unit volume) which previously had energies in the xy plane within the range $p\hbar\omega_c$ to $(p + 1)\hbar\omega_c$, and impress upon every state in the bundle the energy $(p + \frac{1}{2})\hbar\omega_c$. This collection process is shown schematically in Figure 3-61. The stronger the magnetic induction, the more states there are per bundle and the fewer bundles (i.e., Landau levels) there are within a specified range of $(\epsilon - \hbar^2 k_z^2/2m)$.

Viewed over a sufficiently large range of energy or k-space, the average density of states is not grossly affected by a magnetic field; but the detailed distribution is certainly changed. It is this detailed redistribution which explains the dimagnetism of the electron gas and the various phenomena that show an oscillatory dependence on B_z. The density of states as a function of energy for the energy spectrum of Equation 3-176 is

$$g(\epsilon) = \frac{1}{4\pi^2}\left(\frac{2m}{\hbar^2}\right)^{3/2} \sum_{p=0}^{p_{max}} \frac{\hbar\omega_c}{[\epsilon - (p + \frac{1}{2})\hbar\omega_c]^{1/2}} \qquad (3\text{-}178)$$

whereby each Landau level contributes one term in the summation. The value of p_{max} is set by the requirement that the denominator be real; the higher the energy under consideration, the more Landau levels commence below that energy.

From Equations 3-177 and 3-178, the total electron density in the presence of a magnetic field of any magnitude, large or small, can be expressed in the rather unpalatable form

$$n = \frac{1}{4\pi^2}\left(\frac{2m}{\hbar^2}\right)^{3/2} \sum_{p=0}^{p_{max}} \int_0^\infty \frac{\hbar\omega_c \cdot d\epsilon}{\left[1 + \exp\left(\frac{\epsilon - \epsilon_F}{k_0 T}\right)\right]\left[\epsilon - (p + \frac{1}{2})\hbar\omega_c\right]^{1/2}} \qquad (3\text{-}179)$$

This can be written as

$$n = \left[\frac{m^3 k_0 T \omega_c^2}{2\pi^4 \hbar^4}\right]^{1/2} \sum_{p=0}^{p_{max}} F_{-1/2}(x_p/k_0 T) \qquad (3\text{-}180)$$

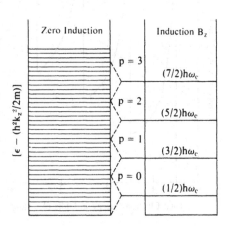

Figure 3-61 An illustration of the magnetic reordering of the distribution of electron states in energy. The ordinate is the energy associated with motion in the plane perpendicular to the field. The magnetic induction B_z causes bundles of $(m\omega_c/\hbar)$ states (per unit volume of crystal) to coalesce into highly degenerate states (subbands, Landau levels) at energies $(p + \frac{1}{2})\hbar\omega_c$. Each of these corresponds with one of the cylinders in k-space shown in Figure 3-59; the space in k-space between the successive cylinders is no longer permitted.

where x_p denotes the quantity

$$x_p = [\epsilon_F - (p + \tfrac{1}{2})\hbar\omega_c] \tag{3-181}$$

and $F_{-1/2}(x_p/k_0 T)$ is a member of the family of Fermi-Dirac integrals introduced earlier in the chapter.[12,13]

When the magnetic induction is small compared with $(\epsilon m/e\hbar)$, the summation in Equation 3-178 can be replaced by an integral. (This is suggested in Problem 3.33.) The density of states and the electron density–Fermi energy relationships then assume the conventional quantized-free-electron forms of Equation 3-44 and Equation 3-46. Of course, if the magnetic induction is rather small, $(\omega_c \tau_m)$ will probably be less than unity, and thermal broadening will further assist in de-emphasizing any magnetic modification of the energy levels and their occupancy.

On the other hand, when B_z is large, and Equations 3-178 through 3-180 must be treated as summations over a limited number of terms, we must observe that $g(\epsilon)$ is infinite at each of the energies $(p + \tfrac{1}{2})\hbar\omega_c$, which is the minimum *total* energy in a new Landau level. The discontinuous nature of $g(\epsilon)$ is demonstrated in Figure 3-62. The properties of the electron gas then depend on the Fermi energy ϵ_F which satisfies the normalization condition Equation 3-179; on how this Fermi energy compares with the minimum energies for nearby Landau levels, and on the relationship of ϵ_F and $\hbar\omega_c$ to the thermal distribution width $k_0 T$.

In order to appreciate the *diamagnetism* of an electron gas, note that a Landau level combines states with energy of motion in the xy plane ranging from $p\hbar\omega_c$ up to $(p + 1)\hbar\omega_c$ into a degenerate set at the energy $(p + \tfrac{1}{2})\hbar\omega_c$. The average energy per electron is unchanged by this reorganization, provided that all these states are occupied. But for a Landau level whose quantum number is not much smaller than $(\epsilon_F/\hbar\omega_c)$, the magnetic reorganization pictured in Figure 3-61 combines

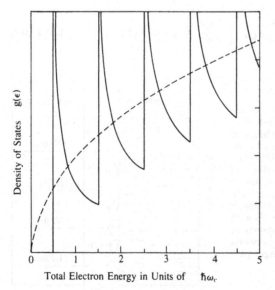

Figure 3-62 The density of states prescribed by Equation 3-178 as a function of energy, in the presence of a large magnetic induction. Note that $g(\epsilon)$ is zero below an energy $(\hbar\omega_c/2)$, and that there is a discontinuous increase in $g(\epsilon)$ at that energy and for every subsequent odd multiple of $(\hbar\omega_c/2)$. The infinity in $g(\epsilon)$ prescribed for the lowest energy of each sub-band is in practice made finite by thermal scattering. The dashed curve shows the conventional density of quantized states (Equation 3-44) in the absence of any magnetic induction.

Density of States $g(\epsilon)$

0   1   2   3   4   5

Total Electron Energy in Units of $\hbar\omega_c$

heavily occupied states from the lower part of the range with less heavily occupied states from higher parts of the range. The result is that the
average energy per *occupied* electron state is increased in the magnetic
field for the uppermost sub-bands. The total electron energy is

$$n<\epsilon> = \frac{1}{4\pi^2}\left(\frac{2m}{\hbar^2}\right)^{3/2} \sum_{p=0}^{p_{max}} \int_0^\infty \frac{\epsilon\hbar\omega_c \cdot d\epsilon}{\left[1 + \exp\left(\frac{\epsilon - \epsilon_F}{k_0T}\right)\right]\left[\epsilon - (p + \frac{1}{2})\hbar\omega_c\right]^{1/2}}$$

(3-182)

where n is as given by Equation 3-179. The increase in average energy
$<\epsilon>$ compared with the value $(3\epsilon_F/5)$, which we deduced in Equation
3-55 for a zero field situation, can be described as a diamagnetic susceptibility; as noted in Section 3.3, Landau deduced that this diamagnetic
susceptibility

$$\chi_{dia} = -(n\mu_0\hbar^2\omega_c^2/8\epsilon_F B_z^2) = -(n\mu_0\mu_B^2/2\epsilon_F) \qquad (3-183)$$

cancels out one-third of the paramagnetic susceptibility produced by
electron spin.

Equation 3-183 cites a magnetic susceptibility with no dependence
on B_z. This follows from Equation 3-179 and 3-182, provided that B_z is
not too large (nor the temperature too small). This remark is made since
the field-invariant Landau susceptibility of Equation 3-183 is calculated
upon the assumption that the width k_0T of the Boltzmann tail to the
Fermi distribution extends over several sub-bands. This is the case if
$(\hbar\omega_c) \equiv (2\mu_B B_z)$ is small compared with k_0T. For a very large field (or a
very small temperature), the Fermi distribution cuts off abruptly within
a single sub-band range, and the average electron energy then depends
on the exact relationship of the Fermi energy to the energies of the next
lower and next upper Landau levels. A change of magnetic field produces oscillatory changes in the magnetic susceptibility, a phenomenon
first reported experimentally by de Haas and van Alphen (1931).

The origins of the de Haas-van Alphen effect (dHvA effect) can readily be appreciated from the discontinuous density of states shown in Figure 3-62. As the strength B_z of magnetic induction increases, the set of
Landau levels depicted as cylinders in the k-space of Figure 3-59 ex-
pands outward. The energies for these sub-bands move upward in Figures 3-60 through 3-62. At a series of critical B_z values, these threshold
energies for sub-bands coincide with the Fermi energy (as dictated by
Equation 3-179 or 3-180. Thus the Fermi energy oscillates as a function
of magnetic field in response to the discontinuities in the density of
states. Figure 3-63 shows the oscillation of ϵ_F as a function of B_z for a supposed temperature of $T = (\epsilon_F/20k_0)$. In the right hand portion of this figure, only the lowest Landau level is permitted any occupancy.

At this point we should remind ourselves once more that the Landau
model in its original form (and as outlined in the last few pages) is one
for a quantum gas of *free* electrons. Should the tensor of effective mass
be consistent with a reasonably spherical Fermi surface for a non-mag-

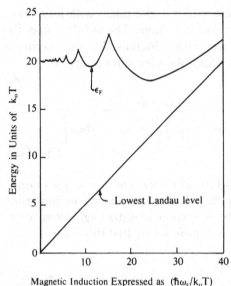

Figure 3-63 is along the right side:

Figure 3-63 A calculated example of the variation of Fermi energy with magnetic induction for a degenerate electron gas. A temperature $T = (\epsilon_F/20k_0)$ is assumed for low-field conditions. Both ordinate and abscissa are expressed in dimensionless units. It should be noted that the cusp-like maxima of the Fermi energy will be slightly rounded in practice by electron scattering. After J. S. Blakemore, *Semiconductor Statistics* (Pergamon, 1962).

netized metal, Landau's model can be used intact by the substitution of an effective mass m^* wherever the free electron mass appears. For a metal in which the Fermi surface is decidedly non-spherical, things are more complicated, but the concept of magnetic sub-bands is still extremely helpful in understanding what goes on.

For each of the critical values of **B** in Figure 3-63, the area of the outermost cyclotron orbit in k-space (i.e., the area of the outermost cylinder in Figure 3-59) is the area of an extremal orbit of Fermi energy. This is the required condition that the Fermi energy should just coincide with the lowest possible energy for one of the Landau levels. Such conditions do not occur at fixed intervals of **B**, as is very obvious from Figure 3-63. Instead, the critical condition is periodic in (1/B). The dHvA oscillatory component of the magnetic susceptibility is controlled by the oscillatory component in the average energy of Equation 3-182, and this has maxima at same linear sequence of values for (1/B). The condition for this sequence can be established as follows.

In the presence of magnetic induction B_z, the area S of an orbit in the $k_x k_y$ plane of k-space must have one of the set of values

$$S = (2\pi m^* \omega_c/\hbar)(p + \tfrac{1}{2})$$
$$= (2\pi e B_z/\hbar)(p + \tfrac{1}{2}) \tag{3-184}$$

This area is π times the radius squared given in Equation 3-174. Now the largest possible value for p (and hence for S) must be that corresponding with the Fermi energy and with zero component of motion in the z-direction. From Equations 3-40 and 3-49, the radius of the Fermi sphere in k-space is $k_F = (3\pi^2 n)^{1/3}$. A circle of this radius has the extremal area

$$S_F = \pi(3\pi^2 n)^{2/3} \tag{3-185}$$

Thus the largest orbit allowed by Equation 3-184 will correspond exactly with the Fermi energy for a highly degenerate electron gas whenever

$$(1/B_z) = (2\pi e/\hbar S_F)(p + \tfrac{1}{2}) \tag{3-186}$$

This condition is repeated at intervals in the reciprocal magnetic induction of

$$P \equiv \Delta(1/B_z) = (2\pi e/\hbar S_F) = (9.55 \times 10^{15}/S_F) \qquad \text{tesla}^{-1} \tag{3-187}$$

where the Fermi cross-section S_F is expressed in m^{-2}. Thus by measuring the period P (in tesla^{-1}) between successive dHvA maxima, the Fermi cross-section can be deduced, and the total electron density found from Equation 3-185.

A simple example of dHvA experimental data is given in Figure 3-64. These data for the alkali metal potassium can be analyzed in terms of an isotropic free electron model, as is suggested in Problem 3.34. For a solid in which the effective mass is anisotropic, the orbits in k-space will not be circular, but the *area* of the orbit is still given by Equation 3-184, and the procedure of Equation 3-187 can still be used to determine S_F. Measurements of the dHvA period for a variety of orientations can then be used to determine the size and shape of the Fermi surface. Thus observe the dHvA trace for metallic silver shown in Figure 3-65. Silver has a Fermi surface similar to copper, as illustrated in Figures 3-51 and 3-52. The smallest Fermi energy orbit which can be performed is one around a "neck" of the Fermi surface where this reaches the B.Z. zone boundary in one of the regions equivalent to $[\tfrac{1}{2}\tfrac{1}{2}\tfrac{1}{2}]$, and the short period oscillations of the magnetic susceptibility shown in Figure 3-65 belong to this "neck" orbit. Superimposed is a slower modulation of the susceptance, which has been ascribed to the beating of two different kinds of orbits, around the main "belly" of the Fermi quasi-sphere. For all but the simplest metals, the interpretation of dHvA data is highly complicated, and the several successes which have been scored rank among the more significant achievements in solid state experimentation during the 1950's and 1960's.

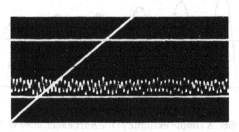

Figure 3–64 An experimental trace of the oscillatory component of magnetic susceptibility (the dHvA effect) in potassium at a temperature of 1.07K. The diagonal line is a superimposed trace of the magnetic induction, which varies from 12.69 T to 13.28 T across the width of the figure. The spacing of successive maxima can be used in determining the area of the Fermi surface in k-space, as suggested in Problem 3.34. This figure is taken from A. C. Thorsen and T. G. Berlincourt, Phys. Rev. Letters, 6, 617 (1961).

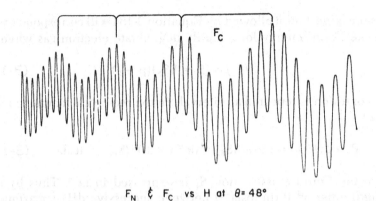

$$F_N \quad \& \quad F_C \quad \text{vs} \quad H \quad \text{at} \quad \theta = 48°$$

Figure 3–65 An example of a more complicated dHvA trace, one of those obtained for silver by A. S. Joseph and A. C. Thorsen, Phys. Rev. **138**, A1159 (1965). The magnetic field is at an angle $\theta = 48°$ away from the cubic [001] direction, in the (110) plane; that is, some 7° away from the [111] direction. Silver has a Fermi surface similar to that of copper (see Figures 3-51 and 3-52), with "necks" touching the zone boundary in the [111] directions of k-space.

In addition to an oscillatory component of susceptibility, a variety of other phenomena[70] in a degenerate electron gas can show a periodic dependence on (1/B) when $(\omega_c \tau_m) \gg 1$. One of these phenomena is illustrated in a very simple form in Figure 3-66. Displayed here is an ex-

---

[70] The various oscillatory phenomena have been described by A. H. Kahn and H. P. R. Frederikse, in *Solid State Physics, Vol. 9*, edited by F. Seitz and D. Turnbull (Academic Press, 1959).

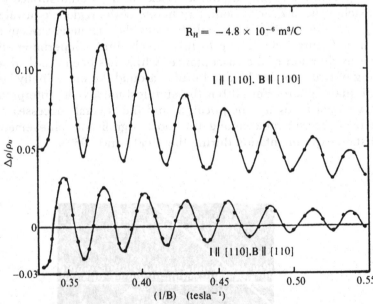

Figure 3–66 A simple example of the SdH effect, after W. M. Becker and H. Y. Fan, *Proceedings 7th International Semiconductor Conference* (Dunod, Paris, 1964). The magnetoresistance ratio ($\Delta\rho/\rho_0$) is plotted as a function of reciprocal magnetic induction for a sample of the n-type semiconductor GaSb at 4.2K. Current and field are perpendicular for the top trace, parallel for the lower trace. The quoted value for the Hall coefficient, $R_H = -4.8 \times 10^{-6}$ m³/C, enables one to deduce the total electron density (Problem 3.35), and to compare this with expectations from the SdH period.

ample of the Shubnikov-de Haas (SdH) effect, which was first reported in 1930 for the semimetal bismuth by L. Shubnikov and W. J. de Haas. The SdH effect is an oscillatory dependence of electrical resistance on magnetic induction (superimposed on a monotonic magnetoresistance which may be much larger than the oscillatory term for some solids and some orientations). As with the dHvA effect, the oscillations of the SdH effect are caused by the movement of Landau levels past the Fermi energy.

For many metals and semiconductors with non-simple Fermi surfaces, interpretation of SdH data is as complex as that of dHvA or other investigatory methods. The data chosen for Figure 3-66 are particularly simple, since they are for an electron population in a band (actually in a semiconductor rather than in a metal) for which the effective mass tensor reduces to a simple scalar. As evidence of the isotropy of the effective mass, it can be noted that the period P is the same for the upper curve (transverse magnetoresistance) as it is for the lower curve (longitudinal magnetoresistance). The calculations suggested in Problem 3.35 verify that the SdH period is consistent with the total electron density inferred from the magnitude of the Hall coefficient in the crystal of Figure 3-66.

In Chapter 4 we shall consider an oscillatory component of magnetoresistance that arises in the *non-degenerate* electron gas of a weakly doped semiconductor. This component is called *magnetophonon resonance*, and arises from the interaction of a Landau level structure, slow electrons in the first Landau level, and optical phonons of energy suitable for boosting an electron into a higher Landau level. Magnetophonon resonance requires a high electron mobility, and thus usually has a very poor chance of competing with other processes in the degenerate electron gas of a metal or a strongly doped semiconductor.

3.6 Superconductivity

THE SUPERCONDUCTING STATE

Low temperature physics as we understand it today may be said to have started with the successful liquefaction of helium at Leiden by H. Kamerlingh Onnes (1908). Onnes promptly used liquid helium in a variety of studies of metals at low temperatures, and observed in 1911 that the electrical resistivity of pure mercury dropped abruptly upon cooling to below 4.2K. To Onnes' surprise, the addition of impurities to the mercury failed to provide any residual resistance at the lowest temperatures, and by 1913 he concluded that "mercury has passed into a new state which on account of its remarkable electrical properties may be called the *superconductive* state." Figure 3-67 recreates the behavior of the electrical resistance in the neighborhood of the *superconducting transition temperature* (or *critical* temperature) T_c, as reported by Onnes.

As a preliminary to the more extensive account of superconductive phenomena in this section, we may remark that a superconductor has an infinite D.C. conductivity but displays normal resistance above a certain frequency. An anomaly in the electronic specific heat (but the absence of a heat of transformation) upon entering the superconductive state shows that this is a second-order transition (in the thermodynamic sense) of the electrons to a more ordered condition. A superconductor is perfectly diamagnetic, and totally excludes magnetic flux; the application of a magnetic field strong enough to penetrate the metal destroys the infinite conductivity. All of these phenomena caused problems of a theoretical nature for nearly half a century, even though some properties could be

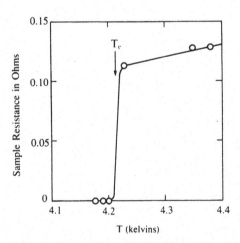

Figure 3–67 The transition from normal to superconducting behavior in mercury, as reported by H. K. Onnes, Leiden Comm., **124c** (1911). The resistance of less than 10^{-5} ohms that Onnes found below the transition temperature T_c corresponded with an electrical conductivity of at least 10^{15} ohm^{-1} m^{-1}. Modern thermometric scales place T_c for mercury at 4.15K.

systematized in thermodynamic and electrodynamic theories. The nature of the electronic reorganization was explained in a microscopic theory developed in 1957 by Bardeen, Cooper, and Schrieffer.[71] We shall follow the usual practice of referring to this as the BCS theory of superconductivity.

The BCS theory showed that interactions between electrons through the virtual exchange of phonons could lead to a *total* energy (kinetic plus potential energy) which was smaller than that of independent electrons in a Fermi distribution, provided that at zero current density electrons occupied pairs of states with opposing momentum and spin. A finite current, once initiated, was supported by the entire electron distribution displaced uniformly from $\mathbf{k} = 0$, and a finite quantity of energy (an energy gap) had to be provided as a single highly unlikely quantity in order to produce any subsequent change in the electron distribution. Thus the BCS theory explained why a current in a superconductor "persists," in addition to rationalizing the other properties of a superconductor and accounting for experiments which suggested the existence of an energy gap in the electron spectrum.

THE OCCURRENCE
OF SUPERCONDUCTIVITY

Superconductivity is a common phenomenon.[72] At the time of writing, superconducting phases have been demonstrated for 25 of the chemical elements. Transition temperatures for a few of these elements are given in Table 3-4. Many alloys and intermetallic compounds have also been shown to be superconductors. A few examples of these are also given in Table 3-4. A binary compound may be a superconductor even though one (or both!) of its constituent elements are not superconducting. As predicted by Cohen,[73] superconductivity has been observed in semiconductor samples with large free electron densities.

It is possible that every pure element becomes superconducting below some sufficiently low critical temperature, though it may take a long time to establish or disprove this possibility. The monovalent alkali and noble metals have been investigated to less than 0.1K without showing any evidence of superconductivity, and it may well be that they are excluded by a need for superconductive electron-phonon-electron

---

[71] J. Bardeen, L. N. Cooper, and J. R. Schrieffer, Phys. Rev. **108**, 1175 (1957). This is the classic exposition of the very difficult concepts of the BCS theory. The books by Lynton, Kuper, and Tinkham cited in the bibliography take these concepts at a rather slower pace.

[72] An extensive account of experimental data on superconductivity through the mid-1950's is given by B. Serin, *Handbuch der Physik, Vol. 15* (Springer, 1956). Criteria for the occurrence of superconductivity are given by B. T. Matthias, T. H. Geballe, and V. B. Compton, Rev. Mod. Phys. **35**, 1 (1963). Alternatively, see B. T. Matthias, *Progress in Low Temperature Physics, Vol. 2*, edited by C. J. Gorter (Wiley Interscience, 1957).

[73] Theoretical prediction by M. L. Cohen, Phys. Rev. **134**, 511 (1964). Experimental reports have been made of superconductivity (with a very small value for T_c) in a number of semiconductors in which a large electron density can be created and in which there is a strong coupling between electrons and lattice vibrations.

TABLE 3-4 SUPERCONDUCTING TRANSITION
TEMPERATURES FOR SELECTED ELEMENTS,
COMPOUNDS, AND ALLOYS°

Element	T_c (K)	Compound	T_c (K)
Technetium	11.2	Nb_3Ge	23.2
Niobium	9.46	Nb_3Ga	20.3
Lead	7.18	Nb_3Sn	18.05
Tantalum	4.48	NbN	16.0
Mercury	4.15	Mo_3Ir	8.8
Indium	3.41	NiBi	4.25
Aluminum	1.19	AuBe	2.64
Cadmium	0.56	$PdSb_2$_	1.25
Titanium	0.40	TiCo	0.71
Iridium	0.14	$AuSb_2$	0.58
Tungsten	0.01	$ZrAl_2$	0.30

° This table predates the work of Bednorz and Müller (1986) which showed
$T_c \approx 30$ K for La-Ba-Cu-O ceramic and earned them the 1987 physics Nobel prize.
Transition temperatures near 100 K (Y-Ba-Cu-O, by Chu *et al.* and others in 1987)
have swept away the arguments of the paragraph below, and suggest much wider
horizons for large-scale superconductor use (cf. pp. 281–4). Much work of consoli-
dation from the exciting announcements of 1986/87 will be needed before the full
implications of the new "high T_c" materials are known.

coupling at low temperatures to be accompanied by strong electron-
phonon coupling (i.e., rather modest electrical conductivity) at higher
temperatures.

Superconductivity is also absent in the ferromagnetic metals, and it
is well known that magnetic impurities have a deleterious effect on
superconductivity in their host solid. This is easy to reconcile with the
startling magnetic behavior of a superconductor. Matthias[72] has developed empirical rules for the appearance of superconductivity, rules
which favor solids in which the number of valence electrons per atom is
at least two but less than nine. Matthias also correlates a high transition
temperature in transition elements with an odd number of valence electrons, and discusses the absence of superconductivity in any of the rare
earth elements following lanthanum. He is also able to correlate the transition temperature T_c with the atomic mass and with the spacings and
configurations of atoms. All of this is connected in a complex manner
with the BCS microscopic theory of superconductivity.

D.C. AND A.C. CONDUCTIVITY

If the D.C. conductivity of a superconductor is indeed infinite, then
a current initiated in a closed loop of superconducting material should
continue forever, without any need for a driving electromotive force. On
the other hand, should a superconducting loop of inductance L have a
finite resistance R, then a persistent current should slowly decline in
accordance with

$$I = I_0 \exp(-Rt/L) \qquad (3\text{-}188)$$

Attempts to detect a decay rate in persistent currents were begun early in the history of superconductivity, and Onnes and Tuyn (1924) concluded that the conductivity of lead was at least 10^{18} ohm^{-1}m^{-1} in the superconducting state, compared with perhaps 10^{11} ohm^{-1}m^{-1} for highly purified lead just above T_c. Numerous experiments since then (such as the one discussed in Problem 3.36) have extended the lower bound for D.C. conductivity of a superconductor. Quinn and Ittner[74] report a conductivity of at least 10^{25} ohm^{-1}m^{-1}.

It is still important to draw a distinction between a *superconductor*, as a solid with infinite D.C. conductivity, and a *perfect conductor*, as a solid in which electrons occupy Bloch states with zero scattering. Bardeen (1956) remarked that it is more fruitful to view a superconductor as an extreme case of diamagnetism rather than as a limiting case of infinite conductivity. At $T = 0$, the conductivity is infinite only up to a limiting frequency, and at a finite temperature (smaller than T_c), there is a small A.C. loss at all frequencies.

The absence of superconductivity at optical frequencies was noted very early in the history of this subject (a superconductor is visibly opaque!), and measurements in the 1930's showed that superconductivity disappeared at a frequency of more than 10^9 Hz but less than 10^{14} Hz. The more recent developments of far infrared techniques have permitted an accurate determination of the frequency dependence in the interesting region for a variety of superconductors. A typical thin-film absorption result is shown in Figure 3-68. The threshold occurs for a

[74] D. J. Quinn and W. B. Ittner, J. Appl. Phys. **33**, 748 (1962).

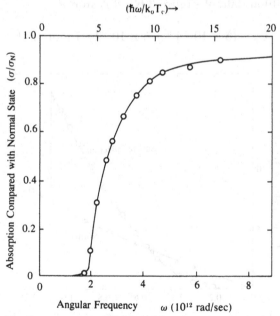

Figure 3–68 Frequency dependence of the electromagnetic absorption by a thin superconducting film of indium. The ordinate is normalized to the strength of absorption in the normal condition. Note that superconductivity is absent above an angular frequency of about $(4k_0T_c/\hbar)$. Data are from D. M. Ginsberg and M. Tinkham, Phys. Rev., **118**, 990 (1960).

photon energy of about $4k_0T_c$, which is to be associated with an energy gap ϵ_g between superconducting and "normal" electron states. Such an energy gap was postulated by F. London (1935) as an integral part of his phenomenological theory of superconductivity, and it lies at the heart of the BCS microscopic theory.

HEAT CAPACITY

There is no heat of transformation associated with the superconducting-normal transition in a metal, but there is an anomaly in the electronic component of the specific heat. An example of this is shown in Figure 3-69. The discontinuity in the specific heat reflects the second-order transition from a relatively disordered (normal) state to a more highly ordered (superconducting) state of lower entropy. Of course, a similar specific heat anomaly occurs when liquid ^{4}He becomes superfluid, and it is quite useful to regard a superconducting metal as a superfluid medium (the bosons involved being electron pairs of opposing spin and momentum according to the BCS theory). That superconductivity is a manifestation of a quantum phenomenon, with electron wave functions which are coherent over macroscopic distances, was apparent from the earlier phenomenological theories.

For a metal in the "normal" state, the total specific heat is the sum of a T^3 term arising from the lattice vibrations (the Debye law term) and one directly proportional to the absolute temperature representing the temperature-dependence of energy for a Fermi electron gas. With the "normal" gallium data of Figure 3-69, this sum is

$$C_p = (6 \times 10^{-5}T^3 + 6 \times 10^{-4}T) \ \text{J mol}^{-1}\text{K}^{-1} \qquad (3\text{-}189)$$

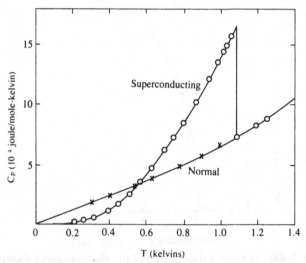

Figure 3–69 Temperature dependence of the specific heat of gallium in the normal condition and the superconducting state, as measured by N. Phillips, Phys. Rev. **134**, 385, (1964). Data for normal Ga below T_c were measured by application of a 0.02 tesla magnetic field.

The contribution of phonons is unchanged in the superconducting state, but the electronic term is much more strongly temperature-dependent. In some superconducting metals, C_{el} varies approximately as T^3 at low temperatures, but for gallium, among other superconductors, the behavior is demonstrably exponential:

$$C_p = [6 \times 10^{-5}T^3 + 9.5 \times 10^{-3} \exp(-1.9/T)] \ \text{J mol}^{-1}\text{K}^{-1} \qquad (3\text{-}190)$$

This exponential dependence suggests that electrons in a superconductor at absolute zero are all in a completely ordered state of zero entropy and specific heat, but that electrons can be thermally promoted across an energy gap (with a width of some $4k_0T_c$) to "normal" states of the system. According to such a two-fluid model, the fraction of electrons in the superconducting (or superfluid) condition would decline to zero at a certain temperature – the critical temperature T_c. This concept is incorporated into the BCS theory, but in a more complicated form, since the energy gap of the BCS model *itself* decreases with increasing temperature.

SUPERCONDUCTIVITY AND MAGNETIC FIELDS

During his early measurements, Onnes discovered that the superconducting state could be quenched reversibly by passing an excessive current through a wire, and it was soon recognized that it was the magnetic field accompanying a current which enforced a return to normal conduction. Superconductivity can flourish only if the external magnetic field is smaller than a critical value H_c, a quantity which varies from a maximum value H_0 at $T = 0$ to zero field at the critical temperature. For many superconductors the temperature dependence is approximately of the form

$$H_c = H_0[1 - (T/T_c)^2] \qquad (3\text{-}191)$$

which is often referred to as *Tuyn's law* (1924, 1929). This "law" is plotted as the solid curve in Figure 3-70, with accompanying experimental data for lead shown to be in tolerable agreement. [Since magnetic fields are usually quoted in the literature in gauss, oersteds, or teslas rather than in amp/meter, it proves convenient in Figure 3-70 and in several succeeding figures and equations to quote the critical condition in terms of the critical *induction* $B_c = \mu_0 H_c$ (teslas) rather than as the applied field intensity H_c (amp/meter).]

The departure of the data for lead in Figure 3-70 from Tuyn's law is magnified by the choice of coordinates in Figure 3-71. This figure shows comparable information for mercury and indium also. The curves illustrate that an accurate description of the relationship between H_c and T requires a polynomial representation.

Figure 3-71 uses reduced variables: magnetic field in units of H_0 and temperature in units of T_c. These and other reduced variables have often

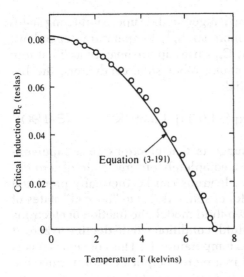

Figure 3-70 Temperature dependence of the magnetic field necessary to quench superconductivity in lead ($T_c = 7.19K$, $B_0 = 0.0803$ tesla). The circles are experimental data, after D. Decker et al., Phys. Rev. **112**, 1888 (1958). The solid curve follows Tuyn's law, Equation 3-191. The deviation between this equation and experiment is amplified in Figure 3-71.

been used in the comparison of data for various metals. Many properties of the superconducting state were systematized in terms of dimensionless coordinate systems prior to the development of a microscopic theory of the phenomenon.

The zero temperature critical field (H_0 or B_0) is generally larger for materials of high critical temperature, but the results for supercon-

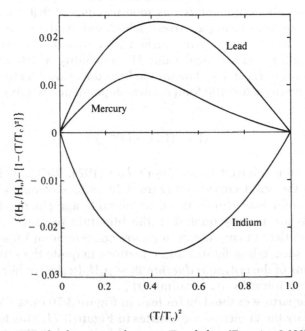

Figure 3-71 The discrepancy between Tuyn's law (Equation 3-191) and experiment for the temperature dependence of the critical magnetic field. The curve for lead magnifies the discrepancy between the curve and the experimental points of Figure 3-70. Note that the discrepancy can be of either sign, but that negative trends, as with indium, are easier to reconcile with the BCS theory. After J. Bardeen and J. R. Schrieffer, Prog. Low Temp. Phys. **3**, 170 (1961).

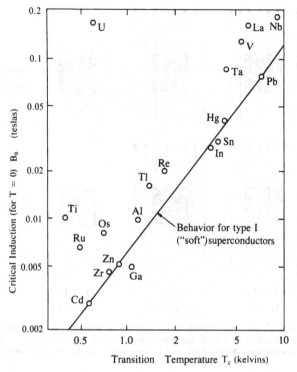

Figure 3–72 Critical magnetic induction B_0 for quenching superconducting behavior at $T = 0$ plotted as a function of T_c for elemental superconductors (data from Lynton).

ducting elements displayed in Figure 3-72 can be seen to have a great deal of scatter. The scatter is much smaller when superconductors are grouped into two categories in accordance with their diamagnetic response curves and when the line in Figure 3-72 is associated only with the "soft" or *Type I superconductors*. These are the superconducting elements in the group of boxes in the right portion of the Periodic Table on the inside cover of this book, and they are the superconductors which are most satisfactorily described by theory. The "hard" or *Type II superconductors*, formed from transition and actinide series elements, and including most superconducting compounds, have a more complicated structure, which is reflected in their magnetic properties. Before discussing the difference between Type I and Type II magnetic behavior, we must reflect on the expected interaction of a magnetic field with a perfect conductor.

If we could place a perfect conductor in an external magnetic field, no lines of magnetic flux would penetrate the sample. Induced surface currents would counteract the effect of the external field. However, a normal conductor in a field (top left in Figure 3-73) which became a "perfect conductor" while in that field should behave as in the top center part of that figure, and upon removal of the external field (top right) would acquire persistent currents to lock in the internal flux. Thus the final state of the cooled "perfect conductor" would depend on the path taken, and transitions would not be reversible.

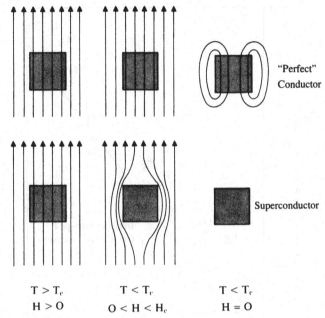

$$T > T_c \qquad\qquad T < T_c \qquad\qquad T < T_c$$
$$H > O \qquad\qquad O < H < H_c \qquad\qquad H = O$$

Figure 3–73 An illustration of the difference between an imaginary "perfect conductor," at the top, and an actual superconductor, in the lower sequence of sketches. As shown by Meissner and Ochsenfeld (1933), a superconductor expels magnetic flux from its interior.

The situation is different for a superconductor, as was shown by measurements of the magnetic field surrounding a metallic sphere by Meissner and Ochsenfeld (1933). They showed that the state of a superconductor is independent of any previous magnetic environment, and that a metal cooled through the superconducting transition temperature expels all flux lines from its interior. (If the external field is larger than H_c, the flux is not expelled because the metal is then *not* superconducting.) Any flux lines expelled from the interior must be offset by an induced persistent surface current. Termination of the external field induces an opposing surface current which cancels the previous one and leaves the superconductor both field-free and current-free. The *Meissner effect* of flux expulsion makes the transition between the superconducting and normal states reversible.

In discussing the Meissner effect, it is important to notice the emphasis on exclusion of **B** from the *interior* of a sample. Unlike an electric field (which cannot penetrate a superconductor at all), a magnetic field can penetrate a superconductor to a depth λ, which is typically of the order of 10^{-8} to 10^{-7} m. According to the phenomenological theory of superconducting electrodynamics developed by H. and F. London (1935), the penetration depth is given by

$$\lambda = (m\varepsilon_0 c^2/e^2 n_s)^{1/2} \qquad\qquad (3\text{-}192)$$

in which n_s denotes the density of electrons in the superfluid condition. At zero temperature, this includes all electrons; but the number of superconducting electrons decreases towards zero as temperature approaches

T_c, to permit infinite field penetration. Measurements of field penetration vary with temperature approximately as[75]

$$(\lambda/\lambda_0) = [1 - (T/T_c)^4]^{-1/2} \qquad (3\text{-}193)$$

in accordance with the predictions of a thermodynamic model for superconductors developed by Gorter and Casimir (1934). (Note again in Equation 3-193 the convenience of dimensionless coordinates for discussion of superconductive phenomena.)

FLUX QUANTIZATION AND
PERSISTENT CURRENTS

The exclusion of magnetic flux from a superconducting body refers of course to a *simply connected body;* that is, to one without holes or the topological equivalent of holes. Let us suppose now that the experiment of Figure 3-73 is performed on a hollow metallic cylinder. Cooling to below T_c will cause magnetic flux to be ejected from the metal but not from the space within the inner diameter of the cylinder. These flux lines *will* be trapped when the external field is turned off and will be maintained by a persistent circulating electrical current in the surrounding cylinder. The magnitude of the trapped magnetic flux decays when (or if) the persistent current decays.

F. London[76] suggested that since superconductivity is manifestly a quantum phenomenon, then the trapped magnetic flux should be quantized in units of $(2\pi\hbar/e) = 4.14 \times 10^{-15}$ Wb. It would be necessary to make observations with metallic cylinders of very small cross-sectional area in order to establish the quantized nature of the flux. Of course, quantization of the flux requires quantization of the total persistent current (see Problem 3.37).

Measurements of the kind suggested by London have since been performed,[77] and it has been possible to isolate the flux associated with a small quantum number. The magnitude of the flux quantum is one-half of the amount postulated by London,

$$\phi = (\pi\hbar/e) = 2.07 \times 10^{-15} \text{ Wb.} \qquad (3\text{-}194)$$

which implies a fundamental charge of $-2e$ for the carriers of a superconducting current. This agrees well with the Cooper pair concept, which lies at the heart of the BCS theory, that a pair of electrons of opposite spin and opposite wave vector form the unit of the superconducting system.

The free energy of a superconducting cylinder is smallest when no persistent current flows, and has a series of minima for the set of persistent currents which correspond with the allowed values of the quantized flux. Thus a continuous slow decline of persistent current is not permitted. The current can be reduced only in macroscopic steps, each of

---

[75] Some of the earliest measurements of field penetration were made by R. B. Pontius, Phil. Mag. **24**, 787 (1937). The temperature dependence of λ in tin was measured by E. Laurmann and D. Shoenberg, Proc. Roy. Soc. **A198**, 560 (1949).

[76] F. London, *Superfluids, Vol. I* (Wiley, 1950).

[77] B. S. Deaver and W. M. Fairbank, Phys. Rev. Letters **7**, 43 (1961); R. Doll and M. Näbauer, Phys. Rev. Letters **7**, 51 (1961).

which requires a highly unlikely degree of cooperation among the possible sources of electron scattering. This is one reason (in addition to the idea of "infinite conductivity") for the temporal stability of induced persistent currents.

DIAMAGNETISM OF TYPE I AND TYPE II SUPERCONDUCTORS

The flux exclusion (Meissner effect) described in connection with the lower sequence of Figure 3-73 occurs as pictured there only up to a limiting field strength which depends on H_c, on the sample shape, and on whether we are discussing a Type I or a Type II superconductor. Even for a Type I superconductor, a sample shape of essentially zero demagnetization factor (such as a long solid rod) must be used if we wish flux exclusion to be total up to the field H_c. (Above this field, normal conduction returns.) For a sample of spherical shape, the finite demagnetization associated with the shape only permits us to exclude *all* flux up to an externally applied field of $(2H_c/3)$.

Let us suppose a sample of a Type I superconductor fabricated as a long solid cylinder. In the presence of an applied field smaller than H_c, this sample sets up circulating surface currents which make $\mathbf{B} = 0$ in the interior. The bulk sample is behaving as a *perfect diamagnetic*, with a negative magnetization $\mathbf{M} = -(\mathbf{H} - \mathbf{B}/\mu_0)$ which varies with field as shown in the "Type I" curve of Figure 3-74. Above the critical field H_c, the conduction of the metal has been forced to become "normal," and $(\mathbf{B}/\mu_0)$ inside the solid will differ from $\mathbf{H}$ outside only to the same small degree as in any other paramagnetic solid.

The second curve in Figure 3-74 shows the contrasting behavior of a Type II superconductor in a magnetic field. This type of material permits some penetration of magnetic flux above a first critical field H_{c1}. For fields larger than H_{c1} but not as large as the second indicated limit H_{c2}, the material is said to be in the *intermediate* state. A "hard" Type II superconductor is one for which H_{c2} is very large. [At the time of writing, Type II superconductors are known in which H_{c2} is as large as 2.8×10^7 amp/meter ($B_{c2} = 35$ tesla) at absolute zero.] Such materials are used now for practical superconducting magnet coils, and are anticipated for use in the generation and distribution of electrical power (see the final

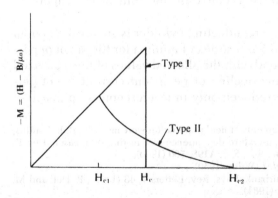

Figure 3-74 The flux exclusion demonstrated in the Meissner effect requires a *perfect diamagnetism* with a magnetization versus field as shown here for a Type I (or "soft") superconductor. In a Type I superconductor, the superconducting state ceases abruptly at the critical field H_c. A *partial* flux exclusion, which is still accompanied by infinite D.C. superconductivity, occurs with Type II superconductors for fields larger than H_{c1} yet smaller than H_{c2}.

sub-section of this chapter), since a current density of 10^{10} A/m^2 can be carried without destruction of the superconducting state.

A Type II superconductor has infinite D.C. conductivity in the intermediate state, though it consists of a mosaic of filaments of "normal" and superconducting materials. Such a mixture of superconducting and non-superconducting material was postulated by Landau (1937), and can be established as needing a lower energy than either a completely normal or perfectly diamagnetic state for materials which satisfy a certain series of inequalities. The condition for the existence of an intermediate state is crudely that the *coherence length* associated with superconducting state wave-functions (Pippard, 1953) should be smaller than the Landau penetration depth λ. For a Type I superconductor, the wave-functions are coherent over distances much larger than the penetration depth, and it takes less energy to create a single external barrier between the magnetic field and the superconductor than it does to permit a filamentary structure.

THE MICROSCOPIC THEORY
OF SUPERCONDUCTIVITY

In the last few pages we have on numerous occasions mentioned the BCS microscopic theory, which represented a major advance in our understanding of the behavior of electrons in a superconductor. This theory came nearly 50 years after the discovery of the phenomenon, and nearly 30 years after Bloch's description of the one-electron quantum states in a normal metal. In the interim there has been a great deal of theoretical interest in the problem of superconductivity, and much of the prior work on phenomenological theories by Gorter, Casimir, London, Ginzburg, Landau, and Pippard has been incorporated into a framework which can now describe superconductivity on both macroscopic and microscopic scales.

Noting that superconductivity is more likely in solids in which ordinary conductivity is not large (i.e., ones in which electrons are frequently scattered by phonons), Fröhlich[78] in 1950 proposed a model for an electron moving through a lattice continuously emitting and reabsorbing virtual phonons. In this model Fröhlich was influenced by some work he had previously done concerning polaron states (electron plus polarization) in ionic crystals. Fröhlich supposed that an electron in a metal perturbs the atoms in its neighborhood, causing them to oscillate; the lattice perturbations then react on the electron. He argued that this kind of interaction might produce a ground state of lower energy than the completely filled Fermi sea of noninteracting electrons, and that such a superconducting ground state would be separated from "normal" conducting states by an energy gap. The distribution of electrons in k-space which would give this new lower energy state was to be calculated by perturbation theory (which, alas, did not work).

The failure of perturbation theory for the Fröhlich model is less significant than his suggestion that if electron-phonon interactions are important, then T_c should be proportional to the Debye θ_D for isotopes of a

---

[78] H. Fröhlich, Phys. Rev. **79**, 845 (1950).

given element. This means that T_c should vary as $M^{-1/2}$, where M denotes the isotopic mass. Unknown to Fröhlich at the time he made this suggestion were two experimental investigations[79] of T_c versus M for isotopes of mercury which showed that T_c *did* vary as $M^{-1/2}$. Similar trends of T_c with isotopic mass have since been reported for other elements, and the demonstration of the *isotope effect* guaranteed that the models of Fröhlich and of Bardeen,[80] relating superconductivity to electron-phonon interactions, would be studied thoroughly.

The important missing link was supplied in the study by Cooper[81] of an electron-phonon-electron interaction which could make the Coulomb repulsion between two electrons smaller (or even absent) for the super-conducting condition than that in the normal state. Cooper considered a sequence of events which starts when an electron deforms the lattice in its vicinity. This constitutes the creation of a phonon, which we imagine to propagate through the crystal. A second electron is then affected by absorbing this phonon. If we denote the initial and final wave-vectors of the first electron as k_a and $k_a{}^*$ and those for the second electron as k_b and $k_b{}^*$, then

$$(k_a - k_a{}^*) = (k_b{}^* - k_b) = k_{phonon} \qquad (3\text{-}195)$$

and we might at first imagine a similar conservation for the *energy* gained by one electron at the expense of another. However, since the phonon exchange is a virtual one, energy is conserved *for the crystal* but the energy of the pair of electrons does not have to be exactly the same as before the exchange. Thus Cooper showed that for a pair of electrons just above the Fermi surface, a bound state could result if the phonon exchange resulted in an attractive interaction, no matter how weak.

Bardeen, Cooper, and Schrieffer[71] then followed up this possibility, showing that favorable conditions for a condensation into bound pairs can exist between electrons of opposing spin which have exactly opposing values of wave-vector for zero current density. Pairing involves electrons in states within an energy range $k_0\theta_D$ of the Fermi energy. The lectures given by Schrieffer, Cooper and Bardeen in Stockholm on the occasion of their receipt of the 1972 Nobel prize in physics (reprinted in the July 1973 issue of "Physics Today") give a fascinating account of how the contributions of the three authors resulted in the BCS theory.

It is thus hypothesized that at absolute zero and with zero current density the ground state of a superconductor is a highly correlated one whereby in k-space the electron states near to the Fermi surface are to the fullest extent possible occupied with pairs of opposite wave-vector and spin. The ground state of all "Cooper pairs" is represented by a single coherent wave-function. A Cooper pair $(k_1 \uparrow, -k_1 \downarrow)$ can, by an exchange of a suitable virtual phonon, go to any other unoccupied pair of positions $(k_2 \uparrow, -k_2 \downarrow)$. The existence of such pairs is strongly implied by the experimental evidence of flux quantization.

[79] E. Maxwell, Phys. Rev. **78**, 477 (1950); C. A. Reynolds *et al.*, Phys. Rev. **78**, 487 (1950).

[80] J. Bardeen, Phys. Rev. **78**, 167 (1950).

[81] L. N. Cooper, Phys. Rev. **104**, 1189 (1956).

A significant energy gap is involved between the energy of a Cooper pair and the energy of two single and unpaired electrons. This energy is many times larger than the energy required only to break up a pair, since the occupancy of a single electron state by an unpaired electron voids the opportunities for any remaining Cooper pair to exchange a phonon which would land either member of the pair in that occupied state. Thus the energy gap which is so apparent from the infrared properties and in the specific heat behavior emerges as a natural property in the BCS approach.

As we have just emphasized, each Cooper pair is symmetric about $k = 0$ for a superconductor *with zero current density*. For a situation with a non-vanishing persistent current, the entire Fermi sphere is shifted bodily (carrying its surrounding energy gap with it), and every Cooper pair has a non-zero net momentum. Unlike the ordinary electrical current in a "normal" conductor, this superconducting current cannot be relaxed by the scattering activity of a phonon or a localized defect. The energy gap stabilizes the Cooper pairs against any small change of net momentum. Thus the infinite D.C. conductivity of a superconductor also emerges as a natural (and essential) consequence of the theory.

The energy gap described by the BCS theory decreases towards zero as the temperature approaches T_c, since the fraction of electrons in paired states decreases with rising temperature. The total separation between states above and below the gap is conventionally described as $2\epsilon_{g0}$ at zero temperature, and as $2\epsilon_{gT}$ for a finite temperature. BCS theory provides an integral equation between ϵ_{gT} and temperature, and in the manner of most equations for superconducting properties this can best be described for the dimensionless $(\epsilon_{gT}/\epsilon_{g0})$ versus (T/T_c). The solid curve in Figure 3-75 shows the mutual dependence of these two dimensionless variables, compared with experimental data for the gap width in three elements. The data in this figure for In and Ta have been obtained by the previously mentioned tunneling technique of Giaever,[62] in which the current voltage characteristics are measured for a metal-insulator-superconductor sandwich. The insulating layer in this sandwich is so thin ($<100\text{Å}$) that electrons can easily tunnel from states in the metal on one side to states in the superconductor on the other. In this manner, electrons of the Fermi energy for a "normal" metal on one side can scan the energy dependence of the density of states in the superconductor. Changes in the applied bias voltage over a range of a few millivolts show up discontinuities in current which correspond with the upper and lower edges of the superconducting energy gap. Such measurements, together with other experimental observations (far infrared absorption, acoustic attenuation, and so forth), provide convincing confirmation that the ground state of a superconductor is separated from the states of "normal" conduction in the manner outlined by Bardeen, Cooper, and Schrieffer.

The three parts of Figure 3-76 give a highly exaggerated picture of the difference between the spectrum and occupancy of states in a normal metal and those in a superconductor. Part (a) considers the density of states at $T = 0$ in the absence of superconductivity (which can be arranged just by applying a magnetic field larger than H_{c0}). This density of

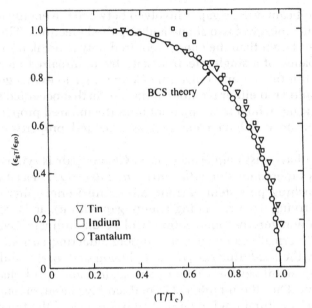

Figure 3-75 Temperature dependence of the superconducting energy gap, with the gap expressed in units of the gap at absolute zero, and the temperature expressed as a fraction of the transition temperature. The curve follows the BCS theory. Data for tin obtained from acoustic attenuation measurements by R. W. Morse and H. V. Bohm, Phys. Rev. **108**, 1094 (1957). Data for indium obtained in electron tunneling experiments by I. Giaever and K. Megerle, Phys. Rev. **122**, 1101 (1961). Data for tantalum also from electron tunneling, by P. Townsend and J. Sutton, Phys. Rev. **128**, 591 (1962).

states is just that of Equations 3-44 and 3-160. The Fermi-Dirac occupancy factor, Equation 3-45, is unity up to energy ϵ_{F0}, and zero for any higher energy.

The superconducting ground state of the BCS theory is pictured for zero temperature in part (b) of Figure 3-76. This shows a zero density of states for energies within $\pm\epsilon_{g0}$ on either side of the Fermi energy, and a piling up of the displaced states on either side of the gap. At T = 0, *all* electrons are in Cooper pairs; none are excited to higher states.

Part (c) of the figure imagines the consequences of a finite temperature less than T_c. The superconducting energy gap ϵ_{gT} is now smaller than ϵ_{g0}. Moreover, not all the electrons are still in superconducting pairs; a fraction are in states above $(\epsilon_{F0} + \epsilon_{gT})$, leaving some unoccupied states ("pair holes") in states below $(\epsilon_{F0} - \epsilon_{gT})$. Finally, the gap decreases to zero and the fraction of paired electrons goes to zero when the temperature reaches T_c.

This picture of the relationship between occupied and unoccupied states in a superconductor is useful in visualizing how experiments such as those of Giaever are able to scan the density of states with respect to energy, by measuring the current-voltage characteristic of a tunnel junction involving a superconductor on one or both sides of the junction. It is not an exact picture in that the BCS ground state is *not* of non-interacting single-particle states which obey Fermi-Dirac statistics, but of interacting electron pairs, with a consequent departure from Fermi statistics.

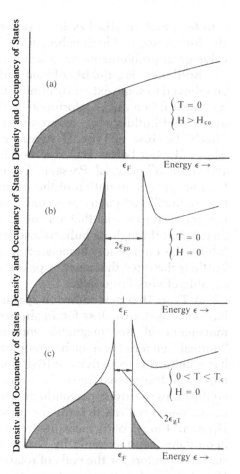

Figure 3-76 A visualization of the difference between the spectrum of states for a normal metal and for a superconductor at very low temperatures. The size of the superconducting energy gap has been grossly exaggerated in parts (b) and (c) of this figure.

APPLICATIONS OF SUPERCONDUCTIVITY

The applications of superconductivity are discussed at length in a number of books, including the ones by Fishlock, Newhouse, Solymar, and Williams cited in the bibliography at the end of this chapter. This sub-section is intended to do no more than indicate the range of applications, present and potential.

The most widespread of all superconducting devices to date is the *superconducting magnet*, which uses a large D.C. current in a coil wound of superconducting wire to generate a high magnetic field for laboratory or industrial use. The disadvantage of requiring a liquid refrigerant in which the magnet coil must be immersed is frequently outweighed by the advantage of having a magnet coil with negligible I^2R losses, capable of producing an induction up to 15 T. Thus superconducting magnets have been possible only since very "hard" Type II superconducting materials became a reality, in which H_{c2} at the operating temperature is as large as possible.[82] Some of these materials are listed at upper right in Table 3-4. They have large values of H_{c2} at

---

[82] J. E. Kunzler, Rev. Mod. Phys. **33**, 501 (1961).

zero temperature [B_{c2} has been reported as high as 35 T for Va$_3$Ga at absolute zero], and high values for T_c. Many of these materials also involve great problems in manufacture, notably extreme brittleness.

Brittleness is a quality of a polycrystalline material that can usually be relieved to some extent by elimination of impurities and by annealing (i.e., heat treatment). Unfortunately, these attempts to ease the task of the magnet builder also compromise the ability to carry a large current. This is because a Type II superconductor in the intermediate state (above H_{c1} but below H_{c2}) is a mosaic of filaments of superconducting and "normal" material. Passage of a current sufficient to make H exceed H_{c1} encourages migration of the magnetic flux vortices associated with the "normal" inclusions, as a consequence of the Lorentz force. Thus it is a practical essential that a Type II material have lattice irregularities distributed through its bulk, whose job it is to "pin" these vortices and inhibit the energy loss associated with vortex motion. A wire which is brittle is the price that must be paid if a superconducting magnet is to be capable of very large fields.

A Type II superconductor operated above the limit H_{c1} inevitably has some hysteresis loss for an alternating current, since periodic small movement of the magnetic vortices associated with filaments of "normal" material cannot be completely prevented. Thus the material has a finite A.C. resistivity at high fields. The larger the frequency, the more I^2R losses that must be anticipated. This limits the potential applications of hard superconductors to frequencies of a few hundred Hz unless one is willing to settle for currents that will keep the field below H_{c1} at all times. With such low frequency operation in mind, work has been done (see the books by Newhouse or by Williams) in using hard superconductors for the coils of rotary alternator generators and motors. Here the disadvantage of refrigeration must be balanced against the advantage of reduction in mass (possibly by as much as a factor of 10) for a given power handling capability. Superconducting materials have also been considered for use in underground power distribution cables, using either a hard Type II material for D.C. transmission, or a Type I material (probably in thin film form) for an A.C. transmission system.

Since "hard" superconductors with large T_c, large H_{c2}, and large current carrying capacity were first reported,[82] there has been intense interest in seeing how large T_c can be made. The two materials cited at the top right of Table 3-4 bracket the boiling point of liquid hydrogen, though this should not be taken to mean that hydrogen rather than helium can be used as the refrigerant in all future applications of superconductors. (The superconducting energy gap goes to zero at $T = T_c$, and so does the capacity to exclude magnetic flux and to carry a lossless supercurrent.) It *will* be possible to consider superconducting devices cooled with liquid hydrogen if T_c can be made substantially larger than 23K. Matthias[83] anticipates the formulation of alloys in which T_c may reach 30K, but he is pessimistic about increases beyond the 25–30K region. There have been high hopes by some writers[84] that organic

[83] B. T. Matthias, Physics Today **24**, No. 8, p. 23 (August 1971).
[84] For example, W. A. Little, Phys. Rev. **134**, A1416 (1964).

charge-transfer complexes with a large conductivity in one direction might result in superconductivity in one direction with a large T_c, but most attempts in this direction have been frustrated by a solid-state phase transition of the material to an insulating form on cooling. It is important to distinguish between those organic charge-transfer salts which are metallic or semimetallic (with a high conductivity in one direction), and those for which the alternations of intermolecular spacing result in a band structure with no partly filled bands. In his classic book (see bibliography) Peierls warned that a partly filled 1-D band can lower its energy by forming a (nonconducting) superlattice sequence of spacings. Thus while a perchlorate salt with an organic (HMTSF) cation is superconducting,[85] the value of T_c is only a little more than 1 K. Scientific interest in 1-D conduction seems certain to yield somewhat higher T_c values in time.

A number of applications of superconductivity center on the use of Type I superconductors, without especial reference to very high transition temperatures or high magnetic fields, but with an interest in high frequencies. One example of this is the superconducting linear accelerator, in which the microwave resonant cavities are lined with a thin layer of lead or niobium, cooled with superfluid helium. A superconducting LINAC at Stanford University[86] operates at 950 MHz, and can achieve a Q value of 10^{10} within the resonant cavities.

Since a magnetic field smaller than the bulk critical field does not appreciably penetrate a Type I superconductor, no resistance is presented to the motion of flux past it. Even the flux which does penetrate to the Landau depth λ of Equation 3-192 enters and leaves in a reversible, lossless mechanism. This provides the basis for a frictionless *superconducting bearing* which can withstand a thrust of 10^4 N/m² or more. Such bearings are contemplated for high speed transit systems. The absence of friction between a Type I superconductor and a magnetic field has also led to the successful demonstration of a *superconducting gyroscope*, with a spinning Type I superconducting sphere levitated in a magnetic field.

Figure 3-67 showed the rather abrupt change of resistance with temperature reported by Ohnes for the superconducting transition of mercury. The years since 1911 have seen many studies of the abruptness of this transition in various materials. The superconducting bolometer[87] is a thin strip of material maintained at temperature T_c by a suitable thermal support system, whereby a small amount of absorbed thermal radiation causes the element to move at least part of the way to the "normal" resistivity. This type of bolometer compares well with other thermal detectors in respect of detectivity and time constant, but the need for highly accurate thermal control has limited the use of this device.

The disparity between the "normal" and superconducting resis-

[85] K. Bechgaard et al., Phys. Rev. Lett., **46**, 852 (1981).
[86] W. M. Fairbank and H. A. Schwettman, Cryog. Eng. News 2, No. 8, p. 46 (1967).
[87] D. H. Andrews, R. M. Milton, and W. de Sorbo, J. Opt. Soc. Amer. 36, 518 (1946). See also D. H. Martin and D. Bloor, Cryogenics 1, 1 (1961).

tance states was suggested by Buck[88] as the basis for a binary switching element called a *cryotron*. This has two windings of superconducting wire (or two thin films), of which the current in one controls the existence or absence of superconductivity in the other. Thus this can serve as a binary logic element or as a memory storage element in a digital computer. Attempts have also been made to use transient currents to magnetically form small "islands" of "normal" material at chosen locations in a continuous superconducting thin film. Such islands would be surrounded with a circulating supercurrent, which could later be detected to serve as the basis for a computer memory. The importance of optimized computer technology to present-day society needs no emphasis.

A number of applications of superconductivity stem directly from the quantum nature of this state of matter. Thus the quantization of magnetic flux within a superconducting ring described by Equation 3-194 has been used[89] as the basis for a sensitive magnetometer, which is also applicable as an absolute A.C. ammeter.

Tunneling junctions of metal-insulator-superconductor (MIS) and superconductor-insulator-superconductor (SIS) sandwich construction have been mentioned earlier in connection with the experiments of Giaever.[62] Experiments with these sandwich junctions have been informative about the density of states in a superconductor and about other properties of the participating solids. Moreover, SIS tunnel junctions have been demonstrated as amplifiers, oscillators, and detectors of microwave electromagnetic energy and of high frequency phonons. The book by Solymar[64] discusses applications of tunnel junctions, including Josephson junctions.

Josephson junctions[63] were mentioned previously as SIS junctions for which the insulating layer is so thin (perhaps 10 Å) that superconducting Cooper pairs can make the transition from one metal to the other. The first striking consequence is that a finite supercurrent may flow through a circuit containing a Josephson junction with zero voltage across the junction! This supercurrent is a sensitive function of magnetic field, and goes through a zero whenever the total magnetic flux in the insulating gap is a multiple of $(\pi\hbar/e)$. Thus a Josephson junction can form the heart of a sensitive magnetometer or galvanometer. When a small voltage V_0 is applied across a Josephson junction, the current oscillates at an angular frequency

$$\omega_0 = (eV_0/\hbar) \tag{3-196}$$

A voltage of 1 μV results in generation of a frequency $(\omega_0/2\pi) = 484$ MHz. A very accurate value for the ratio $(\hbar/e)$ has been obtained by measurement of the frequency associated with a known voltage,[90] and this A.C. Josephson effect has been suggested as the portable standard for intercomparison of international voltage standards.

[88] D. A. Buck, Proc. IRE **44**, 482 (1956).

[89] B. S. Deaver and W. S. Goree, Rev. Sci. Instr. **38**, 311 (1967).

[90] W. H. Parker, B. N. Taylor, and D. N. Langenberg, Phys. Rev. Lett. **18**, 287 (1967). The numerical values in the table on page 490 of this book are taken from Taylor, Parker, and Langenberg, Rev. Mod. Phys. **41**, 375 (1969); these values incorporate the work of the same authors in measuring $(\hbar/e)$ with the Josephson effect.

Problems

3.1 The alkali metals lithium through cesium all crystallize in the B.C.C. structure, with nearest-neighbor spacings of 3.03, 3.71, 4.62, 4.87, and 5.24 Å respectively. The radii of the various monovalent positive ions are 0.68, 0.98, 1.33, 1.48, and 1.67 Å respectively. Show that the fraction of space in the lattice outside the core regions decreases smoothly from 94 per cent in lithium to 82 per cent in cesium. Calculate the number of free electrons per m^3 and the average electronic charge density in metallic sodium. Show by solving Poisson's equation that the electrostatic potential is raised by about 12 volts at the center of a spherical region of 10 Å radius inside a sodium crystal for which the free electron density is uniformly 15 per cent lower than in the crystal as a whole.

3.2 Show that an electron of radius R is likely to travel a distance $(\pi R^2 N)^{-1}$ before the tube it sweeps out intersects one of a set of N points per unit volume distributed in space, provided that R is small compared with $N^{-1/3}$. Discuss what this means in connection with the validity of Equation 3-7.

3.3 Calculate a relationship between electrical conductivity, temperature, and mean free path for copper on the basis of the Drude model. Use the conductivity curve of Figure 3-1 to produce a corresponding curve for λ as a function of temperature. You are given that copper has a nearest-neighbor spacing of 2.55 Å, and you should assume that the electron gas comprises one electron per atom.

3.4 Express the conductivity of Equation 3-33 as an integral with respect to electron speed and as an integral with respect to electron energy for a Boltzmann distribution of speeds and energies. Show that either of these results in Equation 3-34. (You may find it convenient to integrate by parts.) Derive an expression for the scattering mean free time τ_m (in seconds) and for the electron mobility μ (in m²/V s) for the Lorentz model when the mean free path is 100 Å, and plot curves to show how these quantities vary with temperature.

3.5 A spherical body of mass M and charge Q moves through a viscous medium at a constant speed (AQE_x) under the influence of an electric field E_x. A magnetic field is now applied in the z-direction, of induction B_z. Set up the equations of motion in the xy plane, and solve for the radius of curvature and the trajectory when the magnetic field is small. (Hint: the velocity component in the x-direction is virtually unchanged.)

3.6 Imagine a free electron gas for which two-thirds of the electrons move with speed s and the other one-third with speed 2s. Each of these groups can be subdivided again into six equal numbers moving in the positive and negative x-, y-, and z-directions. Discuss the conductivity, Hall effect, and magnetoresistance of this free electron gas for current flow in the x-direction and a magnetic field in the z-direction. If we were to inject one more electron for which the magnetic and Hall fields balanced each other, at what speed must this electron travel in the x-direction?

3.7 Show that the spectrum of allowed states for an electron trapped in a cubical box of side L is the same as that for cyclical boundary conditions. (This problem is for readers who have made some use of wave mechanics on previous occasions.)

3.8 The ideal gas law for a Maxwell-Boltzmann gas is $pV = RT = N_A k_0 T$. Find the corresponding relationship between pressure, specific volume, and total energy for a Fermi-Dirac gas. You will need to show that the "pressure" exerted by an electron gas is two-thirds of the kinetic energy per unit volume.

3.9 From the lattice constants and valencies of the respective metallic elements, you are given that the free electron concentrations in sodium, copper, and aluminum are 2.5, 8.5, and 18×10^{28} m^{-3} respectively. Substantiate the 3.1 eV value for the Fermi energy of sodium quoted in Figure 3-9, and show that electrons of the Fermi energy in the three metals move with speeds of 1.04, 1.57, and 2.01×10^6 m/s respectively. To what extent is the mass of an energetic free electron in aluminum relativistically increased?

3.10 Show by using Equations 3-46, 3-49, 3-50, and 3-52 that the slight downward movement of the Fermi level in a metal upon warming from absolute zero is controlled by Equation 3-53. How large is this reduction for metallic sodium at 300K?

3.11 Calculate the electronic specific heat per m^3 at 750K for the metals sodium, copper, and aluminum (for which useful numbers appear in Problem 3.9). What fraction of the total *molar* specific heat is this for each metal, given that 750K is in each case well above the Debye temperature?

3.12 Show that Equation 3-62 is consistent with the specific heat quoted in Equation 3-59. You should use Equation 3-44 as the basis for defining density of states at the Fermi energy.

3.13 The splitting shown in Figure 3-12(b) between the energy origins of the two portions of the density of states would in fact require an extremely large magnetic induction. For metallic sodium, how large an induction (in teslas) is required to make the energy splitting $2\mu_B\mu_0 H = 2\mu_B B$ equal to $k_0 T$ for room temperature (300K)? How large must this induction be for the splitting to reach 10% of the Fermi energy?

3.14 Compare Equation 3-4 for the classical distribution function with Equation 3-73 for the Fermi-Dirac equilibrium distribution function. The first of these is proportional to n (the volume density of particles) and the second is not. Why is this? Show that f_0 of Equation 3-73 is indeed a *normalized* distribution function.

3.15 Explore the properties of $(-\partial f/\partial \epsilon)$ in a Fermi-Dirac distribution, showing that it is symmetrical and that its integral is unity. How does this function compare with functions customarily used as an analytic representation of the Dirac delta function?

3.16 Repeat the tasks requested in Problem 3.3, this time doing them for copper in terms of the Sommerfeld conduction model.

3.17 Consider a gold crystal to be an F.C.C. array of touching spheres with radius 1.44 Å. Each atom contributes one free electron. The electrical conductivity of pure gold is 5.0×10^7 ohm^{-1}m^{-1} at T = 273K. Now suppose a gold crystal with 0.1 per cent vacancies, randomly distributed. Estimate the mean free path λ_{imp} determined by vacancy scattering alone. From the known conductivity, evaluate the free path for phonon scattering at 273K, and use Matthiessen's rule to determine the revised conductivity.

3.18 Note the relationship (Equation 3-88) between the x-component of speed inside a metal and the much smaller speed after successful thermionic emission. Show that the distribution of x-component emergent velocities is of the form

$$n(u_x)du_x \propto u_x \exp(-mu_x^2/2k_0 T)du_x$$

for an RMS component of $(u_x)_{RMS} = (2k_0 T/m)^{1/2}$. Show that the velocity components in the y- and z-directions are consistent with an average emergent kinetic energy of $2k_0 T$ per electron.

3.19 A pure tungsten filament is used as a source for thermionic emission in an experimental system, and it is found that operation at 2000K produces the desired emission-limited current. What filament temperature would be necessary if a thoriated tungsten filament were used? (Use the parameters of Table 3-3.) Show that if filament temperature is limited by black-body radiation, then a change to thoriated tungsten permits a 44 per cent reduction in filament driving current, assuming that electrical resistivity is directly proportional to the absolute temperature. In what ratio must the driving current be increased again if the thoriated filament gets contaminated, increasing the effective work function by 25 per cent while leaving the Richardson constant unchanged?

3.20 Note that the potential energy expression (Equation 3-100) for field-aided emission is based upon Equation 3-98 rather than the more elaborate Equation 3-99, even though the latter should be more reliable for small values of x. Set up the equation for field-modified potential energy based on Equation 3-99 and show that the location and height of its maximum are not appreciably different from Equation 3-101 with any reasonable electric field. Verify that an electric field of 10^6 V/m will lower the work function by only 0.04 eV/electron, and that the maximum in the potential curve occurs for $x_{max} = 190$ Å with such a field.

3.21 Show that integration of the force expressed by Equation 3-103 over all of the time that a primary electron interacts with a conduction electron in a solid results in a momentum change with magnitude and direction as shown in Equation 3-104. What must be the impact parameter b for an electron of initial kinetic energy 1 eV to have energy 10 eV after a primary of energy 500 eV has passed by?

3.22 Sodium becomes transparent in the ultraviolet, below a wavelength $\lambda = 210$ nm. How well does this compare with the wavelength $\lambda_p = (2\pi c/\omega_p)$ expected from the plasma frequency of Equation 3-108? As a numerical contrast, estimate the minimum average electron density in the Earth's ionosphere, given that broadcast band radio signals up to 3 MHz are customarily reflected from the ionospheric layer back to Earth.

3.23 Figure 3-31 shows that the fourth zone of the square lattice (Figure 3-30) has the same size as the first, and that it can be fitted into the shape of the first. Draw a general curved line which passes across a major portion of the first zone for this lattice, and show how this line is broken up into numerous sections of the exploded fourth zone. This is not a meaningless exercise, for later in the chapter you will be concerned with the progression of an electron through k-space under the influence of electric and/or magnetic fields.

3.24 Try the jigsaw puzzle task of fitting the second through fifth zones of a hexagonal lattice (Figure 3-33) into the hexagonal size and shape of the reduced zone.

3.25 Illustrate the first four zones for a two-dimensional rectangular lattice in which the ratio of atomic spacings in the primitive cell is 2:1.

3.26 For a monatomic lattice which crystallizes in the H.C.P. structure, the Brillouin zone is a hexagonal prism. Determine the dimensions of this prism in k-space for interatomic spacing a. Show that the permitted density of states in k-space is consistent with two electron states per atom for each Brillouin zone.

3.27 Draw the first two Brillouin zones for the two-dimensional lattice with a 2:1 ratio of atomic spacings, as in Problem 3.25. Inscribe a Fermi surface for nearly free electrons, such that the Fermi surface just reaches to the center of the second zone in the reduced zone representation. Draw a picture of the reduced zone for the first and second zones, showing that the degree of electron filling specified is just sufficient to close off empty pockets in the first zone to four fragments.

3.28 Show that the group velocity is twice the phase velocity for any state in a perfectly free electron gas. Determine the group and phase velocities in a system for which velocities are spherically symmetrical and the ϵ–$\mathbf{k}$ relationship takes the form

$$\epsilon = Ak^2 - Bk^4$$

and show that the group and phase velocities are the same when $\epsilon = (2A^2/9B)$.

3.29 Consider a one-dimensional crystal, for which energy varies with wave-vector in accordance with

$$\epsilon = \epsilon_1 + (\epsilon_2 - \epsilon_1)\sin^2(ak_x/2)$$

Populate this band with a single electron, which is not scattered. Discuss the behavior of the effective mass, the electron velocity, and the electron position in real space under the influence of a time-independent electric field, assuming that Bragg reflection occurs at the zone boundaries as expected by Houston. If a = 1 Å, then for how long must a field of 100 V/m be applied to make this electron execute one complete oscillation in space? If the band is 1 eV wide, what is the range of distance covered in this oscillation?

3.30 Consider the band and electron of Problem 3.29 again. Describe the behavior of the electron in a sinusoidally modulated electric field, first placing the electron at the bottom of the band, and then considering a sinusoidal field for an electron at the mid-point on the energy scale.

3.31 Consider an electron with the Fermi energy for sodium, and initial motion in the xy plane. Show that a magnetic induction $B_z = 1$ T will produce cyclotron motion with an orbital radius of 6 μm. How is the area of the orbit in real space (in m²) related to the area of the orbit performed in k-space (in m⁻²)?

3.32 Consider the problem of an extremal cyclotron orbit in the $k_x k_y$ plane of the Fermi surface for a non-isotropic solid. The equation for the Fermi energy is $\epsilon = (\alpha k_x{}^2 + \beta k_y{}^2)$. Apply a magnetic induction B_z, and describe the rate at which k changes during an orbit when α and β are unequal. Describe the orbit performed in real space for this situation.

3.33 Show that when magnetic induction becomes small, the summation of Equation 3-178 can then be expressed as an integral over an infinite range, and that this gives the conventional density of quantum states, Equation 3-44. In the same manner, show that Equation 3-179 reduces to Equation 3-46 in the weak-field limit.

3.34 From the allowed density of states in k-space, show that the Fermi sphere for a cold degenerate electron gas has a cross-section S_F given by Equation 3-185. Determine the spacing P (in tesla⁻¹, or m²/V s) between the successive dHvA peaks of Figure 3-64, and evaluate the corresponding Fermi orbit cross-section S_F in m⁻². Calculate the density n (m⁻³) of electrons forming this electron gas (assuming that the mass tensor reduces to a scalar for potassium), and show that electrons in potassium must have an effective mass very close to the mass of a free electron, since $\epsilon_F = 2.0$ eV in this metal.

3.35 As a companion to Problem 3.34, consider the experimental data for electrons in the semiconductor GaSb in Figure 3-66. Note that the Hall coefficient is cited as $R_H = -4.8 \times 10^{-6}$ m³/C, and from this evaluate the total electron density and the area of an extremal Fermi energy orbit in k-space. Show that this same orbital area is required from the spacing P in reciprocal magnetic field of the SdH effect.

3.36 J. File and R. G. Mills [Phys. Rev. Letters **10**, 93 (1963)] made a solenoid of the superconducting alloy Nb_3Zr, using a tube with a 10 cm diameter, and winding a double layer solenoid of 984 turns of 0.5 mm diameter wire (for a coil length of 25 cm). Assuming that this is long enough to use infinite-length approximations, calculate the solenoid inductance. We are told that the magnitude of a persistent current in this solenoid in the superconducting state decreases by about one part in 10^9 per hour. If we describe this decay by the equation

$$I = I_0 \exp(-Rt/L)$$

then determine an upper limit for the coil resistance, and hence a lower limit for the conductivity in the supercon-ducting state. The latter is reported by File and Mills to be at least 10^{23} ohm^{-1}m^{-1}.

3.37 Consider the first quantum state for a superconducting loop made from a single turn of aluminum wire. The wire has a diameter d = 2 μm, and the loop has a radius R = 20 μm. Assuming that the flux through this loop is given by Equation 3-194, calculate how much current must be circulating around the loop, and the value of **H** or **B** at the center of the loop. Now use the more nearly correct expression

$$\phi = (\pi \hbar/e)[1 + (8\pi\lambda^2/Rd)]$$

for the flux enclosed by a loop whose dimensions are not infi-nite compared with the Landau penetration depth λ. Assume that $n_s = 1.8 \times 10^{29}$ m^{-3} of electrons are in the superconducting state at the temperature of the experiment, which will be true for aluminum near absolute zero. How much difference does the above equation make to your evaluation of current and of the field at the center of the loop?

3.38 This problem will give you some slight idea of how difficult it is to design a superconducting magnet which will generate a large field without destroying its own superconducting state. On an insulating cylindrical form of radius R = 50 mm, wind a Helmholtz coil arrangement (two coils, separated by distance R), each comprising a single turn of lead wire. The wire is of diameter d = 1 mm. Operated immersed in liquid helium at 4.2K, how large a current can be sent through each coil, and how large a value for **H** or **B** does this create at the point on the axis midway between the coils? (Consult Figure 3-70 for the critical field of lead at the operating temperature. This value must not be exceeded at the surface of the metal wire.)

Bibliography

Experimental Metallic Conduction

F. J. Blatt, P. A. Schroeder, C. L. Foiles, and D. Greig, *Thermoelectric Power of Metals* (Plenum Press, 1976).

A. N. Gerritson, *Handbuch der Physik, Vol. 19*, 137 (Springer, 1956).

C. M. Hurd, *The Hall Effect in Metals and Alloys* (Plenum Press, 1972).

D. K. C. MacDonald, *Handbuch der Physik, Vol. 14*, 137 (Springer, 1956).

G. T. Meaden, *Electrical Resistance of Metals* (Plenum Press, 1965).

Free Electron Theory [Several of these, alas, out of print.]

H. Jones, *Handbuch der Physik*, *Vol. 19*, 227 (Springer, 1956).

H. A. Lorentz, *The Theory of Electrons* (Dover, 1952).

N. F. Mott and H. Jones, *Theory of the Properties of Metals and Alloys* (Dover, 1958).

R. Peierls, *Quantum Theory of Solids* (Oxford Univ. Press, 1965).

F. Seitz, *Modern Theory of Solids* (McGraw-Hill, 1940).

A. H. Wilson, *Theory of Metals* (Cambridge Univ. Press, 2nd ed., 1965).

J. M. Ziman, *Electrons and Phonons* (Oxford Univ. Press, 1960).

J. M. Ziman, *Principles of the Theory of Solids* (Cambridge Univ. Press, 2nd ed., 1972).

Collective Electron Phenomena: Plasmas in Solids

H. Haken, *Quantum Field Theory of Solids* (North-Holland, 1976).

M. F. Hoyaux, *Solid State Plasmas* (Pion, 1970).

J. C. Inkson, *Many-Body Theory of Solids* (Plenum Press, 1984).

P. M. Platzman and P. A. Wolff, *Waves and Interactions in Solid State Plasmas* (Academic Press, 1973).

G. Rickayzen, *Greens Functions and Condensed Matter* (Academic Press, 1984).

Band Theory

J. Callaway, *Energy Band Theory* (Academic Press, 1964).

J. Callaway, *Quantum Theory of the Solid State* (Academic Press, 1974), 2 vols.

G. C. Fletcher, *Electron Band Theory of Solids* (North-Holland, 1971).

W. A. Harrison, *Electronic Structure and the Properties of Solids* (Freeman, 1980).

W. A. Harrison, *Solid State Theory* (Dover, 1980).

J. C. Slater, *Handbuch der Physik*, *Vol. 19*, 1 (Springer, 1956).

Fermi Surfaces and Electron Dynamics

A. A. Abrikosov, *Introduction to the Theory of Normal Metals* (Academic Press, 1972).

A. P. Cracknell and K. C. Wong, *The Fermi Surface* (Oxford Univ. Press, 1973).

T. E. Faber, *Introduction to the Theory of Liquid Metals* (Cambridge Univ. Press, 1972).

W. A. Harrison and M. B. Webb (eds.), *The Fermi Surface* (Wiley, 1960).

A. B. Pippard, *Dynamics of Conduction Electrons* (Gordon & Breach, 1968).

Superconductivity and its Applications

J. M. Blatt, *Theory of Superconductivity* (Academic Press, 1964).

C. G. Kuper, *An Introduction to the Theory of Superconductivity* (Oxford Univ. Press, 1968).

E. A. Lynton, *Superconductivity* (Halsted, 3rd ed., 1971).

V. L. Newhouse (ed.), *Applied Superconductivity* (Academic Press, 1975), 2 vols.

R. D. Parks (ed.), *Superconductivity* (Dekker, 1969), 2 vols.

L. Solymar, *Superconductive Tunneling and Applications* (Halsted, 1972).

M. Tinkham, *Introduction to Superconductivity* (Krieger, 1980).

J. E. C. Williams, *Superconductivity and its Applications* (Pion, 1970).

chapter four

SEMICONDUCTORS

It was not feasible to discuss metals in Chapter 3 without mention of the companion topic of electronic conduction in insulating and semiconducting materials on several occasions. Now it is time to consider these solids in some detail.

For a metal, it is possible to vary the *mobility* of charge carriers (by changing the temperature to change the density and spectrum of phonons, or by changing the density of defects in the crystal), but the *number* of charge carriers is fixed. This number of free electrons may be characterized by a fixed electrochemical potential (Fermi energy) or in terms of some other suitable invariant parameter, and most investigative techniques with metals tell us only about the properties of electrons at the Fermi energy.[1]

Things are very different in a semiconductor. The numbers (as well as the mobilities) of charge carriers depend on temperature and on the presence of defects or impurities. At thermodynamic equilibrium it is possible to express the occupancy of all electron states at all energies in terms of a single normalizing parameter or Fermi level, but this Fermi level is a consequence of the overall electrostatic balance and does not have to coincide with the energy of any mobile electrons. Thus in a semiconductor, we have to evaluate the *dependence* of the Fermi level on temperature and flaw concentration.[2] The electrons in a semiconductor cannot be characterized in terms of a single Fermi level when thermodynamic equilibrium is destroyed, and (again in contrast to a

---

[1] Footnote 58 of Chapter 3 commented on some of the experimental techniques which *are* informative about states below the Fermi energy of a metal, notably photoemission and optical reflectance.

[2] The term "flaw" was popularized by Shockley and Read (1952) as a general name for anything which contributes a *localized* energy state capable of donating or accepting at least one electronic charge. Thus a foreign atom (impurity atom) is likely to act as a flaw whether it occupies a regular lattice site or whether it is wedged into an interstitial site. A native defect of any of the types discussed in Section 1.6 will usually have some effect on the electronic spectrum, and will present one or more "flaw" energy levels. The numbers and spectra of flaw levels help in determining the numbers of free electrons and holes in a semiconductor at equilibrium. Flaws also help to manufacture and annihilate free electrons and holes, both at equilibrium and away from equilibrium.

metal) a few additional free electrons in a semiconductor can represent a major departure from thermal equilibrium.

In the first section of this chapter we consider the distribution of electrons over various sets of electron states in thermodynamic equilibrium, for "intrinsic" and "extrinsic" situations. As a reflection of the interests of the writer, this first section has more detail than found in the treatment of some other topics.[3] (The last sub-section can be omitted without loss of continuity with subsequent sections.) Of course, an electron density cannot be measured directly. We measure a property or combination of properties of a solid, and then interpret this data as *inferring* the presence of a certain number of electrons and/or holes. Even so, since the numbers of electrons and holes crucially affect almost anything we choose to measure, it is still useful to survey the influences of flaws and of thermal excitation upon the electronic distribution.

The chapter continues with a section concerning the observable transport phenomena for free electrons in a spatially uniform semiconductor. In most of the discussion of transport, it is assumed that any external field applied for the purpose of making the observation is too weak to cause a serious departure from thermodynamic equilibrium.

The observable properties of most semiconductors are complicated by the complexities of constant energy surfaces in k-space near the top of the valence band and near the bottom of the conduction band. A lengthy discussion of this continuously evolving field would be inappropriate, but Section 4.3 does illustrate some common complexities in terms of well-studied semiconductors. An appreciation for the actual shapes of energy bands in a semiconductor is of great help when we want to know which recovery mechanisms may be dominant in the event of a disturbance from an equilibrium electron population, and Section 4.4 gives an account of generation and recombination mechanisms, as these relate to the dynamics of a semiconductor away from equilibrium. Most applications of semiconductors require a departure from equilibrium.

---

[3] This writer has discussed the distribution of electrons among band states and flaw states for equilibrium and non-equilibrium situations in a homogeneous semiconductor, in the monograph *Semiconductor Statistics* (Pergamon, 1962). This book will be referred to at several places in the narrative for a more extensive account than is warranted here. Thus, *Semiconductor Statistics* discusses the mechanisms of and dynamics of generation and recombination for nonequilibrium situations in much more detail than Section 4.4.

Equilibrium Electron Statistics

DETAILED BALANCE

When a material is in a state of thermodynamic equilibrium, it is possible to describe its thermodynamic condition in terms of a temperature T. Indeed, the term "temperature" has no strict meaning unless the material *is* in equilibrium with its surroundings, though we often speak loosely of the "temperature" of an object which is subject to some mild external stimulus.

The above observations are true for any medium, semiconducting or not, but the need to remind ourselves of them is much more acute for semiconducting solids than with metals. The general shape of the Fermi surface in a metal is preserved regardless of the temperature, and whether or not thermal equilibrium exists. For the small mobile electron and hole densities in a semiconductor, it is important to know whether or not an equilibrium distribution does exist, and the value of the "temperature" parameter if equilibrium does prevail.

At any rate, when thermodynamic equilibrium *does* exist in a semiconducting solid, the temperature T determines the numbers and spectra of electrons, holes, phonons, and photons within the solid. The photon population is distributed in energy in accordance with the Planck radiation law, since these quasi-particles obey Bose-Einstein statistics. The phonon population similarly obeys the Bose-Einstein distribution law (Equation 2-54), operating on a density of phonon modes determined by the interatomic restoring forces in the solid. The distribution of electrons of various potential and kinetic energies is determined by the Fermi-Dirac distribution law (Equation 3-45), and the same Fermi normalizing parameter ϵ_F operates for *all* electron energies.

At any instant, photons from the black body radiation field are being absorbed in the excitation of electrons to states of higher energy. At the same time, an electron which can find unoccupied states of lower energy into which it can drop may do so, creating a photon of appropriate energy in the process. In thermodynamic equilibrium these opposite processes match in overall rate and in spectral detail. Similarly, there is a *detailed balance* in the transformations between phonon energy and electron energy, and in every other process of energy transformation which can be identified within the solid. Such a detailed balance is required by the laws of thermodynamics.

Because of detailed balance we can have confidence that there is a unique electronic energy distribution characterized by a unique Fermi level ϵ_F for material of specified composition at a given temperature T. The value of this Fermi energy, and the distribution of the total electron supply over the complete spectrum of allowed states, are independent of the rates of the various identifiable transition processes, for we are talking about a *time-independent* distribution.

When thermodynamic equilibrium is destroyed, detailed balance is temporarily destroyed. It then becomes a matter of interest to know which physical processes will dominate the attempt to restore equilibrium. For one semiconductor, a relaxation towards equilibrium may occur at a rate dominated by the temporary excess of photon creation over photon absorption. In another solid, a non-equilibrium population of high energy electrons may be thermalized more easily by a temporary excess of phonon creation over phonon absorption. These matters are of importance, and will be mentioned again in Section 4.4. For the remainder of this section, however, it will be assumed that a semiconductor can be studied *at* thermodynamic equilibrium. On the basis of this assumption, we shall discuss equilibrium electronic distributions without regard to the rates at which the various energy transformation processes and their inverses may be able to operate.

INTRINSIC AND EXTRINSIC SEMICONDUCTORS

It was noted in Section 3.4 that intrinsic semiconducting behavior occurs in an insulating solid when $k_0 T$ is large enough to permit some thermal excitation of electrons from the upper part of an otherwise filled band to the lower portion of the next higher (normally empty) band.[4] At that time we introduced the notation of *valence band* for the normally filled band (so called because the electrons which occupy these energy states are the ones involved in the chemical valence bonds between atoms), and of *conduction band* for the next empty band (since any electron in this band is certainly free for conduction). Throughout this chapter we shall signify the numbers of conduction band electrons and valence band holes as follows:

n_0 = density of conduction band electrons for a situation of thermodynamic equilibrium.

n = total density of conduction band electrons (not necessarily at equilibrium).

$n_e = (n - n_0)$ = excess electron density in the conduction band caused by a departure from equilibrium.

p_0 = density of valence band holes in thermodynamic equilibrium.

p = total density of valence band holes.

$p_e = (p - p_0)$ = excess free hole density caused by a departure from equilibrium.

Now if a semiconductor is completely pure and crystallographically perfect, the densities n_0 and p_0 must be equal, for conduction band electrons can then be derived only by excitation of valence band states. (We

---

[4] See the right portion of Figure 3-40 for a very crude visualization of the highest filled (valence) band and lowest empty (conduction) band for silicon as a typical semiconductor. Footnotes 47 and 48 of Chapter 3 cite the papers in which A. H. Wilson proposed the concepts of intrinsic and extrinsic semiconducting behavior.

shall signify by n_i the intrinsic electron-hole density.) A single Fermi level characterizes the energy distribution of both the free holes and the free electrons, and this intrinsic Fermi level is close to the center of the intrinsic gap, as shown in Figure 4-1(a). Later in this section we shall see why the Fermi level must occupy a position so strikingly at variance with metallic materials, but first we must review the relationship between electron density and Fermi energy for a single band.

Only for quite pure semiconducting crystals, and at comparatively high temperatures, does intrinsic excitation dominate a semiconductor. Particularly at low temperatures, we are likely to find that the intrinsic electron-hole pair density n_i is negligibly small compared with the density of *one* kind of free carrier derived by thermal excitation of flaw levels. Thus an intrinsic conductor must inevitably become either *n-type* or *p-type* extrinsic when it is cooled. These two alternatives are shown in parts (b) and (c) of Figure 4-1.

In an *n-type* extrinsic semiconductor, free electrons are more numerous than holes. The notation "n-type" serves as a reminder that the conduction processes in such a conductor are dominated by *negatively* charged mobile carriers, so that phenomena such as the Hall effect and thermoelectric effects (which depend on the first power of the mobile charge) are *negative*. Similarly, these effects are *positive* in a *p-type* extrinsic semiconductor for which *positively* charged free holes are the dominant free carriers. A *donor* impurity is a flaw which can become positively charged, for this releases or donates one or more electrons to the system. An electron so donated may become mobile in the conduction band (though it may just become trapped at another flaw). Similarly, an *acceptor* impurity is a flaw which can assume one or more states of negative charge, thus accepting electronic charges (and perhaps liberating mobile holes). Donors exceed acceptors in an n-type semicon-

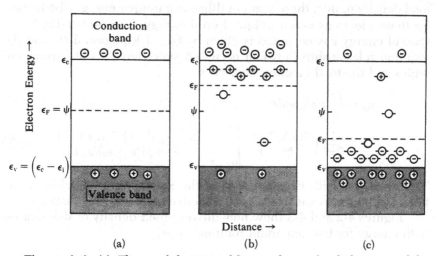

Distance →

(a)   (b)   (c)

Figure 4-1   (a), The equal densities of free conduction band electrons and free valence band holes for an intrinsic semiconductor. The Fermi level required for $n_0 = p_0 = n_i$ is at the energy ψ which is usually near the center of the intrinsic gap. (b), The same semiconductor with a miscellany of localized flaw states added to make $n_0 > n_i > p_0$. This is an n-type conductor, and the Fermi level is higher than ψ. (c), p-type conduction results when the kind(s) of flaws added make the Fermi level lower than ψ, so that $p_0 > n_i > n_0$.

ductor, and are less numerous than acceptors if p-type behavior is exhibited.

The values of n_0 and p_0 for an extrinsic semiconductor depend both on the temperature and on the complete spectrum of localized flaw states, some of which become ionized and others not. Ionization of flaw states leading to an increase of the free electron density also enforces a reduction of the free hole density, and vice versa. As indicated by parts (b) and (c) of Figure 4-1, the value of $n_0 p_0$ is essentially just n_i^2 when we depart from the intrinsic condition, and we shall soon see why this is so.

ELECTRON DENSITY AND FERMI LEVEL FOR A SIMPLE BAND

In order to place the preceding statements about free carrier densities on a quantitative footing, let us first consider how a particular Fermi energy ϵ_F is consistent with a given total conduction electron density n_0 for the simplest possible band. Suppose that this band has a *single* energy minimum for energy $\epsilon = \epsilon_c$ at the center of the Brillouin zone, and suppose that spherical surfaces of constant energy in **k**-space surround the location of this minimum, parabolically spaced as required for a scalar and energy-independent effective mass m_c. Then the density of states for an energy not much larger than ϵ_c is

$$g(\epsilon) = \frac{1}{2\pi^2} \left(\frac{2m_c}{\hbar^2}\right)^{3/2} (\epsilon - \epsilon_c)^{1/2} \qquad (4\text{-}1)$$

which we may compare directly with Equations 3-44 or 3-160.

Now suppose that some of the states in this conduction band are populated at thermal equilibrium by thermal excitation of electrons from lower (flaw and/or valence band) states. At a temperature T, and for a total density n_0 m^{-3}, there is at equilibrium a unique energy distribution for these electrons and a unique Fermi energy. The probability that a state of energy ϵ is occupied is given by $f(\epsilon)$ of Equation 3-45, and the required relationship between occupied state densities at various energies and the total electron density is

$$n_0 = \int_{\epsilon_c}^{\infty} f(\epsilon) g(\epsilon) d\epsilon$$

$$= \frac{1}{2\pi^2} \left(\frac{2m_c k_0 T}{\hbar^2}\right)^{3/2} \int_{(\epsilon_c/k_0 T)}^{\infty} \frac{[(\epsilon - \epsilon_c)/k_0 T]^{1/2} \cdot d(\epsilon/k_0 T)}{1 + \exp[(\epsilon - \epsilon_F)/k_0 T]} \qquad (4\text{-}2)$$

This coincides with Equation 3-46 of the Sommerfeld model for a metallic free electron gas (aside from effective mass renormalization).

Figures 4-2 and 4-3 show how the occupied density of states varies with energy for two important limiting cases:

1. A conduction electron gas is *degenerate* (Figure 4-2) when conduction electrons are numerous and the temperature is rather small. Under these conditions the solution of Equation 4-2 requires that $(\epsilon_F - \epsilon_c) \gg k_0 T$, and the conduction properties are dominated by the varying occupancy of states at the Fermi energy.

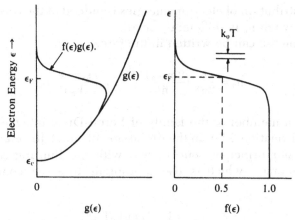

Figure 4-2 A degenerate conduction electron distribution in a semiconductor, as will occur when electrons are very numerous and/or the temperature is small. The total shaded area in the left portion of the figure is n_0 of Equation 4-2, and this equation demands a solution with $\epsilon_F > \epsilon_c$. Thus the Fermi energy and electron density are related by Equation 4-6.

2. An electron gas is *non-degenerate* or *classical* when the total electron density is so small (or the temperature so high) that the two sides of Equation 4-2 balance for a Fermi energy several k_0T lower than ϵ_c. As Figure 4-3 shows, only a minor fraction of the band states are then occupied, even for the lowest states. Since

$$f(\epsilon) = \{1 + \exp[(\epsilon - \epsilon_F)/k_0T]\}^{-1}$$
$$\rightarrow \exp(\epsilon_F/k_0T)\,\exp(-\epsilon/k_0T) \quad \text{for } \epsilon \gg \epsilon_F \qquad (4\text{-}3)$$

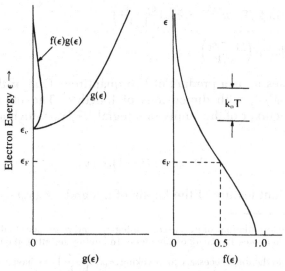

Figure 4-3 A non-degenerate or "classical" solution for Equation 4-2 when the total n_0 is small and/or the temperature is large. The occupancy factor $f(\epsilon)$ has adopted its Boltzmann limiting form for the range of energy in which there is a non-zero density of states, so that the electrons comprising the shaded area form a classical gas. The value of n_0 is now related to the Fermi energy by Equation 4-9.

then the distribution of electron energies is indeed of the classical Boltzmann form when n_0 is sufficiently small.

Equation 4-2 can be written in the form

$$n_0 = \frac{1}{2\pi^2}\left(\frac{2m_c k_0 T}{\hbar^2}\right)^{3/2} F_{1/2}\left(\frac{\epsilon_F - \epsilon_c}{k_0 T}\right) \qquad (4\text{-}4)$$

which uses a member of the family of Fermi-Dirac integrals $F_j(y_0)$ we defined by Equation 3-50. In the discussion of a metallic electron gas in Chapter 3, our principal concern was with the limiting condition of strong degeneracy, which is also appropriate in a semiconductor provided that

$$n_0 \gg \frac{1}{2\pi^2}\left(\frac{2m_c k_0 T}{\hbar^2}\right)^{3/2} \qquad (4\text{-}5)$$

When this inequality is indeed well satisfied, then

$$\left.\begin{array}{l} \epsilon_F \approx \epsilon_c + (\hbar^2/2m_c)(3\pi^2 n_0)^{2/3} \\ n_0 \approx (1/3\pi^2)[2m_c(\epsilon_F - \epsilon_c)/\hbar^2]^{3/2} \end{array}\right\} \qquad (4\text{-}6)$$

from the degenerate asymptotic form Equation 3-52 of $F_{1/2}\left(\dfrac{\epsilon_F - \epsilon_c}{k_0 T}\right)$.

However, the electron population within a band of a semiconductor is often much too small to justify Equation 4-5. In order to deal with these more nearly "classical" situations, it is more convenient to rewrite Equation 4-2 in the form

$$\begin{aligned} n_0 &= 2(m_c k_0 T/2\pi\hbar^2)^{3/2} \int_{(\epsilon_c/k_0 T)}^{\infty} \frac{(2/\pi^{1/2})[(\epsilon - \epsilon_c)/k_0 T]^{1/2} \cdot d(\epsilon/k_0 T)}{1 + \exp[(\epsilon - \epsilon_F)/k_0 T]} \\ &= 2(m_c k_0 T/2\pi\hbar^2)^{3/2} \cdot \mathscr{F}_{1/2}\left(\frac{\epsilon_F - \epsilon_c}{k_0 T}\right) \\ &= N_c \mathscr{F}_{1/2}\left(\frac{\epsilon_F - \epsilon_c}{k_0 T}\right) \qquad (4\text{-}7) \end{aligned}$$

This expresses n_0 as a product of two quantities. One of these is $N_c = 2(m_c k_0 T/2\pi\hbar^2)^{3/2}$, with dimensions of length^{-3}. The other is dimensionless, a member of the family of integrals $\mathscr{F}_j(y_0)$ which are related to the $F_j(y_0)$ by[5]

$$F_j(y_0) = \Gamma(j + 1)\mathscr{F}_j(y_0) \qquad (4\text{-}8)$$

One convenient feature of the family of integrals $\mathscr{F}_j(y_0)$ is that they all

---

[5] These two formulations of Fermi-Dirac integrals, and references to their tabulations, were given in Footnotes 12 and 13 of Chapter 3. In forcing the integral on the first line of Equation 4-7 into the form necessary for invoking $\mathscr{F}_{1/2}\left(\dfrac{\epsilon_F - \epsilon_c}{k_0 T}\right)$, we have used the fact that $\Gamma(3/2) = (\pi^{1/2}/2)$.

asymptotically approach exp (y_0) when y_0 is a large negative quantity. In the particular case of $\mathscr{F}_{1/2}(y_0)$, the asymptotic form can be used with an error of less than 5 per cent provided that $y_0 < -2$. Thus

$$n_0 \approx N_c \exp\left(\frac{\epsilon_F - \epsilon_c}{k_0 T}\right) \Bigg\}$$

$$\text{if } (n_0/N_c) < 0.15 \qquad (4\text{-}9)$$

$$\epsilon_F \approx \epsilon_c - k_0 T \ln(N_c/n_0)$$

The simple interrelationship between electron density and Fermi level expressed by Equation 4-9 is extremely useful in discussions of the many semiconductor situations for which degeneration of the conduction electron gas does not occur. It is possible to use a relationship a little more complicated than Equation 4-9 (but still with an exponential term as the major contributor) in order to deal also with "semi-degenerate" situations of slightly larger total electron density, and with a combination of correction factors $\mathscr{F}_{1/2}(y_0)$ can be described in analytical terms for any value of y_0, positive or negative.[6]

The quantity N_c introduced in Equation 4-7 and used in Equation 4-9 is often called the *effective density of conduction band states*. This term is used since Equation 4-9 shows that for a classical (non-degenerate) distribution, n_0 is just N_c multiplied by the Boltzmann occupancy factor for energy $\epsilon = \epsilon_c$. Thus the *total* electron density (distributed, in fact, over a range of several $k_0 T$ of energy as shown in Figure 4-3) is the same as though the band were replaced by a set of N_c levels all at the energy ϵ_c. In numerical terms

$$\begin{aligned} N_c &= 2(m_c k_0 T/2\pi \hbar^2)^{3/2} \\ &= 4.83 \times 10^{21} (T m_c/m_0)^{3/2} \quad \text{m}^{-3} \end{aligned} \qquad (4\text{-}10)$$

with the temperature expressed in kelvins. Problem 4.1 examines a simple test of the circumstances under which n_0 may be either much larger than N_c (degeneracy) or much smaller than N_c (nondegeneracy), depending on the temperature.

FREE HOLE STATISTICS

For an almost full valence band (see Figure 4-4) of which the uppermost portion can be characterized by an effective mass m_v, the number of

---

[6] The form $\mathscr{F}_{1/2}(y_0) \approx [0.27 + \exp(-y_0)]^{-1}$ correctly conforms with the classical asymptotic form for y_0 strongly negative, and involves an error of no more than 3 per cent for $y_0 \lesssim +1$. For larger positive values of y_0, the form $\mathscr{F}_{1/2}(y_0) \approx (4/3\pi^{1/2})(y_0^2 + \pi^2/6)^{3/4}$ involves a similarly small error and has the correct asymptotic form for very large positive y_0. The considerable literature on attempts since Sommerfeld (1928) to describe Fermi-Dirac integrals in simple analytic forms was reviewed by the present author in *Solid-State Electronics*, **25**, 1067 (1982).

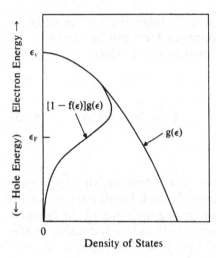

Electron Energy →

ϵ_v

$[1 - f(\epsilon)]g(\epsilon)$

ϵ_F

$g(\epsilon)$

(← Hole Energy)

0

Density of States

Figure 4-4 A pocket of free holes at the top of an otherwise filled valence band. Since the probability that a state of energy ϵ is filled is $f(\epsilon) = \{1 + \exp[(\epsilon - \epsilon_F)/k_0T]\}^{-1}$, then the probability of having an empty state (i.e., a hole) is $[1 - f(\epsilon)]$, which is $\{1 + \exp[(\epsilon_F - \epsilon)/k_0T]\}^{-1}$. For the particular situation illustrated, the total free hole density is large enough to create a degenerate situation, which with hole statistics means that the Fermi level is *lower* than the top of the band.

holes (empty states) at equilibrium for temperature T and Fermi energy ϵ_F is

$$p_0 = \int_{-\infty}^{\epsilon_v} g(\epsilon)[1 - f(\epsilon)]d\epsilon$$

$$= \frac{1}{2\pi^2}\left(\frac{2m_vk_0T}{\hbar^2}\right)^{3/2} \int_{-\infty}^{(\epsilon_v/k_0T)} \frac{[(\epsilon_v - \epsilon)/k_0T]^{1/2}\cdot d(\epsilon/k_0T)}{1 + \exp[(\epsilon_F - \epsilon)/k_0T]} \quad (4\text{-}11)$$

Thus we can choose to write p_0 as either

$$p_0 = \frac{1}{2\pi^2}\left(\frac{2m_vk_0T}{\hbar^2}\right)^{3/2} F_{1/2}\left(\frac{\epsilon_v - \epsilon_F}{k_0T}\right) \quad (4\text{-}12)$$

or

$$p_0 = N_v\mathscr{F}_{1/2}\left(\frac{\epsilon_v - \epsilon_F}{k_0T}\right) \quad (4\text{-}13)$$

where $N_v = 2(m_vk_0T/2\pi\hbar^2)^{3/2}$ is analogous, for the valence band, to N_c in the conduction band. Equations 4-11 through 4-13 have the asymptotic forms

$$p_0 \approx (1/3\pi^2)[2m_v(\epsilon_v - \epsilon_F)/\hbar^2]^{3/2} \quad \text{(degenerate)}$$

$$p_0 \approx N_v \exp\left(\frac{\epsilon_v - \epsilon_F}{k_0T}\right) \quad \text{(non-degenerate)} \quad (4\text{-}14)$$

in complete accordance with the expressions for electrons at the bottom of the conduction band.

It is very convenient that hole statistics can be handled in exactly the same way as electron statistics. (The sole difference is that a *decrease* in electron energy is an *increase* in hole energy.) This happy state of affairs comes about because $f(\epsilon)$ is symmetrical about 50 per cent occupancy at energy ϵ_F. Thus $f(\epsilon_F + \zeta)$ is always the same as $[1 - f(\epsilon_F - \zeta)]$, for any value of ζ.

The reality of the hole (the absence of an electron from an almost full band) as a quasi-particle with statistics and transport properties of its own was mentioned in Section 3.5, and will be taken for granted in this chapter. This subject has received some careful scrutiny from the quantum-mechanical point of view, particularly by Shockley.[7] Shockley concluded that to the extent that *any* single-particle model of conduction in a solid has some validity, then the wave-packet which is a single hole in an otherwise full band has the same kinds of properties as a single electron wave-packet in an otherwise empty band.

INTRINSIC ELECTRON-HOLE PAIR DENSITY

We can now put the equations of the last two subsections to work in determining the value of the intrinsic hole-electron pair density n_i for a semiconductor containing a negligibly small flaw density. The intrinsic semiconductor of Figure 4-5 has different masses, m_c and m_v in the two bands, so that N_c and N_v are numerically different. As in Figure 4-1, we shall use the symbol ψ to denote the value of the Fermi level when the intrinsic condition $n_0 = p_0 = n_i$ is sustained.

Provided that the intrinsic gap ϵ_i is large compared with k_0T, then n_i will be small compared with either N_c or N_v. Thus the neutrality condition

$$n_i = N_c \mathscr{F}_{1/2}\left(\frac{\psi - \epsilon_c}{k_0 T}\right) = N_v \mathscr{F}_{1/2}\left(\frac{\epsilon_v - \psi}{k_0 T}\right) \qquad (4\text{-}15)$$

---

[7] W. Shockley, *Electrons and Holes in Semiconductors* (Van Nostrand, 1950) Chapter 15.

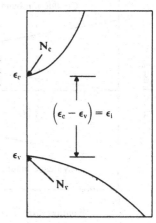

Figure 4-5 An intrinsic semiconductor for which the conduction band mass m_c is appreciably smaller than the valence band mass m_v. This disparity is emphasized by a three-halves power in the ratio of N_v to N_c.

which we can write using Equations 4-7 and 4-13, can be expressed in terms of the non-degenerate limit for each of the Fermi integrals:

$$n_i = N_c \exp\left(\frac{\psi - \epsilon_c}{k_0 T}\right) = N_v \exp\left(\frac{\epsilon_v - \psi}{k_0 T}\right) \qquad (4\text{-}16)$$

The two equalities of Equation 4-16 then permit us to express n_i and ψ separately as functions of temperature for a non-degenerate intrinsic semiconductor. The carrier pair density is

$$n_i = (N_c N_v)^{1/2} \exp(-\epsilon_i/2k_0 T), \qquad n_i \ll N_c, N_v \qquad (4\text{-}17)$$

where ϵ_i is the gap width $(\epsilon_c - \epsilon_v)$.

For the corresponding progression of the Fermi level,

$$\psi = \frac{1}{2}(\epsilon_c + \epsilon_v) + \frac{1}{2} k_0 T \ln(N_v/N_c)$$

$$n_i \ll N_c, N_v \qquad (4\text{-}18)$$

$$\text{or} \quad \psi = \frac{1}{2}(\epsilon_c + \epsilon_v) + \frac{3}{4} k_0 T \ln(m_v/m_c)$$

Thus, unless m_c and m_v happen to be the same, the intrinsic Fermi level ψ is displaced from the center of the intrinsic gap, except at absolute zero (when n_i is indeterminately small). The course of the intrinsic Fermi level for the mass disparity of Figure 4-5 is in the sense shown in Figure 4-6.

The exponential factor in Equation 4-17 almost (but not quite) determines the entire temperature dependence for n_i. A semilogarithmic plot of n_i versus $(1/T)$, as in Figure 4-7(a), has an average slope which is a tolerable first approximation to the value of $(-\epsilon_i/2k_0)$, but the plot has a

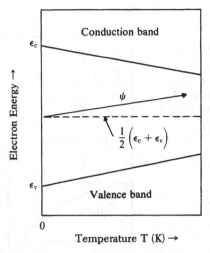

Figure 4-6 The rise with increasing temperature of the intrinsic Fermi level ψ with respect to the midpoint of the gap, for an intrinsic semiconductor in which $m_c < m_v$. The disparity of the densities of states in the two bands will permit equal densities for free electrons and free holes only when ψ is raised above the midpoint. The semiconductor imagined in this figure has an intrinsic gap that decreases upon heating, which drives ψ still closer to the low-mass band. In an extreme case it is possible for an intrinsic semiconductor to become *partly degenerate* in the low-mass band at high temperatures. [This happens in indium antimonide above 400K according to the analysis of Austin and McClymont, Physica **20**, 1077 (1954). See also Figure 4-9 for InSb intrinsic data.]

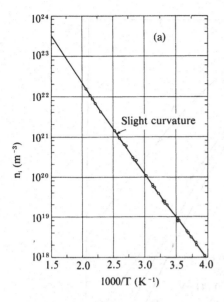
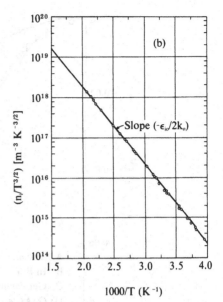

Figure 4-7 (a), Semilogarithmic presentation of intrinsic concentration as a function of reciprocal temperature for germanium. The experimental points are the data of Morin and Maita, Phys. Rev. **96**, 28 (1954). Note the slight curvature caused by the $T^{3/2}$ temperature dependence of $(N_c N_v)^{1/2}$. (b), The same data replotted with $(n_i/T^{3/2})$ as ordinate. This can be interpreted in terms of Equation 4-20. The points shown are based on experimental measurements of the Hall effect.

slight curvature. This curvature can be seen if the plot is viewed from one end with the page tilted. In contrast, it will be seen that the plot of $\ln (n_i/T^{3/2})$ versus $(1/T)$ in the same figure is quite straight, since the non-exponential temperature dependence of $(N_c N_v)^{1/2}$ has been thereby extracted.

Equation 4-17 implies that the straight line plot of Figure 4-7(b) has a slope of $(-\epsilon_i/2k_0)$. This would be true if it were not for the slight complication that the intrinsic gap between the valence and conduction bands of germanium decreases slightly with increasing temperature, as happens for most semiconductors.[8] (Such a decrease of gap width upon warming was shown schematically in Figure 4-6.) A change of energies for the various sets of Bloch states as a function of temperature is to be expected from both the variation of interatomic spacings with T and the temperature-dependence of the amplitude of lattice vibrations. Figure 4-8 illustrates the variation of intrinsic gaps with temperature for some well known semiconductors, and these dependences introduce complications into the interpretation of data for the intrinsic electron-hole pair density.

---

[8] The lead compounds PbS, PbSe, and PbTe are among the minority of semiconductors for which the intrinsic gap becomes wider upon heating (at any rate up to some 600K). For most semiconductors, $(d\epsilon_i/dT)$ is very small at low temperatures and becomes progressively more negative as temperature increases. In a tabulation of $(d\epsilon_i/dT)$ for 48 different semiconductors given by R. H. Bube, *Photoconductivity of Solids* (Wiley, 1960), the three lead salts are the only ones shown with positive values.

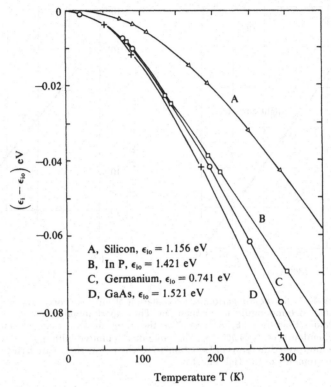

Figure 4-8 Temperature variation of intrinsic gap for some well-known semiconductors. The gap is "direct" for GaAs and InP, but "indirect" for germanium and silicon (see Section 4.3). The data for Ge and Si is from T. P. McLean, *Progress in Semiconductors*, *Vol. 5* (Wiley, 1960). Data for GaAs from M. D. Sturge, Phys. Rev. **127**, 768 (1962). Data for InP from W. J. Turner, W. E. Reese, and G. D. Pettit, Phys. Rev. **136**, A1467 (1964).

Provided that the gap width can be approximated by a linear dependence:

$$\epsilon_i = \epsilon_{io} - \alpha T \qquad (4\text{-}19)$$

within the temperature range of interest, then Equation 4-17 can be rearranged as

$$(n_i/T^{3/2}) = [2(m_c^{1/2}m_v^{1/2}k_0/2\pi\hbar^2)^{3/2}\exp(\alpha/2k_0)]\exp(-\epsilon_{io}/2k_0T) \qquad (4\text{-}20)$$

with the temperature dependence of the right side vested solely in the final exponential factor. Thus a plot of $\ln(n_i/T^{3/2})$ versus $(1/T)$ for a semiconductor can then be quite linear, but the slope is $(-\epsilon_{io}/2k_0)$ rather than $(-\epsilon_i/2k_0)$, while the intercept on the ordinate for $(1/T) \rightarrow 0$ is controlled by the parameter α as well as by m_c and m_v.

Analysis of the germanium data of Figure 4-7 is suggested in Problem 4.2. Even before this analysis is made, it will be clear from a comparison of the germanium curve in Figure 4-8 with Equation 4-19 that ϵ_{io} is not the separation of the valence and conduction bands in this semiconductor for *any* temperature. Comparable or greater difficulties occur in gap width determinations from n_i data for any semiconductor in which ϵ_i must be expressed as a polynomial function of temperature.[9] (This emphasizes the value of optical absorption data as an adjunct to the analysis of n_i data.) The conversion from n_i to ϵ_i is more difficult still if the density-of-states effective mass for one or both bands is a sensitive function of energy or temperature. Among other semiconductors, this complication is prominent in the III-V compound InSb, for which intrinsic data are displayed in Figure 4-9.

The complications for the InSb intrinsic data of Figure 4-9 are increased further by the fact that this is a semiconductor for which the intrinsic gap is relatively small and the ratio (m_c/m_v) is far from unity. The disparity of the densities of states in the two bands forces ψ to be far away from the midpoint of the gap at high temperatures (as may be seen from

---

[9] A relatively simple nonlinear expression for variation of gap width with temperature has been suggested by Y. P. Varshni, Physica **34**, 149 (1967). His suggestion arose from consideration of the two separate physical mechanisms which contribute to a thermal change in a gap. He noted that the thermal contraction of the gap due to the increasing strength of coupling between electrons and lattice vibrations should vary as T^2 when $T \ll \theta_D$, changing to a linear dependence on temperature well above the Debye temperature, when the full spectrum of vibrations is excited. Superimposed on this contraction of the gap is a separate term resulting from lattice dilation. Since the thermal expansion coefficient is small at low temperatures for many solids, and becomes rather more constant at high temperatures, this contribution to $\Delta\epsilon_i$ should also be linear in T for the higher temperatures. Accordingly, Varshni fits (with reasonable success) data for the intrinsic gap in several semiconductors to an expression:

$$\epsilon_i = \epsilon_{i0} - AT^2(T + B)^{-1}$$

However, the values for B which permit the best fits for the various semiconductors are poorly correlated with the Debye temperature from thermal methods.

Figure 4-9 Variation with temperature of the intrinsic free carrier density in indium antimonide, based on the Hall effect and conductivity data of H. J. Hrostowski et al., Phys. Rev. **100**, 1672 (1955). A correct analysis has to take into consideration the temperature dependence of gap width and effective masses, and allow for conduction band degeneracy at high temperatures. The procedure suggested in Problem 4.3, omitting these complications, will thus give values for ϵ_i which differ from the correct values.

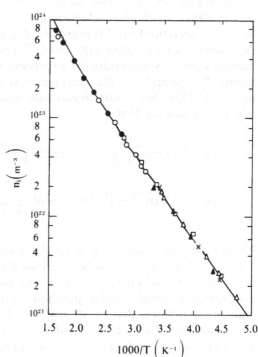

$n_i \left(m^{-3} \right)$

$1000/T \left(K^{-1} \right)$

Equation 4-18), and if the gap is only a few k_0T wide it can well happen that the solution of Equation 4-15 requires ψ to enter the low-mass band. This complication was alluded to in the caption of Figure 4-6 but is deliberately overlooked in the preparation of an (erroneous) intrinsic gap curve for InSb in Problem 4.3.

The warnings of complications in the preceding paragraphs should alert the reader to numerous incorrectly evaluated "intrinsic gaps" in the semiconductor physics literature. As was remarked earlier in connection with Figure 1-54, it is much easier to measure a quantity with an exponential dependence on $(1/T)$ than it is to be sure that the measured slope is the quantity one *thinks* it is.

DEPARTURE FROM THE INTRINSIC SITUATION

For an electrically neutral semiconductor in thermodynamic equilibrium, any difference in the densities of mobile electrons and holes must be matched by a charge $e(n_0 - p_0)$ on flaw levels within the forbidden gap. We shall consider the occupancy statistics for flaw levels in the next subsection. However, without knowing anything about the densities of or nature of flaw states, we can from Equations 4-7, 4-13, and 4-15 construct the statement that

$$n_0 p_0 = n_i^2 \left[\frac{\mathscr{F}_{1/2}\left(\dfrac{\epsilon_F - \epsilon_c}{k_0 T}\right) \mathscr{F}_{1/2}\left(\dfrac{\epsilon_v - \epsilon_F}{k_0 T}\right)}{\mathscr{F}_{1/2}\left(\dfrac{\psi - \epsilon_c}{k_0 T}\right) \mathscr{F}_{1/2}\left(\dfrac{\epsilon_v - \psi}{k_0 T}\right)} \right] \leqslant n_i^2 \qquad (4\text{-}21)$$

The product of n_0 and p_0 is *smaller* than n_i^2 when degeneration occurs either for the free electron gas ($\epsilon_F > \epsilon_c$) or for the free hole gas ($\epsilon_F < \epsilon_v$), since the Fermi integral $\mathscr{F}_{1/2}(y_0)$ increases less rapidly than an exponential for a positive argument. But $n_0 p_0$ *coincides with* n_i^2 for the many extrinsic semiconductor situations which keep the Fermi level within the bounds of the intrinsic gap (i.e., for which $n_0 < N_c$ and $p_0 < N_v$). Thus for a non-degenerate semiconductor (extrinsic or intrinsic) we can say that

$$\left.\begin{aligned} n_0 &= N_c \exp\left(\frac{\epsilon_F - \epsilon_c}{k_0 T}\right) = n_i \exp\left(\frac{\epsilon_F - \psi}{k_0 T}\right) \\[2ex] p_0 &= N_v \exp\left(\frac{\epsilon_v - \epsilon_F}{k_0 T}\right) = n_i \exp\left(\frac{\psi - \epsilon_F}{k_0 T}\right) = (n_i^2/n_0) \end{aligned}\right\} \qquad (4\text{-}22)$$

Then whatever action by flaw levels causes n_0 to be increased must at the same time cause p_0 to be decreased in the same *ratio* (not by the same amount), and *vice versa*. In consideration of the properties of semiconducting materials (and of junction semiconductor devices), it is very convenient to know that a shift of k_0T in the electrochemical potential ϵ_F is accompanied by an increase of one free carrier density by a factor of $e = 2.718$ and by reduction of the other free carrier population in the same ratio. This is of course the "Law of Mass Action" from thermodynamics.

While still refraining from specifying what kinds of flaw states may be operating, let us suppose that $N_r = (n_0 - p_0)$ in a non-degenerate n-type extrinsic semiconductor. Then N_r denotes the *difference* between the densities of flaws which have *lost* an electron and those which have *gained* an electron. The condition

$$p_0 = (n_i^2/n_0) = n_0 - N_r \qquad (4\text{-}23)$$

permits us to express the *majority carrier* density n_0 and the *minority carrier* density p_0 in terms of N_r and the intrinsic condition as follows:

$$\left. \begin{aligned} n_0 &= (N_r/2)[(1 + 4n_i^2/N_r^2)^{1/2} + 1] \\ p_0 &= (N_r/2)[(1 + 4n_i^2/N_r^2)^{1/2} - 1] \end{aligned} \right\} \qquad (4\text{-}24)$$

These two quantities are shown in Figure 4-10, varying from the almost intrinsic situation $[N_r \ll n_i,\ n_0 \approx (N_i + \tfrac{1}{2}N_r),\ p_0 \approx (n_i - \tfrac{1}{2}N_r)]$ to the strongly extrinsic situation $[N_r \gg n_i,\ n_0 \approx N_r,\ p_0 \approx (n_i^2/N_r)]$. Of course, the curves for electron density and hole density must be interchanged if the semiconductor is p-type, in which case holes are the majority carriers and electrons the minority carriers.

The transition from nearly intrinsic conditions, as at the left of Figure 4-10, to completely extrinsic conditions, happens fairly abruptly when an intrinsic semiconductor is cooled. This happens since n_i usually varies rapidly with temperature, while the value of N_r is usually much less sensitive to slight cooling. An illustration of these remarks is provided by the curves of Figure 4-11, which display the majority and minority densities as functions for temperature for two samples of InAs which are intrinsic at the upper end of the temperature range displayed

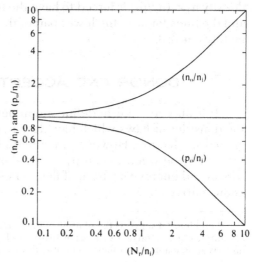

Figure 4–10 Mutual dependence of the free electron and free hole densities (on a scale normalized to the intrinsic density) when an excess of ionizable donor flaws causes electrons to outnumber holes; $N_r = (n_0 - p_0)$. These curves follow Equation 4-24. For the situation shown, electrons are the majority carriers, and holes are the minority carriers.

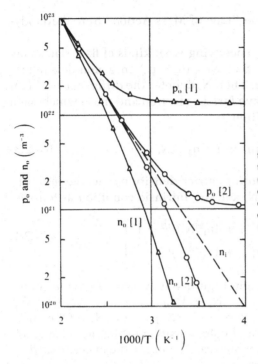

Figure 4–11 Variation with the inverse of the absolute temperature of the majority (hole) and minority (electron) populations for two samples of InAs which become p-type upon cooling. The course of the intrinsic density is shown as the dashed curve.

but become p-type when cooled.[10] The transition temperature is lower for the sample with the smaller density of ionized flaws.

For each sample in Figure 4-11, the product $p_0 n_0$ is the temperature-sensitive but purity-independent quantity n_i^2; while the difference $(p_0 - n_0)$ is the excess of ionized acceptor-like flaws over ionized donor-like flaws, a quantity which varies very little over the temperature range shown in the figure. Almost certainly the majority carrier density for each sample *will* decrease with sufficient cooling (since thermal ionization even of fairly shallow flaw states cannot be maintained at the lowest temperatures), but such a decrease of extrinsic majority density is not apparent for these samples at 250K. Accordingly, the minority carrier density $n_0 = (n_i^2/p_0)$ is forced to have the temperature-dependence of n_i^2, and the lines for n_0 in the lower part of the figure have a slope of approximately $(-\epsilon_i/k_0)$.

DONOR AND ACCEPTOR STATES

It is all very well to define a number N_r as the difference between the densities of flaws which have lost an electron and those which have gained an electron. However, in order to understand how and why a difference N_r between n_0 and p_0 comes about, we must appreciate the nature of and energy spectrum of flaw states. How, then, do donors and acceptors arise?

---

[10] Problem 4.4 suggests the plotting of the Fermi level as a function of temperature for the intrinsic and extrinsic situations of Figure 4-11. The departure of the ratio (p_0/n_0) from unity is accompanied by a movement of ϵ_F downwards from ψ.

Figure 4–12 A rather stylized two-dimensional representation of part of a silicon crystal. Silicon crystallizes in the diamond lattice of Figure 1-31(a), and each atom participates in covalent bonds with its four nearest neighbors. A phosphorus atom is shown substituting for one silicon atom, and this forms a substitutional donor impurity (flaw). The phosphorus atom has one extra electron, not needed for bonding, and the "donor electron" moves in a large orbit of small binding energy.

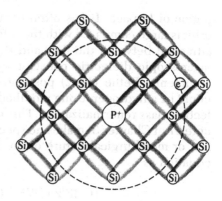

As a simple example of a *donor flaw* in a semiconductor, consider the situation sketched in Figure 4-12 of a silicon crystal with one regular atomic site preëmpted by a phosphorus atom. Each atom in the crystal is shown as establishing four covalent bonds with its four nearest neighbors, which means that a phosphorus atom has an extra electron, one not needed for bonding. This electron is not completely free, for the phosphorus nucleus has one more positive charge than a silicon nucleus; but the extra electron is bound rather weakly to the vicinity of the flaw. Thus we can represent the neutral condition of the flaw (the electron in residence) by a *localized electron state* below the conduction band. Figure 4-13 shows this state at an energy ϵ_d below the bottom of the conduction band, signifying that the electron can be promoted to a (non-localized) Bloch state by receiving additional energy of at least ϵ_d. The donor is ionized and exhibits a localized net positive charge when this happens.

In the case of a phosphorus donor in silicon, the donor binding energy is small, and the orbit of the bound electron is large compared with the interatomic spacing. Nevertheless, a bound donor electron is highly restricted in real space compared with the size of the crystal; thus Figure 4-13 is drawn to show that the bound state extends over a finite

Figure 4–13 Representation of an electron state associated with an electron bound to a donor flaw by an energy within the intrinsic gap. According to the uncertainty principle, the degree of localization necessary for the bound state determines the range of k-space for which the bound state is non-vanishing.

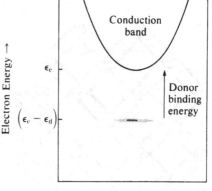

Conduction band

ϵ_c

Donor binding energy

$(\epsilon_c - \epsilon_d)$

Wave Vector **k** $\rightarrow$

region of **k**-space. It has often been argued that when the bound state orbit is large compared with the interatomic distance, then the Coulomb attraction between the flaw and the bound electron is reduced by the macroscopic dielectric constant κ of the semiconductor. Using such arguments, Bethe (1942) postulated that a donor state wave-function might be a hydrogenic one modified by the dielectric constant and by effective mass renormalization. This model (which was referred to in footnote 57 of Chapter 3) makes no specific acknowledgment as to the chemical or metallurgical nature of the flaw site in writing the Schrödinger equation as

$$[(-\hbar^2/2m_c)\nabla^2 - (e^2/4\pi\kappa\varepsilon_0 r)]\psi = \epsilon\psi \qquad (4\text{-}25)$$

Quantum mechanics and the semi-classical Bohr model agree concerning the ground state binding energy and "Bohr radius" for this scaled hydrogenic model of an impurity:

$$\left.\begin{aligned} \epsilon_d &= (m_c e^4/32\pi^2\hbar^2\kappa^2\varepsilon_0^2) \\ a_d &= (4\pi\kappa\varepsilon_0\hbar^2/m_c e^2) \end{aligned}\right\} \qquad (4\text{-}26)$$

There is also a hydrogenic sequence of excited (but not ionized) states.

We shall see that two kinds of complication, one depending on the nature of the band structure and the other on the nature of the impurity, often cause the spatial distribution of bound charge, and the binding energy, to be more complicated than indicated by Equations 4-25 and 4-26. However, before getting involved with the problems and complexities of actual donor flaws, we should verify that *acceptor flaws* are also plausible.

Figure 4-14 portrays in stylized form the simplest kind of substitutional acceptor impurity in silicon, the result of placing a boron atom in a regular silicon site. It might seem natural that one of the four valence bonds to the boron atom itself should go unsatisfied, but in fact the location of the missing bond can move from one interatomic position to another, subject only to the screened Coulomb attraction of the central negative charge. The situation can be described most simply as that of a

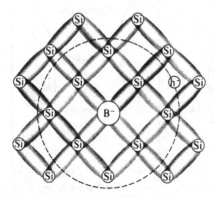

Figure 4-14 A silicon lattice as in Figure 4-12, except that the atom which replaces a silicon atom is now boron, which has one fewer valence electrons than silicon. One valence bond in the vicinity of the boron center must be unfilled when this region of the crystal is electrically neutral. We can think of the changing location of this unfilled bond as the orbital motion of a "bound hole" associated with the boron acceptor.

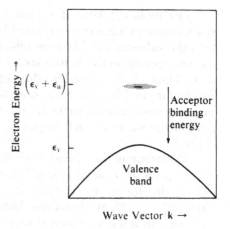

Figure 4–15 Representation of the state associated with a monovalent acceptor flaw. The neutral flaw has a circulating bound hole, and for this condition the *electron state* at energy $(\epsilon_v + \epsilon_a)$ is *unoccupied*. The flaw becomes ionized when an electron is captured, thus filling the flaw level and rendering unoccupied one of the non-localized valence band states. This is not an easy thing to understand at the first reading!

bound hole moving in a state which depends on the dielectric constant and on the effective mass tensor for free holes. The hole can be completely non-localized if it is provided with energy ϵ_a, and so we can represent the neutral state of the acceptor by an unoccupied electron state at energy ϵ_a *above* the top of the valence band. This convention is shown in Figure 4-15. The binding energy ϵ_a and radius a_a for a *hydrogenic* acceptor are given by the expressions of Equation 4-26, except that m_v rather than m_c must be used as the effective mass.

Problem 4.5 concerns a situation of *shallow* donor states (i.e., states of small binding energy and large Bohr radius) under circumstances which *might* make the hydrogenic approach of Equations 4-25 and 4-26 a valid one. However, this model is certainly too simplified for most chemical and metallurgical flaws in most semiconductors.

As an indication of the effects band shapes can have in making the picture of an impurity state more complicated than a scaled hydrogenic one, Kohn[11] remarks that "a donor state may be regarded as a wave-packet which is mainly composed of Bloch waves chosen from near the bottom of the conduction band." Thus the hydrogenic states which result from Equation 4-25 may be reasonable if the effective mass m^* is an energy-independent scalar. But a Schrödinger equation of much greater complexity is necessary when the effective mass is a tensor quantity. Kohn[11] has described the procedures by which this more complicated *effective mass theory* can be established for shallow donors and shallow acceptors in the well known semiconductors germanium and silicon. Only for a handful of semiconductors are the shapes of energy bands known well enough to permit a rigorous calculation of flaw states by the "effective mass" method.

---

[11] W. Kohn in *Solid State Physics, Vol. 5*, edited by F. Seitz and D. Turnbull (Academic Press, 1957). Effective mass theory is complicated for donors in germanium or in silicon, since for each of these semiconductors a surface of constant energy just above ϵ_c has to comprise several prolate ellipsoidal surfaces centered on locations away from the middle of the Brillouin zone. The valence band maximum does occur just at the center of the B.Z., but acceptor states are complicated in both Ge and Si, since *two* valence bands reach a joint maximum simultaneously.

Just about any kind of foreign atom incorporated as an impurity in a semiconductor has a ground state binding energy ϵ_d or ϵ_a which is *larger* than the value calculable from effective mass theory. The extra binding energy, specific to the particular kind of impurity, is called the *chemical shift*. Thus a fully successful theoretical model for flaw states in a semiconductor must be able to account for the chemical shift of each type of impurity atom, and also account for the complete spectrum of excited but not ionized states of each impurity, since each of these is "chemically" shifted.

The origin of the chemical shift is the departure of the attractive potential from $(e^2/4\pi\kappa\epsilon_0 r)$ at small distances from the impurity nucleus. A bound electron in its orbital motion may experience a Coulomb attraction $(e^2/4\pi\kappa\epsilon_0 r^2)$ at distances large compared with the interatomic spacing, but it feels a much stronger attraction at distances so small that the macroscopic concept of a dielectric constant breaks down. This enhanced attraction at small distances has a maximum effect on states of small principal quantum number, and is particularly important for states of s-like symmetry (since the wave-function then does not have a node at the nuclear location).

As a result of effective mass tensor complications, the spectrum of excited states for a flaw may have little resemblance to a scaling of the excited states for a neutral hydrogen atom. Figure 4-16 shows hy-

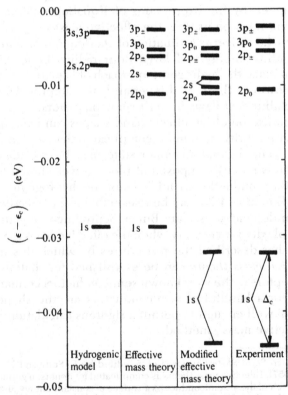

Figure 4–16 The ground state and the first few excited states of a phosphorus donor in silicon, as estimated from a hydrogenic model, from the effective mass theory of Kohn,[11] and as deduced from optical[12] and electrical[13] experiments.

drogenic scaling for a shallow donor in silicon in the left column, with the Kohn effective mass model[11] for any shallow donor in silicon in the next column. The model of this second column has taken into account the complications of the effective mass tensor for the conduction band of silicon (see Figure 3-50 for constant energy surfaces near the bottom of this band), but still makes no specific recognition of what *phosphorus* does to the problem. Note that effective mass asymmetry makes the energy for an n = 2 or n = 3 state depend on the azimuthal and magnetic quantum numbers to a significant extent.

The third column of energy states marked in Figure 4-16 comes from the recognition that a bound electron in its orbital motion experiences a Coulomb attraction at large distances but a stronger attraction at distances so short that the macroscopic concept of a dielectric constant breaks down. The specific nature of the flaw center determines how much the crystal is distorted by its presence, and at what radius the idea of a screened Coulomb attraction breaks down; thus the 2p and 3p states of other donor flaws in silicon look rather like those shown here for phosphorus, but the 1s states are quite different.

An electron in the ground state of a phosphorus donor flaw in silicon has a choice concerning what fraction of time it spends close to the nucleus, and this choice causes a further split of the ground state, as shown in the third column of Figure 4-16 and as supported by the experimental data of the fourth column. The lowest ground state and the p-like excited states have been measured from the wavelengths of optical absorption in a number of experiments, and the energies shown here are those reported by Aggarwal and Ramdas.[12] Transitions to s-like states above the lowest one are optically forbidden, but the energy separation Δ_c up to the "upper ground state" has been inferred from electrical measurements by Long and Myers.[13] We shall refer to their experiment again later.

Complications in the structure of the ground state and in the spectrum of excited states are to be expected for "shallow flaws" in most semiconductors. The problems are even more formidable for deep flaw states, those which have a binding energy which is an appreciable fraction of the entire intrinsic gap. The ground state of a deep flaw is highly localized, and the dielectric constant concept has little validity. One possible modification of the effective mass model which has shown signs of promise for the deeper kinds of flaw state is the *quantum defect technique*, which assigns a non-integral quantum number (the regular quantum number minus a "quantum defect") to each bound state.[14]

These complications become rampant for a *multivalent* flaw, which can bind two or more electrons (or holes) with differing energies. As an extension of the concepts we have discussed relative to phosphorus as a substitutional flaw in a silicon lattice, it can be appreciated that a substitutional sulphur atom in a silicon crystal will be able to release one elec-

[12] R. L. Aggarwal and A. K. Ramdas, Phys. Rev. **137**, A602 (1965).

[13] D. Long and J. Myers, Phys. Rev. **115**, 1119 (1959).

[14] The quantum defect model and other approaches to deeper flaw states are discussed by H. B. Bebb and R. A. Chapman, *Proceedings 3rd International Conference on Photoconductivity*, edited by E. M. Pell (Pergamon, 1971), p. 245.

tron very easily, and another with a little more reluctance. The procedure can be crudely compared with the successive states of ionization for an isolated helium atom. While sulphur in silicon behaves as a *double donor* flaw, an element such as zinc behaves as a *double acceptor*.

More complicated still are the *amphoteric* flaws, which tend to acquire a net positive charge when the Fermi level is very low, yet which pass through neutrality to one or more states of net negative charge when the electrochemical potential is very high. This happens when some noble metal atoms or certain transition element atoms are dissolved in a variety of semiconductors.

OCCUPANCY STATISTICS FOR DONORS AND ACCEPTORS

We have used the Fermi-Dirac occupancy law of Equations 3-45 and 4-3 to describe the probability that *any* Bloch state in a band will be occupied for a given temperature and a given value of $(\epsilon - \epsilon_F)$. Our use of this equation for any state within a band is unimpaired by our knowledge or ignorance concerning whether or not the companion state of the same wave-vector and energy but opposite spin has been occupied. For the wave-functions of Bloch states extend throughout a crystal, and the presence of one electron whose wave-vector has a certain value has no perceptible influence on the availability of the other (opposite spin) state with the same value of **k**; the two electrons will rarely be in the same part of real space. Thus for *band states,* each location in **k**-space provides two distinct and separately tenable states, each with a statistical weight $g = 1$.

As a contrast, consider a localized flaw level such as the ground state of a donor. This level will usually have a statistical weight g which is larger than unity. Thus for the s-like ground state of a "hydrogenic flaw," there are two spin opportunities (up or down), and we should set $g = 2$ for this state. However, *only one of these states can be occupied* by an electron, since the Coulomb repulsion occasioned by the presence of an electron in a bound state of *either* spin possibility would effectively prevent further occupancy.

For a more general type of donor flaw, it may be possible to describe the total flaw wave-function (including orbital and spin alternatives) in g_n ways for the unexcited neutral flaw and in g_i ways when an electron is removed. Later in this section we shall worry about what happens when the probability is appreciable that the flaw is neutral but in an excited state. For the present we exclude this possibility, and consider the relationship between N_{di} ionized flaws and N_{dn} unexcited neutral flaws for which the bound electron is held at an energy $[(\epsilon_c - \epsilon_d) - \epsilon_F]$ relative to the Fermi energy. Thus at temperature T,

$$\left\{\frac{N_{dn}}{N_{di}}\right\} = \left\{\frac{g_n \exp\left[\dfrac{\epsilon_F - (\epsilon_c - \epsilon_d)}{k_0 T}\right]}{g_i}\right\} \tag{4-27}$$

Remembering that excited states are being neglected, we can rearrange Equation 4-27 to express the fraction of the total donor density which has retained an electron in one or other of the available spin possibilities as

$$\left\{\frac{N_{dn}}{N_{dn} + N_{di}}\right\} \equiv \frac{N_{dn}}{N_d} = \frac{1}{1 + (g_i/g_n) \exp\left[\dfrac{\epsilon_c - \epsilon_d - \epsilon_F}{k_0 T}\right]} \qquad (4\text{-}28)$$

In equivalent language, the probability $P_e(\epsilon_c - \epsilon_d)$ that *any one* of a set of donors with a ground state at energy $(\epsilon_c - \epsilon_d)$ has retained its bound electron is

$$P_e(\epsilon_c - \epsilon_d) = \frac{1}{1 + \left(\dfrac{g_i}{g_n}\right) \exp\left[\dfrac{\epsilon_c - \epsilon_d - \epsilon_F}{k_0 T}\right]} \equiv \frac{1}{1 + \beta \exp\left[\dfrac{\epsilon_c - \epsilon_d - \epsilon_F}{k_0 T}\right]}$$

$$(4\text{-}29)$$

The name *spin degeneracy weighting factor* is often applied to the quantity $\beta \equiv (g_i/g_n)$. For a donor type of flaw, β is usually smaller than unity, which has the effect that the occupancy factor $P_e(\epsilon_c - \epsilon_d)$ is larger than 50 per cent when the Fermi energy coincides with the flaw energy.

A similar terminology can be used to express the probability that an acceptor state at energy $(\epsilon_v + \epsilon_a)$ is ionized (i.e., contains an *electron*):

$$P_e(\epsilon_v + \epsilon_a) = \frac{1}{1 + \beta \exp\left[\dfrac{\epsilon_v + \epsilon_a - \epsilon_F}{k_0 T}\right]} \qquad (4\text{-}30)$$

The statistical weighting of neutral or negatively ionized status for an acceptor center makes β a number larger than unity for this type of flaw. The probability of finding a *neutral* acceptor (i.e., one with a bound hole) is

$$P_h(\epsilon_v + \epsilon_a) \equiv [1 - P_e(\epsilon_v + \epsilon_a)] = \frac{1}{1 + \beta^{-1} \exp\left[\dfrac{\epsilon_F - \epsilon_v - \epsilon_a}{k_0 T}\right]} \qquad (4\text{-}31)$$

and as an analog of Equation 4-29 for donor states, we find that $P_h(\epsilon_v + \epsilon_a)$ for an acceptor is larger than 50 per cent when ϵ_F coincides with $(\epsilon_v + \epsilon_a)$.

COMPENSATION

Unless *extreme* care is taken in the purification, preparation, and crystallization of a semiconductor, there is likely to be a significant density of several kinds of donor and acceptor flaws present, with bound states at a variety of energies. At thermodynamic equilibrium, a single normalizing parameter or Fermi energy must be appropriate for a description of all electron states in the crystal at all energies, localized or non-localized.

What happens of course is that any donor flaws with relatively high

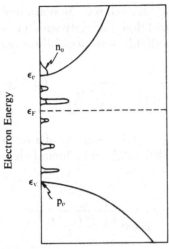

Density of States and Their Occupancy.

Figure 4–17 A compensated semiconductor, containing several kinds of donor flaw and several kinds of acceptor flaw at a variety of energies within the intrinsic gap. Donor states at high energies tend to become ionized, ionizing any acceptor states of lower energy in the process. The densities and degree of compensation of the flaw states will determine how the equilibrium densities n_0 and p_0 are related to the intrinsic situation.

characteristic energies will tend to be emptied, in favor of the acceptance of these by any low lying acceptors present (see Figure 4-17). The presence of acceptor states at low energies is said to *compensate* (partially or totally) the presence of donor states at higher energies. The condition of overall electrical neutrality at equilibrium is

$$(n_0 - p_0) = N_r = \sum_j N_{dj}[1 - P_e(\epsilon_{dj})] - \sum_k N_{ak}P_e(\epsilon_{ak}) \qquad (4\text{-}32)$$

and this is a condition which requires a unique Fermi energy at any specified temperature for a given semiconductor with a given distribution of flaws.

The first term on the right in Equation 4-32 is the aggregate effect of j kinds of donor in losing electrons, and the second term on the right is the aggregate effect of k kinds of acceptor in receiving electrons. The difference between these summations is the quantity we called N_r in Equations 4-23 and 4-24 to indicate the total effect of flaws in enforcing a difference between n_0 and p_0. When we first introduced N_r, we chose not to worry about just how certain flaws could be either neutral or ionized. However, it is now time to see how the degree of occupancy of a set of flaws may vary with temperature, and to observe the resulting effect on the majority free carrier density.

Of course, donor states well below the Fermi energy will make a negligible contribution to Equation 4-32, as will acceptor states appreciably above the Fermi level. Donor states well above ϵ_F and acceptors well below ϵ_F make a significant but temperature-insensitive contribution. It is usually practicable to study the temperature-dependence of occupancy for the flaw states closest in energy to the Fermi energy, assuming that the occupancy of other flaw states is relatively insensitive to the coupled changes of T and ϵ_F. This assumption will cause no appreciable error if all other kinds of flaw states are at least $3k_0T$ higher or lower than the Fermi energy.

FREE ELECTRON DENSITY IN
A COMPENSATED SEMICONDUCTOR

We wish to consider Equation 4-32 in this sub-section, under conditions whereby the occupancy of only *one* band and *one* set of flaw states varies over some relatively broad range of temperature. This can occur for the situation imagined in Figure 4-18, a situation which is dominated by N_d donor flaws at energy ϵ_d below the conduction band. The effect of all flaws other than this set is described as a "net compensator density" N_a. For simplicity it is assumed that the intrinsic gap is very large and that p_0 is zero over the entire temperature range of interest. Equation 4-32 reduces to

$$n_0 = N_d[1 - P_e(\epsilon_d)] - N_a \qquad (4\text{-}33)$$

under these conditions. This can be rewritten as

$$(N_d - N_a - n_0) = N_d P_e(\epsilon_d) = \frac{N_d}{1 + \beta \exp\left[\dfrac{\epsilon_c - \epsilon_d - \epsilon_F}{k_0 T}\right]} \qquad (4\text{-}34)$$

In order to continue the discussion we shall assume that the number of free electrons produced from the flaw levels is small enough so that $n_0 \ll N_c$ at all temperatures. (N_c is the "effective density of conduction band states.") Then we can use the very simple relationship of Equation 4-9 between the free electron density and the corresponding Fermi energy:

$$n_0 = N_c \exp\left(\frac{\epsilon_F - \epsilon_c}{k_0 T}\right) \qquad (4\text{-}35)$$

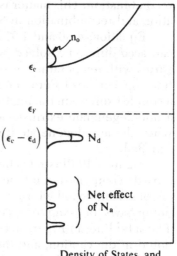

Figure 4-18 Simplified energy state diagram for an n-type extrinsic semiconductor which is controlled by a single set of donor flaws and by the partial compensation of these flaws.

Density of States, and Their Occupancy.

When Equations 4-34 and 4-35 are both valid, we have two conditions relating n_0 and ϵ_F. The joint solution of these provides a quadratic equation either for n_0 or for $\exp(\epsilon_F/k_0T)$. Electing the former alternative, we can say that

$$\frac{n_0(N_a + n_0)}{(N_d - N_a - n_0)} = \beta N_c \exp(-\epsilon_d/k_0T), \quad n_0 \ll N_c \qquad (4\text{-}36)$$

The comparable quadratic equation for the free hole density excited from a set of N_a acceptors of ionization energy ϵ_a, partly compensated by the presence of N_d donors, is

$$\frac{p_0(N_d + p_0)}{(N_a - N_d - p_0)} = (N_v/\beta) \exp(-\epsilon_a/k_0T), \quad p_0 \ll N_v \qquad (4\text{-}37)$$

Equations 4-36 and 4-37 are entirely equivalent in their behavior as functions of temperature and compensation ratio (N_a/N_d), and both equations are widely used. In the following comments, we shall choose to talk about electrons and about Equation 4-36, since this is the situation illustrated by Figure 4-18.

A semiconductor at thermal equilibrium enjoys *detailed balance* between any process of energy transformation and its inverse. Because of transformation processes which interchange electron energy for lattice vibrational energy (phonons) and background black body radiation (photons) within a solid, electrons are constantly being excited to the conduction band, and constantly falling from that band. The rates of *electron generation* and *electron recombination* are equal at thermal equilibrium, and a time-independent n_0 is then observed. If the semiconductor is dominated by compensating sets of flaws, this n_0 will satisfy Equation 4-36. Equation 4-36 for electrons and the companion 4-37 for holes are called the *mass action* equations, because they are written in a way which clearly reflects the law of mass action and the principle of detailed balance in mediating the dynamic balance between generation and recombination. This matter is taken up as part of the discussion of generation and recombination in Section 4.4.

Equations 4-36 and 4-37 are quadratic in n_0 and p_0 respectively, and can accordingly be solved with ease to learn how free carrier density varies with temperature. Inversion of Equation 4-35 then permits us to trace the course of Fermi energy versus temperature. Problem 4.6 gives a simple exercise in tracing the course of n_0 and ϵ_F versus temperature for a non-degenerate extrinsic semiconductor situation, and a reminder of when the non-degenerate limit required for Equation 4-36 is no longer satisfied.

Figure 4-19 illustrates how n_0 varies with reciprocal temperature for a partly compensated n-type semiconductor. The samples in this figure are of silicon doped with arsenic. Each contains about 10^{22} m^{-3} of arsenic donor flaws, but sample (a) contains more compensators than sample (b). The solid lines in Figure 4-19 fit Equation 4-36 with N_d and N_a values as noted in the caption, and the dashed line (c) is the extrapolation for a sample similar to (b) but with *zero* compensation.

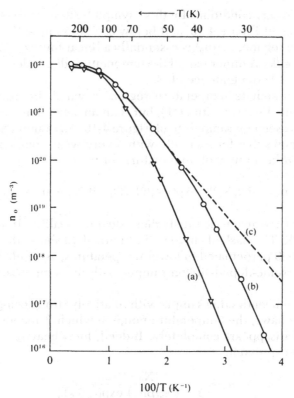

Figure 4-19 The manner in which free electron density varies with temperature for a compensated extrinsic situation governed by Equation 4-36. The data points for samples (a) and (b) are deduced from Hall effect data for silicon samples doped with arsenic donors, and containing smaller amounts of unknown compensating centers. The lines fit Equation 4-36 with the right side set as $2.7 \times 10^{21} T^{3/2} \exp(-615/T)$ m^{-3}, as appropriate for $m_c = 1.08m$, $\beta = \frac{1}{2}$, and $\epsilon_d = 0.053$ eV. A fit for sample (a) requires $N_d = 1.1 \times 10^{22}$ m^{-3} and $N_a = 2 \times 10^{21}$ m^{-3}. For the less heavily compensated sample (b), the required parameters are $N_d = 1.02 \times 10^{22}$ m^{-3} and $N_a = 3 \times 10^{19}$ m^{-3}. Problem 4.7 is concerned with the temperature dependence of the Fermi energy for samples (a) and (b).

As explored in Problem 4.7, the data of Figure 4-19 *may* be slightly affected by occupancy of donor bound states higher than the ground state in part of the temperature range, but this complication can be ignored in the general comparison of the figure with the expectations of Equation 4-36.

Equation 4-36 takes several important limiting forms, which can be observed via Figure 4-19. First we may observe that n_0 approaches a limiting maximum value of $(N_d - N_a)$ for temperatures larger than about $(\epsilon_d/3k_0)$. (This is the situation above about 200K in Figure 4-19.) The high temperature region is commonly called the *exhaustion range* of the system, since no further change in electron density can occur unless the Fermi level moves low enough to start depopulation of other types of flaw or to start massive intrinsic excitation.

For the opposite situation of very low temperatures, Equation 4-36 simplifies to

$$n_0 = [\beta N_c(N_d - N_a)/N_a] \exp(-\epsilon_d/k_0 T), \quad \text{if } n_0 \ll N_a < N_d \qquad (4\text{-}38)$$

and the necessary conditions for this asymptotic form are met below 70K for sample (a) of Figure 4-19, and below 35K for sample (b). In this low temperature regime, $\ln \cdot (n_0)$ is essentially a linear function of $(1/T)$, with a slope of $-\epsilon_d/k_0$. A minor part of the temperature dependence is contributed by the $T^{3/2}$ dependence of N_c.

An intermediate temperature region in which $\ln \cdot (n_0)$ also varies approximately linearly with $(1/T)$, but with an activation energy of only $(\epsilon_d/2)$, can be seen for sample (b) of Figure 4-19. An examination of Equation 4-36 shows that for a sample with a very weak compensation ratio there *should* be a range of temperature for which

$$n_0 = (\beta N_c N_d)^{1/2} \exp(-\epsilon_d/2k_0 T), \quad \text{if } N_a \ll n_0 \ll N_d \qquad (4\text{-}39)$$

and sample (b) manages to satisfy these dual inequalities from about 45K to about 90K. The dashed curve (c) in Figure 4-19 shows the behavior of Equation 4-39 perpetuated to lower temperatures, as would occur for an idealized arsenic-doped-silicon sample with no compensating flaws at all.

For a semiconducting sample with relatively strong compensation of the primary flaws, the temperature range in which Equation 4-39 could be applied disappears completely. Indeed, for a heavily compensated crystal, the expression

$$n_0 \approx \frac{(N_d - N_a)}{1 + (N_a/\beta N_c) \exp(\epsilon_d/k_0 T)} \qquad (4\text{-}40)$$

becomes a tolerable approximation for the *entire* temperature range.

The balance between bound electrons and conduction electrons will change from so-called *reserve conditions* $[n_0 \ll (N_d - N_a)]$ at low temperatures to *exhaustion conditions* $[n_0 \approx (N_d - N_a)]$ at sufficiently high temperatures, whether the compensation in a crystal is small or large. The increase of free electron density upon warming can of course be equally well described as a temperature variation of the Fermi energy, which we may write

$$\epsilon_F = \epsilon_c - k_0 T \ln \cdot (N_c/n_0) \qquad (4\text{-}41)$$

by an inversion of Equation 4-35. It is certainly possible to generate a quadratic equation in $\exp(\epsilon_F/k_0 T)$ by a simultaneous solution of Equations 4-34 and 4-35, and this is discussed in *Semiconductor Statistics*. For most situations it is more practical to find ϵ_F from Equation 4-41, once n_0 has been determined from Equation 4-36 or directly from experimental data. The latter procedure is suggested in Problem 4.7 for exploring the maximum Fermi energy as temperature varies in the samples of Figure 4-19.

The general manner in which ϵ_F varies with temperature when a set of flaws is depopulated is sketched in Figure 4-20. The two limiting forms of behavior are

$$\epsilon_F = (\epsilon_c - \epsilon_d) + k_0 T \ln \cdot [\beta(N_d - N_a)/N_a],$$
$$\text{T small, } n_0 \ll N_a < N_d \qquad (4\text{-}42)$$

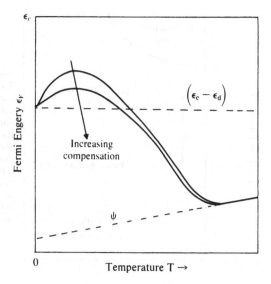

Figure 4-20 The way in which the electrochemical potential ϵ_F must vary with temperature in order to portray the depopulation of a set of partially compensated donors in favor of the conduction band. The resulting free electron density is given by Equation 4-36, and ϵ_F can be described over the entire range of temperature by Equation 4-41. The rising trend at the left of the figure shows the behavior of Equation 4-42, and the subsequent fall of ϵ_F with rising temperature is the exhaustion range of Equation 4-43. (The arrow indicates the shift produced in the curve by adding more compensator centers.) Of course the Fermi level must join the intrinsic Fermi position ψ at sufficiently high temperatures.

and

$$\epsilon_F = \epsilon_c - k_0 T \ln \cdot [N_c/(N_d - N_a)], \quad T \text{ large}, n_0 \approx (N_d - N_a) \quad (4\text{-}43)$$

but no simple and convenient expression is suitable for the entire temperature range. This subject is discussed in *Semiconductor Statistics* in more detail, including the complications which occur when n_0 becomes comparable with N_c and the conditions in the conduction band become "semi-degenerate."

COMPLICATIONS IN THE OCCUPANCY OF FLAW LEVELS

Electrons rise in energy from flaw states to conduction states when any n-type semiconductor is warmed, and eventually intrinsic conduction will take over unless the semiconductor melts first. The last sub-section has described the simplest way in which this thermally-provoked progression can take place. The progression is rather different in many semiconductor/flaw systems, because of a variety of complicating factors. Some of these are mentioned in the present sub-section, which can be omitted by the reader whose main interests lie elsewhere. Future sections on semiconductors do not depend on this material.

Influence of Excited States

One such complicating factor can be the existence of numerous flaws in an excited but not ionized configuration during some part of the temperature range. Figure 4-21 depicts an n-type semiconductor dominated by N_d flaws with N_a assorted compensators, and now we show

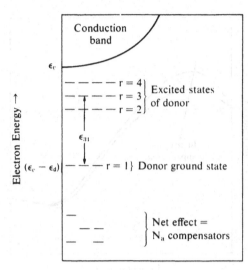

Figure 4-21 A slightly more complicated model than Figure 4-18 for the spectrum of energy states in an n-type semiconductor dominated by a set of N_d donor flaws and N_a assorted compensator flaws. This figure also includes the first few sets of excited states of each donor flaw. When a donor has an electron bound in *any* one of the β_1 ground states or in *any* one of the β_r excited states at energy $(\epsilon_c - \epsilon_d + \epsilon_{rl})$, then Coulomb repulsion prevents occupancy of any of the remaining states by a second electron.

some of the excited states of electrons bound to flaws. A flaw may be neutral with its electron in any one of the β_1 states at energy $(\epsilon_c - \epsilon_d)$, but it can alternatively be neutral with this same electron in one of the β_r states at energy ϵ_{rl} above the ground state. (Here r is an integer which ranges from 2 upwards to represent different types of excited state.)

When excited states are taken into account, the statistical weighting at thermal equilibrium is modified in a way which can make a larger fraction of the flaws neutral for a given Fermi energy. Only if every excited state is at least $3k_0T$ above the Fermi energy can these states be ignored. As discussed in *Semiconductor Statistics*, the weighting effects of the upper bound states modifies Equation 4-36 to read

$$\frac{n_0(N_a + n_0)}{(N_d - N_a - n_0)} = \frac{\beta_1 N_c \exp(-\epsilon_d/k_0T)}{1 + \sum_{r=2}^{\infty} (\beta_1/\beta_r) \exp(-\epsilon_{rl}/k_0T)} \tag{4-44}$$

The denominator of the right side of this equation is larger than unity unless *all* excited states are several k_0T higher than the Fermi energy, in which case the simple analysis of the last subsection is correct. When the denominator of the right hand side of Equation 4-44 is significantly larger than unity, the free electron density is smaller than would have been predicted by Equation 4-36 for the same temperature.

Ground State Split by Crystal Field

It will be remembered from the third and fourth columns of Figure 4-16 that phosphorus donors in silicon are found to have two possible energies for the "1s" ground state, split by an amount Δ_c. The upper ground states are just as effective as excited states for principal quantum number $n > 1$ in providing opportunities for a donor to become excited but not ionized. The effect is once more that n_0 is smaller for a given temperature than would have been the case in the absence of any bound states above the lowest. Equation 4-36 must now be replaced by a modification of Equation 4-44:

$$\frac{n_0(N_a + n_0)}{(N_d - N_a - n_0)} = \frac{\beta_1 N_c \exp(-\epsilon_d/k_0 T)}{1 + (\beta_1/\beta_{ug}) \exp(-\Delta_c/k_0 T) + \Sigma} \quad (4\text{-}45)$$

where the summation sign at the end of the denominator on the right means the same summation as in Equation 4-44. The new term in this denominator is the contribution of the upper ground states.

A ground state split in this manner occurs in many semiconductor/flaw systems in addition to phosphorus-doped-silicon. The experimental data shown in Figure 4-22 happen to be for the Si-P situation, and the curves of this figure show how the upper ground states depress the value of n_0 over a wide temperature range. These states can be ignored at very low temperatures, since then almost all electrons are frozen on the lowest possible bound states. Similarly, the upper and lower ground states (and all excited states) are very sparsely occupied at high temperatures, for all bound states are then overshadowed by the many available states in the conduction band. This produces the exhaustion situation of $n_0 \simeq (N_d - N_a)$. It is for the intermediate temperatures that Equations 4-36 and 4-45 predict different things, because in the latter case there is a partial occupancy of the upper 1s states at the expense of the conduction band. Long and Myers[13] interpreted their experimental (Hall effect) data for n_0 in Figure 4-22 in terms of Equation 4-45 and were able to deduce a value for Δ_c which agrees tolerably well with optical measurements of the same quantity.[12]

Thus the distribution of a given number of electrons between a set of flaws and the conduction band can become complicated if excited

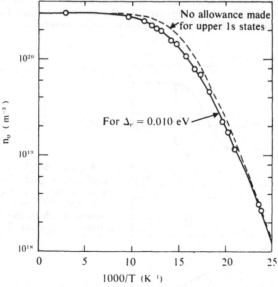

Figure 4-22 Temperature dependence of the free electron density in a partially compensated crystal of phosphorus-doped silicon, after D. Long and J. Myers, Phys. Rev. **115**, 1119 (1959). The solid line shows a fit to Equation 4-45 for $N_d = 6.9 \times 10^{20}$ m^{-3}, $N_a = 3.8 \times 10^{20}$ m^{-3}, and $\epsilon_d = 0.0435$ eV, $\Delta_c = 0.010$ eV. The dashed curve is the expectation of Equation 4-36 for the same values of N_d, N_a, and ϵ_d.

states become appreciably occupied, or if the crystal field splits the degeneracy of the ground states. A similar splitting of the ground state can occur in a large magnetic field, or if a semiconductor is strained in an anisotropic manner. For most (but not all) semiconductors, an isotropic stress will not make a major change in the structure of flaw levels.

Effect of Two or More Types of Flaw

Complications of a different character must be considered if two or more kinds of flaws make varying contributions to the free electron population when the temperature is changed. Thus suppose that an n-type semiconductor has N_{d1} donor flaw states of ionization energy ϵ_{d1}, and N_{d2} completely distinct donor flaw states of a rather larger binding energy ϵ_{d2}. There will usually also be some density N_a of miscellaneous compensating flaws, and for an n-type semiconductor we require that N_a be smaller[15] than $(N_{d1} + N_{d2})$. Then the distribution of the available electrons at thermal equilibrium will satisfy

$$n_0 = N_c \exp(\epsilon_F/k_0T) = N_{d1}[1 - P_e(\epsilon_{d1})] + N_{d2}[1 - P_e(\epsilon_{d2})] - N_a \quad (4\text{-}46)$$

provided that $n_0 \ll N_c$ and that excited states of the various flaws can be neglected. Equation 4-46 is formally a cubic for either n_0 or $\exp(\epsilon_F/k_0T)$, but it simplifies to a quadratic if the separation between the two flaw energies is large compared with k_0T. Figure 4-23 sketches the typical

---

[15] It is mandatory that N_a should be smaller than $(N_{d1} + N_{d2})$ if the semiconductor is to be dominated by the donors at all. However, we can apply the stricter requirement that $N_a < N_{d1}$ if the behavior sketched in Figure 4-23 is to be seen. For if compensator centers are more numerous than the shallower set of donors, the Fermi level will be "locked" to the *lower* donor energy at low temperatures. The situation is then formally that of a single set of N_{d2} donors with $(N_a - N_{d1})$ effective compensating centers.

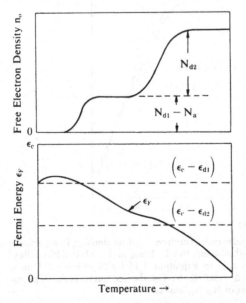

Figure 4-23 A schematic of the variation of n_0 and ϵ_F with temperature for an extrinsic n-type semiconductor controlled by two separate donor species, of which the shallower set is partially compensated.

manner in which n_0 will increase upon warming (accompanied by a temperature variation of electrochemical potential) for two sets of flaws which are well separated in energy. One set of flaws becomes "exhausted" before appreciable ionization of the second set can commence.

Multivalent Flaws

We have already commented that some kinds of flaw have the ability to donate or accept more than one electron. The temperature-dependence of the extrinsic free electron density when multivalent donors control a semiconductor has some similarity to the model just discussed (with two sets of separate monovalent donors), but there are also interesting differences.

Figure 4-24 symbolizes the electronic spectrum for divalent donors. The left column of this figure shows the situation when the Fermi level is near to the conduction band edge. Then each donor is neutral and holds two moderately weakly bound electrons with a *combined* binding energy of $(\epsilon_1 + \epsilon_2)$. These electrons may in practice have similar orbits. However, two separate binding energies ϵ_1 and ϵ_2 are marked in Figure 4-24 because the *flaw* must receive energy ϵ_1 in order that it can become singly ionized, and the second electron then moves in an orbit of binding energy ϵ_2. Most divalent donor flaws will be in this singly ionized condition when $\epsilon_1 < (\epsilon_c - \epsilon_F) < \epsilon_2$, as indicated in the second column of Figure 4-24. Each singly ionized donor offers empty electron states at energy $(\epsilon_c - \epsilon_1)$ with statistical weight of β_1^{-1}. (Remember that $\beta < 1$ for a donor-like flaw.)

When the Fermi level is forced still lower, through compensation or by increased temperature, each donor becomes doubly ionized, and β_2^{-1} electron states per donor are now offered at the energy $(\epsilon_c - \epsilon_2)$. This is illustrated by the third column of Figure 4-24. Note that no states are now shown at the upper level, since these states *do not exist* until the

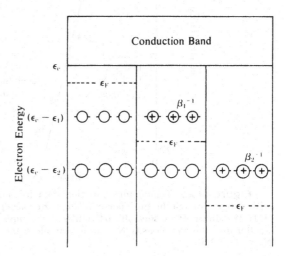

Figure 4–24 The three possible charge configurations for a set of divalent donor flaws in a semiconductor, showing the location of the Fermi energy consistent with each state of ionization.

flaw has been populated with a first electron. As soon as this first electron has been received, a second electron can be captured only into one of the upper states which will then appear.

In general, there will be excited states as well as a ground state for capture by either a singly or doubly charged flaw. However, it is not proper to construe the placement of an electron at energy $(\epsilon_c - \epsilon_1)$ as a valid excited state of the singly ionized configuration; the true excited states for the condition of single ionization will usually be quite different.

The equations for the distribution of electrons over a conduction band, a set of multivalent donor flaws, and miscellaneous compensator centers are unwieldy, but they can usually be broken down into tractable fragments when the possible flaw energy states are separated from each other by several $k_0 T$. (*Semiconductor Statistics* discusses some of the solvable situations.) The simplest situation of multivalent flaws is that of *divalent donors*, with the three states of charge discussed in connection with Figure 4-24. For this case the conduction electron density must equal $n_0 = (2N_2 + N_1 - N_a)$, where N_a is the number of compensator centers, and $N = (N_0 + N_1 + N_2)$ is the total number of donors, of which N_0 are neutral, N_1 are singly ionized, and N_2 are doubly ionized. The situation is controlled by the states at ϵ_1 when $(n_0 + N_a)$ is smaller than N, and is controlled by the deeper level when a combination of high temperature and/or strong compensation makes $N < (n_0 + N_a) < 2N$.

The last sentence is illustrated by the curves in Figure 4-25 for the free hole density in two samples of germanium controlled by divalent

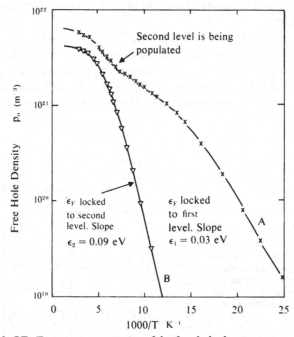

Figure 4–25 Temperature variation of the free hole density in germanium crystals made p-type through the presence of divalent zinc acceptors. Data from W. W. Tyler and H. H. Woodbury, Phys. Rev. **102**, 647 (1956). For sample A the compensator density N_d is smaller than the zinc density N_a, but for sample B the situation must be one for which $N_a < N_d < 2N_a$.

zinc *acceptors*. Sample *A* is rather weakly compensated, and none of the zinc acceptors are doubly ionized at low temperatures. The free hole density is augmented at higher temperatures by receipt of a second free hole from each acceptor in this sample.

However, the more severe compensation for sample *B* prevents *any* of the zinc acceptors from achieving a neutral condition at low temperatures. The mixture of singly ionized and doubly ionized acceptors at all temperatures is demonstrated by the larger slope ϵ_2 for the data of sample *B*.

Amphoteric Flaws

An *amphoteric flaw* is one which is most likely to be neutral when the Fermi level lies at some energy ϵ_0 *within* the intrinsic gap. For a lower value of the electrochemical potential, the flaw is likely to lose one or more electrons (the attributes of a donor center), while for a Fermi energy higher than ϵ_0 the center may add one or more states of negative ionization to behave as an acceptor. This behavior is exhibited for gold centers in germanium, and the spectrum of known flaw states for this center is shown in Figure 4-26. In strongly p-type germanium, gold can act as a single donor, but for strongly n-type germanium each gold atom can be as much as a triple acceptor. The mathematics necessary for a description of how many flaws are in each state of charge is an extension of the set of equations used for "ordinary" multivalent flaws.

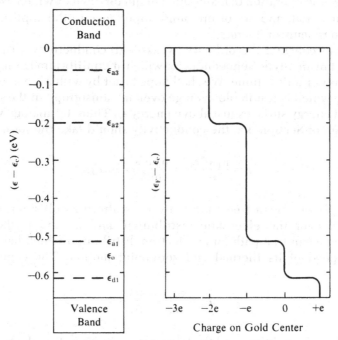

Figure 4–26 The energy levels of the amphoteric gold flaw in germanium. The right portion of the figure shows how the charge per gold atom depends on the Fermi energy at a fairly low temperature. Based on the information of W. C. Dunlap, in *Progress in Semiconductors, Vol. 2* (Wiley, 1957) p. 165.

4.2 Electronic Transport in a Semiconductor

MEAN FREE PATH, MOBILITY, AND CONDUCTIVITY

Probably the most common measurement made of the properties of a semiconducting solid is the electrical conductivity. This measurement depends on the number of mobile charge carriers (electron, holes, or both), on their distributions of thermal velocities, and on the change from an equilibrium distribution provoked by an applied electric field. The theory used is customarily based on the Boltzmann transport equation approach of Section 3.2.

It is always safe to assume that any applied electric field in a metal produces a shift in the distribution of electron speeds which is a small perturbation of the equilibrium distribution. The same assumption is made in considering "ordinary" or "quasi-equilibrium" conductivity in a semiconductor under the influence of a modest field. However, it is possible to apply an electric field in a semiconductor which is large enough to produce a radical change in the velocity distribution. The resulting conduction (which is mentioned again later in the section) is then said to be controlled by *hot carriers*. High field (hot carrier) effects influence the operation of many semiconductor devices which use single crystal material, and are of profound importance to the applications of amorphous semiconductors.

Let us consider the ordinary (weak-field) conductivity arising from transport in an n-type semiconductor which at equilibrium has n_0 mobile electrons per unit volume. We shall hope to get by with a scalar effective mass m_c which is a suitable average over any anisotropy in the shape of constant energy surfaces just above energy ϵ_c. Then, by analogy with the formalism of Section 3.2, the conductivity should take the form

$$\sigma = \frac{n_0 e^2 \tau_m}{m_c} = \frac{n_0 e^2 \bar{\lambda}}{m_c \bar{s}} = n_0 e \mu_n \qquad (4\text{-}47)$$

Here τ_m is the mean free time between scattering collisions, suitably averaged over the electronic distribution, and $\bar{\lambda} = \bar{s}\tau_m$ is the corresponding mean free path for an electron distribution which has $\bar{s}$ as the mean speed of its thermal and zero-point motion. The combination

$$\mu_n = \frac{e\tau_m}{m_c} = \frac{e\bar{\lambda}}{m_c \bar{s}} \qquad (4\text{-}48)$$

is the *electron mobility*, and the transport properties of an electron gas in a semiconductor can be expressed most conveniently in terms of the single quantity μ_n. The mobility is the ratio of drift speed (in m/s) to applied field (in V/m), and accordingly is quoted in units of $m^2/V \cdot s$. This unit is the same as $tesla^{-1}$.

For holes in a p-type semiconductor, the transport can be described in terms of the equivalent *hole mobility* μ_p. With the *ambipolar conduction* of an intrinsic or near-intrinsic conductor, both electron and hole mobilities appear in the expression

$$\sigma = e(n\mu_n + p\mu_p) \tag{4-49}$$

for the total electrical conductivity.

The manner in which electron mobility behaves as a function of the temperature depends on (1) whether or not the electron gas is a degenerate one, (2) the specific relationship between energy and wave-vector for the lowest conduction band states of the semiconductor (in a way we try to account for by employing some "average" effective mass if the mass tensor is actually anisotropic), and (3) what processes dominate the scattering. These processes are likely to include scattering by acoustic phonons, optical phonons, neutral and ionized flaws, and dislocations. When two or more scattering processes have comparable rates, we *should* calculate the mobility (and hence the conductivity) by determining the quantity

$$\tau = [(1/\tau_a) + (1/\tau_b) + (1/\tau_c) + \cdot \cdot \cdot]^{-1} \tag{4-50}$$

as a function of electron energy, and then performing the averaging operations with respect to electron speed or energy in order to determine the correct τ_m. In practice, it is found that the error is not too grave if each scattering process is considered separately as being responsible for a velocity-averaged mean free path, these then being combined after the fashion

$$\bar{\lambda} = [(1/\lambda_a) + (1/\lambda_b) + (1/\lambda_c) + \cdot \cdot \cdot]^{-1} \tag{4-51}$$

This latter (and slightly dubious) procedure really amounts to combining "mobilities" from the various scattering processes with all others neglected in the manner

$$\mu_n = [(1/\mu_a) + (1/\mu_b) + (1/\mu_c) + \cdot \cdot \cdot]^{-1} \tag{4-52}$$

Only in the occasional and particularly scrupulous calculation does one find that scattering processes have been combined in accordance with Equation 4-50 rather than with Equation 4-52.

Figures 4-27 through 4-29 show how the mobility of holes (Figure 4-27) or of electrons (Figures 4-28 and 4-29) varies with temperature for three typical semiconductors in single crystal form. The shapes of these curves are determined by the competition of various scattering processes, a competition which is affected by temperature. We shall have occasion to refer to these figures again as the various scattering mechanisms are discussed.

Degenerate conditions occur when n_0 is large and the temperature

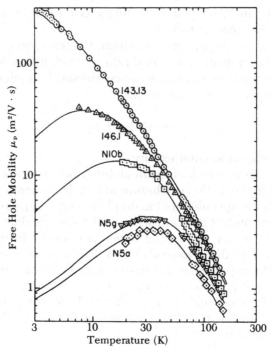

Figure 4–27 Mobility as a function of temperature for free holes in germanium crystals doped with copper acceptors (and with much smaller densities of compensating donors). The copper density ranges from $\sim 10^{19}$ m^{-3} for sample 143.13 to $\sim 10^{22}$ m^{-3} for sample N5a. After P. Norton and H. Levinstein, Phys. Rev. **B.6**, 470 (1972).

is rather small. We might expect the conduction process under these conditions to resemble that in a metal, in which case μ_n would have the temperature-dependence of the electrical conductivity curve shown in Figure 3-1. This stretches the analogy between a metal and a degenerate semiconductor a little too far, since electrons in a semiconductor are much more at the mercy of static imperfections.

For a degenerate electron distribution in a semiconductor, the mean free paths and times of interest are those for electrons within one or two k_0T of the Fermi energy, and the important speed for substitution in the denominator of Equations 4-47 and 4-48 is the Fermi velocity $s_F = [2(\epsilon_F - \epsilon_c)/m_c]^{1/2}$. It is possible to calculate how μ_n *ought* to vary with temperature for such a degenerate distribution in which the dominant scattering is by phonons, ionized flaws, and so forth.[16] However, in practice the mobility in a semiconductor sample with very numerous free electrons usually has a very uninteresting behavior, with little change of mobility over a wide range of temperature. Some examples of this are the two almost completely flat characteristics for InSb samples XS22 and XS25 in Figure 4-29.

There is a more interesting diversity in the patterns of mobility-temperature behavior when electrons in a semiconductor have a non-

---

[16] Electron scattering for degenerate electron distributions is discussed for example in the books by Smith and by Putley, cited in the bibliography at the end of this chapter.

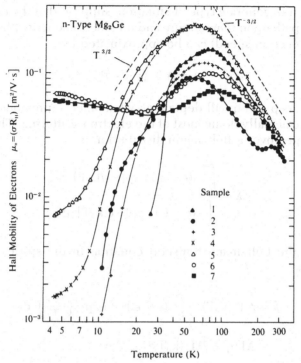

Figure 4–28 Temperature dependence of Hall mobility for electrons in single crystals of n-type Mg₂Ge. Data from P. W. Li, S. N. Lee, and G. C. Danielson, Phys. Rev. **B.6**, 442 (1972). The dashed lines show the conventional expectations for lattice scattering ($T^{-3/2}$) and for scattering by ionized impurities ($T^{3/2}$). The "Hall mobility" for several samples drops anomalously at low temperatures because of impurity conduction.

degenerate (Boltzmann) distribution. When the electron density is small enough and/or the temperature is large enough to validate a nondegenerate approach, then the speed $\bar{s}$ to be used in Equations 4-47 and 4-48 is the quantity

$$\bar{s} = (8k_0T/\pi m_c)^{1/2} \tag{4-53}$$

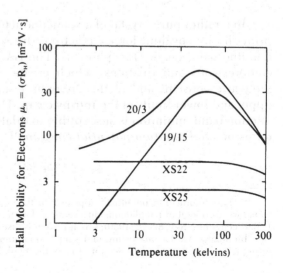

Figure 4–29 The mobility for electrons in n-type samples of InSb, as deduced from a combination of conductivity and Hall effect by E. H. Putley, Proc. Phys. Soc. **73**, 280 (1959). Samples XS22 and XS25 have degenerate electron populations, with $n_0 \sim 10^{22}$ m⁻³. For sample 20/3, $n_0 \sim 3 \times 10^{20}$ m⁻³, and the total density of ionized flaws which can participate in scattering is not much larger. Sample 19/15 has a comparable free electron density, but is more heavily compensated and thus has more ionized flaws to reduce the low temperature mobility.

For this non-degenerate type of situation, suppose that the dependence of the free path upon temperature and on electron energy (that is, on the square of electron speed) can be approximated by

$$\bar{\lambda} = A \ T^p (\epsilon - \epsilon_c)^q \qquad (4\text{-}54)$$

The numbers p and q will depend on what objects (phonons, flaws, and so forth) are actually doing most of the electron scattering. Then the free path averaged over a Boltzmann distribution is

$$\bar{\lambda} = \frac{\displaystyle\int_{\epsilon_c}^{\infty} \lambda(\epsilon - \epsilon_c) \exp[-\epsilon/k_0 T] \cdot d\epsilon}{\displaystyle\int_{\epsilon_c}^{\infty} (\epsilon - \epsilon_c) \exp[-\epsilon/k_0 T] \cdot d\epsilon} \qquad (4\text{-}55)$$

for use in the Boltzmann transport equation in the spirit of Equation 3-33. Thus

$$\bar{\lambda} = A T^p (k_0 T)^{-2} \int_{\epsilon_c}^{\infty} (\epsilon - \epsilon_c)^{q+1} \exp\left(\frac{\epsilon_c - \epsilon}{k_0 T}\right) \cdot d\epsilon$$

$$= A \Gamma(q + 2) T^p (k_0 T)^q \qquad (4\text{-}56)$$

with a temperature-dependence which specifically involves both p and q. Inserting the temperature dependences of Equations 4-53 and 4-56 into Equation 4-48, the mobility should have the form

$$\mu_n = B \ T^{(p + q - 1/2)} \qquad (4\text{-}57)$$

where a knowledge of the constant B is contingent upon the establishment of a suitable microscopic model for electron scattering.

LATTICE SCATTERING

In a rather pure crystal of a semiconductor which is bound together primarily by covalent forces, electrons are scattered predominantly by longitudinal acoustic (LA) phonons. The LA modes produce a series of compressions and dilations, which create a local modulation of the dielectric constant and of the energies ϵ_c and ϵ_v which describe the upper and lower limits of the intrinsic gap. The extent to which the conduction band minimum is susceptible to dilation is usually described in terms of a *deformation potential constant*[17]

$$\epsilon_1 = -V(\partial \epsilon_c/\partial V) \qquad (4\text{-}58)$$

---

[17] The "deformation potential" approach to the scattering of electrons by compressional waves (longitudinal phonons) in a semiconductor was proposed by W. Shockley and J. Bardeen, Phys. Rev. **80**, 72 (1950); this approach is discussed in detail in Shockley's book (see bibliography). An interesting and simple quantum-mechanical picture of the scattering process is described in Section 9.11 of the book by McKelvey (see bibliography).

which is arranged to have the dimensions of an energy. When ϵ_1 is non-zero, any LA phonons of appropriate wave-vector will create peaks and troughs in the minimum conduction band energy, any one of which can result in the scattering of an electron. When an electron encounters a series of compressions and rarefactions in such a direction that there is an angle ϕ between the electron direction and the normal to the compressional planes, then the most probable angle of scattering is $(\pi - 2\phi)$.

The LA phonons which can scatter an electron must have wavelengths at least as large as the electron wavelength. Thus a simple calculation shows that scattering of an electron with energy k_0T by an LA phonon involves a major change in momentum direction but a very small change of electronic energy. (This is confirmed in Problem 4.8.) These relatively long wavelength, low energy phonons remain in ample supply down to temperatures well below the Debye temperature of the solid. At very low temperatures (which are of reduced interest for lattice scattering because other effects interfere when T is small), the scattering by LA phonons becomes anisotropically favorable for scattering in the forward direction, but for most of the temperature range it was shown by Shockley and Bardeen[17] that LA phonons would scatter electrons isotropically. These authors showed that the free path between collisions could be expressed in terms of the deformation potential by

$$\lambda = \frac{4\rho \, v_{LA}^2}{k_0T} \left(\frac{\hbar^2}{2m_c\epsilon_1} \right)^2 \qquad (4\text{-}59)$$

for the rather idealized situation of a conduction band in which the effective mass is isotropic and energy-independent. In this equation, v_{LA} denotes the speed of the long-wave LA phonons and ρ is the density of the solid.

A comparison of Equations 4-54 and 4-59 indicates that LA phonon scattering corresponds with the choices $p = -1$ and $q = 0$ in Equation 4-54. Accordingly, from Equation 4-57 we expect to find a *lattice-scattering mobility*

$$\mu_L = BT^{-3/2} \qquad (4\text{-}60)$$

With this simple kind of scattering, the quantity B can easily be evaluated in terms of the quantities of Equations 4-53 through 4-59. An example of such an evaluation is suggested in Problem 4.9.

A mobility varying as some negative power of temperature is seen for the purer samples at high temperatures in each of Figures 4-27 through 4-29. The variation is close to $T^{-3/2}$ for the electrons in Mg_2Ge and InSb, but is nearer $T^{-1.9}$ for the holes in germanium sample 143.13. Departures from a precisely $T^{-3/2}$ dependence for lattice scattering mobility occur to a greater or lesser extent in almost every semiconductor. These arise from a variety of competing effects, including the nonspherical shapes of constant-energy surfaces for conduction electrons in semiconductors such as silicon or germanium, the energy dependence of effective mass in a semiconductor such as InSb, scattering caused by shear-type acoustic vibrations (TA phonons), and scattering of the more energetic electrons at high temperatures by optical branch phonons. The latter species of phonon can of course produce a large change in the electron energy whether or not there is much of a change in the direction of motion.

IONIZED FLAW SCATTERING

It will be noted in Figures 4-27 through 4-29 that the free carrier mobility in a non-degenerate sample does not continue to rise without limit on cooling. Indeed, for a number of the samples illustrated, the mobility begins to decrease again at the lower temperatures. This is expected to happen in a crystal when the phonon population is depleted by cooling and the mean free path or time is controlled by *ionized-flaw scattering*.

Each ionized flaw in a crystal is a stationary positive or negative charge which can deflect the path of a passing electron, as sketched in Figure 4-30. The orbit executed will be a hyperbolic one whether the interaction is an attractive or repulsive one, and the mathematical arguments required are exactly those used by Rutherford in his model for alpha-particle scattering by atomic nuclei.[18] The Rutherford expression for scattering angle must be scaled for the dielectric constant κ of the medium.

The mean free path associated with Rutherford scattering does not depend explicitly on the temperature of the system but is proportional to the square of electron energy. Inserting these facts into Equation 4-57, we can see that the resulting mobility μ_I should vary as $T^{3/2}$. This is essentially correct, but there is a complication we must consider.

The *maximum* scattering angle in the interaction of an electron with an ionized flaw is 180°, when the electron scores a direct hit on the charged center. Conwell and Weisskopf[19] suggested that the *lower* angle

---

[18] This was originally described by E. Rutherford, Phil. Mag. **21**, 669 (1911). The derivation of the Rutherford scattering equation appears in any textbook on modern and atomic physics.

[19] E. M. Conwell and V. F. Weisskopf, Phys. Rev. **77**, 388 (1950).

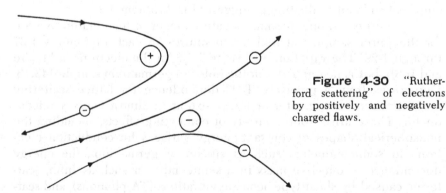

Figure 4-30 "Rutherford scattering" of electrons by positively and negatively charged flaws.

limit of scattering for use in the Rutherford integrals for total scattering cross-section be the angle

$$\phi_{min} = (Ze^2 N_i^{1/3} / \pi \kappa \varepsilon_0 m_c s^2)$$ (4-61)

for an electron of speed s. Here N_i is the density of ionized flaws, each of which has an effective charge $\pm Ze$. Equation 4-61 represents the fact that an electron can never be a distance larger than $(8N_i)^{-1/3}$ away from an ionized flaw, and sets the effective Coulomb force between electron and flaw as zero for that distance.[20]

In the form adopted by Conwell and Weisskopf, the Rutherford scattering result for electron mobility is

$$\mu_I = \frac{(64/N_i)(\kappa\varepsilon_0/Z)^2(\pi/m_c)^{1/2}(2k_0 T/e^2)^{3/2}}{\ln \cdot [1 + (12\pi\kappa\varepsilon_0 k_0 T/Ze^2 N_i^{1/3})^2]}$$ (4-62)

when the integrations are performed with respect to impact parameter (i.e., with respect to scattering angle) and with respect to a Boltzmann distribution of electron speeds. For temperatures which are not *too* small, the logarithmic term in the denominator plays a rather insignificant role in the temperature-dependence of mobility, and

$$\mu_I = \frac{CT^{3/2}N_i^{-1}}{\ln \cdot [1 + DT^2]}$$ (4-63)

has essentially a $T^{3/2}$ dependence as previously remarked. However, note that the quantity called D in Equation 4-63 is inversely proportional to the two-thirds power of the scattering center density. Thus when ionized flaws are rather numerous, μ_I will tend to vary as $T^{-1/2}$ at the lowest temperatures. Problem 4.9 looks at this low temperature region.

The mobility is influenced by ionized-flaw scattering for a number of the samples in Figures 4-27 through 4-29. The $T^{3/2}$ dependence cited in Equation 4-63 is seen very clearly in sample 19/15 of Figure 4-29, and shows up with rather less clarity for several of the samples in Figures 4-27 and 4-28. The quantity plotted for the samples of Figure 4-28 is the "Hall mobility" (σR_H), and this no longer reflects the true behavior of μ_n at very low temperatures for several samples because of impurity conduction (which will be discussed shortly).

OTHER SOURCES OF SCATTERING FOR ELECTRONS

An interpretation of curves such as those of Figures 4-27 through 4-29 is complicated by the miscellany of departures from a perfectly

---

[20] The Conwell-Weisskopf procedure for handling the small-angle scattering when the trajectory of an electron has a large *impact parameter* relative to an ionized flaw is not the only possible one. An alternative procedure, based on screening by free carriers of the ionized impurity Coulomb field, was developed in the 1950s by H. Brooks and C. Herring, and also by R. B. Dingle. Models for ionized impurity scattering were reviewed by D. Chattopadhyay and H. J. Queisser, *Rev. Mod. Phys.*, **53**, 745 (1981).

periodic lattice which *can* scatter electrons of thermal speeds, and by the complexities of band structures that permit additional methods for rearranging crystal momentum.

As an example of the latter class of effects, electrons in the conduction band of a semiconductor such as silicon or germanium occupy several low-energy "valleys" at locations in the reduced Brillouin zone which are related by the crystal symmetry. (The six equivalent valleys for electrons in silicon were shown in Figure 3-50.) The effective mobility of an electron in such a semiconductor is likely to be considerably affected at the higher temperatures by *inter-valley scattering*, as has been discussed by Herring.[21]

Another scattering mechanism for electrons and holes, particularly at reasonably high temperatures and in partially ionic lattices, is that effected by *optical mode* phonons. Whereas acoustic mode phonons mostly have an energy which is very small compared with that of the electrons they scatter (as explored in Problem 4.8), optical mode phonons have a large energy $\hbar\omega_{opt}$ regardless of their wave-vector. Optical phonon scattering (often called *polar mode* scattering) is especially important for a solid with a polar (partially or completely ionic) lattice, since an optical phonon creates an electric dipole field in such a solid.

We can expect polar mode scattering to be unimportant for low temperatures, but it can become predominant in some materials when k_0T becomes comparable with the formation energy $\hbar\omega_{opt}$ of optical phonons. Petritz and Scanlon[22] discuss a model for this scattering in which mobility varies as $\exp(\hbar\omega_{opt}/k_0T)$.

An electron in a polar solid sets up a spatial distribution of lattice polarization about it; each ion is displaced to a small extent by the presence of the electron nearby. This *polarization cloud* surrounding each electron has stored electrostatic energy. Thus when the electron moves through the crystal, its polarization cloud must accompany it. The movement of an electron plus its polarization cloud through a polar lattice is called the motion of a *polaron*.[23] It was at one time speculated by Landau (1933) and by Pekar (1946) that an electron might be self-trapped by the set of deformations it creates in an ionic crystal, so that a polaron would be completely immobile. This drastic view of the polaron is no longer maintained, but it is expected that the effective mass will be large (and the mobility correspondingly small) when the energy states of an electron are perceptibly affected by polaron formation.

In both ionic and covalent crystals, electrons should also be scattered by neutral flaws, for which a simple model has been suggested by Erginsoy.[24] In adapting a model for the scattering of slow electrons by the neutral atoms of a gas, Erginsoy noted that an atom in which the out-

---

[21] C. Herring, Bell System Tech. Journal **34**, 237 (1955).

[22] R. L. Petritz and W. W. Scanlon [Phys. Rev. **97**, 1620 (1955)] discuss polar mode scattering for lead sulphide. This paper gives references to several earlier alternate models for scattering by optical phonons.

[23] The considerable literature concerning polarons in ionic crystals can be reached via the lectures reproduced in *Polarons and Excitons*, edited by C. G. Kuper and G. D. Whitfield (Plenum, 1963).

[24] C. Erginsoy, Phys. Rev. **79**, 1013 (1950).

ermost electron has a wave-function of characteristic radius a_0 will present a target of cross-section

$$\sigma_e \approx (20\hbar a_0/sm_c) \tag{4-64}$$

for an electron of effective mass m_c and speed s. [It must be assumed that the electron has a kinetic energy $(m_c s^2/2)$ which is considerably smaller than the first ionization energy of the atom.] Thus for N_n neutral scattering centers per unit volume, the collision time is

$$\tau_m \equiv (s\sigma_e N_n)^{-1} = (m_c/20\hbar a_0 N_n) \tag{4-65}$$

which dictates a mobility of

$$\mu_N = (e/20\hbar a_0 N_n) \tag{4-66}$$

This neutral-flaw limited mobility does not depend on the temperature, and the scattering efficiency is independent of the electron speed distribution. In practice, neutral-flaw scattering is particularly effective at low temperatures for the faster moving electrons in the thermal distribution, since the ionized-flaw scattering process is weighted in favor of scattering slow electrons. The 1972 data of Norton and Levinstein for mobility of holes in Cu-doped-Ge (see Figure 4-27) suggest that "deep" copper acceptors are less effective as neutral scattering centers than the simple Erginsoy model would predict.

In addition to all the forms of scattering that occur when objects of atomic size interrupt the periodicity of the lattice, electrons can be scattered by objects which are extensive in one or more dimensions. A dislocation line running through a crystal is not very effective in scattering if it is electrically uncharged, but can be quite effective if it consists of a line of electrically charged acceptors.[25] Grain boundaries within a polycrystalline sample certainly affect the mobility of mobile carriers, and may inhibit electron motion from one microcrystal to another if there is a potential barrier at the interface. Because of this, the properties of a semiconductor in single crystal form can be described with much more confidence than the same material in the form of a polycrystalline ingot, polycrystalline film, or pressed powder.

HALL EFFECT AND MAGNETORESISTANCE

Suppose that electrons in the conduction band of a semiconductor must move under the combined influences of an electric field, a magnetic field, and some effective scattering mechanism. If the conduction band enjoys an isotropic scalar effective mass, then the argument can proceed via the Boltzmann transport equation to the result

$$\mathbf{E} = \sigma^{-1}\{\mathbf{J} - [\sigma R_H]\mathbf{J} \times \mathbf{B}\} \tag{4-67}$$

which was noted previously as Equation 3-37 in the classical Lorentz model of a metal. According to this model, the Hall coefficient (which

---

[25] A model for scattering of electrons by an *electrically charged* crystal dislocation was proposed by W. T. Read, Phil. Mag. **45**, 775, 1119 (1954).

determines the magnitude of the response perpendicular to both current and field) is

$$R_H = \frac{-\langle \tau^2 \rangle}{\langle \tau \rangle^2 n_0 e} = \frac{-r}{n_0 e} \qquad (4\text{-}68)$$

The dimensionless quantity r is called the *Hall factor*. It depends on the combination of scattering processes effective under the prevailing conditions and on how the free time between collisions varies with electron energy. The Hall factor is usually fairly near to unity. Thus r = 1 when all electrons move at the same speed, as is true for a classical Drude model or for a degenerate electron gas (in which all members move with the Fermi velocity). We find that r = $(3\pi/8)$ = 1.18 when the electron gas is non-degenerate and scattering is exclusively by LA phonons. When there is appreciable scattering by ionized flaws, r can exceed 1.9 for an isotropic band; on the other hand, r can be as small as 0.7 if the constant energy surfaces for an electron are perturbed sufficiently from a spherical form. The books by Smith and by Putley in the bibliography discuss some of the criteria for a determination of r.

From Equations 4-47 and 4-68, the combination of conductivity and Hall coefficient is $(-\sigma R_H) = r\mu_n \equiv \mu_H$, a quantity often called the *Hall mobility*. The mobility plotted in both Figures 4-28 and 4-29 is actually the Hall mobility $\mu_H = (-\sigma R_H)$, and not the *conductivity mobility* or *drift mobility* $\mu_n = (\mu_H/r)$ itself. For some semiconductors it has been possible to measure the drift mobility independently, but it is often necessary to rely exclusively on Hall mobility data.

The dimensionless quantity $(B\mu_H)$ is sometimes called the *Hall angle*, since the tangent of this angle is the ratio of the transverse to longitudinal electric fields when a transverse magnetic field is applied. (This is reviewed in Problem 4.10.)

For electrons in a semiconductor, as for electrons in the Lorentz classical model of a metal, the conductivity in a magnetic field *should* decrease in accordance with 3-39. Thus a semiconductor should display a maximum magnetoresistance (proportional to B^2) when **B** and **J** are perpendicular, and magnetoresistance should disappear when **B** and **J** are parallel. Things are rarely that simple in practice. For many semiconductors there is still a very healthy magnetoresistance in the longitudinal configuration because of anisotropy in the tensor of effective mass or anisotropy in the scattering time.[26] The complexities of competing scat-

---

[26] In the theory of magnetoresistance there are some interesting complications which are appropriate for a band in which an energy just above the minimum must be described by a set of ellipsoidal energy surfaces in k-space (such as those of Figure 3-50). The relevant theory is discussed in the books by Smith and by Putley cited in the bibliography. The early measurements of "anomalous" magnetoresistance in germanium [G. L. Pearson and H. Suhl, Phys. Rev. **83**, 768 (1951)] and in silicon [G. L. Pearson and C. Herring, Physica **20**, 975 (1954)] merged beautifully with the OPW band calculations of Herman [Phys. Rev. **93**, 1214 (1954)] in predicting complicated band structures which subsequent measurements have confirmed. The magnetoresistive behavior of germanium and silicon is reviewed by M. Glicksman in *Progress in Semiconductors, Vol. 3* (Wiley, 1958).

tering mechanisms may make the resistance rise more rapidly or more slowly than a strict B^2 dependence. And of course for large magnetic fields and weak scattering there is a possibility of seeing the oscillatory Shubnikov-de Haas effect, as in Figure 3-66. These complications are to be welcomed rather than mourned, for they have provided much insight into the details of how electrons are distributed as functions of energy and wave-vector in the bands of a solid.

MULTI-BAND AND INTRINSIC TRANSPORT

The electrical conductivity in an intrinsic semiconductor is the sum of the separate conductivities from the two bands:

$$\sigma_i = en_i(\mu_n + \mu_p) \qquad (4\text{-}69)$$

In a semiconductor which is not quite intrinsic but for which both bands make a conduction contribution, this must be generalized to

$$\sigma = e(n_0\mu_n + p_0\mu_p) \qquad (4\text{-}70)$$

where n_0 and p_0 are related by the requirement that their product is n_i^2 (for non-degenerate statistics) and that their difference is any difference between densities of ionized donors and ionized acceptors. For an n-type semiconductor we have seen in Equation 4-24 that these requirements lead to

$$n_0 = (p_0 + N_{di}) = \{(N_{di}/2) + [(N_{di}/2)^2 + n_i^2]^{1/2}\} \qquad (4\text{-}71)$$

where N_{di} symbolizes the excess of ionized donors over ionized acceptors. For a p-type semiconductor we can similarly write

$$p_0 = (n_0 + N_{ai}) = \{(N_{ai}/2) + [(N_{ai}/2)^2 + n_i^2]^{1/2}\} \qquad (4\text{-}72)$$

The latter equations have been written because they are appropriate for an analysis of conductivity curves such as those of Figure 4-31. Three of the InSb samples shown in this figure become n-type upon cooling; the other four become p-type when cooled from the intrinsic range. Note that since $(\mu_n/\mu_p) \equiv b$ is considerably larger than unity for this semiconductor, the conductivity required by Equations 4-70 and 4-72 in the intermediate temperature range is *smaller* for the p-type samples than that for an intrinsic sample at the same temperature.

In two-band transport, the Hall effect is more complicated than the conductivity since the Lorentz force encourages both the electron and hole populations to be displaced towards the same direction. (If the electric field causes electrons to drift from left to right, and the magnetic field encourages a clockwise deflection, then holes must be drifting from right to left in the electric field, and are deflected in a counter-clockwise manner by the magnetic field.) Thus electrons and holes attempt to cancel the contributions of each other towards a Hall effect (see Problem 4.11) and

$$R_H = \frac{r(p_0\mu_p{}^2 - n_0\mu_n{}^2)}{e(p_0\mu_p + n_0\mu_n)^2} \tag{4-73}$$

Examples of the kinds of curves generated by this equation can be seen in Figure 4-32, which shows the same InSb samples as in Figure 4-31. It will be seen that p-type samples have a positive Hall coefficient at low temperatures but a negative coefficient above the temperature for which $n_0\mu_n{}^2 = p_0\mu_p{}^2$. Moreover, the negative branch of the Hall coefficient for these formerly p-type samples goes beyond the intrinsic curve, just as the conductance behavior of these same samples dipped below the intrinsic line.

The reader can learn a lot about conduction in a semiconductor and the interplay of hole-dominated and electron-dominated phenomena from an analysis of curves such as these; a modest attempt in this direction is suggested in Problem 4.12.

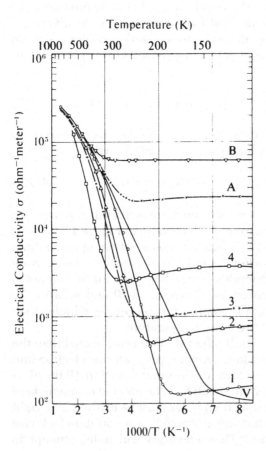

Figure 4-31 Electrical conductivity expressed logarithmically as a function of reciprocal temperature for samples of indium antimonide. Samples V, A, and B are n-type. Samples 1, 2, 3, and 4 are p-type at low temperatures. After O. Madelung, *Physics of III-V Compounds* (Wiley, 1964).

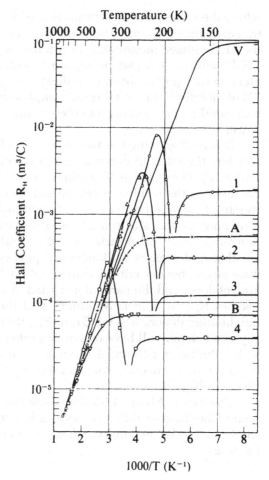

Figure 4-32 Hall coefficient as a function of reciprocal temperature for the seven samples of InSb whose conductivity curves appear in Figure 4-31. After O. Madelung, *Physics of III-V Compounds* (Wiley, 1964).

IMPURITY CONDUCTION

In all of our discussion of flaws to this point, we have characterized them as having localized wave-functions for bound states of finite binding energy. Thus one would expect that cooling of an extrinsic semiconductor towards absolute zero would result in the recapture of *every* mobile electron (or hole) to give an infinite resistivity. This picture is correct *if* the average spacing of flaws is very large compared with the radius a_d of the flaw ground state wave-function. However, the behavior at low temperatures is different if there is any possibility that electrons may travel from one flaw to another without entering the conduction band. Such transport is known as *impurity conduction*.

When impurity conduction is able to occur among the donors of an n-type semiconductor, the total conductivity can be expressed approximately as

$$\sigma = n_0 e \mu_n + \sigma_2 \exp(-\epsilon_2/k_0 T) + \sigma_3 \exp(-\epsilon_3/k_0 T) \qquad (4\text{-}74)$$

where the second and third terms relate to two different kinds of electron transport among donors. These impurity conduction processes also affect the Hall voltage measurable in a transverse magnetic field. Thus, the "Hall mobility" σR_H when impurity conduction contributes to the total electron transport is not a true indication of μ_n. This effect resulted in the "Hall mobility" curves for some samples in Figure 4-28 which dropped very rapidly upon cooling into the impurity conduction range of temperature.

Before discussing the two processes of impurity conduction, let us consider the effect of donor-donor proximity upon the donor ionization energy ϵ_d. There are at least three reasons why strong doping should decrease ϵ_d: the residual potential energy of an electron with respect to its parent donor if it cannot travel far without encountering other donors; screening by the electron gas which accompanies a large density of ionized donors; and polarization of neutral centers. The consequences are that ϵ_d decreases as donors are pushed into closer proximity, and disappears above a critical concentration N_{crit} which is specific to the kind of donor and the particular semiconductor. When the donor density reaches N_{crit}, it is usually found[27] that the *average* distance between neighboring donors is some $5a_d$ to $6a_d$; there is a strong probability that any given donor will have one or two other donors within a distance $3a_d$ to $4a_d$. Problems 4.13 and 4.14 are concerned with separation of donors in an n-type semiconductor, and the application of this subject to the value for N_{crit}.

The data in Figure 4-33 show how the ionization energy for arsenic donor flaws in germanium varies as a function of the ionized donor density N_{di}. It can be seen that the data are fitted quite well by an equation of the form

$$\epsilon_d = \epsilon_{do}[1 - (N_{di}/N_{crit})^{1/3}] \qquad (4\text{-}75)$$

which makes the reduction of ionization energy inversely proportional to the distance between *ionized* flaws.

For a flaw density which exceeds N_{crit}, no thermal activation is necessary for the electrons associated with the flaws, and the semiconductor has become an *impurity metal*. In such a heavily doped semiconductor, the free carrier density and probably the electrical conductivity are quite insensitive to temperature, except for intrinsic conduction as a possibility at a sufficiently high temperature. The samples of lowest resistivity in Figure 4-34 fall into this category.

---

[27] The "average donor spacing" and "nearest donor-donor distance" would be the same only if donors were situated on a precise super-lattice; each donor would then have 12 neighbors at the same distance $(2/N_d^2)^{1/6}$. Since impurities are randomly distributed in any real semiconductor, each donor interacts mostly with the one to three other donors likely to be found with the range $(3/4\pi N_d)^{1/3}$. Both Mott and Hubbard have considered the distance likely to cause collapse of ϵ_d and the onset of a quasi-metallic state when $a_d N_{crit}^{1/3} \simeq 0.25$. Literature on this topic is summarized by N. F. Mott, *Metal-Insulator Transitions* (Taylor and Francis, 1974).

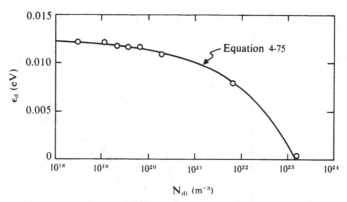

Figure 4-33 The ionization energy for arsenic donor flaws in germanium, plotted as a function of the density N_{di} of *ionized* flaws in the temperature range for which the fit was made. [After P. Debye and E. M. Conwell, Phys. Rev. **93**, 693 (1954).]

The curves in Figure 4-34 for antimony concentrations smaller than N_{crit} show that progressive cooling is at first accompanied by a rapid rise of the electrical resistivity (i.e., a fall of conductivity) since electrons are forced to leave the conduction band in favor of localized states. But the reduction of conductivity with falling temperature becomes much more gradual when the second or third term on the right of Equation 4-74 controls the low temperature conduction. The respective opportunities for the mechanisms of these two terms depend greatly on the densities of majority impurities and of compensator centers. These mechanisms have been discussed (within the context of a model involving shallow impurities) by Davis and Compton[28] in the light of their detailed measurements (as exemplified by Figure 4-34) with antimony-doped germanium.

The term $\sigma_3 \exp(-\epsilon_3/k_0T)$ is the *hopping conduction* of electrons tunneling from neutral to ionized donors in a partially compensated semiconductor. Hopping conduction does not by itself produce a large conductivity, but with some shallow impurity systems it can be strikingly dominant at low temperatures when the average spacing $<r>$ between participating donors is as large as $10a_d$ to $20a_d$. Tunneling takes place with the donor ground states spread over an energy range $\epsilon_3 \simeq (e^2/4\pi\kappa\varepsilon_0<r>)$ as a consequence of the random Coulomb fields arising from ionized donors and ionized compensating acceptors. Thus the tunneling transitions have to be phonon-assisted and display a thermal activation energy. If tunneling to *more remote* but *energetically favorable* donors becomes popular at very low temperatures, then Austin and Mott[29] have argued that the hopping conductivity term should convert to the form $\sigma_3' \exp(-A/T^{1/4})$.

Samples E3 through F3 in Figure 4-34 show the $\sigma_3 \exp(-\epsilon_3/k_0T)$

[28] E. A. Davis and W. D. Compton, Phys. Rev. **140**, A2183 (1965).
[29] I. G. Austin and N. F. Mott, Advanc. Phys. **18**, 41 (1969).

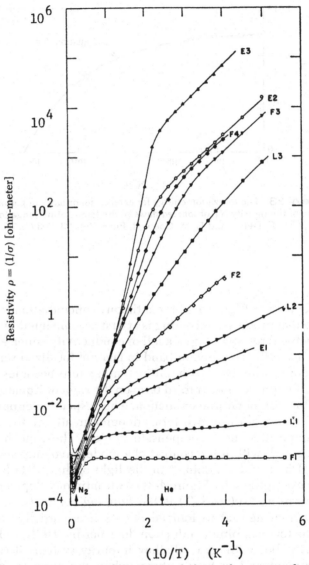

Figure 4-34 Temperature dependence of electrical resistivity for germanium samples doped primarily with antimony, in an amount ranging from $\sim 7 \times 10^{21}$ m^{-3} for sample E3 to $\sim 2 \times 10^{23}$ m^{-3} for sample F1. [After E. A. Davis and W. D. Compton, Phys. Rev. **140**, A2183 (1965).]

hopping behavior below about 5K, and this term probably dominates the conduction of sample L3 just below the range shown in the figure. For samples L3 through E1, the density of donors is large enough for the $\sigma_2 \exp(-\epsilon_2/k_0T)$ mechanism to be in control for most of the temperature range in Figure 4-34. This is impurity conduction of a type which requires closer proximity of the participating donors but does not need acceptors present to keep the donors partly ionized. The energy ϵ_2 is believed to be associated with the energy necessary to place a second electron onto a donor which previously had one (i.e., was neutral). Thus the "band" in which conduction takes place is one of negatively charged donors (a D^- band). Some alternative explanations for the ϵ_2 mechanism have been proposed, but the concept of conduction in D^- states still seems to be the most likely.[28]

All states associated with the donors become a completely non-localized part of the conduction band when $N_d > N_{crit}$. However, for the localized states that exist when $N_d < N_{crit}$, an important distinction should be drawn between conduction of electrons which receive excitation to the conduction band and conduction of electrons which travel from one impurity to another. We have used the Boltzmann equation to describe the transport of electrons in non-localized Bloch states, and discussed the relationship between a measured electron mobility and the equivalent mean free time and mean free path between scattering collisions. All of this is different for transitions from one localized state to another.

Thus the "mobility" measured in an impurity conduction situation is apt to be very small. This would infer a mean free path comparable with or smaller than the size of an atom, except that the conventional idea of a mean free path is invalid. The general motion of electrons from one end of a sample to the other in the presence of an electric field is then in the nature of a *percolation* process.[30] Random variations in donor-donor separation mean that some paths for an electron are *connected* from one end of the sample to another, while other microscopic regions of the solid fail to achieve connectivity with one or both current contacts. We shall encounter percolation again when amorphous semiconductors are discussed, and then the division into connected and disconnected regions will be determined by electron energy.

One consequence of percolation current flow is that only part of the total volume contributes to the low temperature D.C. hopping conduction when $N \ll N_{crit}$. The contributing volume, and the conductivity, can thus increase dramatically when measurements are made with an alternating current. Figure 4-35 shows an example of this, for weakly doped silicon, and here $\sigma_\omega \propto \omega^{0.8}$. Comparable differences between D.C. and A.C. conductivity are seen in the hopping currents of some amorphous semiconductors and other "low mobility" solids, which we turn to in the next sub-section.

---

[30] The difference between diffusion and percolation is discussed by V. K. S. Shante and S. Kirkpatrick, Advanc. Phys. **20**, 325 (1971). The relevance of percolation for electron motion is treated by J. M. Ziman, J. Phys. C., **1**, 1532 (1968) among others.

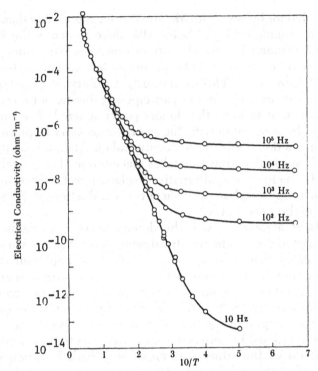

Figure 4–35 D.C. and A.C. conductivity for weakly doped and weakly compensated n-type silicon, showing a dramatic enhancement of the opportunities for low temperature hopping conduction at large alternating frequencies. From M. Pollak and T. H. Geballe, Phys. Rev. **122**, 1742 (1961).

CONDUCTION IN LOW MOBILITY SOLIDS

Apart from low temperature impurity conduction in semiconductors which otherwise demonstrate conduction by highly mobile electrons and/or holes at higher temperatures, there are several classes of solids for which the mobility is small at all temperatures. The preceding remarks concerning hopping conduction and percolation are relevant to many of these solids at room temperature.

The smallness of conductivity and of mobility arises for differing reasons in the differing classes of solids: ionic crystals, organic semiconductors, transition metal oxides and sulphides, amorphous semiconductors, etc. Immediately apparent applications of materials in these classes have been exploited from time to time, but the underlying physical principles have received much less attention than the theory of "easier" solids. This attitude is now changing. After decades of quasi-neglect, research on many of these materials has become popular in the 1970's, with a rapidly growing literature.[31]

---

[31] See *Conduction in Low-Mobility Materials, Proceedings 2nd International Conference* (eds. N. Klein, D. S. Tannhauser and M. Pollak) (Barnes & Noble, 1971) for a review of most classes of low mobility solids.

The low mobility seen in many ionic solids is partly the consequence of polaron formation and partly due to the presence of many localized defects which can trap a moving carrier. In a transition metal oxide, the electrons for conduction are derived from the partially filled 3d states of the isolated metallic atoms, and these are too tightly bound to the parent atoms for formation of a "band" in the Bloch sense of totally non-localized states. Thus an electron must endeavor to hop from one atom to another. Some transition element oxides and sulphides make a transition[32] to a *metallic* state upon warming, since enough overlap then exists to permit the non-localization of the states into bands – of which the "3d" band is only partly full.

Organic Semiconductors

Organic semiconductors have been studied for many years, mostly in the form of pressed polycrystalline powders (which is not ideal for understanding the basic properties) and occasionally in single crystal form. An organic semiconductor is a good example of a *molecular solid*, in that each molecule has strong covalent bonds between its constituent atoms, but molecules are assembled into a solid with usually no more than van der Waals forces. A stronger interaction of molecules occurs when an organic material is encouraged to polymerize, but if the example of polyethylene in Figure 3-21 is anything to judge by, a large conductivity is not an automatic result for a highly cross-linked organic solid.

The distressing truth must be admitted that most organic solids are very poor conductors, with room temperature conductivities widely reported[33] as small as 10^{-17} ohm^{-1}m^{-1}. Anthracene is one of the most extensively studied molecular solids, and here the room temperature conductivity is 10^{-12} ohm^{-1}m^{-1} in the favorable direction for current flow and considerably less in other directions. Conduction in the typical organic solid occurs by hopping and/or tunneling, and the conductivity frequently varies in accordance with

$$\sigma = \sigma_0 \exp(-\Delta\epsilon/k_0 T) \tag{4-76}$$

over a wide temperature range. The activation energy $\Delta\epsilon$ often appears to be associated with mobility variation rather than with temperature dependence of the electron supply.

One group of organic solids which lead to possibilities for a much higher conductivity comprises the *charge-transfer complexes*, in which one type of molecule acts as a donor (D) while another type also present in the crystal acts as an acceptor (A). For a complex in which these ingredients are present in a 1:1 ratio, the ground state wave-function is a

---

[32] See N. F. Mott, *Metal-Insulator Transitions* (Taylor and Francis, 1974); and also N. F. Mott and E. A. Davis, *Electronic Processes in Non-Crystalline Solids* (Oxford Univ. Press, 2nd ed., 1979).

[33] A source of older information on many weakly conducting organics is F. Gutmann and L. E. Lyons, *Organic Semiconductors* (Krieger, 1983). More recent developments are covered by M. Pope and C. E. Swenberg, *Electronic Processes in Organic Crystals* (Oxford Univ. Press, 1982).

resonance between ionic and non-ionic forms of the donor-acceptor relationship

$$\psi = \alpha\psi_0(A:D) + \beta\psi_1(A^-:D^+) \tag{4-77}$$

It was noted in Section 1.1 that such a resonance between ionic and covalent configurations leads to a stronger bond. Here, resonance between ionic and van der Waals bond configurations inevitably leads to a stronger bond than the van der Waals mechanism could produce by itself. It also leads to a closer packing than usually found in molecular crystals, which opens up the possibility for overlap and broadening of levels into bands. The bands are still narrow, with large effective masses and small mobilities, but the possibilities for conduction are much improved compared with that of a single organic compound. Tight packing occurs particularly with complexes resulting in the formation of a linear columnar arrangement of stacked donor and acceptor units. This is a feature of a number of the charge-transfer complexes formed by tetracyanoquinodimethane (TCNQ) with organic, inorganic, and metallic donor entities.[34] Figure 4-36 illustrates the structure of TCNQ, and the way an extra electronic charge is distributed. Some TCNQ salts are metallic or semimetallic, but the instability of a columnar lattice against a Peierls distortion (see page 283) often results in a semiconductor situation. This also happens in charge-transfer salts involving other recently-synthesized organic acceptor radicals.

Compacted samples of TCNQ complexes frequently have a conductivity which varies in the manner of Equation 4-76 over a wide temperature range; single crystals have the same temperature dependence and additionally show a massive anisotropy of conductivity. Siemons *et al.*[35] show data for $Cs_2(TCNQ)_3$ which give an excellent fit to Equation 4-76 over 13 decades of resistance [$\sigma_0 = 4 \times 10^4$ ohm^{-1}m^{-1}, $\Delta\epsilon = 0.33$ eV, for a room temperature conductivity of 0.1 ohm^{-1}m^{-1}]. Other TCNQ complexes have shown a room temperature conductivity as large as

---

[34] Originally reported by R. G. Kepler, P. E. Bierstedt and R. E. Merrifield, Phys. Rev. Letters **5**, 503 (1960).

[35] W. J. Siemons, P. E. Bierstedt and R. G. Kepler, J. Chem. Phys. **39**, 3523 (1963).

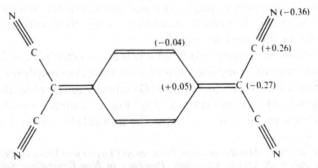

Figure 4–36 Structure of TCNQ. The numbers in parentheses show the anticipated charge in units of e for each atom (and atoms of equivalent position) when the molecule is in the acceptor charge state TCNQ$^-$, according to J. G. Vegter *et al.* in the proceedings of the 1971 "low mobility" conference (footnote 31).

10^4 ohm^{-1}m^{-1} in the favorable direction, often with a value 100 times smaller in directions perpendicular to this.

The easiest thing to measure in an organic semiconductor is the electrical conductivity. This has sometimes been supplemented by measurements of magnetic susceptibility or of electron spin resonance (E.S.R.). Attempts to measure mobility by conventional techniques have usually proved frustrating. This is not too surprising, for our concepts of such things as Hall effect, magnetoresistance, and thermoelectric power are based on the Boltzmann equation approach to perturbation of a mobile electron gas, and the ideas of mean free path and mean free time between collisions. All of these ideas are invalid for a low-mobility material in which transitions occur only by hopping or tunneling. Hall voltages have often been sought in organic conductors, sometimes with total failure and often with ambiguous results. Thus complexes of TCNQ with N-methylphenazinium can yield either negative or positive Hall voltages,[36] each indicative of a "Hall mobility" with an absolute magnitude of less than 10^{-6} m^2/V·s.

Another interesting type of conductor based on organic chemistry is the "doped polymer", of which doped polyacetylene[37] has provided some of the most interesting information. Here one deals not with a single crystal, but a polymer whose degree of orientational ordering depends on the skill in growing a (CH)$_x$ film. Undoped, this is highly resistive, but highly conducting n-type and p-type forms can be achieved by doping with alkalis, halogens, etc.

Amorphous Semiconductors[38]

Amorphous solids have been mentioned in Chapter 1 in connection with their structural features: the lack of long range order, even though each atom *may* have just the conventional number of nearest neighbors at the conventional distance. The electronic properties of amorphous or glassy and disordered systems are our concern in these paragraphs. Many studies of electrical characteristics for amorphous solids have been made in the past, but these have often been hampered by lack of control over the conditions of preparation and the resulting structure.

The most important amorphous semiconductor in past years has been the vitreous phase of selenium, the active material in many xerographic copying machines. Amorphous materials with semiconducting properties exist as elements, as binary compounds, and as ternary and even quaternary alloys involving Si and Ge from Group IV; P, As, Sb, and Bi from Group V; and the chalcogenide elements O, S, Se and Te from Group VI of the Periodic Table. These are the main participants, though some transition metal oxides can exist in an amorphous semiconducting form, and other elements (as in CdAs$_2$Ge) may be incorporated with members of the group just listed.

---

[36] A. R. Blythe and P. G. Wright, Phys. Letters **34A**, 55 (1971).

[37] See for example J. C. W. Chien, *Polyacetylene : Chemistry, Physics, and Materials Science* (Academic Press, 1984).

[38] See the books by Adler *et al.*, Mott and Davis, and Tauc, in the bibliography at the end of this chapter.

An "ideal covalent glass" has been defined as a random network with no long range order but excellent short range order. Such a glass should have no structural defects (such as voids), and should permit local valence requirements to be satisfied for every atom—i.e., there should be no dangling bonds. Probably the nearest approaches to this ideal are amorphous films of silicon and germanium prepared by vacuum evaporation onto a cool substrate. The electron diffraction data of Moss and Graczyk shown in Figure 1-47 indicate that each silicon atom has four silicon neighbors at the same distance as in crystalline silicon. Moreover, both amorphous silicon and germanium can be deposited as films with a density near to that of the respective crystals. There can be few voids in an amorphous Ge film with a density 97% of that for a Ge crystal.

The intrinsic gap ϵ_i (as measured optically) is slightly smaller for a very high quality amorphous Si or Ge film than in the corresponding crystalline material. Amorphous layers of greater disorder show a gap which is smaller and less distinct. Other electronic properties show a much greater difference between crystal and "ideal glassy" structures. Thus Figure 4-37 shows four "scans" of the density of states in the germanium valence band carried out by photoemission. The upper curve for a layer evaporated onto a substrate held at 260°C shows the van Hove singularities in the density of states for the valence band of crystalline germanium. These singularities are not evident in any of the three lower curves for amorphous germanium. Such an amorphous film may be crystallized by subsequent warming, and then the van Hove singularities of the crystal Ge band appear. It must be concluded that there *is* a $g(\epsilon)$ for valence band states in amorphous Ge, but it does not resemble $g(\epsilon)$ for crystalline Ge.

Figure 4-38 shows how the electron drift mobility and electrical conductivity vary with reciprocal temperature for a 1.3 μm evaporated

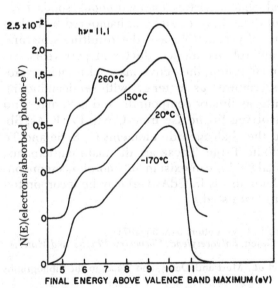

Figure 4-37 The energy spectrum of photoemitted electrons from germanium layers, using 11.1 eV photons to survey the density of states in the valence band. The various curves (with staggered zeros on the ordinate) are for different substrate temperatures as the layers were evaporated; the lower three are amorphous, while the layer formed at 260°C is crystalline. From C. G. Ribbing, D. T. Pierce, and W. E. Spicer, Phys. Rev. **B.4,** 4417 (1971).

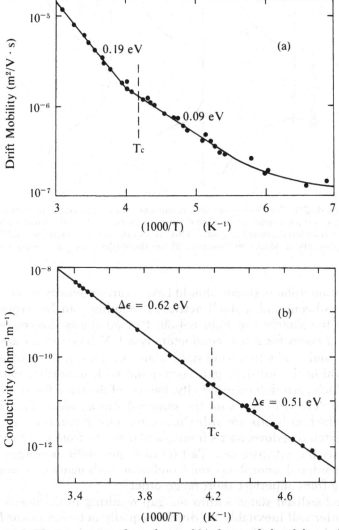

Figure 4–38 Temperature dependence of (a) electron drift mobility, and (b) electrical conductivity for an amorphous silicon film. After P. G. Le Comber and W. E. Spear, Phys. Rev. Letters **25**, 509 (1970).

film of amorphous silicon. The values of $\Delta\epsilon$ quoted in part (b) refer to a comparison with Equation 4-76, and the activation energies marked on the mobility curve similarly are the values needed for comparison with $\mu \simeq A \exp(-\epsilon_\mu/k_0T)$. Since $\epsilon_i = 1.1$ eV for crystalline silicon, and $n_i \sim \exp(-\epsilon_i/2k_0T)$ from Equation 4-17, there is a superficial resemblance between the slopes of the two portions of the conductivity line in Figure 4-38(b) and the slope of intrinsic conduction in an ordered silicon lattice. However, both conductivity and mobility for this amorphous silicon are smaller by a factor of 10^4 than for an intrinsic crystal. Moreover, the mobility of Figure 4-38(a) has a thermally activated behavior unknown in pure crystalline silicon.

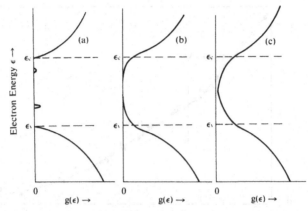

Figure 4-39 The density of states versus energy for (a) a crystalline semiconductor, (b) the same semiconductor as an "ideal covalent glass," and (c) the same material as a much more disordered amorphous solid. This view of states in an amorphous solid is based on the arguments of Mott (see footnote 38 for the underlying references) and others.

Why amorphous silicon should have a carrier density no larger than intrinsic silicon, and a small activated mobility, can be explained in terms of the sketches of Figure 4-39. Part (a) shows the conventional density of states for a semiconductor crystal. Valence states are shown below ϵ_v and conduction band states above ϵ_c. A few pockets of localized states within the intrinsic gap correspond with impurities and native flaws which perturb the periodicity. Part (b) of the same figure shows the expected density of states for the same solid as an *ideal glass*. Valence and conduction bands are still shown, but now a continuum of $g(\epsilon)$ is shown extending downwards from ϵ_c and upwards from ϵ_v into what was previously the intrinsic gap. Part (c) of Figure 4-39 envisages a much more disordered amorphous semiconductor, with many dangling bonds resulting from imperfect short range order.

The localized states within the gap resulting from imperfect short range order will inevitably be divided equally between *donor-like* and *acceptor-like* states. Thus the disorder of an amorphous semiconducting element, compound, or alloy has an automatically *self-compensating* effect in creating electron and hole traps in equal numbers. The Fermi energy ϵ_F is "pinned" to the central portion of the gap as a result. Chemical impurities are unlikely to encourage movement of ϵ_F far from the midpoint of the gap since most impurity atoms do not become donors or acceptors; their different valence requirements are met by changing the local coordination.

Does the term "gap" still mean anything, particularly if the disorder is so great that there is a finite $g(\epsilon)$ all the way across? Yes, apparently it does. There is still a massive rise in optical absorption coefficient when $\hbar\omega$ for incident photons exceeds $(\epsilon_c - \epsilon_v)$ and permits electron-hole pair creation. Photons of lesser energy do not create *mobile* pairs. For an electron in a state above ϵ_c is mobile, even though we can not use a Bloch function to describe its wave-function. However, a state for which $\epsilon_v < \epsilon < \epsilon_c$ is localized, and an electron can travel only by hopping or tunneling from one such state to another at a nearby location and comparable energy. Thus ϵ_c and ϵ_v are called the edges of a *mobility gap* for the amorphous semiconductor.

Data such as those of Figure 4-38 can be interpreted in the light of Figure 4-39(b). The conductivity seen is that of *intrinsic* amorphous silicon, since whatever few impurities were evaporated with the silicon will have taken up coordination configurations compatible with their number of valence electrons. The equality of donor-like and acceptor-like states then results in "pinning" the Fermi level to the center of the gap. For temperatures below the indicated T_c, transport occurs by hopping among the localized sites within the gap. Since this is assisted by optical-mode phonons[39] of about 0.07 eV energy, it is not unreasonable that mobility should vary as $\simeq \exp(-0.09/k_0 T)$ in this hopping range. For higher temperatures, the mobility rises more rapidly because electrons *can* be thermally excited to above ϵ_c, though their rate of progress is severely limited by frequent falls into empty localized states above ϵ_F but below ϵ_c. Such states serve as *traps* for electrons. The ratio of "free time" to "trapped time" varies with temperature to give a mobility dependence of $\exp(-0.19/k_0 T)$.

Figure 4-40 pictures a two-dimensional amorphous semiconductor for energies below, at, and above the "mobility edge" ϵ_c. For the low energy of part (a), there are a few small isolated regions in which electrons may move freely, but there is no connection between these regions. Current can flow from one end of the sample to the other only by electron hopping. The intermediate energy of part (b) reveals some connected channels which permit long-distance percolation without hopping. For the still higher energy of part (c), only a few isolated pockets remain which are not contributing to the overall current flow.

---

[39] What, if anything, does the term "optical phonon" mean for an amorphous solid, since there is no periodic lattice and no Brillouin zone? Even so, a long wave optical phonon is a state of motion for which neighboring atoms move 180° out of phase with each other. Each atom still has nearest neighbors in an amorphous solid; thus the energy $\hbar\omega_{opt}$ is valid in describing oscillations of the nearest-neighbor bond length.

Figure 4–40 Percolation theory applied to the concept of a mobility edge in an amorphous semiconductor, after R. Zallen and H. Scher, Phys. Rev. **B.4**, 4471 (1971). Parts (a), (b), and (c) show the regions (shaded) of unblocked conduction paths for three successively higher electron energies.

The ability of an amorphous semiconductor to conduct is thus dependent on getting electrons into states above ϵ_c (or holes into states below ϵ_v). As a consequence, just about any amorphous solid has a conductivity that increases on warming. Moreover, experiments on the current-voltage characteristics of thin films of amorphous material in the Te:As:Si:Ge quaternary system show that a strong electric field can precipitate a localized highly conducting state.

Chalcogenide glasses such as Te:As:Si:Ge alloys show promise for switching[40] and memory applications in electronic systems by virtue of the current-voltage characteristics found in thin films. For a film a few μm thick, the current at small fields is a sensitive function of temperature, and is non-linear with applied voltage. This is the "OFF" state of the switch, with perhaps a 10^7 ohm resistance at room temperature. If a field of 10^7 V/m is exceeded, a transition takes place to the conducting "ON" state with a resistance of a few tens of ohms. The switching can apparently be repeated very many times without significant damage to the film.

The switching of chalcogenide glass films has sometimes been attributed to a thermal runaway of the film over part of its area, and this may well happen with thicker films. Investigation[41] with selected pulse sequences for very thin films shows that the breakdown is then primarily electronic rather than thermal. It seems plausible that in such a case one or more filaments (a few μm in diameter, supporting a current density of perhaps 10^{10} A/m^2) manage to establish a *non-equilibrium* electron distribution, whereby virtually all the traps within the gap become saturated, and current within a filament is sustained by a dense electron-hole plasma.[42] Each filament is surrounded with material containing unsaturated traps, and this keeps the filament self-trapped in position and regulated in size.

Departures from Ohm's Law occur in crystalline as well as in amorphous semiconductors, as we must consider in the next sub-section.

HOT ELECTRONS

From our previous discussions of electronic drift motion superimposed upon a much more massive thermal motion, it is clear that the drift velocity of conduction band electrons in a semiconductor ought to increase with the strength of an applied electric field, at any rate up to a certain point. It is reasonable to suppose that a limiting drift velocity would be reached in a sufficiently large field, and this supposition is confirmed by experimental curves such as the one for electrons in germanium which appears in Figure 4-41. We shall discuss this curve, observing that a similar analysis is possible for many other semiconductors.

---

[40] D. Adler *et al.*, *J. Appl. Phys.*, **51**, 3289 (1980).

[41] H. K. Henisch *et al.*, *J. Non-Cryst. Solids*, **8–10**, 415 (1972); and D. Adler *et al.*, *Rev. Mod. Phys.* **50**, 209 (1978).

[42] This may be compared with the space-charge-limited current flow studied for many years in trap-filled insulating photoconductors. See A. Rose, *Photoconductivity of Solids* (Wiley-Interscience, 1963), Ch. 4.

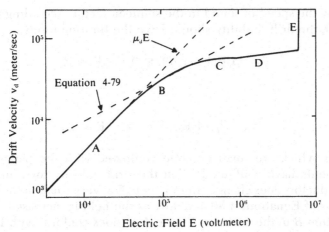

Figure 4–41   Drift velocity as a function of electric field for electrons in germanium at 300K. After J. B. Gunn, J. Electronics **2**, 87 (1956). The five regions are described in the text.

Five characteristic regions are marked on the curve of Figure 4-41. These can be described as follows.[43] Region A is the ordinary low field regime in which the drift velocity is directly proportional to the field strength, so that Ohm's law is obeyed. In this low field region, the many collisions of each electron with the phonon population keep the average electronic energy independent of time, and indicative of an equilibrium distribution.

We must remember, however, that each electron-phonon collision makes a quite small change in the energy of the electron. As the applied electric field is increased, it becomes progressively more difficult for an electron to dispose of the energy it picks up from the electric field, and the electronic energy distribution starts to become "warm." A warm electron gas can still hopefully be described by a Boltzmann distribution of speeds,

$$dn \propto s^2 \exp(-m_c s^2/2k_0 T_e) \tag{4-78}$$

but the effective electron temperature T_e is higher than the lattice temperature T, and is a function of the applied field strength. An analysis[43] of the steady state balance between the rate of energy absorption by the "warm" electrons and the rate at which they can lose energy by LA phonon scattering suggests that the drift velocity should vary with field E in accordance with

$$v_d = \frac{2\mu_d E}{[2 + \sqrt{4 + (E/E_1)^2} \,]^{1/2}} \tag{4-79}$$

Here μ_d denotes the low-field drift mobility, and E_1 is a parameter con-

---

[43] An extensive review of the theory and experiment for high field conduction in germanium and silicon is given by J. B. Gunn in *Progress in Semiconductors, Vol. 2* (Wiley, 1957). See also E. M. Conwell, *High Field Transport in Semiconductors* (Academic Press, 1967).

trolled by the speed of sound in the semiconductor. According to Equation 4-79, the drift mobility should have the limiting forms

$$v_d = \mu_d E(1 - E^2/32E_1{}^2), \qquad E \ll E_1 \qquad (4\text{-}80)$$

and

$$v_d = 2\mu_d(EE_1)^{1/2}, \qquad E \gg E_1 \qquad (4\text{-}81)$$

for fields which are small or large compared with the parameter E_1. Experiments have confirmed[43] that the drift velocity for a number of semiconductors does change (from $v_d = \mu_d E$ at very small fields) through the forms of Equations 4-80 and 4-81 as the field is increased.

Section B of the curve in Figure 4-41 does conform with Equation 4-81, but some other mechanisms evidently must intervene to further subdue the rise of v_d with applied field. The region of saturated velocity shown as part C in the figure occurs because *hot electrons* now have enough energy to excite optical phonons. Soon after an electron is warmed by the field to an energy exceeding $\hbar\omega_{opt}$ it is likely to excite an optical mode, which takes away almost all of its energy. Thus in region C most of the conduction is carried on by electrons just before they have collected enough energy from the field to make them susceptible to this energy-loss process. A further increase of the field, as in part D of the curve, results in a very modest increase of current density since a few of the electrons then manage to maintain an energy larger than $\hbar\omega_{opt}$ for part of the time.

The final portion of the curve in Figure 4-41 shows the *carrier multiplication* which occurs as an *avalanche breakdown* process. At a critical field, some electrons have managed to struggle past energy $\hbar\omega_{opt}$ to an energy sufficient for the generation of additional hole-electron pairs in the semiconductor, and the current then soars with very little change in applied voltage (part E). The avalanche breakdown mechanism sets the upper limit of permissible voltage in most semiconductor devices employing p-n junctions.

In an extrinsic semiconductor at temperatures low enough to depopulate the conduction band in favor of flaw sites, an avalanche breakdown process can occur when any remaining free electrons acquire enough energy from an electric field to permit the impact ionization of neutral flaws. For shallow flaw levels, this extrinsic kind of dielectric breakdown can occur in a field as small as 500 V/m. The field necessary becomes larger, of course, for "deeper" flaw states.

Let us now return to the variation of drift velocity with field when flaw ionization is *not* involved. By the term "drift velocity," we customarily mean $v_d = (E\sigma/en)$ for an ensemble of n electrons accelerated by a field E. Figure 4-41 shows an apparently monotonic increase of v_d with E for electrons in germanium, but the increase is not necessarily monotonic for all carrier populations. Indeed, by a suitable choice of crystal orientation (and some assistance from uniaxial compression), it is possible to make v_d for electrons in germanium decrease slightly above a certain field strength, as demonstrated in the curves of Figure 4-42. A

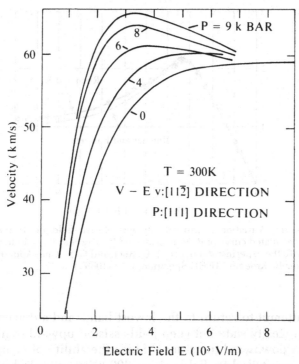

Figure 4–42 Electron drift velocity as a function of electric field for germanium at 300K, with and without uniaxial compression applied in the [111] direction. The field is in the [11$\bar{2}$] direction. The measurements were made by a "time-of-flight" technique, in which electrons were injected into and drifted through high-purity p-type germanium. From A. Neukermans and G. S. Kino, Phys. Rev. **B.7**, 2693 (1973).

much more prominent *retrograde velocity* or *negative differential conductivity* (NDC) is shown in the gallium arsenide curve of Figure 4-43, and for GaAs no anisotropic distortion of the lattice is necessary to make the effect a large one.

The rather startling reduction in effective conduction per électron which occurs for high fields in GaAs is a consequence of the sequence of conduction bands for that semiconductor. Figure 4-44 shows that the lowest energy minimum in the conduction band of gallium arsenide is at the zone center, and all electrons reside here at thermal equilibrium. This favored minimum at $\mathbf{k} = (000)$ has a low effective mass and a correspondingly high mobility. However, there are other conduction band minima (of larger mass and lower mobility) at higher energies; and Figure 4-44 shows that the first and most important are at $\mathbf{k} = (\frac{111}{222})$, at eight zone-boundary locations.

The existence of the upper minima is unimportant at thermal equilibrium or for a small applied electric field. However, these valleys can become populated at the expense of the central band minimum as a result of scattering processes in a high electric field. Thus one consequence of the large field is that some electrons are forcibly transferred from band states at the zone center (where they enjoyed a low effective mass and high mobility) to band states of these upper "valleys" where their effective mass is much larger and the mobility smaller. They will,

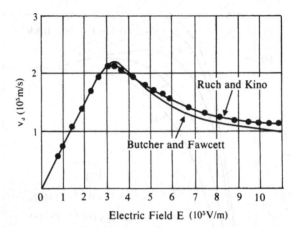

Figure 4-43 Variation of drift velocity with electric field for electrons in gallium arsenide. The calculated curve of P. N. Butcher and W. Fawcett, Phys. Lett. **21**, 489 (1966) is compared with the experimental data of J. G. Ruch and G. S. Kino, Appl. Phys. Lett. **10**, 40 (1967). [After H. Kroemer, IEEE Spectrum **5**, 47 (1968).]

of course, attempt to return to the lowest band, and a dynamic balance will exist in steady state between field-assisted upward transitions and spontaneous downward transitions. Since the ability of many electrons to conduct is curtailed for fields above 300 kV/m in GaAs by this forced electron transfer, the observed curve for $v_d = (E\sigma/ne)$ decreases as E increases. For a *sufficiently* large field, v_d must start to increase again.

The conduction band for germanium will be discussed in Section 4.3, and we shall see there that the lowest band has four equivalent "valleys." The small NDC shown for germanium in Figure 4-42 arises from a field-provoked distribution of electrons among these valleys. NDC has been detected in a number of semiconductors, and it has been predicted[44] that some ternary intermetallic alloys should be more ef-

---

[44] See, for example, M. P. Shaw, H. L. Grubin, and P. R. Solomon, *The Gunn–Hilsum Effect* (Academic Press, 1979).

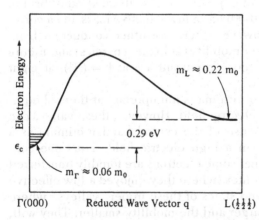

Figure 4-44 The lowest conduction band for gallium arsenide, showing the single minimum at the zone center, and the course of energy versus reduced wave vector to one of the subsidiary minima at $L(\frac{111}{222})$. See also Figure 4-48 for more of the GaAs band scheme.

ficient than GaAs in exhibiting NDC.

What happens when the *average* electric field in a slab of semiconductor corresponds with a situation of retrograde electron velocity? J. B. Gunn[45] showed experimentally that an appropriate voltage applied to a slab of GaAs causes the material to form a sequence of low-field and high-field *domains*, such that (dv_d/dE) is positive in both kinds of domain. A high-field domain forms at one terminal, moves through the sample (at a speed of approximately v_d for the low-field domains), and disappears at the other terminal. The current carried by the semiconductor slab oscillates at a frequency which is the inverse of the domain transit time, and this is the basic form of the *Gunn effect*, an effect with many applications.[46]

Since the domain velocity is of the order of 10^5 m/s in GaAs, a sample of this semiconductor with a current path of length some 5 μm should result in the generation of 20 GHz oscillations, and this is close to the maximum frequency currently possible with domain-motion Gunn devices. A somewhat higher frequency is possible with the oscillatory mode referred to as LSA (limited space-charge accumulation) which does not involve full domain propagation.

This brief mention of the Gunn effect should remind us that a meaningful discussion of many semiconductor phenomena is not possible without a recognition of the complications of band structure for real semiconductors. This forms the topic for the next section.

---

[45] J. B. Gunn, Solid State Comm., **1**, 88 (1963); IBM J. Res. Devel. **8**, 141 (1964).

[46] Various forms of "transferred electron" device are discussed in the book by Shaw *et al.* (Footnote 44), as well as in the texts by Sze and by Milnes (see the General References on page ix).

Band Shapes in Real Semiconductors

THE IDEALLY SIMPLE SEMICONDUCTOR BAND STRUCTURE

For discussions of the *densities* of free and bound electrons and holes, and of the corresponding Fermi energy, it makes no difference whether the effective mass tensors for the bands really do reduce to scalar quantities, or whether they are just replaced by "density of states" scalar equivalents. However, many effects in semiconducting materials depend quite crucially on the actual shapes of constant-energy surfaces in three-dimensional k-space. We have had some hints of this from transport processes, and shall be concerned in this section with phenomena which are much more demonstrative of the simplicity or complexity of a band structure.

Simply as a matter of convenience, let us suppose a semiconductor with an ideally simple band structure which chooses to crystallize in a lattice whose *translational* symmetry is F.C.C. (This supposition allows us to include any solid with a rocksalt, diamond or zincblende lattice, as well as F.C.C. solids.) Then our bands can be expressed as sets of constant energy surfaces within the reduced Brillouin zone of Figure 4-45.

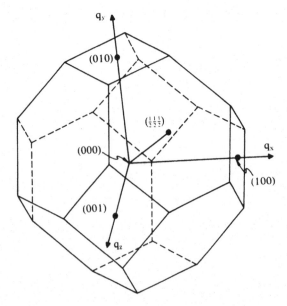

Figure 4–45 The reduced zone for a solid which crystallizes in a lattice whose translational symmetry is face-centered-cubic, including the diamond and zincblende structures of Figure 1-31. Each coordinate must be multiplied by $(2\pi/a)$ to give the actual magnitude of the component of k (where a is the unit cube edge). For any location intermediate between (000) and (100) there are five additional locations equivalent through symmetry. Similarly, there are a total of eight equivalent points for any location describable as (ttt). For a location in the zone which is not on any of the principal axes of symmetry, there are 47 other equivalent locations which must have the same set of eigenenergies.

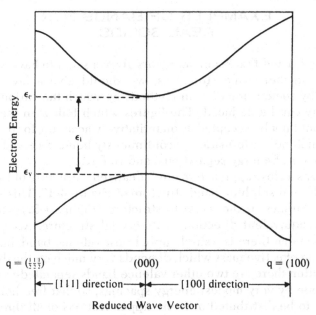

Figure 4–46 The ideally simple band structure for a semiconductor.

Later in the section we can demonstrate bands which are far from simple within the same B.Z. shape.

The "ideally simple" band structure is pictured in Figure 4-46. A single conduction band minimum is supposed at $k = (000)$, and it is supposed that a spherical constant-energy surface concentric upon the zone center can be drawn for any energy a little higher than ϵ_c. This ensures that the conduction electron effective mass m_c is a scalar. This mass is also energy-independent if $(\epsilon - \epsilon_c)$ is required to increase in proportion to k^2. Figure 4-46 shows energy increasing monotonically all the way to the B.Z. boundary in both cubic and diagonal directions, and the energies at the two kinds of boundary limit are each so much higher than ϵ_c that the exact band shape is unimportant for free electrons.

There must obviously be other completely unoccupied bands at higher energies than the lowest conduction band, but it is assumed that these are high enough not to interfere with the postulated simplicity of the band shown. Similarly, the semiconductor must have a number of completely full bands, but Figure 4-46 shows only the highest of these. This highest valence band is assumed to have a form comparable with (but inverted with respect to) the conduction band, with a maximum at (000), and states for slightly lower energies characterized by a scalar and energy-independent mass m_v.

This picture is the one we have adopted throughout the last two sections, apart from occasional admissions concerning the complexities of transport phenomena for *real* bands. It must now be conceded that no semiconductor has yet been found which proved (after exhaustive study) to have a band structure in the vicinity of the intrinsic gap as simple as that drawn in Figure 4-46.

EXAMPLES OF BANDS FOR
REAL SOLIDS

Having stated that semiconductors always seem to have some complications in their band structures, we should give a few examples, followed by some comments on methods by which the various forms of complexity can be deduced. The figures which follow in the next few pages should not be accepted as quantitatively accurate, for this is a subject in which new information is continuously being received to update our estimates of energy separations, and so forth.

Perhaps a close approach to our "idealized" structure of Figure 4-46 is the cadmium sulphide band structure of Figure 4-47. This solid ordinarily crystallizes in the wurtzite structure (Figure 1-32) which has a single nonequivalent direction in its crystal structure. Accordingly, it can be seen that there is a single uppermost valence band, but that this band has an effective mass which depends very much on the direction of **k**. In addition there are two other valence bands separated from the uppermost one by very modest energy spacings, so that free holes in CdS are likely to be distributed over the upper sections of all three valence bands at high temperatures. As a result of all these factors, free holes in CdS have quite a small mobility. The conduction band behavior, however, is fairly simple, and there is little mass anisotropy for this band.

A rather more complicated band structure, but one for which the valence and conduction band extrema still both occur at the zone center, is shown in Figure 4-48. For these bands in gallium arsenide, the V1 and V2 valence bands are *degenerate*[47] at their upper extremum. Indeed, any theory of the valence bands in such a material requires all three upper-

---

[47] In this sentence the word *degenerate* is being used in the spectroscopic sense of "two or more states which happen to have the same energy." In footnote 16 of Chapter 3 we commented that the word *degenerate* is used both for this purpose *and* for the purpose of describing a Fermi-Dirac condensation of electrons with a "degenerated" electronic specific heat.

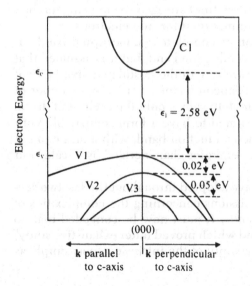

Figure 4-47 The three uppermost valence bands and the lowest conduction band for CdS, a solid which crystallizes in the wurtzite structure of Figure 1-32. Note in particular the strong mass anisotropy of the uppermost valence band.

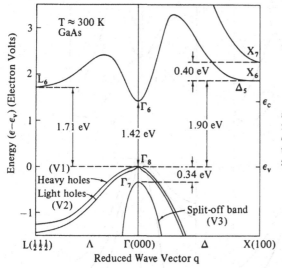

Figure 4-48 The bands just above and just below the intrinsic gap for gallium arsenide. This material crystallizes in the zincblende structure of Figure 1-31(b). The Brillouin zone is the polyhedron of Figure 4-45.

most valence bands to be degenerate at the center of the zone until *spin-orbit interaction*[48] is taken into account. The V3 valence band in Figure 4-48 is known as the *split-off* band, and the energy separation marked as 0.36 eV is the spin-orbit splitting.

The V3 band in gallium arsenide is low enough to remain unpopulated by free holes distributed between the uppermost states of the V1 and V2 bands. The V1 band, with a large density of states (i.e., a large effective mass), is popularly called the *heavy hole band,* and in a similar manner the V2 band is called the *light hole band.* From the similarity in curvature of the $\epsilon - k$ curves in the left and right portions of the valence band structure within Figure 4-48, it might be surmised that the light and heavy holes would at least enjoy scalar effective masses within their respective bands. However, this illusion is destroyed by more detailed analysis. A surface of constant energy is warped from a spherical form, pushed in along [100] directions for the heavy holes and along [111] directions for the light holes. In addition, there are theoretical reasons for suspecting that the V1 band reaches its highest energy at a set of equivalent locations in **k**-space just a small distance out from **k** = (000).

The conduction bands of GaAs are less complex than the valence bands but still do not conform with the "idealized" situation of Figure 4-46. The lowest conduction band minimum, marked as Γ_6, does have a scalar effective mass; but energy increases above ϵ_c less rapidly than k^2, so that the magnitude of the effective mass m_c must increase with energy. At thermal equilibrium the upper conduction band valleys marked L_6 do not contain any free electrons, but electrons can be excited from Γ_6 into L_6 in a large electric field, as discussed in Section 4.2. If contours of

---

[48] Spin-orbit interactions take place within isolated atoms as well as in a solid. The phenomenon is the interaction between the magnetic field set up by a spinning electron and the magnetic field produced by the orbital motion of the same electron. The importance of this interaction in splitting energy bands which would otherwise be degenerate was shown by R. J. Elliott, Phys. Rev. **96**, 266 (1954).

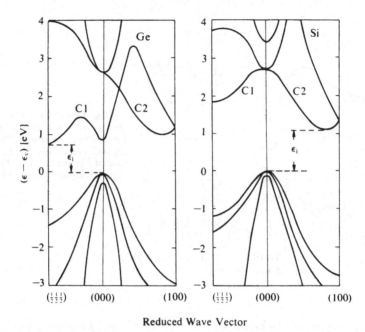

Reduced Wave Vector

Figure 4-49 A semi-quantitative picture of electron energy versus reduced wave vector for the three uppermost valence bands and the first few conduction bands of germanium and silicon. The bands are drawn here as suggested by F. Herman et al. in *Quantum Theory of Atoms, Molecules, and the Solid State*, edited by P-O Löwdin (Academic Press, 1966) p. 381.

constant energy are inscribed in the Brillouin zone for the GaAs L_6 valleys, an energy just above the bottom of the band is described by a set of eight prolate half-ellipsoids, resembling those shown for Ge in Figure 4-50.

Of course, gallium arsenide and the homologous III-V compounds have other conduction bands, and Figure 4-48 shows only the lowest two of these. The higher ones are not involved in quasi-equilibrium studies of the semiconductor, but they still influence the optical and photoemissive properties in the visible and ultraviolet parts of the spectrum. When we consider the band structure for a solid which is crystallographically similar to GaAs, but in which the periodic potential has a different Fourier structure, the ordering of the bands can well be different. As an example, Figure 4-49 shows the bands just above and just below the intrinsic gap for the elemental semiconductors silicon and germanium. The diamond lattice for each of these elements is just the zincblende structure of GaAs with two identical atoms as the basis.

The first thing to be observed in Figure 4-49 is that the ordering of the valence bands is essentially the same as for the GaAs valence bands in Figure 4-48. The spin-orbit splitting energy is quite small for both germanium and silicon (0.28 eV in germanium and only 0.04 eV in silicon). However, in all three cases there is a "heavy hole band" with an extremum essentially at (000) and energy surfaces somewhat perturbed from spherical form for lower energies, which dominates most properties of a p-type crystal.

The important difference between the structure of Figure 4-48 and the two structures of Figure 4-49 concerns the location(s) in k-space of the lowest possible conduction band energy. For germanium the conduction band does not reach its lowest energy at the zone center (as in GaAs), but rather at the eight locations on the zone boundary at $q = [\frac{1}{2}\frac{1}{2}\frac{1}{2}]$; conduction electrons will preferentially select states in these eight regions of the zone. In silicon the lowest conduction band energy is achieved at the six locations for which $q = [\frac{3}{4}00]$.

For a band similar to the conduction band described for silicon or germanium, the *density of states* $g(\epsilon)$ and the *density-of-states mass* m_ℓ must take account of the contributions from a set of ellipsoids instead of a single energy minimum. We have already shown in Figure 3-50 the set of six ellipsoids which could describe an energy just above ϵ_c in silicon. Suppose that a conduction band occurs in $\mathfrak{N}$ equivalent locations *within* the reduced zone. Suppose also that we can draw a set of ellipsoidal surfaces for any slightly higher energy, and that $[\hbar^2(\partial^2\epsilon/\partial k^2)^{-1}]$ is equal to m_ℓ along the longitudinal axis of an ellipsoid and equal to m_t along each of the transverse axes of an ellipsoid. Then the combined effect of $\mathfrak{N}$ ellipsoids is a total density of states

$$g(\epsilon) = \frac{\mathfrak{N}}{2\pi^2}\left(\frac{2}{\hbar^2}\right)^{3/2}(\epsilon - \epsilon_c)^{1/2}(m_\ell m_t^2)^{1/2} \qquad (4\text{-}82)$$

for a density-of-states mass

$$m_c = \mathfrak{N}^{2/3}(m_\ell m_t^2)^{1/3} \qquad (4\text{-}83)$$

in Equation 4-1, and so forth.

The above remarks are all very well for the conduction band of silicon where $\mathfrak{N}$ minima occur *within* the reduced zone. For the conduction band of germanium, the minimum energy occurs at eight locations *on* the zone boundary, and a slightly higher energy can be represented by eight *half-ellipsoidal* surfaces in the reduced zone (Figure 4-50). An extremum and its partner a full reciprocal lattice vector away combine to produce a complete ellipsoidal surface, and $\mathfrak{N} = 4$ for insertion into Equations 4-82 or 4-83 in this particular case. Problem 4.16 considers two other simple examples of this principle.

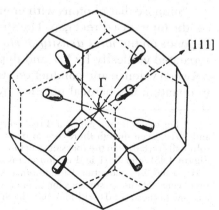

Figure 4–50 Surfaces of constant energy in the B.Z. for an energy just above ϵ_c in germanium. After D. Long, *Energy Bands in Semiconductors* (Wiley-Interscience, 1968).

DIRECT AND INDIRECT
INTRINSIC TRANSITIONS

The bibliography at the end of this chapter cites sources for details on the band structures in a variety of semiconductors, and a lengthy repetition here would serve little purpose. From the preceding examples, we can see that the location(s) in k-space of the maximum valence band energy ϵ_v may or may not correspond to the location(s) in k-space of the minimum conduction band energy ϵ_c. When such a coincidence *does* occur (whether this be at the center of the zone or elsewhere) the semiconductor is said to have a *direct intrinsic gap*. Thus CdS and GaAs both have direct gaps, as do many other semiconductors, including a number for which ϵ_v and ϵ_c occur at the *same* set of locations in k-space *away from* the center of the zone.[49] A semiconductor for which the highest valence band energy does *not* occur at the same place or places in k-space as the lowest conduction band energy is similarly said to have an *indirect* intrinsic gap. It will be clear from Figure 4-49 that both germanium and silicon are indirect gap materials, and numerous other materials are known in this category.

We shall discuss the features of direct gap materials first and come back to the indirect gap classification later. The significance of the term "direct gap" is that a photon of the special energy $\hbar\omega = \epsilon_i = (\epsilon_c - \epsilon_v)$ can excite an electron from the top of the filled band *directly* to one of the states at the very bottom of the conduction band, in a transition which is vertical in k-space as sketched in Figure 4-51(a). Any photon with an energy somewhat larger than ϵ_i can participate in a direct (vertical) transition such as those shown as dashed arrows in Figure 4-51(a); the free hole and free electron so produced share the excess energy $(\hbar\omega - \epsilon_i)$ in such a manner that the hole and electron have the same value of k. As we may readily verify (Problem 4.17), for simple bands the electron and hole will each have a wave-vector

$$\mathbf{k} = (1/\hbar)[2m_r(\hbar\omega - \epsilon_i)]^{1/2} \qquad (4\text{-}84)$$

with respect to the wave-vector for the extremum location. The quantity $m_r \equiv [m_c m_v (m_c + m_v)^{-1}]$ in Equation 4-84 is known as the *joint density of states effective mass*.

Suppose that photons with energy $\hbar\omega \geq \epsilon_i$ are incident upon a semiconductor with a direct gap. The strength of absorption for these photons depends on the *joint density of states* for valence and conduction states separated vertically by $\hbar\omega$, and depends also on the degree of coupling between valence and conduction states. This latter quantity is described in quantum-mechanical notation by an *interband matrix element*.

---

[49] For example, the gap is a direct (vertical) one for the lead compounds PbS, PbSe, and PbTe, even though there are no band maxima or minima at the zone center. These solids all crystallize in the rocksalt structure (which once again has the B.Z. shape shown in Figure 4-45), and SdH and cyclotron resonance results suggest that the valence band maxima and the conduction band minima coincide at the eight locations $q = [\frac{1}{2}\frac{1}{2}\frac{1}{2}]$. Thus a set of constant energy surfaces for an energy just below ϵ_v *or* for an energy just above ϵ_c in any one of these solids will look like the array of semi-ellipsoids in Figure 4-50.

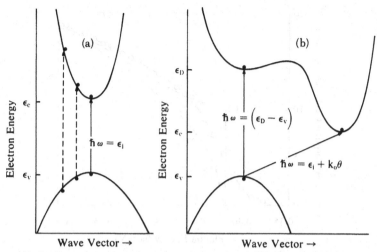

Figure 4-51 Energy versus wave-vector for (a), a direct gap semiconductor, and (b), an indirect gap semiconductor. A direct transition is vertical on such a diagram, as is appropriate for a transition induced by a photon alone (which has a wave-vector of typically one ten-thousandth of the zone width). A vertical transition produced by a photon with more energy than ϵ_i starts from a state below ϵ_v and proceeds vertically to a state of the same k higher than ϵ_c as shown by the dashed lines in part (a). The indirect (non-vertical) transition of part (b) is phonon-assisted to achieve momentum conservation. Any direct transition in such a solid requires more energetic photons.

Now if the transition from ϵ_v to ϵ_c does not violate any spectroscopic selection rules,[50] it is said to be an *allowed transition,* and the matrix element is nonvanishing for photon energy $\hbar\omega = \epsilon_i$. In such a case, it might seem reasonable to expect that the matrix element would be essentially independent of the transition energy (at least for photon energies not too much larger than ϵ_i), and that the optical absorption coefficient α_i would have the spectral dependence only of the joint density of states:

$$\alpha_i = A(\hbar\omega - \epsilon_i)^{1/2}, \qquad \hbar\omega \geq \epsilon_i \qquad (4\text{-}85)$$

Curve (b) in Figure 4-52 is plotted to conform with Equation 4-85, and it can be seen that such an expression is a tolerable though not perfect fit to the experimental data of curve (a) in that figure.

One reason for the imperfection of the fit is that the arguments leading to Equation 4-85 neglect the electron-hole interaction which may result when $(\hbar\omega - \epsilon_i)$ is not very large. Since the electron and hole of a photoexcited pair are created in the same region of real space, they

---

[50] Optical selection rules are less rigorous in solids than they are in isolated atoms, since a state is rarely a "pure" s-state or p-state or d-state. However, some transitions are nominally forbidden from the very top of a filled band to the very bottom of an empty band; the matrix element for such a transition becomes vanishingly small. Even so, a transition between such a pair of bands usually becomes progressively "less forbidden" as we consider vertically separated pairs of states further away from the wave-vector k_0 of the extremum location. In many such cases, the matrix element coupling the lower and upper states is proportional to $|k - k_0|$ or to $|k - k_0|^2$.

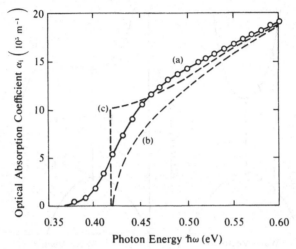

Figure 4–52 Spectral dependence of the optical absorption coefficient for a semiconductor with a direct energy gap. Curve (a) shows experimental data for a film of lead sulphide 1.33 μm thick (deposited epitaxially on a NaCl single crystal), as measured at room temperature by V. Prakash (Thesis, Harvard University, 1967). This data is compared with the behavior plotted in curve (b), dictated by Equation 4-85 for a situation in which complete screening obviates any interaction between the free electron and the free hole generated by a photon. A comparison can alternatively be made with curve (c), which conforms with Equation 4-86 for the Elliott model[52] of optical absorption with electron-hole interaction.

exert a Coulomb force upon one another,[51] an influence which is screened to some extent by other charges in the vicinity (as represented by an appropriate dielectric constant). The effect of the electron-hole interaction ceases when systematic motion and scattering processes separate the hole and the electron, but the influence upon the spectral dependence of absorption is found by Elliott[52] to take the form

$$\alpha_i = B\{1 - \exp[-C(\hbar\omega - \epsilon_i)^{-1/2}]\}^{-1}, \qquad \hbar\omega \geq \epsilon_i \qquad (4\text{-}86)$$

where B and C are constants.

In Equation 4-86, α_i rises abruptly as a step function when $h\omega = \epsilon_i$, and eventually merges with the behavior of Equation 4-85 for higher energies, as demonstrated by curve (c) of Figure 4-52. We may regard

---

[51] The Coulomb attraction between an electron and a hole can also lead to a hydrogenic series of *bound* configurations in which the two objects revolve about their common center of mass with a quantized angular momentum. Such an entity is an *exciton*, the solid state analog of a positronium "atom." Excitons, postulated by Frenkel in 1931, can be created by the absorption of light. The photons appropriate for this task must have an energy just a little smaller than ϵ_i. How much smaller is explored in Problem 4.18. The theory of excitons is discussed in the short book *Excitons*, by D. L. Dexter and R. S. Knox (Wiley-Interscience, 1965).

[52] The model for optical absorption including the unscreened interaction of electrons and holes was first described by R. J. Elliott, Phys. Rev. **108**, 1384 (1957). Models for "allowed" and "forbidden" type transitions with and without the electron-hole interaction correction are reviewed by T. P. McLean in *Progress in Semiconductors, Vol. 5* (Wiley, 1960).

Equation 4-85 as a limiting case of Equation 4-86 for perfect electro-static screening of the electron-hole Coulombic interaction.

A comparison of the PbS experimental data in Figure 4-52 with the theoretical curves of Equations 4-85 and 4-86 shows that electron-hole Coulombic interaction in this solid is partly but not completely screened by the high dielectric constant. Curve (a) in Figure 4-52 also shows a characteristic which is found at the absorption edge of any direct gap insulator or semiconductor: a gradual increase of α_i with increasing photon energy when $\hbar\omega$ is smaller than the nominal gap width, rather than a step function start of absorption when $\hbar\omega = \epsilon_i$. As first noted by Moser and Urbach[53] for silver halides, the first portion of the absorption curve in a direct gap solid rises in the manner

$$\alpha_i \propto \exp[\beta(\hbar\omega - \epsilon_i)], \qquad \hbar\omega < \epsilon_i \qquad (4\text{-}87)$$

even if it conforms with the Elliott model of Equation 4-86 for larger photon energies. A variety of explanations have been suggested for this *Urbach rule* behavior below what ought to be the threshold photon energy, and probably a variety of mechanisms contribute to varying extents in the different classes of semiconductors and insulators.

A semiconductor such as germanium or silicon has an *indirect intrinsic gap*, since the conduction band minima are in parts of k-space remote from the valence band maximum. Direct optical transitions are shown in the right portion of Figure 4-51 as a possibility when photons of energy exceeding $(\epsilon_D - \epsilon_v)$ are used, but electron-hole pairs of the minimum creation energy can not be produced in a one-step (direct) optical process. However, as suggested by Bardeen et al.,[54] an *indirect* or *non-vertical* transition can occur in such cases via an intermediate virtual state, whereby photon absorption is accompanied by either the creation of or annihilation of a lattice vibrational phonon. This model requires a minimum photon energy of $(\epsilon_i + k_0\theta)$ for an indirect transition which creates a hole-electron pair *and* a phonon of energy $k_0\theta$ and momentum $\hbar(k_c - k_v)$, a process which can take place at any temperature. The alternative process, in which a photon is absorbed and a phonon of momentum $\hbar(k_c - k_v)$ is *also* absorbed in pair creation, can proceed from the lower threshold energy $(\epsilon_i - k_0\theta)$; but such a mechanism will be attenuated at low temperatures when the needed phonons are no longer readily available. When the Bose-Einstein statistics for the

---

[53] F. Moser and F. Urbach, *Phys. Rev.* **102**, 1519 (1956) reported exponential absorption edges below the intrinsic threshold for the direct gap insulators AgCl and AgBr. The name *Urbach rule* has remained for similar exponential edges in other direct gap solids. For some solids (including the original ones studied by Moser and Urbach) the steepness parameter is inversely proportional to the temperature. For other solids, temperature has little effect on the steepness, which implies that not all exponential edges have the same origin. The large local electric fields in an insulator or semiconductor arising from the random distributions of donors and acceptors have been suggested as the origin of exponential absorption edges by M. A. Afromowitz and D. Redfield, *Proc. IX Intl. Conf. Phys. Semiconductors* (Nauka, Leningrad, 1968), p. 98.

[54] J. Bardeen, F. J. Blatt, and L. H. Hall, in *Photoconductivity Conference* (Wiley, 1956). The subject matter of indirect absorption edges as well as direct edges is reviewed by McLean (footnote 52).

phonon state population are taken into consideration, the optical absorption coefficient for an indirect transition should have the form

$$\alpha_i = \frac{A}{\hbar\omega}\left[\frac{(\hbar\omega - \epsilon_i - k_0\theta)^2}{1 - \exp(-\theta/T)} + \frac{(\hbar\omega - \epsilon_i + k_0\theta)^2}{\exp(\theta/T) - 1}\right] \qquad (4\text{-}88)$$

when only one branch of the vibrational spectrum is involved. In general, phonons from several branches of the vibrational spectrum can provide the needed $(k_c - k_v)$, each with its characteristic θ. The absorption coefficient is then a summation of pairs of terms for phonon creation and phonon annihilation processes. At low temperatures, each of the phonon annihilation terms suffers a severe loss of intensity, while the phonon creation terms remain.

The relative strengths of direct and indirect optical absorption processes in a semiconductor are illustrated by the data for germanium in Figure 4-53. Since the (indirect) intrinsic energy gap is not very much smaller than the direct transition threshold for this particular solid, it is easy to see from this figure that direct processes overshadow indirect processes by a factor of at least 100.

When the first intrinsic optical absorption edge of a semiconductor or insulator *is* direct, we have to locate only one conduction band

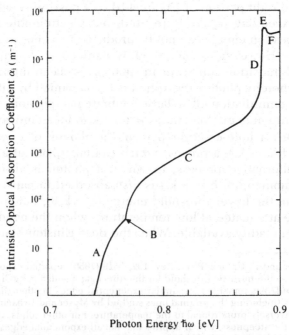

Figure 4–53 The intrinsic optical absorption edge for germanium at T = 20K, after T. P. McLean, *Progress in Semiconductors, Vol. 5* (Wiley, 1960). In region A the only band-to-band process possible is the indirect one which involves simultaneous absorption of a photon and a phonon. B is the onset energy $(\epsilon_i + k_0\theta)$ for the indirect process of photon absorption accompanied by phonon creation, and this process continues to dominate through region C. At D the energy of a photon can start to execute direct transitions at k = (000); an exciton peak is seen (E), followed by direct free electron-hole pair formation at F.

minimum to know (1) other minima by symmetry and (2) that valence band maxima occupy the same places in the zone. On the other hand, if the first intrinsic edge is indirect, a knowledge of the phonon spectrum in the solid can give us a rough idea of the magnitude of $(k_c - k_v)$ (as verified by the crude calculations suggested in Problem 4.19) and can sometimes give us enough clues to deduce the locations of the maxima and minima.

TESTS FOR THE COMPLEXITIES
OF BAND STRUCTURES

Optical absorption is only one of the techniques by which the complexities of bands in semiconductors and insulators can be sorted out. A number of other techniques should receive brief mention here. Tests for the shapes of constant energy surfaces just above ϵ_c and just below ϵ_v, and tests for the symmetry (and hence locations) of band extrema, may be informative about one band only or about the combined properties of the valence and conduction bands. Tests for the complexities of a single band include:

1. The directionality of magnetoresistance. The application of this technique to the complexities of bands in germanium and silicon is discussed, for example, by Glicksman.[26]

2. The directionality of piezoresistance (the change of resistance produced by a uniaxial strain). This effect is comparable with magnetoresistance in its sensitivity to the anisotropy of a constant energy surface in k-space. As with magnetoresistance, results are easy to get, but interpretations require great skill and care.

3. Anisotropy of low-temperature cyclotron resonance. This has been a most useful analytic tool in semiconductor crystals for which careful purification has produced a mean free time long enough to make $(\omega_c \tau_m) > 1$ without having to use an extremely large magnetic field and a very large frequency in the microwave or infrared[55] regions. Cyclotron resonance in semiconductors differs from cyclotron resonance in metals, since the radiofrequency penetration depth in a semiconductor is large compared with the cyclotron orbital diameter, and the strong harmonic effects of the Azbel-Kaner arrangement do not appear. Figure 4-54 shows a typical cyclotron resonance trace for a germanium sample, with resonances for photo-excited electrons and photo-excited holes, and weak evidence of harmonic resonances. Measurements as a function of orientation in germanium demonstrate the ellipsoidal form of electron constant-energy surfaces and the departures from spherical energy surfaces for heavy and light holes.

4. Oscillatory behavior of the magnetoresistance, with any anisotropy of this indicating anisotropies of the constant energy surfaces in

---

[55] If the mean free time τ_m is extremely small for the semiconductor, then cyclotron resonance is impossible without a large magnetic field and a high excitation frequency. Elucidation of the warped shape of energy surfaces for the valence band of tellurium required a 15 T field, and 195 μm radiation from a HCN laser (Button, Lax, *et al.*, 1970).

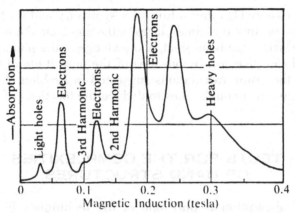

Figure 4-54 A typical cyclotron resonance trace for a semiconductor sample. This curve was recorded by R. N. Dexter, H. J. Zeiger, and B. Lax, Phys. Rev. **104**, 637 (1956), for germanium at 4K in a microwave field at 23GHz. Both electrons and holes are excited at this temperature by weak illumination, and the various possible modes of resonance are observed as the superimposed magnetic field is scanned.

k-space. The name *Shubnikov-de Haas* effect (SdH effect) was used in Section 3.5 to describe the oscillatory magnetoresistance of a cold, degenerate electron gas in a metal or a heavily doped semiconductor. A magnetic field causes the reorganization of the density of states into a series of magnetic sub-bands (as described in connection with Figures 3-59 through 3-62) with Landau level spacing $\hbar\omega_c$. If SdH oscillations are to be seen, n_0 must be large enough and T small enough so that the Fermi surface is well defined, with $k_0 T < \hbar\omega_c$; then magnetoresistance goes through an oscillation whenever a change of magnetic field causes a Landau level to pass through the Fermi energy. The situation previously chosen as an example in Figure 3-66 and Problem 3.35 was for the especially simple situation of a spherical Fermi surface centered on $\mathbf{k} = (000)$. For such a situation, the SdH period P (which is reciprocally dependent on the Fermi surface cross-section S_F) does not depend on orientation. However, if electrons (or holes) are condensed within a number of ellipsoidal or other non-spherical surfaces within the zone, such as those of Figures 3-50 and 4-50, then an anisotropy of the SdH period will give information about the prolateness of each ellipsoid and the angular placement of ellipsoids.

5. Oscillatory magnetoresistance can also be seen in a semiconductor under quite different circumstances. This is the phenomenon of *magnetophonon resonance*, predicted by Gurevich and Firsov (1961) and experimentally pioneered in large measure by the work of Stradling.[56] Magnetophonon resonance is seen in a weakly doped semiconductor for which the conduction electron gas is non-degenerate, and requires the presence of long wave longitudinal optical mode phonons, of energy $\hbar\omega_{LO}$ and small wave-vector. This latter requirement means that the temperature must not be so low that the supply of optical phonons is completely dried up, yet the temperature must not be so high that the density of states pictured in Figure 3-62 is thermally washed out.

[56] R. A. Stradling and R. A. Wood, J. Phys. C., **1**, 1711 (1968); **3**, L94 (1970). This topic is reviewed at length by P. G. Harper, J. W. Hodby and R. A. Stradling, Repts. Prog. Phys., **36**, 1 (1973).

The essentials of the Landau levels in Figure 3-62 will be preserved if $\omega_c \tau_m \gg 1$. The action of the *phonons* in magnetophonon resonance is now in scattering electrons from the lowest Landau level to higher ones, especially for the set of magnetic fields that satisfy the condition

$$\hbar\omega_{LO} = N\hbar\omega_c = (N\hbar eB/m^*), \qquad N = 1,2,3, \ldots \qquad (4\text{-}89)$$

Thus the magnetoresistance has a component periodic in reciprocal field.[57] Stradling points out that the oscillatory term can be detected most easily by plotting the second derivative of sample resistance, (d^2R/dB^2), against B or $(1/B)$. This is done for the data of Figure 4-55. These happen to be simple data for the isotropic conduction band of InSb (suggested for analysis in Problem 4.20). However, it is possible to detect and analyze more complicated and orientation-dependent sets of oscillations[58] when the band under study has multiple anisotropic energy surfaces.

6. The de Haas-van Alphen (dHvA) effect of oscillatory magnetic susceptibility can be used in a semiconductor, as in a metal, to determine the size and shape of the Fermi surface for a cold, degenerate situation. The effect is rather a small one for most semiconductors, and thus tends to be used to greater advantage in metals and in *semimetals*, materials for which a tiny overlap of bands produces pockets of electrons and holes in equal numbers. The overlap in magnesium (Figure 3-40) is so large that this element is emphatically metallic. A much smaller overlap is found in the elements of Group VB of the Periodic Table. Thus the

---

[57] The magnetoresistance periodic in (1/B) sounds like SdH resonance, though the principle underlying magnetophonon resonance is quite different. Magnetophonon resonance is more like cyclotron resonance, with microwave photons replaced by optical mode phonons as the exciting agents.

[58] See for example: R. A. Stradling *et al.*, *Proc. X Intl. Semiconductor Conf.* (A.E.C., 1970), p. 369.

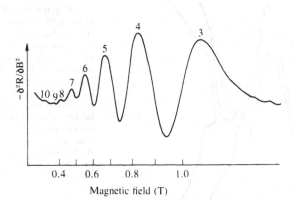

Figure 4-55 Second derivative of resistance (d^2R/dB^2) versus magnetic field B for electrons in the conduction band of InSb at T = 77K. The oscillations show magnetophonon resonance, in this case for electrons in an isotropic band. From R. A. Stradling, J. Phys. E 5, 736 (1972). Stradling says that these data were taken by an Oxford undergraduate during the final year laboratory course. Analysis is suggested in Problem 4.20.

number of electrons (and holes) per atom ranges from 4.6×10^{-3} in arsenic to only 1.1×10^{-5} in bismuth. Though this is much smaller than in a normal metal, the carrier densities are large enough to make the dHvA effect workable, and it has been used with success to explore the size and anisotropy of the hole and electron Fermi surfaces in these semimetals. Figure 4-56 (and the front cover of the book) shows the Fermi surface anticipated for holes in arsenic, based on a pseudopotential band calculation[59] with parameters adjusted to fit observed dHvA data. For antimony, it is believed that the thin connecting regions disappear, leaving six isolated pockets which are still markedly distorted from ellipsoidal form.

Tests of a semiconductor which involve the combined properties of two bands include of course an optical absorption analysis, as detailed in the last sub-section, and a number of other techniques involving the interactions of photons with a solid. Since a semiconductor is not transparent for photon energies larger than its first direct gap energy, transitions from filled bands to empty states of considerably higher energy cannot be observed by optical transmission (except for very thin film samples). The higher energy spectral range is frequently studied by means of the *optical reflectivity*, which may have a significant change in its derivative for any energy which corresponds with a van Hove singularity in the joint density of states.[60] We have already pointed out in Sections 3.3 and

[59] P. J. Lin and L. M. Falicov, Phys. Rev. **142**, 441 (1966).

[60] There are many recent papers on this topic, but a recommended starting point on this subject is a paper by H. R. Philipp and H. Ehrenreich [Phys. Rev. **129**, 1550 (1963)], which reports reflectivity data for some semiconductors, and discusses the analysis of information on the real and imaginary parts of the dielectric constant by Kramers-Kronig methods. The optical properties can be associated with specific filled bands and specific empty bands.

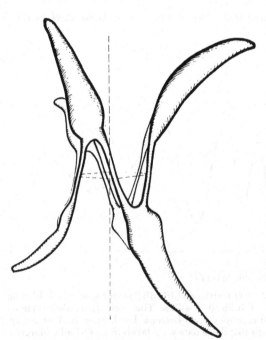

Figure 4-56 A perspective view of the "crown-shaped" Fermi surface for holes in arsenic, based on the pseudopotential calculations of Lin and Falicov[59], guided by experimental dHvA data. Arsenic has a crystal structure which is basically F.C.C., distorted along the cube diagonal to produce a trigonal lattice. The figure shows the preferred axis about which three-fold rotation symmetry exists. From W. M. Lomer and W. E. Gardner, Progress in Materials Science **14**, 143 (1969).

3.5 that photoemission is another valuable analytical procedure for exploring large-photon-energy transitions in which the final state of the electron is higher than the work function surface barrier.[61]

A combination of the fine details of the lowest conduction band states and highest valence band states can be obtained from measuring *magnetoabsorption*, the change of the optical absorption edge characteristic in a magnetic field. Such measurements have been brought to a fine art by Lax and his co-workers.[62]

The process of electron-hole recombination (about which we must say a little more in the next section) can be quite revealing about the directness or otherwise of an intrinsic transition. Efficient *luminescence* (radiative recombination) is quite unlikely unless the transition is a direct one. The process known as *Auger recombination,* or *three-body recombination* is also highly unlikely unless the conduction band and valence band extrema coincide.

---

[61] For a review of photoemission methods applied to nonmetallic solids, see W. E. Spicer and R. C. Eden in *Proceedings 9th International Semiconductor Conference* (Moscow, Academy of Sciences, 1968), p. 65.

[62] See, for example, B. Lax in *Proceedings 9th International Semiconductor Conference* (Moscow, Academy of Sciences, 1968), p. 253.

4.4 Excess Carrier Phenomena

Many interesting phenomena occur in a semiconductor when the distribution of electrons over the permitted states is perceptibly disturbed from equilibrium. Such departures from thermodynamic equilibrium are encountered in any bipolar semiconductor device, as well as in many situations involving homogeneous semiconductor samples.

The description given in this section of non-equilibrium phenomena is a brief one. Much more extensive accounts can be found in the literature cited at the end of the chapter, including the books by Ryvkin, Bube, and by this author.

DETAILED BALANCE

Consider the semiconductor of Figure 4-57, with a total of p holes/m³ in the upper portions of the valence band, and n electrons/m³ in the lower part of the conduction band. The adjustment of electrons over the various energy states is a continuous and dynamic one, but we know that in *thermal equilibrium*

$$n = n_0 = N_c \exp\left(\frac{\epsilon_F - \epsilon_c}{k_0 T}\right) \qquad (4\text{-}90)$$

and

$$p = p_0 = N_v \exp\left(\frac{\epsilon_v - \epsilon_F}{k_0 T}\right) \qquad (4\text{-}91)$$

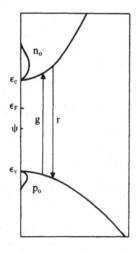

Figure 4–57 Under conditions of thermodynamic equilibrium, the electron and hole densities n_0 and p_0 can both be described in terms of a single Fermi level ϵ_F through Equation 4-22 or Equations 4-90 and 4-91. The equilibrium values n_0 and p_0 result from processes of generation, but are completely independent of the magnitudes of the equal and opposite rates g and r. Nor do n_0, p_0, and ϵ_F depend at all on the *nature* of the generation/recombination processes. The nature and efficiency of the creation and annihilation processes becomes important as soon as thermodynamic equilibrium is disturbed.

regardless of the nature or efficiency of the processes of energy transformation. This can be so only if thermal equilibrium is a situation for which the generation rate exactly equals the recombination rate for each separate process of energy transformation and for each energy interval. Then the *total* generation rate g equals the *total* recombination rate r. What we are saying is that each process has an inverse process with the *same probability* under equilibrium conditions. This is a form of statement for the *principle of detailed balance.*

Thus, as a specific example, photons from the black body environment of our equilibrium semiconductor are absorbed at a certain rate *in toto* and *in detail* in the creation of hole-electron pairs. This matches the rate at which the semiconductor in equilibrium creates photons by a radiative recombination process of electron-hole annihilation. Moreover, the radiative recombination must produce photons with the same spectrum as the absorbed photons if the second law of thermodynamics is to be upheld.

NON-EQUILIBRIUM STATISTICS AND QUASI-FERMI LEVELS

Suppose that the free carrier densities of Figure 4-57 are now modified by an external means to make $np \neq n_0 p_0$. Then in an attempt to recreate a balanced situation generation will exceed recombination if $np < n_0 p_0$. Similarly, recombination will exceed generation if $np > n_0 p_0$, a more common kind of perturbation from equilibrium.

In either event, it is still possible to describe the *total* densities n and p by equations

$$n = N_c \exp\left(\frac{\phi_n - \epsilon_c}{k_0 T}\right) \tag{4-92}$$

and

$$p = N_v \exp\left(\frac{\epsilon_v - \phi_p}{k_0 T}\right) \tag{4-93}$$

where ϕ_n and ϕ_p are normalizing parameters with the dimensions of energy which (following Shockley) we shall call the *quasi-Fermi levels* for electrons and holes respectively. ϕ_n and ϕ_p coalesce to the Fermi energy ϵ_F for thermodynamic equilibrium, but $\phi_n > \epsilon_F > \phi_p$ when $np > n_0 p_0$, as indicated in Figure 4-58.

It should be noted that Equations 4-92 and 4-93 describe the *total* densities of electrons and holes for a non-equilibrium situation, but do not tell us about the distribution of the n electrons or the p holes among the available states. It is important to be aware of this cautionary remark, since the distributions of electrons and/or holes in velocity space may depart from the Maxwell-Boltzmann form of Equations 3-4 and 3-5 if many additional carriers are dumped into the band in a short time with a radically different velocity distribution. Phonons and lattice flaws will

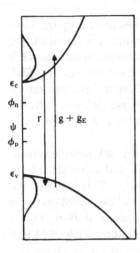

Figure 4–58 The natural rates of generation and recombination no longer balance *in toto* or in detail when thermal equilibrium is disturbed by externally provoked generation or by carrier injection from a region of different composition. The total conduction electron density can now be described by Equations 4-92 or 4-94, which reduce to Equations 4-22 or 4-90 when equilibrium returns. Similarly, Equations 4-22 and 4-91 must be generalized to Equations 4-93 or 4-95 for the free hole density.

help to thermalize the added carriers; thus the quasi-Fermi level is able to characterize the total number *and* the velocity distribution of the total band population provided that the *carrier lifetime* is large compared with the mean free time τ_m between scattering events. The lifetime is the average time spent by an added carrier before it is removed from the band, and we shall be concerned shortly with how large or small this may be for specific energy transformation processes.

By use of Equation 4-16, the electron and hole densities away from equilibrium can be written in terms of the intrinsic Fermi energy ψ and the intrinsic concentration n_i:

$$n = n_i \exp\left(\frac{\phi_n - \psi}{k_0 T}\right); \qquad (\phi_n - \psi) = k_0 T \ln(n/n_i) \qquad (4\text{-}94)$$

and

$$p = n_i \exp\left(\frac{\psi - \phi_p}{k_0 T}\right); \qquad (\psi - \phi_p) = k_0 T \ln(p/n_i) \qquad (4\text{-}95)$$

These generalize Equation 4-22 to non-equilibrium situations. Thus

$$np = n_i^2 \exp\left(\frac{\phi_n - \phi_p}{k_0 T}\right) \qquad (4\text{-}96)$$

which means that

$$e^{-1}(\phi_n - \phi_p) = (k_0 T/e) \ln(np/n_i^2) \qquad (4\text{-}97)$$

The quantity on the left side of Equation 4-97 can be construed as an electromotive force driving toward the restoration of equilibrium. We shall see later that $\nabla\phi_n$ and $\nabla\phi_p$ are associated with the total current density caused by electrons and holes respectively in a semiconductor, as a consequence of two kinds of current flow: field-provoked drift motion and diffusion motion.

Let us consider how a departure from equilibrium affects the quantities ϕ_n and ϕ_p in relation to the intrinsic location ψ and the equilibrium location ϵ_F. Suppose first a sample which is intrinsic at equilibrium, so that $\epsilon_F = \psi$. When intense illumination or some other non-equilibrium influence creates extra hole-electron pairs, it is probable that n and p will continue to be virtually the same.[63] Then ϕ_n will be as far above ψ as ϕ_p is below this indicator of the intrinsic state: $(\phi_n - \psi) = (\psi - \phi_p) = k_0 T \ln(n/n_i)$.

As a contrast, consider now a sample which is strongly n-type, $n_0 \gg p_0$ so that ϵ_F is many $k_0 T$ higher than ψ. When this sample is illuminated, the *fractional* change is much more significant for *minority carriers* (holes) than for *majority carriers* (electrons). Thus ϕ_n will move very little until the added electron density approaches the magnitude of n_0. However, ϕ_p will move rapidly for any added hole density, will lie below ψ as soon as $p = (p_0 + p_e)$ exceeds n_i, and will be as far below ψ as ϕ_n is above this point by the time the density of excess pairs reaches n_0. Problem 4.21 looks at a simple exercise of this principle, which is very relevant to the flow of current in a biased p-n junction diode or transistor.

When the electron and hole densities are larger than those for equilibrium, the natural processes of recombination speed up in an attempt to redress this situation. They speed up just because there are more electrons and holes around to be presented with the possibilities for recombination. Suppose now that we have some means at our disposal for making n_0 and p_0 *less than* at equilibrium. Under these conditions of *carrier exclusion* or *carrier depletion*,[64] recombination is slowed down, but generation proceeds at the same pace as usual, in trying to restore the normal carrier densities. For a semiconductor sample in which carrier depletion has been effected, the quasi-Fermi-level ϕ_p for holes is *higher* than that for electrons, as required by Equations 4-92 through 4-95.

Any space charge established by carrier depletion will be accompanied by an electrostatic potential (see Problem 4.22) in accordance with Poisson's equation. Thus the reductions of hole density and electron density must be essentially the same unless there is some simultaneous adjustment of the charge bound to flaws. This means that carrier depletion can be carried out to best advantage with material which is nearly intrinsic to start with.

The central region of a p-n junction suffers severe depletion when biased in the reverse (i.e., blocking or high resistance) direction.

---

[63] Probable, but not certain. A massive distribution of flaws within the gap may just happen to make $n_0 = p_0 = n_i$ at equilibrium, even though these flaws are more adept at capturing electrons rather than holes. Under strong illumination, these flaws would quickly capture a larger fraction of the added electrons rather than of the added holes, to make $n_e \equiv (n - n_0)$ smaller than $p_e \equiv (p - p_0)$.

[64] This process was first reported by A. F. Gibson, Physica **20**, 1058 (1954) with the name "carrier extraction," though "exclusion" and "depletion" have been preferred descriptions in subsequent accounts. The theory of carrier depletion is discussed by W. T. Read, Bell. Syst. Tech. J. **35**, 1239 (1956). Depletion in a bulk semiconductor sample requires contacts which will extract but not inject carriers. A voltage pulse then leads to extraction, which sweeps the length of the sample as a shock wave.

CONTINUITY EQUATIONS AND
CARRIER LIFETIMES

Suppose that conduction band electrons are created at a rate g_E as a result of some external influence. In addition to this, there are natural rates of generation, g, and recombination, r, resulting from a variety of energy transformation mechanisms. Then if we concentrate on the free electron population and its dependence on time, we must be able to express an electron *continuity equation* in the form

$$\frac{\partial n}{\partial t} = (g - r) + g_E + e^{-1}\nabla \cdot \mathbf{I}_n \qquad (4\text{-}98)$$

Here the last term represents the tendency of excess electrons to move elsewhere if their distribution is not uniform.

Now it is plausible to assume that $(r - g)$ will be crudely proportional to the excess electron density $n_e \equiv (n - n_0)$ when this excess density is not unduly large. It is convenient to define the *electron lifetime* τ_n by

$$\tau_n = \frac{n_e}{(r - g)} \qquad (4\text{-}99)$$

recognizing that a lifetime so defined will probably not be entirely independent of n_e. Thus in a semiconductor situation for which spatial dependences can be ignored, the electron continuity equation reduces to

$$\frac{dn_e}{dt} = \frac{dn}{dt} = g_E - n_e/\tau_n \qquad (4\text{-}100)$$

and for a situation of steady state excitation, the excess electron lifetime is just (n_e/g_E). During the transient decay which follows a period of externally stimulated generation, the time constant of decay is $\tau_n = -n_e(dt/dn)$ provided that the prior generation was reasonably uniform in space.[65]

A continuity equation for holes can be formulated in the same manner,

$$\frac{\partial p}{\partial t} = (g' - r') + g_E' - e^{-1}\nabla \cdot \mathbf{I}_p \qquad (4\text{-}101)$$

---

[65] Of course the picture of a semiconductor in which excess generation is completely homogeneous and the resulting recombination is spatially independent, is an ideal which can be approached with varying degrees of success. For a semiconductor sample of finite dimensions, there will inevitably be recombination on the surfaces as well as in the bulk, and the spatial distribution of excess electrons can then be described as a summation of Fourier components, each of which is influenced by the surface to a different extent and has a different effective lifetime. When externally provoked generation is spatially non-uniform (as it must be to some extent for any kind of ionizing radiation or injecting contact), the high order Fourier modes of short lifetime come into play in a manner which automatically tries to make the excess electron distribution *less* non-uniform. The literature pertaining to Fourier decomposition analysis of spatially dependent situations is discussed in Chapter 10 of *Semiconductor Statistics*.

For a situation of spatial uniformity this becomes the ordinary differential equation

$$\frac{dp_e}{dt} = \frac{dp}{dt} = g'_E - p_e/\tau_p \qquad (4\text{-}102)$$

expressed in terms of the hole lifetime $\tau_p = p_e(r' - g')^{-1}$. Holes and electrons have equal lifetimes if generation and recombination are both exclusively band-to-band processes: $g = g'$, $r = r'$, $n_e = p_e$. Hole and electron lifetimes are different if only one carrier population is changed from equilibrium, as will happen if the semiconductor is illuminated with low energy photons that excite electrons from flaw states. Hole and electron lifetimes are also generally different if generation of electrons and holes is a band-to-band affair but recombination occurs principally by the Hall-Shockley-Read mechanism, which involves the successive capture of free electrons and free holes by localized flaw centers.

DYNAMICS OF EXTRINSIC
GENERATION-RECOMBINATION[66]

Before we discuss the generation-recombination processes which take place band-to-band, let us consider the non-equilibrium dynamics of a single carrier population — say the excess population n_e of a conduction band. The model to be used is based on Figure 4-18, for a partially compensated n-type semiconductor with donor flaws of ionization energy ϵ_d. Since the conduction band states are sparsely occupied, the recombination rate should be proportional to the total free electron density $n = (n_0 + n_e)$ and to the number $\beta^{-1}(n + N_a)$ of available states on ionized donors. We can write this rate as

$$r = (\bar{v}\bar{\sigma}/\beta)n(n + N_a) \qquad (4\text{-}103)$$

where $\bar{v}$ is the average speed of a free electron (thermally averaged for a Boltzmann distribution) and $\bar{\sigma}$ is the similarly averaged capture cross-section that an empty donor state presents for any free electron.

The opposing rate of generation is determined by $N_{dn} = (N_d - N_a - n)$, the number of electrons on donors, and by the thermal environment. Provided the conduction band is not degenerated, it will not depend on the small existing state of occupancy in that band. Thus we can write

$$g = A(N_d - N_a - n) \exp(-\epsilon_d/k_0T) \qquad (4\text{-}104)$$

since the effect of the thermal environment is bound to be expressed by the ratio of ϵ_d to k_0T. The quantity A will depend on the density of conduction band states, but *not* on N_d, N_a, or n.

---

[66] This topic is discussed at some length since it can be associated with several worthwhile problems (4.23 through 4.25) of moderate difficulty. The sub-section may be omitted without serious loss of continuity.

Since $g = r$ at equilibrium, Equations 4-103 and 4-104 tell us that

$$\frac{n_0(n_0 + N_a)}{(N_d - N_a - n_0)} = (A\beta/\bar{v}\bar{\sigma}) \exp(-\epsilon_d/k_0 T) = n_1 \qquad (4\text{-}105)$$

In this way, dynamic arguments of generation and recombination recreate the equation previously derived as Equation 4-36 and given the name *mass action equation* at that time. We can now see the origin of that name. The quantity $(A/\bar{v}\bar{\sigma})$ is just N_c. Since each side of Equation 4-105 has the dimensions of volume^{-1}, it is convenient to write the right side as the "mass action density" n_1. In terms of n_1, the generation rate away from equilibrium is

$$g = (\bar{v}\bar{\sigma}/\beta)n_1(N_d - N_a - n) \qquad (4\text{-}106)$$

Substitution of Equations 4-103 and 4-106 into Equation 4-99 (with the aid of 4-105) results (see Problem 4.23) in an electron lifetime

$$\tau_n = (\beta/\bar{v}\bar{\sigma})[N_a + 2n_0 + n_1 + n_e]^{-1} \qquad (4\text{-}107)$$

suitable for insertion into the continuity Equation 4-100.

The most strongly temperature dependent quantity in Equation 4-107 is usually n_1, though there is a $T^{1/2}$ dependence of the average electron speed $\bar{v}$, and for some kinds of flaw the capture cross-section $\bar{\sigma}$ can be a very sensitive function of temperature.

Whether $\bar{\sigma}$ varies with temperature, and to what extent, depends on the physical basis for electron capture. How is the recombination energy disgorged? There are three possibilities, plus combinations of the three:

(a) The energy can be released as a photon, perhaps with phonons as well.

(b) The energy can be given as additional kinetic energy to another free electron.

(c) A number of phonons can be created, perhaps simultaneously or perhaps in a timed sequence.

Extrinsic luminescence is the process whereby the recombination energy appears as a photon, maybe with one or two phonons as well. The efficiency of this process usually does not depend on temperature. For most flaws, the efficiency of this *radiative recombination* is very small, and process (c) is dominant. There is a great practical interest in semiconductor-flaw systems for which extrinsic luminescence *can* be efficient. Some of those vigorously studied include luminescent transitions from the conduction band of some III-V and II-VI compounds to flaw states well below the middle of the gap. Visible luminescent displays (as used in digital meters, electronic calculators, etc.) can use radiative transitions whether they are band-to-band or band-to-flaw.

A process whereby the recombination energy appears as added kinetic energy for another electron is the extrinsic version of *Auger recombination*, a name given in view of the similar process in atomic physics. Auger recombination is the inverse of the process of impact ionization, whereby a fast electron loses much of its energy in ionizing a donor. In a semiconductor for which Auger recombination is actually dominant, the formulation of the recombination rate r in Equation 4-103 must be modified to $r = Bn^2(n + N_a)$, since the "capture cross-section" is itself proportional to the electron density. Problem 4.24 considers the resulting changes in the expression for electron lifetime. Since the Auger rate varies as n^2, this process can usually be ignored in a semiconductor when n is small.

In many capture processes by flaws, the perturbation of the lattice created by the presence of the flaw provides an easy coupling to lattice vibrations. Sometimes a *multi-phonon emission* takes care of all of the recombination energy at once. In other cases, a *phonon cascade*[67] occurs, whereby phonons are emitted one after another as the electron progresses through the spectrum of bound excited states of the flaw, finally reaching the ground state. This process has a strong temperature dependence, since at high temperatures an electron has a good chance of escaping from a shallow excited state instead of proceeding downward. Figure 4-59 shows the temperature dependence of $\bar{\sigma}$ for phosphorus donor in silicon, and it will be noted that the cross-section for low temperatures is enormous compared with the size of an atom. The capture seen in Figure 4-59 is aided by the Coulomb attraction of an ionized

---

[67] The phonon cascade process was first analyzed by M. Lax, Phys. Rev. 119, 1502 (1960). There have been many subsequent elaborations of this model.

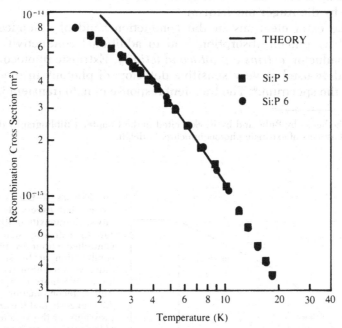

Figure 4–59 Capture cross-section as a function of temperature for ionized phosphorus donors in silicon. Capture is believed to take place accompanied by the emission of a cascade of phonons, essentially as envisaged by Lax.[67] Data from P. Norton, T. Braggins, and H. Levinstein, Phys. Rev. Letters **30**, 489 (1973).

donor for a conduction electron, and much smaller cross-sections are seen for capture of electrons or holes by neutral centers.

The dynamics of generation and recombination for phonon-assisted transitions are the same as for photon-assisted events. Thus (unlike Auger recombination), the dependence of τ_n on n_e is that of Equation 4-107.

Suppose that we use steady illumination to create a steady-state non-equilibrium situation in an n-type extrinsic semiconductor. Let there be a steady flux of I_0 photons per square meter per second, and let radiative generation be determined by I_0 and by the photo-ionization cross-section σ_I of each neutral donor. For simplicity, let the generation rate be uniform in space, which will be true if the sample thickness is small compared with $(N_d - N_a - n)\sigma_I$. There is then a monotonic relation between I_0 and the resultant n_e for this steady state, as sketched in Figure 4-60. Problem 4.23 has a second part which is concerned with identification of the regions in which n_e varies as I_0, then as $I_0^{1/2}$, with an eventual saturation. The form of the curve shown in Figure 4-60 is for a weakly compensated sample at a low temperature, provided that n_e has the dependence of Equation 4-107 (i.e., that radiative or phonon recombination processes dominate). The region of quadratic dependence shown in the figure shrinks and disappears either for high temperatures or for strong compensation. Incidentally, if radiative recombination *is* the dominant mechanism, the behavior for large I_0 and large n_e becomes complicated by stimulated emission processes, a topic we shall take up in connection with band-to-band radiative transitions.

As Problem 4.24 examines, the variation of n_e with I_0 is more complicated than shown in Figure 4-60 if recombination is predominantly by the Auger mechanism.

The extra electrons in the conduction band of a semiconductor caused by photon absorption lead to additional conductivity for the semiconductor, *extrinsic photoconductivity*. Extrinsic photoconductors are widely used as very sensitive detectors of photons in the infrared part of the spectrum.[68] The transient response of n_e to transient illumina-

---

[68] The books by Bube and by Ryvkin cited in the Chapter 4 Bibliography discuss the practical aspects of extrinsic photoconductors in detail.

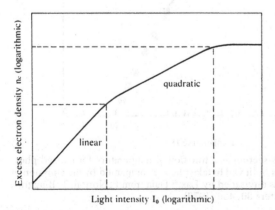

Light intensity I_0 (logarithmic)

Excess electron density n_e (logarithmic)

quadratic

linear

Figure 4–60 Variation of steady state excess electron density with intensity of illumination for an extrinsic semiconductor controlled either by phonon recombination or by spontaneous radiative recombination. Things are a little more complicated for Auger recombination! Problem 4.23 suggests that the analytic expressions for the various portions of the curve be evaluated, and the intensities found for the end of the linear region and the onset of saturation.

tion is usually of concern in these practical applications. Thus, suppose that excess generation caused by light is some arbitrary function $g_E(t)$ of time. How will $n_e(t)$ respond? The answer can be complicated if τ_n is a function of n_e (as in Equation 4-107), but let us suppose a small enough departure from equilibrium so that $\tau_n = \tau_0$, a constant. A simple integral can now relate $n_e(t)$ to g_E for all times prior to t.

Thus, let us write g_E for all times as a sequence of instantaneous generation events, each described by a delta function. The generative activity at time t_0 is $N\delta(t - t_0)$, let us say. Then Equation 4-100 requires that

$$n_e(t) = N \exp[(t_0 - t)/\tau_0] \qquad (4\text{-}108)$$

of the electrons created at t_0 have still survived at time t. Since Equation 4-100 is linear for our supposedly constant lifetime, the solution for $n_e(t)$ when g_E has been a continuous or discontinuous function of time previously is just the sum of terms of the type in Equation 4-108, or

$$n_e(t) = \int_{-\infty}^{t} g_E(t_0) \exp[(t_0 - t)/\tau_0]dt_0 \qquad (4\text{-}109)$$

Thus, when the lifetime does *not* depend on n_e, and Equation 4-109 is valid, it can be a simple matter to find the time dependence of n_e for a generation function which is an isolated pulse (rectangular, triangular, delta function, etc.) or for a series of pulses. Problem 4.25 is concerned with the responses to rectangular pulses and to sinusoidally varying generation. The problem goes on to consider what happens when $(r - g)$ varies as n_e^2 for a large disturbance of equilibrium (as is true in Equation 4-107). Transient decay is then hyperbolic for a large disturbance, becoming exponential as n_e becomes small enough to restore the constancy of τ_n.

The dynamics of generation-recombination have been discussed at considerable length for extrinsic situations, not because they are more important than the dynamics of band-to-band generation-recombination, but because they serve as a simplified introduction to the more complicated concepts of the latter topic.

GENERATION AND RECOMBINATION FOR ELECTRON-HOLE PAIRS

At the risk of being repetitious, it is remarked again that consistency with the laws of thermodynamics and the principle of detailed balance requires that every carrier generation mechanism have a recombination mechanism which does just the opposite. Now we are concerned with those reciprocal pairs of processes which create or destroy electron-hole pairs. For every kind of energy transformation process, the rates of these reciprocal processes at equilibrium must match in detail for every kind of energy transformation process. However, when $n \neq n_0$ and $p \neq p_0$, every kind of energy transformation process must report some kind of imbalance between the rates of generation and recombination.

Suppose, for example, that excess electron-hole pairs are created in a semiconductor sample by illumination. Then we have excess electrons and holes which were radiatively generated. This enforces an increase in every kind of recombination rate, but *radiative* recombination may or may not play a major role, depending on the semiconductor, its temperature, and flaw distribution. It is important to note that the recombination mechanism which dominates (r − g) in striving for a return towards equilibrium depends on the *semiconductor,* not on what caused the original disturbance of equilibrium.

This author has discussed the various kinds of recombination mechanisms which are physically possible in a semiconductor elsewhere at some length (*Semiconductor Statistics*). The reader is referred to that volume for detailed descriptions and associated mathematical formulations of the processes listed below.

Radiative Recombination

Radiative recombination occurring from band to band is just the inverse of intrinsic optical absorption. This is a very weak process for an indirect gap solid, since optical absorption is weak in these solids. However, radiative recombination can be a very efficient process for a direct gap semiconductor, and the *luminescence* (the emitted recombination radiation) for some direct gap semiconductors can be fairly efficient. When n and p are made very large as a result of some externally produced influence, the spontaneous radiative recombination is significantly augmented by *stimulated* radiative recombination. Under suitable conditions the stimulated recombination can become sufficiently dominant to produce *lasing* conditions.

In discussing the extrinsic generation-recombination processes of the last sub-section, we were concerned about mechanisms for energy transformation, but did not appear to worry about momentum conservation. This conservation law could be relaxed because the large mass of a flaw enables it to absorb momentum changes of an electron with ease. Now we must insist on momentum as well as energy conservation in considering radiative transitions between valence and conduction band states. Thus transitions are possible between states of energy ϵ_p in the valence band in Figure 4-61 and those conduction states of energy ϵ_n which have the same wave-vector. The semiconductor in Figure 4-61 is shown to depart from equilibrium, since separate quasi-Fermi-levels ϕ_n and ϕ_p are indicated. The relative occupancy of states at ϵ_n is

$$f_n = \{1 + \exp[(\epsilon_n - \phi_n)/k_0 T]\}^{-1} \qquad (4\text{-}110)$$

while the fraction of states at ϵ_p *not occupied* by electrons is

$$(1 - f_p) = \{1 + \exp[(\phi_p - \epsilon_p)/k_0 T]\}^{-1} \qquad (4\text{-}111)$$

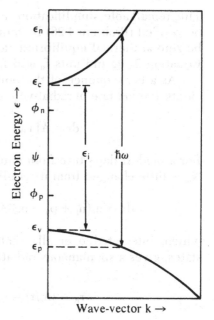

Figure 4–61 Model of a direct gap semiconductor for discussion of radiative transitions between valence states of energy ϵ_p and the conduction states of the same wave-vector and of energy ϵ_n.

Now the net rate of radiative transfer between ϵ_p and ϵ_n is the algebraic sum of three terms:

$$dr = dr_{sp} + dr_{st} - dg_{st} \qquad (4\text{-}112)$$

where dr_{sp} is the rate at which electron-hole recombination takes place spontaneously, with energy released as photons of energy $\hbar\omega = (\epsilon_n - \epsilon_p)$. The third term, dg_{st}, is the rate at which generative events are *stimulated* by the presence of suitable photons in the semiconductor, while dr_{st} is the rate at which *downward* transitions are stimulated by the presence of a photon field. The law of mass action can be used in writing these three rates as

$$\left. \begin{array}{l} dr_{sp} = Af_n(1 - f_p) \\ dr_{st} = Bf_n(1 - f_p)N_\omega \\ dg_{st} = Bf_p(1 - f_n)N_\omega \end{array} \right\} \qquad (4\text{-}113)$$

where N_ω is the number of photons per mode of energy $\hbar\omega$ [and thus must reduce to $<n>$ of Equation 2-54 at thermal equilibrium], and the quantities A and B contain information about densities of states at ϵ_n and ϵ_p, and about the matrix element connecting such states. Einstein[69] showed that upward and downward stimulated transitions must have the same coefficient B, and also demonstrated that

$$A = B \qquad (4\text{-}114)$$

---

[69] A. Einstein, Z. Physik. **18**, 121 (1917).

This remarkable simplification in the quantities of Equation 4-113 can be verified (Problem 4.26) by requiring the rate dr of Equation 4-112 to be zero at thermal equilibrium, a situation which makes $N_\omega \to <n>$ of Equation 2-54, and puts f_n and f_p in terms of the same Fermi energy.

As a consequence of the equality of the Einstein A and B coefficients, the *net* rate of radiative recombination away from equilibrium is

$$dr = A[f_n(1 - f_p) - (f_p - f_n)N_\omega] \qquad (4\text{-}115)$$

For a modest departure from equilibrium, $(f_p - f_n)$ is almost unity, and N_ω is little changed from its equilibrium value $<n>$. Then

$$dr \simeq n_e(n_0 + p_0 + n_e)(A/N_cN_v) \exp[-(\hbar\omega - \epsilon_l)/k_0T] \quad (4\text{-}116)$$

which, integrated over all combinations of valence and conduction states, gives a spontaneous radiative lifetime

$$\tau_R \equiv (n_e/r) = \tau_{R0}\left[1 + \left(\frac{n_e}{n_0 + p_0}\right)\right]^{-1} \qquad (4\text{-}117)$$

with a small-modulation value

$$\tau_{R0} = [N_cN_v/Ak_0T(n_0 + p_0)] \qquad (4\text{-}118)$$

Thus, when direct radiative transitions control a semiconductor, the lifetime of electrons and holes at small modulation has essentially the temperature dependence of $(n_0 + p_0)^{-1}$. The lifetime will vary rapidly with temperature when the semiconductor is intrinsic, but will vary little with temperature over a range for which the majority extrinsic carrier density does not change. From Equation 4-117, we can see that a lifetime independent of n_e can be enjoyed if n_e is small compared with *either* n_0 or p_0, but that recombination is quadratic in n_e for a stronger departure from equilibrium, just as was found for the radiative lifetime in an extrinsic situation (Equation 4-107).

When the excess carrier densities are large, stimulated recombination becomes an important factor, and a *population inversion* of $f_p > f_n$ results in an excess of stimulated recombination over stimulated emission. This can be seen by writing Equation 4-115 differently. The rate of spontaneous recombination is $dr_{sp} = Af_n(1 - f_p)$, and the net rate of downward minus upward radiative transitions can be expressed as

$$dr = dr_{sp}\left[1 - N_\omega\left\{\exp\left[\frac{\hbar\omega - (\phi_n - \phi_p)}{k_0T}\right] - 1\right\}\right] \qquad (4\text{-}119)$$

Thus, when $(\phi_n - \phi_p)$ is larger than the energy separation of the states (which is the same as saying that f_n is smaller than f_p), the term inside the braces in Equation 4-119 *adds* to the gross rate of recombination, to an extent which is proportional to N_ω. This is the basis of lasing in a semiconductor. Enough electron-hole pairs must be created to make

$(\phi_n - \phi_p)$ larger than the intrinsic gap as a necessary but not sufficient condition for lasing. A successful laser will result only if the conditions for optical gain are also satisfied.[70]

The necessary population inversion can be achieved in some semiconductors by injection into the transition region of a p-n junction. For semiconductors in which p-n junctions cannot be fabricated, lasing has been initiated by brief periods of massive electron-hole generation with electron beams and photon beams.[71]

Semiconductor lasing is highly unlikely unless the semiconductor has a direct gap, with the strong coupling of valence and conduction bands possible for such a gap. Radiative recombination is much weaker for an indirect gap material because phonons must assist each transition. Thus GaAs lases at a wavelength of some 0.9 μm, but GaP (with an indirect gap) is unable to lase in the visible part of the spectrum. Visible luminescent devices are of great interest for optical display systems,[72] and the efficiency of spontaneous radiative recombination can often be sufficient for these uses. Other applications of luminescence are not feasible without the greatly increased efficiency of a lasing system.

Radiative recombination can be a two-step process, the first step being formation of an electron-hole exciton,[51] with subsequent radiative annihilation of the exciton. It was predicted by Keldysh (1968) and verified experimentally by Pokrovsky et al.[73] that a large exciton population accumulated in a semiconductor as a result of massive pair creation can under appropriate conditions condense into discrete spherical droplets of degenerate electron-hole plasma a few μm in diameter. The radiative lifetime of such droplets is long enough in an indirect gap semiconductor (such as Ge or Si) at low temperatures so that the droplets can be observed by optical techniques such as Rayleigh scattering.

Auger Recombination

A direct gap semiconductor (in which it is possible for radiative recombination to be a very efficient process) is also likely to have some recombination of the *Auger* or *three-body* type. This process is the inverse of impact ionization; a hole and an electron recombine and donate the annihilation energy to a third particle, either an electron or a hole. Such an event must occur in a manner which conserves both energy and momentum, and the probability for such processes is enhanced if the ratio (m_c/m_v) is considerably smaller than or larger than unity. (See Problem 4.27.) We shall not go into a detailed formulation of band-to-

---

[70] The conditions whereby stimulated radiative recombination can become a major factor in a semiconductor are discussed by G. Lasher and F. Stern, *Phys. Rev.* **133**, A553 (1964).

[71] H. Kressel, in *Lasers, Vol. 3* (edited by A. K. Levine and A. J. DeMaria) (Dekker, 1971) gives an extensive discussion and bibliography of semiconductor lasers.

[72] E. E. Loebner, *Proc. I.E.E.E.* **61**, 837 (1973) surveys the past and likely future of electroluminescent solids for visible display systems.

[73] Y. Pokrovsky, A. Kaminsky, and K. Svistunova, *Proc. 10th Intl. Conf. on Physics of Semiconductors* (A.E.C., 1970), p. 504. See also C. Benoît à la Guillaume *et al.*, *Phys. Rev.* **B.5**, 3079 (1972); **B.7**, 1723 (1973).

band Auger recombination here, and refer the interested reader to *Semi-conductor Statistics*.

For a direct gap semiconductor, the excess carrier lifetime permitted by the combination of radiative and Auger recombination is likely to be quite small (in the range 10^{-10} to 10^{-7} s depending on the equilibrium hole and electron densities). However, the mutual satisfaction of energy and momentum conservation required for Auger recombination is much more difficult to accomplish in an indirect gap semiconductor. Similarly, radiative recombination is very slow in an indirect gap solid, since the radiative process is the inverse of the rather inefficient indirect absorption process. Thus the calculated excess carrier lifetime for an indirect gap semiconductor might range from 10^{-2} to 10^6 sec if we had to depend on band-to-band recombination processes only.

Hall-Shockley-Read (HSR) Recombination

Despite the above forecasts of a very large band-to-band lifetime for an indirect gap semiconductor, carrier lifetimes in the range of 10^{-7} to 10^{-3} seconds are far more typically measured in crystals of well known indirect gap solids. The recombination in these materials is completely dominated by two-step processes involving accidental (and sometimes deliberate) localized flaw centers. Presumably, any crystal of a direct gap solid also has some contribution to the recombination from these flaw-aided processes, but they have less chance to be dominant because the direct processes are so efficient.

It was realized separately by Hall[74] and by Shockley and Read[75] that a flaw could act as a capture agent in promoting electron-hole recombination, and the so-called HSR model has proved to be valid under a variety of conditions. When a flaw captures first an electron and then a hole, it has returned to its original state of charge and can start the process all over again. The HSR model can become quite complicated even when recombination occurs via a single set of flaws at one energy within the gap, as discussed in detail in *Semiconductor Statistics*. Consideration of flaws which can capture several holes (or electrons) one after another before having to start the capture of the opposite carrier species leads to much greater complexities in the analysis.[76] The formal treatment of a semiconductor with several different kinds of capturing flaw has not been seriously carried beyond the initial speculations of Kalashnikov,[77] though considerable *qualitative* insight can be obtained about what kinds of capturing agents will play the major role in a wide gap semiconductor or insulator with a spectrum of flaw states distributed over most or all of the gap.[78]

The flaw states which facilitate electron-hole recombination are usually not the ones which are major contributors to the carrier density at

[74] R. N. Hall, Phys. Rev. **83**, 228 (1951); Phys. Rev. **87**, 387 (1952).
[75] W. Shockley and W. T. Read, Phys. Rev. **87**, 835 (1952).
[76] C. T. Sah and W. Shockley, Phys. Rev. **109**, 1103 (1958).
[77] S. G. Kalashnikov, J. Tech. Phys. U.S.S.R. **26**, 241 (1956).
[78] A. Rose, Concepts in Photoconductivity and Allied Problems (Interscience, 1963); J. G. Simmons and G. W. Taylor, Phys. Rev. **B.4**, 502 (1971).

equilibrium. Flaw states are most efficient at recombination when they are nearer to the middle of the gap. A flaw which acts as a recombination center typically has capture cross-sections for holes and electrons which are very different, and this often leads to trapping problems, whereby $n_e \neq p_e$.

A derivation of the expressions for lifetimes in the HSR model with and without trapping will not be given here. It is simply noted that when the flaw density is small enough to keep n_e and p_e essentially the same, then the hole lifetime τ_p equals the electron lifetime τ_n. This lifetime has the form

$$\tau_p = \tau_n = \frac{\tau_0(n_0 + p_0) + \tau_\infty n_e}{(n_0 + p_e + n_e)} \qquad (4\text{-}120)$$

where τ_0 and τ_∞ are the limiting values of lifetime for very small and very large disturbances of equilibrium (and are each inversely proportional to the density of recombination centers).

Even though flaw states at a variety of energies may play some minor role in the recombination process, a single type of flaw (as envisaged in the derivation of Equation 4-120) often is much more important than all other types combined. Thus, it is essential to avoid all but the smallest traces of contamination with copper if germanium is to be produced with a large lifetime, as is desired for alloy type PNP germanium transistors. In contrast, the transistor manufacturer who wants to make silicon switching transistors with a rapid switching action deliberately allows his silicon to be infused with atoms of gold, a very effective recombination center.

The lifetime, in conjunction with the abilities of carriers to drift and to diffuse, determines how far *minority carriers* can move in a semiconductor before they disappear via recombination. Next we must see what makes electrons and holes move in a semiconductor.

DRIFT AND DIFFUSION CURRENTS

In a metallic conductor, we are accustomed to thinking of a current as resulting solely from an applied field. Diffusion currents, caused by a spatial dependence of the free carrier density, are negligibly small in a metal. This is not so in a semiconductor, and both kinds of currents must be considered. Let us discuss the current carried by the electron population in a semiconductor crystal.

The electron drift current depends on electric field, electron density, and electron mobility:

$$\mathbf{J}_f = -e\mathbf{E}n\mu_n = n\mu_n \nabla \epsilon_c \qquad (4\text{-}121)$$

This is expressed in the manner shown at the right of this equation since an electric field tilts the bands as a function of distance, to make $\nabla \epsilon_c = -e\mathbf{E}$.

Now the diffusion current which must be added to $\mathbf{J}_f$ depends on the gradient of electron density and the diffusion coefficient of the electrons:

$$\mathbf{J}_d = eD_n \nabla n \qquad (4\text{-}122)$$

As shown by Einstein[79] in his classic analysis of Brownian motion, diffusion coefficient and mobility are related by

$$D = (k_0 T \mu / e) \tag{4-123}$$

which means that

$$J_d = k_0 T \mu_n \nabla n \tag{4-124}$$

Thus the total current density due to all electrons is

$$J_n = (J_f + J_d) = \mu_n [k_0 T \nabla n + n \nabla \epsilon_c] \tag{4-125}$$

If we now write the total electron density in the form of Equation 4-92, then ∇n can be expressed in terms of $\nabla \phi_n$. Equation 4-92 requires that

$$\nabla n = (n/k_0 T)[\nabla \phi_n - \nabla \epsilon_c] \tag{4-126}$$

and from Equations 4-125 and 4-126 the current density reduces to the very simple form

$$J_n = n \mu_n \nabla \phi_n \tag{4-127}$$

Thus $e^{-1}\nabla \phi_n$ serves as an *effective field* in producing an uncompensated electron current. Provided that $\nabla \phi_n$ is zero everywhere (as must happen in thermal equilibrium) there is no *net* current of electrons from one place in an inhomogeneously doped crystal to another, though there may be substantial equal and opposite diffusion and drift currents. As soon as thermal equilibrium is violated, the opportunities for diffusion and drift currents to exactly balance each other are likely to be lost, and there will be a net flow of electrons. Similar considerations apply to the drift and diffusion of holes.

Suppose that extra electrons are placed in the conduction band of a semiconductor under circumstances which make their subsequent movement controlled by diffusion rather than by electric field. Eventually all will suffer recombination, and the fraction which cover a distance L by random walk progression before suffering recombination is $\sim \exp(-L/L_n)$. The characteristic length L_n here is the *electron diffusion length*

$$L_n = (\tau_n D_n)^{1/2} \tag{4-128}$$

The effective range of holes under a diffusive influence is similarly $L_p = (\tau_p D_p)^{1/2}$.

The lengths L_n and L_p are of great importance for the operation of semiconductor devices which depend on injection of minority carriers into portions of a semiconductor crystal. The basic device of this type is the p-n junction, for which Shockley[80] outlined the principle of operation which is still accepted. Figure 4-62 sketches a p-n junction at thermal equilibrium, with the bending of the bands required to make the Fermi

---

[79] A. Einstein, Z. Physik. **17**, 549 (1905).

[80] W. Shockley, Bell Syst. Tech. J. **28**, 435 (1949). Also discussed by Shockley in *Electrons and Holes in Semiconductors* (Van Nostrand, 1950).

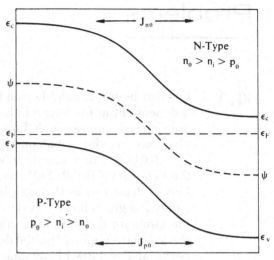

Figure 4-62 Idealized one-dimensional picture of the tilting of the bands through a p-n junction at thermal equilibrium. Since $\nabla \epsilon_c$ is non-zero, there is a large static electric field in the junction region, perhaps 10^6 V/m or more. This causes drift motion of electrons to the right and of holes to the left. $J_n = J_p = 0$ everywhere, since diffusion of electrons to the left and of holes to the right cancels out the drift currents.

Distance through junction

level the same everywhere. Two equal and opposite current densities (J_{p0}) are indicated for diffusion and drift of holes, and two similarly cancelling current densities (J_{n0}) for drift and diffusion of electrons.

We can reason about the approximate magnitudes of J_{n0} and J_{p0} as follows. Electron drift current comes from electrons which happen to be generated on the p-side of the junction within an electron diffusion length of that junction, and are thereby able to diffuse to the junction and slide down the slope. This leads us to expect that

$$\left. \begin{array}{l} J_{n0} \simeq (en_i^2 L_n/p_0 \tau_n) = (en_i^2/p_0)(D_n/\tau_n)^{1/2} \\ J_{p0} \simeq (en_i^2 L_p/n_0 \tau_p) = (en_i^2/n_0)(D_p/\tau_p)^{1/2} \end{array} \right\} \qquad (4\text{-}129)$$

So much for equilibrium. If a voltage is applied to the junction to make the p-side negative compared with the n-side, thus increasing the barrier height, then the diffusion currents are sealed off. A net current density $(J_{n0} + J_{p0})$ flows, resulting from generation of electrons and holes within a diffusion length or so of each side of the barrier. This current is essentially independent of voltage once $V > (k_0T/e)$, and in a well-made p-n junction the current in the high resistance (or blocking) direction is very small.

A voltage in the opposite direction lowers the barrier height, and permits diffusion to carry large numbers of electrons into the p-side and holes into the n-side. This is the phenomenon of carrier injection, which leads to bipolar transistor operation if injected carriers can be captured by yet another separately doped region of the crystal. The relation between total net current and applied voltage is thus nominally

$$J = (J_{n0} + J_{p0})[\exp(eV/k_0T) - 1] \qquad (4\text{-}130)$$

Any real p-n junction has a characteristic which departs from Equation 4-130 for a variety of complicating reasons. The interested reader has an immense body of literature available concerning the principles of junction transistors, integrated circuits, and innumerable other devices. The bibliography at the end of this chapter should serve as a starting point for reaching this literature.

Problems

4.1   Overlap between neighboring flaws results in a temperature-independent n_0 for n-type indium antimonide if the electrons are derived from very shallow donor flaws. The conduction electrons in this semiconductor have an effective mass $m_c = 0.014m$. For a sample in which $n_0 = 10^{22}$ m^{-3}, show that the Fermi level is 0.012 eV above ϵ_c at absolute zero. Calculate how ϵ_F decreases as the sample warms to room temperature, and plot a graph for ϵ_F relative to ϵ_c. (If you wish, you may use the forms for the Fermi integral given in Footnote 6 to cover the awkward region of partial degeneracy. You should find that $\epsilon_F = \epsilon_c$ at T $\simeq$ 140K.) Now imagine the sample at T = 0 again, and find the period P in tesla^{-1} between successive maxima in SdH oscillations.

4.2   Determine the slope and intercept of the straight line through the intrinsic data for germanium in part (b) of Figure 4-7. These numbers are to be interpreted in terms of Equation 4-20. Assume that $m_c = 0.56m$ is the Ge conduction band scalar effective mass *for density-of-states purposes*, and that the corresponding valence band parameter is $m_v = 0.37m$. You should then find that compatibility of your measured intercept with Equation 4-20 requires a temperature dependence $\alpha = (-d\epsilon_i/dT) = 3.7 \times 10^{-4}$ eV/K. Compare your deduced $\epsilon_i = (\epsilon_{i0} - \alpha T)$ with the optically obtained curve of Figure 4-8.

4.3   It is generally agreed from various kinds of measurements that InSb has an intrinsic gap of $\epsilon_i = 0.18$ eV at 300K, and that $(d\epsilon_i/dT) \simeq -2.8 \times 10^{-4}$ eV/K. You are asked here to deduce the gap of InSb from the n_i data of Figure 4-9, but the assumptions you are asked to make will prevent you from getting the right answer. Thus, compute the gap which satisfies Equation 4-17 for four or five temperatures in Figure 4-9. Assume that $m_c = 0.014m$ and $m_v = 0.4m$. Assume also that m_c and m_v do not depend on temperature (which is not really a good assumption for m_c), and that Equation 4-17 is good for all temperatures. (Electron degeneration at high temperatures violates this last assumption.) Your analysis is a warning about the caution with which data for a new semiconductor must be approached.

4.4   The intrinsic gap in indium arsenide has been measured by J. R. Dixon and J. M. Ellis [Phys. Rev. **123**, 1560 (1961)] from the photon energy at which there is an onset of intense optical absorption. This optical gap has been shown by Varshni (foot-note 9) to obey

$$\epsilon_i = 0.426 - 3.16 \times 10^{-4}T^2(93 + T)^{-1} \text{ eV}$$

as a function of temperature. Given that $m_c = 0.022m$ and $m_v = 0.4m$, plot ϵ_c and ψ as functions of temperature for the range 250K to 500K, using ϵ_v as your origin of energy. You should be able to measure values of the ratio (p_0/n_i) for various temperatures from Figure 4-11 with sufficient accuracy to be able to show the course of ϵ_F for each of these two InAs samples below the extrinsic transition temperature.

4.5 Consider a *hydrogenic* donor, for which the Coulomb attraction toward the impurity center of the bound circulating electron is supposed to be inversely proportional to the dielectric constant κ at all distances. Show that the ground state (angular momentum $\hbar$) of a Bohr model then has energy and radius as given in Equation 4-25. You should find that $\epsilon_d = 0.0133$ eV and $a_d = 51$ Å for hydrogenic donors in cadmium telluride, since this semiconductor has a simple conduction band minimum of mass $m_c = 0.11m$ at the center of the zone, and a dielectric constant $\kappa = 10.6$.

4.6 Solve Equation 4-36 numerically for a series of temperatures from 10K to 100K for two hypothetical samples of CdTe ($m_c = 0.11m$) that contain hydrogenic donors ($\beta = \frac{1}{2}$, $\epsilon_d = 0.0133$ eV) of the type discussed in Problem 4.5. $(N_d - N_a) = 10^{22}$ m^{-3} for each of these samples, but one has $N_a = 10^{21}$ m^{-3} compensators while the other has only $N_a = 10^{19}$ m^{-3}. You should find that a semilog plot of n_0 versus $(1/T)$ will resemble Figure 4-19. Now use Equation 4-41 to find how ϵ_F varies with temperature for each sample; the results should crudely resemble Figure 4-20. [This problem is a natural for solution on a computer.]

4.7 Infrared spectroscopic observations by R. L. Aggarwal and A. K. Ramdas, reported in Phys. Rev. **137**, A602 (1965), show that arsenic donors in silicon have a split 1s state, similar to that illustrated for phosphorus donors in Figure 4-16. With arsenic, they find $\Delta_c = 0.022$ eV. Bearing this in mind, use Equation 4-41 to plot curves for the Fermi energy as functions of temperature in the arsenic-doped-silicon samples (a) and (b) of Figure 4-19. Do you find that for either sample the maximum of ϵ_F is high enough to cause appreciable population of the upper 1s states in a certain range of temperature? Should this visibly impair the fit to Equation 4-36 which was used for the solid curves in Figure 4-19?

4.8 Given a semiconductor in which $m_c = 0.25m$, calculate the energy, momentum, wave-vector, and de Broglie wavelength for a conduction band electron of the mean speed for $T = 300K$. Determine the same quantities for an LA phonon which travels at a speed 5000 m/sec and has a wavelength twice as large as the electron. Show that the conservation of energy and crystal momentum permits such a phonon to scatter the electron through an angle of 29° while changing the electron energy by less than 2 per cent.

4.9 Electrons in the conduction band of the semiconductor described in Problem 4.8 are scattered by LA phonons with an effectiveness we can describe by a deformation potential constant $\epsilon_1 = 5$ eV. If the relative density of this material is 5, then show that the mean free time between collisions is $\tau_m \sim 5 \times 10^{-13}$ sec at room temperature, and that the lattice scattering mobility is $\mu_L \simeq 2000T^{-3/2}$ m²/V·s. At what temperature is this equal to the ionized flaw mobility μ_i calculated from Equation 4-62 for 10^{20} m^{-3} of monovalent ionized flaws, if the dielectric constant $\kappa = 10$? Show how the mobility which combines lattice and ionized-flaw scattering varies with temperature, pointing out the temperature below which the logarithmic component of μ_i affects the temperature dependence of the mobility.

4.10 A current density J_x is carried by the electron population in a semiconductor, while a magnetic field B_z is applied. Thus there is a resultant electric field we can describe by components E_x and E_y. If all the electrons have the same drift velocity, show that $E_y = E_x \tan(B_z\mu_H)$. What will happen if the sample is not several times longer than its width?

4.11 Consider an intrinsic semiconductor in which all electrons have one drift velocity and all holes have another, so that their mobilities μ_n and μ_p are different. Show that when current flows and a transverse magnetic field is applied, a Hall field must develop in the direction of $\mathbf{J} \times \mathbf{B}$, and that the magnitude of this will in steady state balance the combined Lorentz forces on both electrons and holes provided that

$$R_H = \frac{-1}{n_ie}\left(\frac{\mu_n - \mu_p}{\mu_n + \mu_p}\right)$$

4.12 Make an attempt to analyze some of the indium antimonide samples for which conductivity and Hall effect data appear in Figures 4-31 and 4-32. Assume that the Hall factor $r = 1$ and that the ionized flaw density $(n_0 - p_0)$ or $(p_0 - n_0)$ does not vary over the temperature range studied. Consult the curve of Figure 4-9 for n_i in InSb if you wish, though it should not be necessary. Show that if the ratio $b \equiv (\mu_n/\mu_p)$ is insensitive to temperature, then the negative peak of the Hall coefficient (for a sample which is p-type at lower temperatures) will occur when $n_0\mu_n = p_0\mu_p$. Try to describe the behavior of sample V, one of the n-type samples of lower resistivity, and one of the p-type samples.

4.13 Assume that donor flaws in a semiconductor are distributed in a regular array which *maximizes* the donor-donor spacing for a given N_d. What two kinds of array will do this? Verify that the donor-donor separation is then $(2/N_d{}^2)^{1/6}$ as mentioned in Footnote 27. Alternatively, now suppose that donors are *randomly distributed* in space. Show why it is then reasonable to expect that any donor will have one or more other donors within a distance $(3/4\pi N_d)^{1/3}$.

4.14 Given that germanium has a dielectric constant $\kappa = 16$, use a hydrogenic model for arsenic donors in germanium to deduce the radius a_d of the ground state wave-function for such donors. [Use the value of ϵ_{do} shown for such donors at infinite separation in Figure 4-33.] From the value of N_{di} at which ϵ_d drops to zero, show that impurity metal behavior occurs if the *average spacing* of donors is less than about $6a_d$, or if the *closest* donor-donor spacing is less than about $3.2a_d$.

4.15 This rather difficult problem looks at the feasibility of creating thermal breakdown in an amorphous semiconductor, simplified by arranging the geometry so that equipotentials are isothermals. Heat flows outward from the symmetrical center along current lines. Imagine two small hemispheres of an amorphous semiconductor, of electrical conductivity σ and thermal conductivity κ. Put ohmic electrodes on the flat faces, and press the curved faces together. When a current is passed from one electrode to another, the region near the central point of contact heats up, and the resistance is the *spreading resistance* on both sides of this contact. With room temperature T_0, suppose the maximum temperature at the contact point of the hemispheres is T_m when the end-to-end applied voltage is V_0. Then show that

$$\int_{T_0}^{T_m} (\kappa/\sigma)dT = (V_0{}^2/8)$$

If $\kappa = (\kappa_0/T)$ and $\sigma = \sigma_0 \exp(-\Delta\epsilon/k_0 T)$, this integral can be expressed as a difference of *exponential integrals* $\overline{\mathrm{Ei}}(x)$ [a tabulated function] for $x = (\Delta\epsilon/k_0 T_0)$ and $x = (\Delta\epsilon/k_0 T_m)$. If $\kappa_0 = 10^4$ watt/meter, $\sigma_0 = 10^4$ ohm^{-1}meter^{-1}, $\Delta\epsilon = 0.3$ eV, and

$T_0 = 300K$, then show that an applied voltage $V_0 = 250$ volts will cause T_m to reach about 350K, while a voltage increase to $V_0 = 290$ volts will create an essentially infinite temperature rise at the central contact.

4.16 Suppose a solid which crystallizes in a simple cubic lattice, and draw the Brillouin zone for this solid. Imagine that the minimum conduction band energy occurs at a location on the B.Z. boundary in (a) the [100] direction; (b) the [110] direction; (c) the [111] direction. Describe how many complete closed surfaces must be necessary to describe an energy just above ϵ_c, and what kinds of distortion from spherical form are possible in these cases. Now suppose another solid with a hexagonal lattice (for which the B.Z. is also hexagonal), and let the minimum conduction band energy occur at a location in the corner of this zone. How many complete closed energy surfaces now exist for an energy just above ϵ_c?

4.17 Show that when a photon is absorbed in a direct gap semiconductor, an electron and a hole share the excess energy $(\hbar\omega - \epsilon_i)$ in a manner which gives them each the wave-vector of Equation 4-84. Given that InSb has a band structure like the GaAs one of Figure 4-48, and that $m_c = 0.014m$, $m_v = 0.4m$, and $\epsilon_i = 0.18 \, eV$ at $T = 300K$ for InSb, then show how much energy from a 0.50 eV photon is converted into free electron kinetic energy and how much is converted into free hole energy. If we include the momentum of the photon itself in the equation for conservation of momentum, then by what percentage do the electron and hole momenta differ?

4.18 A bound state of a simple direct gap exciton comprises (in Bohr theory terms) the combined orbital motion of an electron with mass m_c and a hole of mass m_v in a medium of dielectric constant κ, such that the total angular momentum about the center of mass is a multiple of $\hbar$. Produce an expression for the binding energy of the ground state of such an exciton and for the family of excited states. You may express this conveniently in terms of the joint density of states effective mass m_r. Calculate the binding energies of the ground state and the first excited state for a direct exciton in indium arsenide ($m_c = 0.022m$, $m_v = 0.4m$, $\kappa = 11.6$), and show that at room temperature the photon energy necessary to create an exciton in the ground state is 99.40 per cent of the energy necessary to produce a free electron and free hole at rest. (See Problem 4.4 concerning the gap width in InAs.) Is an exciton in this semiconductor likely to be stable at room temperature?

4.19 Note in Table 2-1 that longitudinal sound waves of low frequency travel through germanium at a speed of 5400 m/s. Estimate the angular frequency for LA modes that correspond with the $[\frac{1}{2}\frac{1}{2}\frac{1}{2}]$ zone boundary, since these modes can presumably be created or destroyed in indirect optical transitions in Ge. Calculate what you would expect the difference $2k_0\theta$ to be between the threshold energies for phonon-absorbing and phonon-creating transitions if this was the only branch of the phonon spectrum involved, and compare your results with what you see in Figure 4-53. In employing Equation 2-33 remember that Ge has a diatomic basis with both atoms of the same mass.

4.20 It is reported from Raman scattering that long wave LO phonons in InSb have a frequency of 5.85 THz. These phonons interact with electrons in InSb to produce the magnetophonon resonance shown in Figure 4-55. Plot the value of (1/B) for each maximum in this experimental curve against the integer number printed, and verify that the relationship is linear, as anticipated from Equation 4-89. From the slope, given the character of LO phonons quoted above, determine the effective mass m_c for conduction electrons in InSb. How does your value for (m_c/m_0) compare with the number previously used in Problem 4.17?

4.21 Consider sample 1 of Figure 4-11. If you have worked Problem 4.4, then you already know how the equilibrium Fermi energy ϵ_F varies with temperature for this material. Show now in a series of little sketches the positions of the quasi-Fermi-levels ϕ_n and ϕ_p for this sample at 300K and again at 400K for excess pair densities of 10^{20}, 10^{21}, and 10^{22} m^{-3}. Suppose now that part of this sample is illuminated at 300K to maintain a steady-state excess pair density of 10^{22} m^{-3}. The transition region to unilluminated material is 100 μm thick. Discuss what controls the current flow of holes and of electrons in this transition region.

4.22 Suppose that carrier exclusion is to take place within a sample in the form of a cube 1 mm on a side. The intrinsic density is $n_i = 10^{19}$ m^{-3}, and $N_r = (n_0 - p_0)$ [viz. Equations 4-23 and 4-24] is 10^{18} m^{-3}. Complete carrier depletion should then result in $n = N_r, p = 0$. Calculate the electrostatic potential at the center of the sample if the electron density is in fact depleted to $n = 9.95 \times 10^{17}$ m^{-3}.

4.23 Verify that the electron lifetime of Equation 4-107 does result from inserting the generation and recombination rates of Equations 4-103 and 4-106. You will need to use Equation 4-105 to achieve cancellation of unwanted terms. Suppose now that there is a steady-state photon flux of I_0 photons per meter^2

per second, and that the ability of neutral donors to respond is represented by a photo-ionization cross-section σ_I. Determine the form of the resulting steady-state n_e versus I_0 for a situation in which $n_0 \ll N_d$ and $N_a \ll N_d$. Substantiate the shape of the curve in Figure 4-60, and determine the values of I_0 and n_e for which the slope on a log-log plot changes.

4.24 Suppose an extrinsic semiconductor for which Auger recombination controls the return rate of electrons to the flaw levels. Retrace the path of Equations 4-103 through 4-107, with recombination now required to vary as $n^2(n + N_a)$, and show that the lifetime must vary as $[(n_0 + n_e)(N_a + n_e)]^{-1}$ for temperatures low enough to make $n_0 \ll N_a < N_d$. This semiconductor is now excited with photons, so that generation occurs radiatively while recombination is still by the Auger process. For a photon flux I_0, show that n_e will vary as I_0 for a small flux, followed by regions of varying as $I_0^{1/2}$, $I_0^{1/3}$, and finally saturation.

4.25 Consider a semiconductor for which the lifetime τ_0 does not depend on n_e. Discuss the time dependence of n_e when excess generation g_E is a repeated rectangular function of time, both for a small and for a large ratio of "off" time to "on" time. What is the response to $g_E = A[1 + B \sin(\omega t)]$, and for what frequency are g_E and n_e 45° out of phase? Now suppose that the lifetime varies as $\tau_n = \tau_0(1 + n_e/C)^{-1}$ (which is the functional dependence of lifetime on n_e in Equation 4-107). This rules out the use of Equation 4-109 when n_e is large. However, you will not need Equation 4-109 to find how n_e varies with time in response to illumination in the form of a large rectangular pulse. What is the decay of n_e in response to a very powerful delta function pulse of illumination?

4.26 Show that the Einstein A and B coefficients are equal, as asserted in Equation 4-114, by requiring the total radiative rate dr to vanish at equilibrium, and substituting equilibrium values into the quantities dr_{sp}, dr_{st} and dg_{st} of Equation 4-113. Thereby, show that when $N_\omega \simeq \langle n \rangle$, and the quasi-Fermi-levels are well inside the intrinsic gap, then the rate dr for non-equilibrium is given by Equation 4-116.

4.27 Consider the two kinds of Auger recombination that can take place band-to-band between a conduction band of mass m_c and a valence band of mass m_v. The extrema are both at the zone center. Show how energy and momentum can both be conserved, in exciting either a highly energetic hole or a highly energetic electron in the course of destroying a hole-electron pair. Why is the probability of this kind of recombination enhanced when m_c and m_v are appreciably different?

Bibliography

Semiconductors in General
A. F. Gibson (ed.), *Progress in Semiconductors* (Wiley), Vol. 1 (1956) through Vol. 9 (1965).
P. S. Kireev, *Semiconductor Physics* (Mir, 1975).
J. P. McKelvey, *Solid State and Semiconductor Physics* (Krieger, 1982).
T. S. Moss (ed.), *Handbook on Semiconductors* (North-Holland, 1980–82), 4 vols.
B. K. Ridley, *Quantum Processes in Semiconductors* (Oxford Univ. Press, 1982).
K. Seeger, *Semiconductor Physics* (Springer, 2nd ed., 1982).
W. Shockley, *Electrons and Holes in Semiconductors* (Van Nostrand, 1950).
R. A. Smith, *Semiconductors* (Cambridge Univ. Press, 2nd ed., 1978).
R. K. Willardson and A. C. Beer (eds.), *Semiconductors and Semimetals* (Academic Press), from Vol. 1 (1966) through Vol. 21 (1984) and continuing.

Equilibrium Electronic Distribution
J. S. Blakemore, *Semiconductor Statistics* (Dover, 1987).
E. Spenke, *Electronic Semiconductors* (McGraw-Hill, 1958).
(Also, in the above-cited Kireev, McKelvey, and Shockley books.)

Crystalline Semiconductor Materials
H. R. Huff (ed.), *Semiconductor Silicon 1981* (Electrochemical Soc., 1981).
D. R. Lovett, *Semimetals and Narrow Bandgap Semiconductors* (Pion, 1977).
O. Madelung, *Physics of III–V Compounds* (Wiley, 1964).
T. P. Pearsall (ed.), *GaInAsP Alloy Semiconductors* (Wiley, 1982).
M. Pope and C. E. Swenberg, *Electronic Processes in Organic Crystals* (Oxford Univ. Press, 1982).
B. Ray, *II–VI Compounds* (Pergamon Press, 1969).
J. L. Shay and J. H. Wernick, *Ternary Chalcopyrite Semiconductors* (Pergamon Press, 1975).

Impurity Effects in Semiconductors
J. S. Blakemore and S. Rahimi, *Models for Midgap Centers in Gallium Arsenide*, in *Semiconductors and Semimetals, Vol. 20* (Academic Press, 1984).
J. Bourgoin and M. Lannoo, *Point Defects in Semiconductors* II: *Experimental Aspects* (Springer, 1983).
W. M. Bullis and L. C. Kimmerling (eds.), *Defects in Silicon* (Electrochemical Society, 1983).
V. I. Fistul', *Heavily Doped Semiconductors* (Plenum Press, 1969).
M. Jaros, *Deep Levels in Semiconductors* (Hilger, 1982).
M. Lannoo and J. Bourgoin, *Point Defects in Semiconductors* I: *Theoretical Aspects* (Springer, 1981).
R. D. Larrabee, *Neutron Transmutation of Semiconductor Materials* (Plenum Press, 1984).
S. Mahajan and J. W. Corbett (eds.), *Defects in Semiconductors* II (North-Holland, 1983).
A. G. Milnes, *Deep Impurities in Semiconductors* (Wiley-Interscience, 1973).
A. M. Stoneham, *Theory of Defects in Solids* (Oxford Univ. Press, 1975).
P. D. Townsend and J. C. Kelly, *Colour Centres and Imperfections in Insulators and Semiconductors* (Crane, Russak, 1973).

Transport in a Crystalline Semiconductor
A. C. Beer, *Galvanomagnetic Effects in Semiconductors* (Academic Press, 1963).
E. M. Conwell, *High Field Transport in Semiconductors* (Academic Press, 1967).

B. R. Nag, *Electron Transport in Compound Semiconductors* (Springer, 1980).
E. H. Putley, *The Hall Effect and Related Phenomena* (Butterworth, 1960).

Transport in Insulators and Non-Crystalline Materials
D. Adler, B. B. Schwartz, and M. C. Steele (eds.), *Physical Properties of Amorphous Materials* (Plenum Press, 1985).
M. H. Brodsky (ed.), *Amorphous Semiconductors* (Springer, 1979).
M. Cutler, *Liquid Semiconductors* (Academic Press, 1977).
M. A. Lampert and P. Mark, *Current Injection in Solids* (Academic Press, 1970).
N. F. Mott, *Metal-Insulator Transitions* (Taylor and Francis, 1974).
N. F. Mott and E. A. Davis, *Electronic Processes in Non-Crystalline Materials* (Oxford Univ. Press, 2nd ed., 1979).
J. J. O'Dwyer, *The Theory of Electrical Conduction and Breakdown in Solid Dielectrics* (Oxford Univ. Press, 1973).
J. Tauc, *Amorphous and Liquid Semiconductors* (Plenum Press, 1974).

Energy Bands in Semiconductors
M. Cardona, *Modulation Spectroscopy* (Academic Press, 1969).
D. Long, *Energy Bands in Semiconductors* (Wiley-Interscience, 1968).
J. C. Phillips, *Bands and Bonds in Semiconductors* (Academic Press, 1973).
(See also the "Band Theory" list at the end of Chapter 3.)

Optical and Non-Equilibrium Phenomena
N. G. Basov, *Optical Properties of Semiconductors* (Plenum Press, 1976).
J. S. Blakemore, *Semiconductor Statistics* (Dover, 1987).
R. H. Bube, *Photoconductivity of Solids* (Wiley, 1960).
C. D. Jeffries and L. V. Keldysh (eds.), *Electron-Hole Droplets in Semiconductors* (North-Holland, 1983).
J. I. Pankove, *Optical Processes in Semiconductors* (Dover, 1975).
A. Rose, *Concepts in Photoconductivity and Allied Problems* (Wiley, 1963).
S. M. Ryvkin, *Photoelectric Effects in Semiconductors* (Plenum Press, 1964).

Processes in Semiconductor Device Structures
(See also the "Solid State Electronics" list on page ix.)
I. Brodie and J. J. Murray, *The Physics of Microfabrication* (Plenum Press, 1982).
H. C. Casey and M. B. Panish, *Heterostructure Lasers* (Academic Press, 1978), 2 vols.
D. A. Fraser, *Physics of Semiconductor Devices* (Oxford Univ. Press, 3rd ed., 1983).
H. L. Grubin, K. Hess, G. J. Iafrate, and D. K. Ferry (eds.), *The Physics of Submicron Structures* (Plenum Press, 1984).
H. K. Henisch, *Semiconductor Contacts* (Oxford Univ. Press, 1984).
H. Kressel (ed.), *Devices for Optical Communication* (Springer, 2nd ed., 1982).
A. G. Milnes and D. Feucht, *Heterojunctions and Metal-Semiconductor Junctions* (Academic Press, 1972).
L. Solymar and D. Walsh, *Lectures on the Electrical Properties of Solids* (Oxford Univ. Press, 3rd ed., 1984).
G. B. Thompson, *Physics of Semiconductor Laser Devices* (Wiley, 1980).
W. C. Till and J. T. Luxon, *Integrated Circuits: Materials, Devices, and Fabrication* (Addison-Wesley, 1982).
E. W. Williams and R. Hall, *Luminescence and the Light Emitting Diode* (Pergamon Press, 1978).

chapter five

DIELECTRIC AND MAGNETIC PROPERTIES OF SOLIDS

This chapter gives a brief account of some of the phenomena on an atomic scale which contribute to the macroscopically observable dielectric and magnetic properties of solids. The discussion of certain topics has been abbreviated or curtailed for two reasons, and a bibliography of more extended accounts is provided at the end of the chapter as a supplement to the limited or descriptive coverage given here.

One reason for compressing dielectric and magnetic phenomena (which are two extremely active fields of solid state research) into a single chapter is that the subjects we shall discuss here have a rather different emphasis than the subject matter of the previous four chapters. Up to this point we have been almost continuously concerned with the consequences of periodicity in real space and in k-space. To be sure, the long range order of a crystalline lattice is significant in any discussion of ferroelectric, ferromagnetic, or antiferromagnetic behavior, but most of the topics in this chapter are less directly involved with the existence of a strictly periodic k-space.

I have also tried to be brief, and thus far from comprehensive, on many of the aspects of dielectric and magnetic behavior in order to hold the total length of this book to a reasonable length for a one semester course (recognizing that there will always have to be selection of topics from within each of the five chapters). Accordingly, several of the topics are dealt with on a purely descriptive basis, relying on the bibliography for further source material.

The separate courses on electromagnetic theory taken by all physics students deal at length and in detail with the electrostatics, magnetostatics, and electrodynamics of continuous media. A modest knowledge of this subject matter is presumed here, in starting with Maxwell's electromagnetic field equations and moving rapidly to a use of the local field corrections. This permits us to assess the effect of the polarizations on an atomic scale which lead to a given response for an applied field. The first section of the chapter then concentrates on dielectric susceptance and its origins in dipole orientation and polarizability. A brief mention follows of piezoelectric, ferroelectric, and antiferroelectric phenomena.

The second section turns to magnetic susceptance, reviewing first the universal weak diamagnetic response and the paramagnetic behavior of uncoordinated magnetic dipoles. Ferromagnetism is discussed in terms of the molecular field and its interpretation as an exchange force, and the latter formulation then provides the framework for a qualitative picture of antiferromagnetic and ferrimagnetic behavior. The third and final section, added for this edition, surveys the principal features of nuclear magnetic resonance (NMR) and of electron spin resonance (ESR). The latter phenomenon is equally well known by the acronym EPR (for electron paramagnetic resonance), and the two names are equally well justified. For consistency, the ESR designation is used throughout Section 5.3.

Dielectric Properties **5.1**

THE BASIS LAWS
OF ELECTROMAGNETISM

In S.I. units,[1] the electric field vector **E** and the electric induction or displacement vector **D** are related by

$$\mathbf{D} = \varepsilon_0 \mathbf{E} + \mathbf{P}$$
$$= \varepsilon_0(1 + \chi_e)\mathbf{E} = \varepsilon_0 \kappa \mathbf{E} \tag{5-1}$$

for an isotropic dielectric medium. Here **P** is the polarization per unit volume, χ_e is the electric susceptibility, and κ is the relative permittivity, or dielectric constant; $\varepsilon_0 = (10^7/4\pi c^2) = 8.8542 \times 10^{-12}$ F/m is the permittivity of free space. It is remarked that Equation 5-1 is written in the form appropriate for an isotropic medium, since **P** and **E** can be in different directions for an anisotropic solid. The susceptibility and permittivity are then tensor quantities.

The analogous relationship between magnetic field intensity **H** and the corresponding magnetic induction **B** is

$$\mathbf{B} = \mu_0(\mathbf{H} + \mathbf{M})$$
$$= \mu_0(1 + \chi_m)\mathbf{H} = \mu_0 \mu_m \mathbf{H} = \mu \mathbf{H} \tag{5-2}$$

Here **M** is the magnetization, χ_m is the magnetic susceptibility, μ_m is the relative permeability, and μ is the absolute permeability; $\mu_0 = 4\pi \times 10^{-7}$ H/m is the permeability of free space. Since we shall need the symbol μ in Section 5.2 to denote magnetic dipole strength, the response of a magnetic material will usually be expressed in terms of either the magnetization **M** or the magnetic susceptibility χ_m.

Equations 5-1 and 5-2 are consistent with the S.I. (rationalized MKSA) expressions for the four Maxwell field equations

$$\nabla \cdot \mathbf{D} = \rho \tag{5-3a}$$

$$\nabla \times \mathbf{E} = -(\partial \mathbf{B}/\partial t) \tag{5-3b}$$

$$\nabla \cdot \mathbf{B} = 0 \tag{5-3c}$$

$$\nabla \times \mathbf{H} = \mathbf{J} + (\partial \mathbf{D}/\partial t) \tag{5-3d}$$

---

[1] Much of the experimental literature on dielectric phenomena, and even more particularly on magnetic phenomena, is expressed in gaussian units. The theory of electromagnetism is treated in S.I. units in some intermediate level texts, including W. B. Cheston, *Elementary Theory of Electric and Magnetic Fields* (Wiley, 1964).

where **J** is the electric current density and ρ is the electric charge density. We shall be more concerned with the magnetic aspects of Equations 5-2 and 5-3 in Section 5.2. For the next few pages we need to concentrate on dielectric properties. In so doing, we shall assume that a solid whose dielectric properties are of interest has a magnetic susceptibility $\chi_m \ll 1$.

MACROSCOPIC AND MICROSCOPIC VIEWS OF DIELECTRIC RESPONSE

Equation 5-3a is of course fully consistent with Coulomb's law

$$\mathbf{E} = (q\mathbf{r}/4\pi\varepsilon_0 r^3) \tag{5-4}$$

for the electric field caused by an isolated charge q at a distance **r** in empty space. [Within a homogeneous dielectric medium this field is reduced to $\mathbf{E} = (q\mathbf{r}/4\pi\kappa\varepsilon_0 r^3)$.] In a discussion of dielectric susceptibility, we are more commonly involved with dipoles rather than isolated charges. For the dipole of Figure 5-1, we define the *dipole moment* **p** by

$$\mathbf{p} = q\mathbf{R} \tag{5-5}$$

and it is a popular and not too arduous problem in electrostatics to show that the field **E** resulting from a dipole is

$$\mathbf{E}(\mathbf{r}) = \frac{3(\mathbf{p} \cdot \mathbf{r})\mathbf{r} - r^2\mathbf{p}}{4\pi\varepsilon_0 r^5} \tag{5-6}$$

in empty space. The proof of this is suggested in Problem 5.1 as a modest exercise in vector geometry.

Any asymmetrical molecule which is composed of atoms with an appreciable difference of electronegativities has a permanent electric dipole moment. Figure 5-2 illustrates two simple examples. Higher

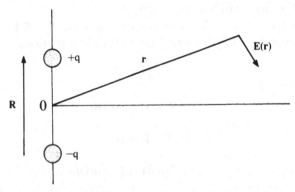

Figure 5–1 An electric dipole constitutes equal and opposite charges q and −q separated in space. If they are separated by distance R, then the dipole moment is **p** = qR. At a location **r** with respect to the center of the dipole, the electric field **E**(r) is given by vector addition of the two Coulomb law contributions. As we can verify (Problem 5.1), the result is Equation 5-6.

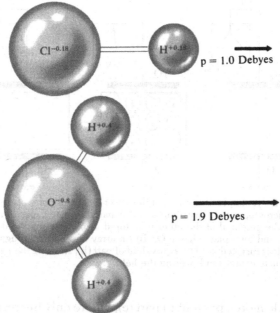

p = 1.0 Debyes

p = 1.9 Debyes

Figure 5-2 Simple examples of molecules with a permanent electric dipole moment. For an H_2O molecule, the moment is directed from the oxygen ion to the midpoint of the line between the two hydrogen ions; the effective ionic charge deduced from the dipole moment and the dimensions fits with Pauling's scale of electronegativity difference and relative ionicity (Figures 1-11 and 1-12). For HCl, the ionicity to be inferred from the dipole moment is considerably smaller than expected from the electronegativity difference between hydrogen and chlorine. Note that the dipole moments of these molecules are expressed in units of the c.g.s. unit, the *Debye* = 10^{-18} esu-cm = 3.33×10^{-30} C m. This unit is associated with the difference of atomic electronegativities, as discussed in connection with Equation 1-21.

order moments also can be involved with complicated and asymmetrical molecules, but dipole moments are the ones of interest to us at the present. The dielectric response of a medium is also concerned with the creation of *induced* electric dipoles. The polarization **P** of Equation 5-1 is the total dipole moment per unit volume, or

$$\mathbf{P} = \sum_n \mathbf{p}_i = \sum_n q_i \mathbf{R}_i \tag{5-7}$$

and it is the existence of dipoles of atomic dimensions which leads to a macroscopic difference between the electric displacement and the electric field.

This is represented in the simplest possible fashion by the parallel plane capacitor of Figure 5-3. We suppose that charges $+Q$ and $-Q$ reside on the plates of this capacitor and that the negative plate is grounded. In part (a) of the figure, the space between the plates is empty, and an electrometer records a potential $V_0 = (Qd/\varepsilon_0 A)$, for an electric field $E_0 = (Q/\varepsilon_0 A)$ between the plates. (It is assumed that fringing field effects can be ignored because of a guard ring arrangement which is not illustrated.) The insertion of a slab of dielectric, as shown in part (b), causes a decrease of potential and of average field by a factor of κ. The

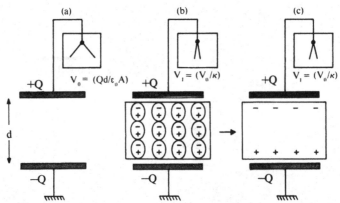

Figure 5–3. A simple illustration of the effect electric dipoles of atomic dimensions have on the macroscopic field distribution. (a) A parallel plate capacitor, with one plate grounded, and the potential of the other monitored. A plate is supposed of area A, with plate spacing d, and total plate charge Q. (b) An array of dipoles throughout the volume when a dielectric is inserted. (c) The equivalent of part (b), showing the *net* charges which partially offset plate charges in lowering the field.

reason is made more apparent in part (c), where only the residual *surface* induced charges resulting from the volume polarization are shown. A dipole moment per unit volume **P** results in a charge density per unit area of

$$q_s = -\mathbf{P} \cdot \mathbf{n} \tag{5-8}$$

where **n** is a unit vector normal to the plane of the surface. We must now use this result again in determining the field at the center of a cavity surrounded with polarized dielectric.

In both dielectric theory and magnetic theory it is often necessary to be able to calculate the *local effective field* at an atom or an ion, as determined by the polarization everywhere else. This is customarily handled by a method suggested by Lorentz; the atomic site in question is enclosed within an imaginary closed surface, and the contributions of dipoles within the surface are calculated one by one, while the contribution of everything outside the surface is treated by a continuum approximation. Thus for an atom in a dielectric sample under the influence of externally imposed charges, we might write the local effective field as a sum of four terms:

$$\mathbf{E}_{loc} = \mathbf{E}_0 + \mathbf{E}_{dep} + \mathbf{E}_{surf} + \mathbf{E}_{dip} \tag{5-9}$$

In this equation $\mathbf{E}_0$ is the field due to externally applied charges, and $\mathbf{E}_{dep}$ accounts for the depolarizing effects (if any) of induced surface charges on the outer surface of the dielectric slab. $\mathbf{E}_{surf}$ is the field produced at the center of an imaginary cavity by the polarization-induced surface charges over the face of the surface which bounds such a cavity. $\mathbf{E}_{dip}$ is the field at the center of this cavity due to the discrete dipoles distributed over all atomic sites in the cavity except for the central one.

The geometry assumed in Figure 5-3 endows the sum $(\mathbf{E}_0 + \mathbf{E}_{dep})$ with the value $\mathbf{E}_1$, whose magnitude is just (V_1/d). For our purposes we shall deal with the local field problem by assuming the same simple geometry for depolarization effects due to remote outer surfaces. We draw Figure 5-4, imagining a plane, parallel-sided dielectric slab with an inner cavity. Accordingly,

$$\mathbf{E}_{loc} = \mathbf{E}_1 + \mathbf{E}_{surf} + \mathbf{E}_{dip} \qquad (5\text{-}10)$$

The effect of the surface charge density of Equation 5-8 must be integrated over the boundary of this cavity in the determination of $\mathbf{E}_{surf}$, and if we choose for simplicity a spherical cavity, then a rather straightforward integration (Problem 5.2) gives us

$$\mathbf{E}_{surf} = (\mathbf{P}/3\varepsilon_0) \qquad (5\text{-}11)$$

An additional reason for proposing a spherical cavity is that the term $\mathbf{E}_{dip}$ then vanishes for cubic lattices.[2] Thus in a cubic lattice the local field according to the Lorentz method of evaluation is

$$\mathbf{E}_{loc} = \mathbf{E}_1 + (\mathbf{P}/3\varepsilon_0) \qquad (5\text{-}12)$$

which is likely to be considerably larger than the applied field. Indeed, by applying a rearranged form of Equation 5-1,

$$\mathbf{P} = \varepsilon_0(\kappa - 1)\mathbf{E} \qquad (5\text{-}13)$$

---

[2] For non-cubic lattices, the procedure for evaluation of the local field is less straightforward. Values for the contribution $\mathbf{E}_{dip}$ made by dipoles inside a supposed simply shaped cavity have been worked out for tetragonal and hexagonal lattices by H. Mueller, Phys. Rev. **47**, 947 (1935) and Phys. Rev. **50**, 547 (1936). The manner in which all dipoles within a spherical cavity in a cubic dielectric have a zero effect on the local field at the center of the cavity is discussed by J. D. Jackson in *Classical Electrodynamics* (Wiley, 1962) page 116.

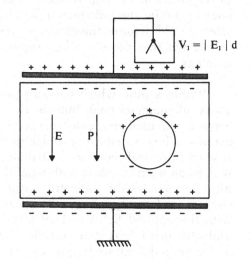

Figure 5–4 Geometry for determination of the local field at an atomic site in a polarized dielectric. For a cubic lattice, the field $\mathbf{E}_{dip}$ is zero at the center of a spherical cavity, and $\mathbf{E}_{surf} = (\mathbf{P}/3\varepsilon_0)$ for a spherical boundary.

to Equation 5-12, we can write the local field at an atomic site in a cubic (isotropic lattice) as

$$\mathbf{E}_{loc} = \left(\frac{\kappa + 2}{3}\right) \mathbf{E} \qquad (5\text{-}14)$$

Of course, things are a great deal more complicated in less isotropic solids, but we can learn many useful things from the behavior of isotropic materials. In the next subsection, we shall return to Equations 5-12 through 5-14 in developing the *Clausius-Mosotti* relation between the polarizability of atoms or ions and the macroscopic dielectric constant of an isotropic (i.e., cubic) medium.

SOURCES OF DIELECTRIC POLARIZABILITY

In discussing how a solid can have a dielectric response which differs from that of empty space, we must consider three forms of polarization on an atomic scale. These are:

1. Partial or complete alignment of the dipole moments of polar molecules with the local electric field. As we have seen from the examples in Figure 5-2, any asymmetrical molecule composed of atoms with a difference of electronegativity has a permanent dipole moment. An electric field encourages rotation of these groups for participation in the electric displacement, a process referred to as *dipole orientation* or *paraelectric response*. This desire to align all permanent dipoles with an applied field is partially frustrated by thermal vibrations, as we shall discuss in terms of the Langevin response function.

2. The inducement of dipoles by relative movement of positive and negative ions in a partially ionic solid under the influence of an electric field. This mechanism is usually called *ionic polarization*.

3. The third kind of contribution, and the only one which occurs within *every* dielectric solid, is the process of *electronic polarization*. This consists of the displacement of electrons in an atom relative to the nucleus, under the influence of an electric field. It can be said that an electric field deforms the electron shells around every nucleus, regardless of whether there are any superimposed changes in internuclear spacings.

Within a solid which contains permanent dipoles, all three of the above phenomena contribute to the polarizability (and hence to the macroscopic dielectric constant) at low frequencies, though to differing extents. When we consider dielectric response at high frequencies, the susceptibility will contain contributions from only those mechanisms which can keep in phase with a rapidly varying electric field. Thus for high frequencies the dielectric constant is complex, the sum of a real component (corresponding with polarization of the dielectric in phase with the applied field) and an imaginary part (which represents the dielectric loss inherent in a mechanism which lags in phase behind that of the applied field) for frequencies at which dispersion occurs. The real and imaginary parts of the complex dielectric constant for any medium (insulating or otherwise) are connected by an extremely valuable pair of integral relationships known as the *Kramers-Kronig dispersion relations*.[3]

Figure 5-5 shows how the real and imaginary parts of the dielectric constant might behave for a solid which has permanent dipoles capable of rotation, and in which ionic as well as electronic polarization can be induced. The component of polarization arising from dipole orientation is not able to keep up with the rapid alternations of the field for frequencies larger than about 10^{10} Hz. Figure 5-5 shows a corresponding decrease in the real part of the dielectric constant in the microwave part of the spectrum. The imaginary part of the dielectric constant is non-zero (i.e., there is an appreciable dielectric loss factor) in this same frequency region, since dipoles attempt to respond but have a motion which is seriously lagging in phase behind the applied field. At rather higher frequencies, the amplitude of dipole rotation is negligibly small.

A further reduction in the real part of the dielectric constant (accompanied by a second spectral region of dielectric loss) occurs around 10^{13} Hz for a solid which is completely or partly ionic in its bonding. This is the reststrahlen frequency region discussed earlier in Section 2.3 and illustrated by Figures 2-14 through 2-16. Above the reststrahlen

---

[3] For an original reference to the Kramers-Kronig relationships, see H. A. Kramers, *Collected Scientific Papers* (North-Holland, 1956). Their relationship to the dielectric and optical properties of solids is discussed in Chapters 7 and 8 of *The Physics of Solids* by F. C. Brown (Benjamin, 1967). In addition to the value of these dispersion relations in insulating solids, they can also be applied to the reflectivity spectra of metallic and semiconducting solids in determining critical energies for van Hove singularities in the density of electron states.

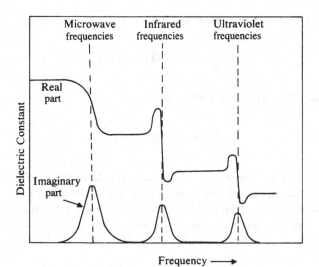

Figure 5-5 The dependence upon applied frequency of the real and imaginary parts of the complex dielectric constant in an ionic solid containing permanent dipoles. The imaginary part of the dielectric constant (or dielectric loss factor) is large whenever the frequency reaches the upper limit for a dielectric response mechanism. This happens for dipole orientation in the microwave range, for ionic polarizability in the infrared, and for bound electron displacements in the ultraviolet.

frequency, ionic movements can no longer keep up with the alternations of the applied field. For higher frequencies, only the induced electronic polarization makes the real part of the dielectric constant larger than unity, and this mechanism also cannot keep up with the field for frequencies much larger than 10^{15} Hz. From this frequency upwards, a solid has a dielectric constant very close to unity.

We shall find it useful to be able to express the real part of the dielectric constant in terms of the polarizabilities on an atomic scale, a task which can be done quite simply for a cubic (and thus isotropic) substance. The argument is based upon the slightly questionable approach that it is the local field $\mathbf{E}_{loc}$ of Equation 5-12 which polarizes the atoms or ions of a dielectric. Such a proposition obviously cannot be entirely correct, since $\mathbf{E}_{loc}$ is derived by removal of the atom whose polarization is to be discussed; and yet there must be some reaction of the induced dipole moment on the local field. Even so, let us define the polarizability α_j of the j th type of atom in the solid by quoting its dipole moment as

$$\mathbf{p} = \alpha_j \mathbf{E}_{loc} \qquad (5\text{-}15)$$

Then the total polarization per unit volume is

$$\mathbf{P} = \mathbf{E}_{loc} \sum_r N_j \alpha_j \qquad (5\text{-}16)$$

where the summation is over the various types of polarizable atoms, and N_j is the volume density of the j th type.

We can now employ Equation 5-14 to express this total polarization in terms of the *applied* field:

$$\mathbf{P} = \left(\frac{\kappa + 2}{3}\right) \mathbf{E} \sum_r N_j \alpha_j \qquad (5\text{-}17)$$

If we rearrange this equation as

$$\sum_r N_j \alpha_j = 3\mathbf{P}/(\kappa + 2)\mathbf{E} \qquad (5\text{-}18)$$

and employ Equation 5-13, then we have

$$\sum_r N_j \alpha_j = 3\varepsilon_0 \left(\frac{\kappa - 1}{\kappa + 2}\right) \qquad (5\text{-}19)$$

Equation 5-19 is known as the *Clausius-Mosotti relation* between atomic polarization and macroscopic dielectric constant. We must remember that this relation has been derived for a solid which crystallizes in a cubic lattice, and that the relationship between atomic and macroscopic parameters is considerably less straightforward for less isotropic solids.

DIPOLE ORIENTATION RESPONSE

The polarizability which arises from the rotation of permanent dipoles is discussed here in the conventional manner which is particularly appropriate for gases and molecules, since polar molecules in solids are not usually completely free to rotate in response to an electric field. The orientation process in a polar solid should more correctly be described in terms of major jumps in the angular placement of molecules from one stable set of orientations to another stable set which is more nearly parallel to the direction of the field. We shall return shortly to a consideration of the angular placements permitted by crystal symmetry, and the relaxation time which controls the jump frequency. First, however, let us determine the dielectric susceptibility supposing that molecules *can* rotate freely. The approach we shall use was developed for (and is still used in) the theory of paramagnetism by orientation of magnetic dipoles.

Consider a medium with N permanent electric dipoles per unit volume, each of dipole moment p. In the absence of an external field, these dipoles will be in random directions. Now suppose that a static field **E** attempts to create some order among the dipoles. The polarization per unit volume is

$$P_{dp} = \sum_N p \, \cos\theta = Np <\cos\theta> \tag{5-20}$$

where θ is the angle between each dipole and the applied field.

The ordering efforts of the field are partially frustrated by the thermal agitation of all particles in the medium, an agitation which we expect to be able to describe by a Boltzmann distribution of energies at ordinary temperatures. Then the expectation value for $<\cos\theta>$ is a ratio of two Boltzmann integrals,

$$<\cos\theta> = \frac{\int_0^\pi 2\pi \, \sin\theta \, \cos\theta \, \exp(-U/k_0 T) \cdot d\theta}{\int_0^\pi 2\pi \, \sin\theta \, \exp(-U/k_0 T) \cdot d\theta} \tag{5-21}$$

where the potential energy U of a dipole at angle θ in the applied field E is

$$U = -\mathbf{p} \cdot \mathbf{E} = -pE \, \cos\theta \tag{5-22}$$

The integrals of Equation 5-21 can be handled most conveniently if we make the substitutions

$$\left.\begin{aligned} y &= (-U/k_0 T) = +(pE/k_0 T) \cos\theta \\ w &= (pE/k_0 T) \end{aligned}\right\} \tag{5-23}$$

In these terms, Equation 5-21 becomes

$$<\cos\theta> = \frac{\int_{-w}^w (y/w) \, \exp(y) \cdot dy}{\int_{-w}^w \exp(y) \cdot dy} \tag{5-24}$$

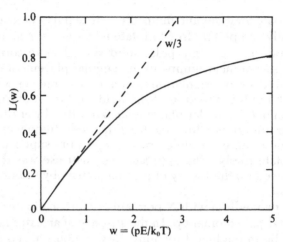

Figure 5–6 The Langevin function, L(w), used in the description of dipole orientation both for paraelectric and paramagnetic behavior.

The numerator integrates by parts very simply to give

$$<\cos\theta> = \left[\frac{1 + \exp(-2w)}{1 - \exp(-2w)} - (1/w)\right] = [\coth w - (1/w)] \equiv L(w)$$

$$(5\text{-}25)$$

The function L(w) defined by Equation 5-25 was introduced by Langevin (1905) in creating the theory for paramagnetic susceptibility, and we shall encounter it again in Section 5.2. In using the Langevin function for paraelectric susceptibility, we may note that the function has a limiting value of unity for large w [i.e., when the applied field E is large compared with (k_0T/p)]. As indicated in Figure 5-6, the low field approximation for the Langevin function is

$$<\cos\theta> \equiv L(pE/k_0T) \simeq (pE/3k_0T), \qquad E \ll (k_0T/p) \quad (5\text{-}26)$$

and as established in Problem 5.3, this low field limiting form of the function is appropriate up to quite large electric fields at ordinary temperatures. The magnitude of electric field which is necessary to invalidate Equation 5-26 at ordinary temperatures is large enough to produce dielectric breakdown in most solids.[4]

---

[4] The intrinsic dielectric strength of an insulating solid is some 10^9 V/m under ideal conditions. A larger electric field than this will inevitably lead to free electron multiplication phenomena (avalanche breakdown) or thermal breakdown, since there are *some* free electrons and *some* sources for Joule heating in any insulating solid. Even in the total absence of free electrons at equilibrium, there is a critical field above which electron-hole pairs are created by the Zener process of tunneling across the large intrinsic gap of the insulator. For any practical insulator, the breakdown field is invariably smaller than the "intrinsic" value because any slight inhomogeneities place a tremendous strain on small regions of the solid. The various breakdown phenomena are discussed by J. O'Dwyer in *The Theory of Electrical Conduction and Breakdown in Solid Dielectrics* (Oxford University Press, 1973).

Accordingly, the total polarization per unit volume caused by reorientation of permanent dipoles is

$$P_{dp} = Np^2E/3k_0T \qquad (5\text{-}27)$$

for a contribution towards the susceptibility of

$$\chi_{dp} = (P_{dp}/\varepsilon_0E) = Np^2/3\varepsilon_0k_0T \qquad (5\text{-}28)$$

This contribution is small compared with unity for a gas, but may be comparable with unity for a polar liquid or for a solid in which dipole reorientation is possible. Thus in a material with $N = 10^{28}$ polar molecules per cubic meter, each of permanent moment 1 Debye unit, then $\chi_{dp} \approx (300/T)$.

Of course, *induced* ionic dipoles and the inevitable electronic polarization also affect the static dielectric constant, which must appear in the form

$$\kappa = 1 + \chi_{dp} + \chi_i + \chi_e \qquad (5\text{-}29)$$

with χ_{dp} given by Equation 5-28. Since χ_i and χ_e do not depend on temperature, we can determine the magnitude of the moment for the permanent dipoles by measuring the static (or low frequency) dielectric constant as a function of temperature.

What happens in practice for a real polar solid is illustrated by the low frequency dielectric constant data of Figure 5-7, from just above the melting point of hydrogen sulphide to well below the melting point. The portion of the total dielectric constant derived from dipole orientation *does* increase on cooling, but in a rather irregular fashion, since the dipoles do not rotate with perfect freedom; rather do they have the abil-

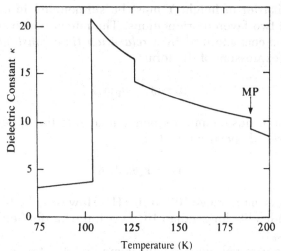

Figure 5-7 Temperature dependence of the dielectric constant of H_2S, measured at low frequencies (5 kHz). After C. P. Smyth and C. S. Hitchcock, J. Am. Chem. Soc. **56**, 1084 (1934).

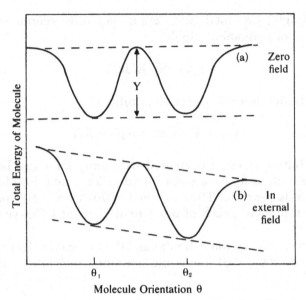

Figure 5–8 (a), A possible curve for the energy of a polar molecule in a solid as a function of its orientation, when there is no external influence. (b), Modification of the relative energies of the two minima by an electric field.

ity to jump from one to another of a small set of orientations permitted by the lattice. The ability for these discrete jumps in orientation is apparently lost below about 103K.

Suppose that the energy of a polar molecule in a crystal (with no external field applied) has the form shown in the upper curve of Figure 5-8. There are two equilibrium positions, at angles θ_1 and θ_2 with respect to a certain direction along which we shall later want to apply a field. For any other orientation of the molecule, an unreasonably large energy is necessary, and a barrier of height Y must be surmounted in order to jump between the two favored orientations. The rate at which such jumping occurs can be characterized by a *relaxation time constant* τ, which is given by an expression of the form

$$\tau = (2\pi\nu_D)^{-1} \exp(Y/k_0 T) \tag{5-30}$$

Here ν_D is an intrinsic jump frequency related to the upper limit of the lattice vibrational spectrum, that is,

$$\nu_D \approx k_0 \theta_D / 2\pi\hbar \tag{5-31}$$

Typically, ν_D is in the range 10^{11} to 10^{13} Hz. However, the time constant τ will be considerably longer than 10^{-11} sec if the barrier height Y is much larger than $k_0 T$.

When an external electric field is applied, the two energy minima are no longer equivalent, and transitions will favor the population of one at the expense of the other. Provided that the field is applied for long enough to permit the establishment of a steady state, the population ratio for orientations 1 and 2 will be

$$\frac{N_1}{N_2} = \exp\left[\frac{\cos\theta_2 - \cos\theta_1}{k_0 T/pE}\right] \tag{5-32}$$

from Equation 5-22 and the Boltzmann distribution law. Population differences such as $(N_2 - N_1)$ establish the bulk polarization, which we have previously described in terms of unhindered rotation in Equation 5-27.

Now suppose that we apply an electric field at an angular frequency which is large enough to jeopardize the jumping response of permanent dipoles (i.e., a frequency comparable with τ^{-1} of Equation 5-30) yet not so large that the induced ionic and electronic polarization cannot keep up with the time-dependence of the field. As first shown by Debye (1929), it is now necessary to write the permittivity or dielectric constant as a complex quantity. The dielectric constant must now have a form

$$\kappa = (\kappa_1 + i\kappa_2) = A + \frac{B}{1 - i\omega\tau} \tag{5-33}$$

The *induced* ionic and electronic polarization is frequency independent and loss-free for the frequency range currently under consideration, and accounts for $(A - 1)$ in Equation 5-33. The hindered rotation of permanent dipoles contributes the second term on the right of Equation 5-33; thus the quantity B depends on the density of permanent dipoles and on their ability to rotate, just as χ_{dp} of Equation 5-28 expressed this for unhindered rotation of dipoles in a fluid.

Equation 5-33 requires real and imaginary parts of the total dielectric constant as follows:

$$\kappa_1 = \left[A + \frac{B}{1 + \omega^2\tau^2}\right] \tag{5-34}$$

and

$$i\kappa_2 = \frac{i\omega\tau B}{1 + \omega^2\tau^2} \tag{5-35}$$

Equations 5-34 and 5-35 are usually referred to as the *Debye equations*. The decrease of the real dielectric constant κ_1 with increasing frequency, centered on a frequency which is the reciprocal of the time constant in Equation 5-30, is sketched as the first drop of the upper curve in Figure 5-5. Beneath this in Figure 5-5 is the first peak in the imaginary part, corresponding with the behavior of Equation 5-35.

For measurements at a given frequency, the effects of a limited jump frequency become apparent by the decrease of κ_1 upon cooling, coupled with a temperature range of appreciable dielectric loss. The duality between the frequency dependence at a given temperature and the temperature dependence at a given frequency is explored further in Problem 5.4.

As a final comment on the orientation of permanent dipoles, we must admit that the preceding discussion was deliberately couched in terms of the paraelectric susceptibility, and was *not* converted to a polarizability by using the Clausius-Mosotti relation. The reason is that the Lorentz procedure for calculating an "effective local field" is appropriate for induced dipoles, but not for orientation of permanent ones. This happens since Equation 5-14 was evaluated assuming that every induced dipole has the same direction as the applied field, whereas the large permanent dipoles in a polar solid point in a variety of directions. The *local field* which is *effective* in dipole orientation is likely to be inappreciably different from the externally applied field.

Next we must return to the situations of induced electronic and ionic dipoles, for which the Lorentz local field corrections *are* appropriate.

ELECTRONIC POLARIZABILITY

In contrast to the relaxation type response of permanent dipoles in a dielectric, the dielectric responses of induced dipoles are resonant phenomena. Consider an electron which is displaced under the influence of a sinusoidally varying external field which we shall represent through the Lorentz local field of Equation 5-14. For a restoring force constant of β, the equation of motion is

$$N \qquad m\frac{d^2x}{dt^2} + \beta x = eE_{loc} = eE_{amp}\exp(i\omega t) \qquad (5\text{-}36)$$

This is the equation of motion for a harmonic oscillator of natural angular frequency $\omega_0 = (\beta/m)^{1/2}$, and so

$$m\frac{d^2x}{dt^2} + m\omega_0^2 x = eE_{loc} \qquad (5\text{-}37)$$

The conventional solution for the amplitude of forced oscillatory motion at angular frequency ω is

$$x_{amp} = \frac{eE_{amp}}{m(\omega_0^2 - \omega^2)} \qquad (5\text{-}38)$$

and this corresponds with a dipole moment of peak amplitude $p = ex_{amp}$ or an electronic polarizability (from Equation 5-15) of

$$\alpha_e = (ex/E_{loc}) = \frac{e^2}{m(\omega_0^2 - \omega^2)} \qquad (5\text{-}39)$$

The contribution of this process towards the dielectric constant is the same for all frequencies much smaller than $(\omega_0/2\pi)$. Indeed, in the visible (optical) frequency region, induced electronic polarization is the only dielectric mechanism which can make the dielectric constant κ and the optical refractive index $n = \sqrt{\kappa}$ depart from unity.

The magnitude of the refractive index can be used to determine the electronic polarizability and hence the resonant frequency ω_0. In the visible spectral region, the Clausius-Mosotti relation of Equation 5-19 can be written

$$\alpha_e = (3\varepsilon_0/N_e)\left(\frac{\kappa - 1}{\kappa + 2}\right) = (3\varepsilon_0/N_e)\left(\frac{n^2 - 1}{n^2 + 2}\right) \tag{5-40}$$

Typical numbers for a real solid are a refractive index of $n = 1.7$ with a density of outer shell electrons $N_e = 10^{29}$ m^{-3}, and this yields $\alpha_e \approx 10^{-40}$ F m^2. Substitution into Equation 5-39 then suggests that $\omega_0 \sim 1.7 \times 10^{16}$ rad/sec.; this is equivalent to a wavelength of 110 nm, in the vacuum ultraviolet part of the electromagnetic spectrum.

Equation 5-39 predicts a singularity in the polarizability at the natural resonant frequency ω_0, but the actual dielectric constant does not reach either $+\infty$ or $-\infty$. The singularity in the behavior is avoided because of dielectric loss, which we have overlooked in setting up Equation 5-36. Since an oscillating electron is constantly being accelerated or decelerated, it must radiate energy at a rate proportional to (dx/dt). Loss also occurs because of inelastic collisions experienced by electrons, and the probability of these can also be gauged by a term proportional to (dx/dt). Thus the equation of motion for an electron should contain a *frictional term* characterized by a *dissipation constant* γ. We can arrange things so that this quantity γ will have the dimensions of a frequency. The full equation of motion is then

or

$$\left. \begin{array}{l} m\dfrac{d^2x}{dt^2} + m\gamma\dfrac{dx}{dt} + \beta x = eE_{loc} \\[4ex] m\dfrac{d^2x}{dt^2} + m\gamma\dfrac{dx}{dt} + m\omega_0^2 x = eE_{loc} \end{array} \right\} \tag{5-41}$$

This once again has a solution for x which varies sinusoidally in sympathy with a sinusoidally driven local field, but now in the form

$$x = \frac{eE_{loc}}{m(\omega_0^2 - \omega^2 + i\gamma\omega)} \tag{5-42}$$

Thus the electronic polarizability is a complex quantity,

$$\alpha_e = (ex/E_{loc}) = \frac{e^2}{m(\omega_0^2 - \omega^2 + i\gamma\omega)} \tag{5-43}$$

We shall now find it convenient to invoke the Clausius-Mosotti equation (5-19) in order to evaluate the complex dielectric constant. This can easily be rearranged to give

$$\kappa = \kappa_1 + i\kappa_2 = \left[1 + \frac{e^2 N_e}{\varepsilon_0 m(\omega_0^2 - \omega^2 + i\gamma\omega) - (e^2 N_e/3)}\right] \tag{5-44}$$

as is suggested in Problem 5.5. Evidently the presence of a large density N_e of polarizable electrons has the effect of shifting the resonant frequency to the value

$$\omega_1 = [\omega_0{}^2 - (e^2 N_e/3 m\epsilon_0)]^{1/2} \tag{5-45}$$

In terms of this parameter ω_1, the real (in phase) and imaginary (out of phase) parts of the dielectric constant are

$$\kappa_1 = \left[1 + \frac{(e^2 N_e/m\epsilon_0)(\omega_1{}^2 - \omega^2)}{(\omega_1{}^2 - \omega^2)^2 + \gamma^2\omega^2}\right] \tag{5-46}$$

and

$$i\kappa_2 = i \left[\frac{(e^2 N_e/m\epsilon_0)\gamma\omega}{(\omega_1{}^2 - \omega^2)^2 + \gamma^2\omega^2}\right] \tag{5-47}$$

These equations show the kind of behavior illustrated for the ultraviolet part of the spectrum in Figure 5-5, whereby the real part of the dielectric constant undergoes a final oscillation and settles down to a value very near to unity above the frequency ω_1. Problem 5.5 explores the behavior of the dielectric constant curves in the vicinity of the damped resonant phenomenon.

IONIC POLARIZABILITY

For an ionic solid, the dielectric contribution of induced ionic dipoles can be discussed in a manner quite similar to that for electronic polarization. Some aspects of this subject were mentioned in Section 2.3, in connection with the interaction of electromagnetic waves and lattice vibrations at the reststrahlen frequency ω_t.

Consider an ionic solid with N_e polarizable electrons per cubic meter and N_i polarizable ion pairs. Then the Clausius-Mosotti equation tells us that the static dielectric constant κ_0 is related to the polarizabilities by

$$3\epsilon_0 \left(\frac{\kappa_0 - 1}{\kappa_0 + 2}\right) = N_i\alpha_i + N_e\alpha_e \tag{5-48}$$

Now, as in Equation 2-43 of Section 2.3, let the symbol κ_∞ signify the dielectric constant at a frequency much too high for any response from induced ionic dipoles, yet for which the electronic response is unimpaired. Then

$$3\epsilon_0 \left(\frac{\kappa_\infty - 1}{\kappa_\infty + 2}\right) = N_e\alpha_e \tag{5-49}$$

Thus if both κ_0 and κ_∞ are known, the ionic polarizability can be found from the difference of Equations 5-48 and 5-49:

$$\alpha_i = (3\varepsilon_0/N_i)\left[\left(\frac{\kappa_0 - 1}{\kappa_0 + 2}\right) - \left(\frac{\kappa_\infty - 1}{\kappa_\infty + 2}\right)\right] \tag{5-50}$$

This quantity is typically a few times 10^{-40} F m^2 in an ionic solid. For example, there are 2.2×10^{28} ion pairs per cubic meter in NaCl, and the dielectric constant changes from 5.6 well below the reststrahlen frequency to 2.25 well above that frequency. These numbers, substituted into Equation 5-50, yield $\alpha_i = 3.8 \times 10^{-40}$ F m^2 for NaCl.

For ion pairs of masses M_+ and M_-, the equation of motion for electrically induced forced oscillations is

$$\left(\frac{M_+ M_-}{M_+ + M_-}\right)\left[\frac{d^2x}{dt^2} + \gamma_i \frac{dx}{dt} + \omega_t^2 x\right] = eE_{loc} \tag{5-51}$$

where, as with Equation 5-41, γ_i represents a damping term and ω_t is the natural resonant frequency (the long wave limiting frequency of transverse optical modes) as determined by the force constants of the solid. The effective field E_{loc} represents any externally applied field and the effects of both the electronic and ionic polarizations. As with the purely electronic response of Equation 5-41, the solution for x is a complex one at frequencies that are at all comparable with ω_t. Including both forms of polarization, the Clausius-Mosotti equation gives a complex dielectric constant

$$\kappa(\omega) = \kappa_\infty + \frac{(\kappa_0 - \kappa_\infty)\omega_t^2}{(\omega_t^2 - \omega^2 + i\gamma_i\omega)} \tag{5-52}$$

This is the expression previously quoted as Equation 2-43. It can be decomposed into real and imaginary parts in just the same way as has been done for the electronic response of Equation 5-44. Thus the change of the real dielectric constant with increasing frequency is a damped oscillatory one analogous to the electronic response dependence of higher frequencies. As a companion to this, a solid in which ionic dipoles can be induced shows a strong peak in the imaginary component $i\kappa_2$ centered on ω_t, which gives the well known optical properties for that spectral range.

PIEZOELECTRICITY AND ELECTROSTRICTION

Some crystalline solids become electrically polarized when they are elastically strained. As a converse phenomenon, these same solids become strained when exposed to an external electric field in certain directions. This *piezoelectricity* is a property which is possible only for ionic solids which crystallize in structures lacking a center of inversion.

We can illustrate the latter statement by considering a molecule of a hypothetical ionic solid which at equilibrium presents three electric dipoles of equal magnitude at 120° intervals, as in Figure 5-9(a). This molecule has the symmetry of the point group 3m of Figure 1-19. The net dipole moment of the molecule vanishes, by symmetry. Yet a net dipole moment will appear if the molecule (along with the rest of the solid) is stretched or compressed along a direction parallel to, or perpendicular to, one of the three vertices. Similarly, an electric field applied parallel to one of the three vertices will enforce a distortion of each molecule. This produces a contraction or elongation of the crystal parallel to the field direction, accompanied by a length change of the opposite sign in the transverse direction. However, an electric field applied perpendicular to the direction of a vertex (such as the horizontal direction in the figure) will not change the crystal dimensions, because this last-mentioned field is perpendicular to a reflection plane of symmetry.

Of the 32 point groups which can be envisaged in three dimensions, 21 lack a center of inversion symmetry. One of these has so high a degree of symmetry in other respects that piezoelectricity is excluded, which leaves 20 types of point group symmetry which are compatible with piezoelectric behavior. Only for a relatively minor fraction of the solids utilizing these 20 point groups is a piezoelectric effect large enough to be measurable. Table 5-1 lists a few of the better known piezoelectric materials, generally in a decreasing order of piezoelectric coefficients. A comparison with Table 5-2 shows that several of these solids are ferroelectric below a transition temperature T_c.

The piezoelectric coefficients of quartz are quite small, but the great mechanical and thermal strength of this solid have made it pre-eminent for highly stable piezoelectric oscillators. These devices control the frequencies of radio transmitters, and of quartz clocks and (more recently) quartz wristwatches.

For other applications of piezoelectricity, the highest possible efficiency is a very desirable attribute. In view of this, Rochelle salt has been used for many years as a sensitive transducer in inexpensive phonographs. Newer models are more likely to use a shaped ceramic block

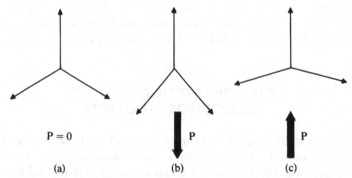

Figure 5–9 Reactions of a piezoelectric molecule to strain. (a) Molecule in equilibrium in an undistorted crystal. (b) Net polarization developed by vertical tension or horizontal compression. (c) Net polarization developed by vertical compression or horizontal tension.

TABLE 5-1 SOME REPRESENTATIVE PIEZOELECTRIC
SOLIDS

Material	Chemical Formula	Relative Strength of Piezoelectric Coefficients
Rochelle Salt	$NaKC_4H_4O_6 \cdot 4H_2O$	Very large
Barium Strontium Titanate	$Ba_xSr_{1-x}TiO_3$	Large
Lead Zirconium Titanate (PZT)	$Pb_xZr_{1-x}TiO_3$	Large
Ammonium Dihydrogen Phosphate (ADP)	$NH_4H_2PO_4$	Large
Potassium Dihydrogen Phosphate (KDP)	KH_2PO_4	Moderate
Ethylene Dihydrogen Tartrate	$C_6H_{14}N_2O_6$	Moderate
Tourmaline	$(FeCrLiNaK)_4Mg_{12}B_6Al_{16}H_8Si_{12}O_{63}$	Small
α Quartz	SiO_2	Small

of barium strontium titanate or PZT, since the latter materials are quite efficient and additionally are resistant to heat and moisture. Much larger ceramic blocks of barium strontium titanate or of PZT are used as agitators in ultrasonic cleaning baths and as transmitters and receivers in underwater sonars.

The deformation of a piezoelectric solid is linear in the applied electric field to a first approximation, and the polarization induced by stress is proportional to the strain achieved. For *any* ionic solid, piezoelectric or not, there is a *much smaller* contraction in an electric field, which varies as the *square* of the field. This more universal phenomenon of *electrostriction* arises from the failure of Hooke's law in describing the changes of interionic distances with applied field. Thus we can expect that a solid which shows an appreciable electrostrictive effect will have lattice vibrational properties which are strongly influenced by anharmonic effects.

FERROELECTRIC BEHAVIOR

For an ionic solid in the ferroelectric state, the center of positive charge of the crystal does not coincide with the center of negative charge even in the absence of any applied field. Thus there is a bulk electrical polarization whether an external field exists or not.

In order that ferroelectric behavior can be even possible, it is necessary that the crystal structure be one that lacks a center of inversion symmetry (which is the minimum requirement for piezoelectric behavior). In addition it is necessary that the crystal structure have a single nonequivalent direction. Of the 20 classes of point group which can qualify for piezoelectric properties, 10 can also meet the second requirement. Thus a ferroelectric material will be piezoelectric, but the converse need not apply.

As a general requirement for the establishment of a ferroelectric state, we may say that an *ordered* array of *induced* ionic dipoles must be able to have a smaller total energy (even in the absence of a field) than the crystal has with zero dipole moment. This requires a large density of extremely polarizable ions, as well as a temperature which is not too large. Ferroelectric behavior disappears above a characteristic temperature, or *Curie point*, T_c, since above this temperature the amplitude of thermal vibrations is strong enough to prevent the formation of an ordered dipole array. A solid which is ferroelectric at low temperatures becomes paraelectric above the Curie point, with a susceptibility which is often found to vary as $(T - T_c)^{-1}$.

Table 5-2 lists a few examples of well known ferroelectric crystalline solids, their characteristic temperatures, and values for the spontaneous polarization density in the absence of a field. As noted in the table, spontaneous polarization disappears again for Rochelle salt below a second characteristic temperature, but this feature is not shared by many ferroelectrics. The final column of Table 5-2 lists values for the spontaneous bulk polarization P_s in the S.I. unit of coulomb/meter^2. How much charge separation is needed to produce the amounts of polarization indicated can be visualized if we note that 10^{29} ion pairs per cubic meter with a 0.25 Å separation of the positive and negative charge distributions would produce a polarization $P_s = 0.4$ C/m^2. All of the polarizations reported in Table 5-2 are smaller than this, some by a factor of 100.

Thus consider the source of the spontaneous polarization of BaTiO$_3$. This compound crystallizes in the perovskite structure of Figure 1-36, which is cubic above the Curie temperature of 393K. The transition to a ferroelectric state is signaled by a transition to a tetragonal distortion of cubic below T_c, in which the unit cell elongates by about 1 per cent along one pseudo-cubic direction (the c-axis) and contracts by about 0.5 per cent along each of the axes perpendicular to this direction. The tetragonal distortion alone produces no polarization; but what in fact happens (at low temperatures, at any rate) is that the crystal energy is smaller when the sublattice of all barium and titanium cations is shifted

TABLE 5-2 SOME REPRESENTATIVE FERROELECTRIC COMPOUNDS[°]

Solid	Chemical Formula	T_c(K)	P_s(C/m^2)
Tri-Glycine Sulfate	(NH$_2$CH$_2$COOH)$_3$·H$_2$SO$_4$	322	2.8×10^{-2} at 275K
KDP	KH$_2$PO$_4$	123	4.7×10^{-2} at 100K
Deuterated KDP	KD$_2$PO$_4$	213	5.5×10^{-2} at 100K
Barium Titanate	BaTiO$_3$	393	2.6×10^{-1} at 300K
Strontium Titanate	SrTiO$_3$	32	3.0×10^{-2} at 4.2K
Potassium Niobate	KNbO$_3$	710	3.0×10^{-1} at 600K
Ammonium Sulfate	(NH$_4$)$_2$·SO$_4$	223	4.5×10^{-3} at 220K
Rochelle Salt	NaKC$_4$H$_4$O$_6$·4H$_2$O	296 (upper) 255 (lower)	2.5×10^{-3} at 275K

[°] Data from F. Jona and G. Shirane, *Ferroelectric Crystals* (Pergamon, 1962).

upwards or downwards along the c-axis with respect to the oxygen sub-lattice, than when the two sub-lattices are in perfect registry. The relative displacement of the two sub-lattices is approximately 0.1 Å according to diffraction experiments,[5] which is sufficient to account for the gross magnitude of the bulk polarization.

For a given direction of the c-axis in a tetragonal $BaTiO_3$ crystal, the polarization P_s can be either "up" or "down," depending on the direction in which one sub-lattice has translated itself with respect to the other. The point to note is that every titanium (and barium) ion has two locations for which its energy is a minimum, and that these are separated by an energy barrier centered on the location which would correspond with a straightforward tetragonal distortion of Figure 1-36. Only above the Curie point does this barrier disappear and the lowest energy correspond with a single location for each ion. The double-minimum curve of energy versus position is a consequence of the local field E_{loc} which each ion experiences.

Explanations for the origin of ferroelectricity vary a great deal from one family of ferroelectric solids to another, but the idea of an energy versus position curve with a double minimum is typical. For KDP and a number of related compounds, certain hydrogen atoms or deuterium atoms are involved in a hydrogen bond as well as a more conventional bond, and ferroelectricity is associated with displacement of the hydrogen-bonded protons (or deuterons) from locations which would give zero net polarization to one or the other of a set of lower energy positions.

In order to illustrate how energy can be smaller in the presence of induced dipoles than it is without them, let us rearrange the Clausius-Mosotti relation

$$[N_i\alpha_i + N_e\alpha_e] = 3\varepsilon_0 \left(\frac{\kappa - 1}{\kappa + 2} \right) \tag{5-53}$$

into the form

$$(\kappa - 1) = \frac{3[N_i\alpha_i + N_e\alpha_e]}{3\varepsilon_0 - [N_i\alpha_i + N_e\alpha_e]} \tag{5-54}$$

Then the dielectric constant becomes infinite (leading to the possibility of spontaneous polarization) provided that

$$[N_i\alpha_i + N_e\alpha_e] \geq 3\varepsilon_0 \tag{5-55}$$

Equation 5-55 includes the "greater than" option as well as an equality, since an excessive polarizability will never rule out ferroelectricity. If

---

[5] The various experiments in structural crystallography which have established the magnitude of this shift are described in Chapter 4 of *Ferroelectric Crystals* by F. Jona and G. Shirane (Pergamon, 1962). The displacements are actually rather more complicated, since the titanium ions, barium ions, and oxygen ions all move by differing amounts in setting up one direction of polarization or the other.

the polarizability is more than sufficient to make the denominator vanish in the right side of Equation 5-54, then anharmonic effects will set an upper limit to the ionic excursions.

We have implied that when the conditions for spontaneous polarization are ripe, the energy density of the polarization field is larger than the energy necessary to create dipoles, whether or not an external field is applied. Let us suppose that a solid contains N_i polarizable ion pairs, each of which is responsive to a local field

$$E_{loc} = E_1 + (P/3\varepsilon_0) \qquad (5\text{-}56)$$

following the terminology of Equation 5-12. For charges $+q$ and $-q$ separated by distance x, the dipole moment is

$$p = qx = \alpha_i E_{loc} = (\alpha_i F/q) \qquad (5\text{-}57)$$

where α_i denotes the polarizability and F is the restoring force trying to bring together the two charges of the dipole. The work involved in *creating* N_i such dipoles per unit volume is

$$
\begin{aligned}
W_1 &= N_i \int F \cdot dx \\
&= (N_i q^2/\alpha_i) \int x \cdot dx \qquad (5\text{-}58) \\
&= (N_i q^2 x^2/2\alpha_i) = (N_i p^2/2\alpha_i) = (P^2/2N_i\alpha_i)
\end{aligned}
$$

This must be set against the energy of the electrical displacement

$$
\begin{aligned}
W_2 &= \int E_{loc} \cdot dP \\
&= \int [E_1 + (P/3\varepsilon_0)] \cdot dP \qquad (5\text{-}59) \\
&= (P^2/6\varepsilon_0) + \int E_1 \cdot dP
\end{aligned}
$$

Thus the net energy associated with a polarized dielectric can be expressed as

$$(W_2 - W_1) = (P^2/2N_i\alpha_i)[(N_i\alpha_i/3\varepsilon_0) - 1] + \int E_1 \cdot dP \qquad (5\text{-}60)$$

The terms on the right of Equation 5-60 have been arranged in this particular way in order to emphasize that the net energy is positive even for zero external field in a ferroelectric solid, for which the inequality of Equation 5-55 is satisfied. Thus it is energetically favorable for an array of highly polarizable dipoles to create themselves in an ordered fashion. At sufficiently high temperatures the thermal vibrations of the lattice frustrate this ordering attempt, and the magnitude of the spontaneous polarization moves smoothly towards zero as the temperature approaches T_c from below.

For a temperature just above T_c, the spontaneous polarization of a ferroelectric material has disappeared, but the solid has a very large dielectric constant. Thus κ can range up to 6000 for some $(Ba,Sr)TiO_3$ ceramics used in capacitors of small size but relatively large capacitance. Other applications of ferroelectrics require operation below T_c and make use of the static polarization. For example, the condenser microphone has experienced a rebirth of popularity, with a polarized ferroelectric taking up most of the space between the plates. The electro-optic characteristics of ferroelectrics such as $BaTiO_3$ and KDP have been used for modulating and deflecting laser beams both inside and outside the optical cavity. At one time it was hoped that single crystal $BaTiO_3$ might form the basis for random access computer memory storage. In this anticipated application, a thin single crystal plate of $BaTiO_3$ would have "bits" stored by causing P to be upwards or downwards as desired at designated locations; this sign would be read non-destructively when desired, and would be reversed by applying a sufficiently large field in the opposite direction. This application proved more difficult than anticipated because of the domain structure of $BaTiO_3$ (to be discussed next). Meanwhile, alternative computer memory systems have been developed, based on the permanent ferrimagnetism of ferrites and on bistable conduction in semiconductor device arrays.

FERROELECTRIC DOMAINS

A large single crystal of a ferroelectric solid need not necessarily demonstrate macroscopic signs of spontaneous polarization, since a crystal is likely to consist of an array of *domains* with different directions of polarization. Most typically, a domain will be surrounded by neighboring domains in which the direction of P_s is reversed. This is the situation of 180° *domain walls*. When an external electric field is applied, the domain walls may move to permit enlargement of domains in which P_s is reasonably parallel to E_1, while domains in which P_s is more nearly antiparallel to E_1 will tend to shrink. The arrangement of domains has several similarities with the magnetic domain structure of a ferromagnetic medium, though there are also important differences. In contrast to the 750 Å thickness of a Bloch wall between magnetic domains (and the relatively small amount of energy associated with the domain wall), the wall separating one ferroelectric domain from another is only one or two atomic spacings thick and has a large energy density associated with it.

Figure 5-10 illustrates a more complicated case of domain patterns in a ferroelectric crystal, once again with tetragonal $BaTiO_3$ as the example. In the crystal seen here, there are "180° domain walls" between regions of antiparallel polarization, but there are also "90° domain walls" between sections of (almost) perpendicular polarization. The origin of a 90° wall in this material can be understood if we remember that $BaTiO_3$ undergoes a slight tetragonal elongation of one of its cubic axes on cooling through the Curie point. For a large crystal, this could easily happen to one of the cubic axes in part of the crystal and to either one of the other two cubic axes in other portions of the crystal. Extension of these two forms of aberration would lead to a highly strained boundary along a pseudo-(110) plane, with the directions of polarization for each kind of domain making an angle of 45° with this boundary. As noted by Jona and Shirane[5], the tetragonal distortion of the lattice forces the angle between the polarization directions of domains separated by a "90° wall" to be actually 89°24′.

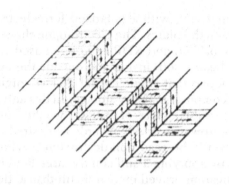

Figure 5-10 Ferroelectric domains in a crystal of tetragonal barium titanate, illustrating 180° and 90° domain walls. From J. A. Hooton and W. J. Merz, Phys. Rev. **98**, 409 (1955).

Comparable complications of the microscopic domain structure appear with many other ferroelectric materials.

ANTIFERROELECTRIC BEHAVIOR

We have seen that a ferroelectric material has induced, ordered dipoles below its transition temperature in order to minimize its energy. Another group of *antiferroelectric* solids also has induced, ordered dipoles below a characteristic temperature, but can demonstrate no spontaneous bulk polarization because each dipole is *antiparallel* to its neighboring dipoles. The structural requirements for this class of behavior are the same as for the genuine ferroelectric state, and well-known antiferroelectric materials are isomorphous with some of the ferroelectrics we have mentioned. Thus sodium niobate ($NaNbO_3$) and lead zirconate ($PbZrO_3$) both crystallize in the perovskite structure, but an antiparallel arrangement of adjacent lines of dipoles permits a lower total energy below a certain temperature than either the fully parallel arrangement of $BaTiO_3$ or $KNbO_3$, or of a state with *no* induced dipoles. Similarly, we find that potassium dihydrogen phosphate (KDP) is ferroelectric, but that the isomorphous ammonium dihydrogen phosphate (ADP) is antiferroelectric.

Magnetic Properties of Solids   5.2

Some of the terms needed for a discussion of magnetic properties have already been defined in Equations 5-2 and 5-3. In contrast to dielectric response, the magnetic response of most solids is dominated by the orientation of permanent dipoles, and we shall be much concerned with the *magnetic susceptibility* χ_m and the *magnetization* $M = \chi_m H$ of a set of permanent dipoles in an applied field. Since induced dipoles are quite unimportant in most solids, a "magnetic polarizability" is not usually defined.

MAGNETIC DIPOLE STRENGTHS

According to Ampère's law, the magnetic moment of a current loop is the product of the loop area and the current. Thus for an orbiting electron in a circular path of radius r and angular frequency ω, the magnetic dipole moment is

$$\mu_{or} = -\tfrac{1}{2}er^2\omega \tag{5-61}$$

Now the relationship between ω and r for an electron in motion about an atom is constrained by the quantum limitation that the orbital angular momentum must be a multiple of $\hbar$. This requires that the magnetic moment associated with orbital motion must be a multiple of the *Bohr magneton*

$$\mu_B = (\hbar e/2m) = 9.274 \times 10^{-24} \text{ J/T} \tag{5-62}$$

The angular momentum associated with the spin of an electron can be characterized by a spin quantum number $s = \pm\tfrac{1}{2}$. This spinning motion has an associated magnetic moment, which is customarily written

$$\mu_{sp} = gs\mu_B \tag{5-63}$$

where g is a quantity called the *spectroscopic splitting factor,* or sometimes just the g-factor. Since $g = 2.0023$ for a free electron, the magnetic moment of a spinning electron is almost exactly one Bohr magneton.

The total angular momentum for a multi-electron atom is obtained by vector addition of the components of orbital and spin angular momentum. This is thoroughly discussed in the standard texts on atomic structure.[6] For the present we shall assume that the vector combination

---

[6] The Russell-Saunders coupling of spin and orbital motion and the Landé splitting factor are discussed in any text on atomic physics, including the books by Bockhoff and by French and Taylor cited in the General References list at the beginning of this book.

proceeds in the manner described by Russell and Saunders (1925). In this scheme, the individual orbital quantum numbers combine vectorially to give a gross orbital angular momentum $\hbar L$, and the various spins to give a total spin angular momentum $\hbar S$. These in turn combine to give a total angular momentum for the electronic system of the atom which is

$$\hbar J = \hbar L + \hbar S \qquad (5\text{-}64)$$

The manner in which electrons are assigned to various quantum states within a partially filled subshell is determined by the Pauli principle and by *Hund's rules*.[7] A *magnetic atom* has a partially filled sub-shell and thus has some unpaired electrons; such an atom has a non-vanishing permanent magnetic moment, as required for paramagnetism, ferromagnetism, antiferromagnetism, or ferrimagnetism.

If we neglect the difference between 2.0023 and 2.0000 for the g-factor of a single electron, then the total *magnetic moment* of an atom or ion is

$$\mu = \mu_B(L + 2S) \qquad (5\text{-}65)$$

This moment precesses around the direction of **J**, and it is customary to employ the terminology of Landé (1923) in writing this moment as

$$\mu = g\mu_B J \qquad (5\text{-}66)$$

Here

$$g = 1 + \frac{J(J + 1) + S(S + 1) - L(L + 1)}{2J(J + 1)} \qquad (5\text{-}67)$$

is the *Landé splitting factor* or *g-factor* for an electron system in which the orbital and spin systems are coupled in the Russell-Saunders manner.

A very small magnetic moment is associated with the spin angular momentum of an atomic nucleus. This is measured in units of the *nuclear magneton*

$$\mu_n = (\hbar e/2M_p) = 5.051 \times 10^{-27} \quad \text{J/T} \qquad (5\text{-}68)$$

which is smaller than the Bohr magneton in the ratio of the proton mass to electron mass. The magnetic moment of the proton itself is 2.793 nuclear magnetons, and a moment numerically smaller than this is found for most heavier nuclei. Nuclear moments can almost always be neg-

---

[7] Hund's rules state that electrons placed in a partially filled sub-shell will arrange themselves (subject to the Pauli principle) in a way which will maximize S, and in a way which will maximize L once maximum S has been obtained. Then $J = (L - S)$ for a shell which is less than half full (with all spins parallel), and $J = (L + S)$ for a subshell which is more than half full. This appears in a logical framework in the books by Anderson and by Born noted in footnote 6.

lected[8] compared with electronic components in discussing static magnetization, though we shall see in Section 5.3 that nuclear magnetic moments control the NMR spectrum and affect the hyperfine structure of an ESR spectrum.

DIAMAGNETISM

The most interesting forms of magnetism are the strong forms which arise in solids including at least some magnetic atoms, and we wish to return to these subjects without undue delay. However it is appropriate first to at least mention the universal diamagnetic response of each orbiting electron. For *every* electron in *every* atom does have an induced diamagnetic response to an applied magnetic field; but the magnetic moments so induced are very weak. Diamagnetism is present but is overshadowed unless every atom within the solid has all of its electrons in paired states in order to exclude permanent dipoles.

In order to examine diamagnetism, we can conveniently study a model of an atom (or ion core) for which all electrons *are* in paired states (recognizing that a similar weak response would also occur for magnetic atoms). In the absence of a magnetic field, the electron motions are spherically symmetrical; there is zero angular momentum, zero circulating current about the nucleus, and zero magnetic moment. If a magnetic field **H** is now applied, the corresponding value of **B** determines the Lorentz term $ev \times B$ for the force on each electron. *Lenz's law* describes the interaction by saying that when magnetic flux changes in a circuit, then a current is induced which opposes the change in flux. The magnetic force rebalances centrifugal and centripetal forces so that the orbital frequency of an electron with orbital magnetic moment parallel to the field is slowed down, while the orbital frequency for an electron with orbital magnetic moment antiparallel to the field is speeded up.

The magnitude of this change in the orbital frequency can be determined from simple classical considerations (Problem 5.7); it is called the *Larmor precession frequency*

$$\Delta\omega = eB/2m \tag{5-69}$$

The effect of this precession adds up in the same direction for all Z electrons of an atom as being equivalent to a circulating current

$$I = -(Ze\Delta\omega/2\pi) = -(Ze^2B/4\pi m) \tag{5-70}$$

Now, we have already cited Ampère's law that a current loop has a magnetic dipole moment which is the product of circulating current and loop area. Thus the induced current of Equation 5-70 must induce a magnetic dipole.

---

[8] A notable exception to this statement is the nuclear paramagnetism of solid hydrogen, which was measured by B. Laserew and L. Schubnikow, Phys. Z. Sowjetunion **11**, 445 (1937). The very feeble paramagnetism of the protons could be seen for this solid because the electron system is diamagnetic.

Let the mean square distance of electrons from the nucleus of this atom be $<r^2>$. Then the mean square displacement in the plane perpendicular to **B** is $(2<r^2>/3)$, and the effective area of the current loop is $(2\pi<r^2>/3)$. The consequent atomic dipole moment is

$$\mu_{dia} = (2\pi I <r^2>/3) = -(Ze^2<r^2>B/6m) \qquad (5\text{-}71)$$

This is a negative quantity, and thus the resulting bulk diamagnetic susceptibility $\chi_{dia} = (+M/H) \simeq (+\mu_0 M/B)$ is also negative. Since the absolute value of the dimensionless quantity χ_m is much smaller than unity for any diamagnetic substance, $(\mu_0 M/B)$ is an excellent replacement for (M/H) in obtaining an expression for χ_{dia}. For a material with N similar atoms per unit volume, it can be seen from Equation 5-71 that

$$\chi_{dia} = -(\mu_0 N Z e^2 <r^2>/6m) \qquad (5\text{-}72)$$

The same result is obtained by a quantum-mechanical derivation, as shown for example in the book by van Vleck cited in this chapter's bibliography.

Equation 5-72 suggests a diamagnetic susceptibility $\chi_{dia} \approx -10^{-5}$ for typical densities and orbital sizes of electrons in a solid. This is comparable with diamagnetic susceptibilities measured in practice, as may be seen from the last five entries in Table 5-3. Diamagnetism is also shown by the Group VIII elements (in solid as well as gaseous form), and by solid compounds comprising positive and negative ions with closed-shell electronic configurations.

More complicated models for diamagnetism, including the diamagnetism of molecules in which atoms are connected via electron-pair bonds, are described by van Vleck.

TABLE 5-3 SOME EXAMPLES OF PARAMAGNETISM AND DIAMAGNETISM

Material	Paramagnetic Susceptibility χ_m (for T = 300K)	Diamagnetic Susceptibility χ_m
Platinum	$+2.9 \times 10^{-4}$	
Chromium	$+2.7 \times 10^{-4}$	
Niobium	$+2.6 \times 10^{-4}$	
Aluminum	$+2.2 \times 10^{-5}$	
Lithium	$+2.1 \times 10^{-5}$	
Calcium	$+1.9 \times 10^{-5}$	
Sodium	$+9.1 \times 10^{-6}$	
Oxygen (at NTP)	$+1.9 \times 10^{-6}$	
Nitrogen (at NTP)		-5.0×10^{-9}
Silicon		-4.2×10^{-6}
Copper		-9.5×10^{-6}
Gallium		-2.3×10^{-5}
Mercury		-2.9×10^{-5}
Bismuth		-1.6×10^{-4}

THE FORMS OF PERMANENT
DIPOLE RESPONSE

Four different kinds of magnetic behavior which involve permanent dipoles in a solid are illustrated in Figure 5-11. We shall have to return to these kinds of solid in a little more detail, but it is useful to remind ourselves at the outset of the qualitative consequences of the possible kinds of ordering.

Paramagnetic behavior occurs when the magnetic moments of the various atoms are uncorrelated in the absence of a magnetic field. Of course the dipoles tend to become aligned in a magnetic field, and the magnetization is governed by something similar to a Langevin function. The situation is actually a little more complicated because of quantum restrictions on the component of angular momentum in the direction of the applied field,[9] but the weak-field susceptibility is still of the *Curie Law* type:

$$\chi_m = C/T \tag{5-73}$$

For an *antiferromagnetic* solid at low temperatures the total energy of the crystal in the absence of an external magnetic field is lowest when dipoles of opposing magnetic moment alternate. This arrangement is very stable at low temperatures, and the susceptibility in an applied field is small. [Thus the quantity $(1/\chi_m)$ illustrated in part (b) of Figure 5-11 becomes large at low temperatures.] When the temperature rises, the efficiency of this dipole-dipole interaction decreases and the susceptibility increases, until the spins become "free" at the Néel temperature T_N to respond to a field. For still higher temperatures the behavior is paramagnetic, and the susceptibility follows a modified Curie law

$$\chi_m = \frac{C}{(T + \theta)} \tag{5-74}$$

A *ferromagnetic* solid is ordered with parallel spins below the Curie temperature T_c, a situation which results in a spontaneous magnetization M_s. The magnitude of this bulk polarization decreases to zero at the Curie point, and the paramagnetic susceptibility for the disordered spin system at higher temperatures obeys the *Curie-Weiss law*

$$\chi_m = \frac{C}{(T - T_c)} \tag{5-75}$$

The low temperature ordering in a *ferrimagnetic* material is similar to that of an antiferromagnetic material, but the two opposing spin systems have magnetic moments of unequal magnitude, and a net spontaneous magnetization results as the lowest energy state of the system. This magnetization declines to zero magnitude when the solid is

---

[9] The formulation of paramagnetic response in terms of a Langevin function parallels our use of this function for a simple picture of the response of permanent electric dipoles in a polar solid. With *paraelectric* solids there were complications in practice because of the limited dipole orientations allowed by the crystal lattice. In a *paramagnetic* situation the complications arise from a completely different cause (the quantization of angular momentum) and force the Langevin function to be replaced by a *Brillouin function*.

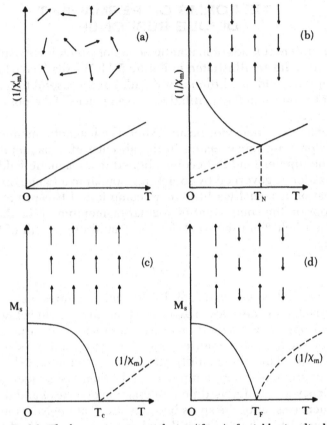

Figure 5–11 The low temperature ordering (if any) of neighboring dipoles, and the consequent behavior of spontaneous magnetization and/or susceptibility, for (a), paramagnetism, (b), antiferromagnetism, (c), ferromagnetism, and (d), ferrimagnetism. After S. Chikazumi, *Physics of Magnetism* (Wiley, 1964).

warmed to the Curie point T_f, and the behavior is once again paramagnetic at higher temperatures. The paramagnetic susceptibility of a ferrimagnetic solid has a more complicated temperature dependence than any of the other three forms illustrated in Figure 5-11.

PARAMAGNETISM

Suppose we have a solid containing N magnetic atoms per unit volume, each with a magnetic moment μ given by Equation 5-66. The lowest energy state of this solid in the absence of a magnetic field is assumed to be one of randomly oriented moments. When a magnetic field **H** is now applied, the energy of a dipole is $-\mu \cdot \mathbf{B}$ and an increase of **H** and **B** should produce preferential orientation of the dipole moments with the field. By analogy with the discussion of electric dipole orientation (see Equations 5-20 through 5-28), it might seem plausible to expect a magnetization

$$M = N\mu L(\mu B/k_0 T) \tag{5-76}$$

Here L(w) is the Langevin function previously cited as Equation 5-25. We know that $L(w) \simeq (w/3)$ when $w \ll 1$. Thus for a combination of applied field and ambient temperature which justifies a *weak-field* approximation, Equation 5-76 connotes a paramagnetic susceptibility

$$\chi_m = (M/H) \simeq (\mu_0 M/B) = C/T \quad [\mu B \ll k_0 T] \tag{5-77}$$

where

$$C = (\mu_0 \mu^2 N/3k_0) \tag{5-78}$$

Equation 5-77 is known as *Curie's Law*, and the quantity C is the *Curie constant* of the solid. A straight line is drawn in Figure 5-11(a) to indicate this kind of behavior. With $N \simeq 5 \times 10^{28}$ magnetic atoms per cubic meter, each of moment one Bohr magneton, the susceptibility of Equation 5-77 is $\chi_m \simeq (0.13/T)$, fairly typical of the numbers encountered in practice. The upper set of entries in Table 5-3 shows the paramagnetic susceptibility for some of the elements.

The Curie law is obeyed quite well by numerous paramagnetic solids, though for other solids there are deviations at low temperatures. Such deviations are to be expected as the rule rather than the exception, since paramagnetic orientation is complicated by the rules of quantum mechanics concerning the permitted quantization of angular momentum in space with respect to a designated axis.

According to quantum mechanics, the total angular momentum vector of an atom has a magnitude $\hbar \sqrt{J(J+1)}$, but the component which can be aligned with the axis of a magnetic field must be one of the set $\hbar m_J$, where the azimuthal quantum number m_J is a member of the set $J, (J-1), (J-2), \ldots, (1-J), -J$. Then the maximum value of $\hbar m_J$ is smaller than the magnitude of the total angular momentum vector, and there must be some transverse component.

For each value of m_J, the total magnetic moment μ of Equation 5-66 has a component $g\mu_B m_J$ aligned with the field axis (g is the Landé splitting factor of Equation 5-67). Such a component of magnetic moment aligned with the field axis requires a potential energy

$$U_{J,m} = -g\mu_B m_J B \tag{5-79}$$

in magnetic induction B, and it is this set of energies which must be considered with respect to the Boltzmann energy $k_0 T$ in determining the gross magnetization for a given temperature and given magnetic field. A very simple example of this is used as the basis for Problem 5.8. For the general case, the magnetic moment per unit volume is

$$M = N \left[\frac{\sum_{-J}^{J} (g\mu_B m_J) \exp(g\mu_B m_J B/k_0 T)}{\sum_{-J}^{J} \exp(g\mu_B m_J B/k_0 T)} \right] \tag{5-80}$$

This we choose to write as

$$M = Ng\mu_B J B_J(y) \qquad (5\text{-}81)$$

in terms of a dimensionless variable $y \equiv (g\mu_B J B/k_0 T)$ and the *Brillouin function* $B_J(y)$. Some manipulation of the summations in Equation 5-80 can reduce the Brillouin function to the form

$$B_J(y) = \left\{ \left(\frac{2J+1}{2J}\right) \coth\left[\frac{(2J+1)y}{2J}\right] - \left(\frac{1}{2J}\right) \coth\left(\frac{y}{2J}\right) \right\} \qquad (5\text{-}82)$$

and this is asymptotic to the Langevin function of Equation 5-25 for very large J. Since J is finite for any real atom, only a limited number of orientations are permitted with respect to the field direction, and none are coincident with this field direction.

The Brillouin function varies from zero when the applied field is zero to unity for infinite field.[10] Thus the saturation magnetization of a paramagnetic solid is $M_{max} = Ng\mu_B J$. Measurements such as those of Figure 5-12, for very low temperature and up to rather large fields, show that the magnetization in a paramagnetic salt does vary in the manner predicted by an appropriate Brillouin function.

Under weak field conditions, the Brillouin function is asymptotic to

$$B_J(y) \approx [y(J+1)/3J], \qquad y \ll 1 \qquad (5\text{-}83)$$

---

[10] Brillouin functions for various orders are tabulated by J. S. Smart in *Effective Field Theories of Magnetism* (Saunders, 1966).

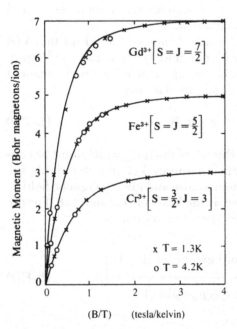

Magnetic Moment (Bohr magnetons/ion)

$$Gd^{3+}\left[S = J = \frac{7}{2}\right]$$

$$Fe^{3+}\left[S = J = \frac{5}{2}\right]$$

$$Cr^{3+}\left[S = \frac{3}{2}, J = 3\right]$$

x T = 1.3K
o T = 4.2K

(B/T)   (tesla/kelvin)

Figure 5-12 Magnetic moment versus applied field for paramagnetic ions incorporated into insulating salts. The abscissa combines field and temperature as (B/T). Gd^{3+} was measured in gadolinium sulphate octahydrate, Fe^{3+} in ferric ammonium alum, and Cr^{3+} in potassium chromium alum. The curves shown are Brillouin functions for appropriate J (or for appropriate S). [Data reported by Warren Henry, Phys. Rev. **88**, 559 (1952).]

and the paramagnetic susceptibility has a Curie law behavior:

$$\chi_m = (M/H) \simeq (\mu_0 M/B)$$
$$= [\mu_0 \mu_B^2 g^2 NJ(J+1)/3k_0 T] \qquad (5\text{-}84)$$

provided that $B \simeq \mu_0 H$ is small compared with $(k_0 T/g\mu_B J)$. Paramagnetic susceptibilities are usually small enough so that replacement of $\mu_0 H$ with B is a good approximation. Comparison of Equation 5-84 with the Curie law, Equations 5-73 and 5-77, shows that the Curie constant is

$$C = [\mu_0 \mu_B^2 g^2 NJ(J+1)/3k_0] \qquad (5\text{-}85)$$

It may be concluded from a comparison of the Curie constants of Equations 5-78 and 5-85 that each magnetic atom behaves as though it had an *effective dipole moment* of

$$\mu_{eff} = g\mu_B \sqrt{J(J+1)} \qquad (5\text{-}86)$$

as far as its *weak-field* response is concerned.

Tests of paramagnetic behavior can be carried out most conveniently with nonconducting compounds which include transition elements. (Measurements on the transition metals themselves would be complicated with the Pauli paramagnetism of the free electrons.) Such measurements have been made on an extensive scale with compounds including transition elements of the iron group (partially filled 3d subshell) and the rare earth group (4f subshell); more scanty evidence is available for the palladium group (4d subshell), the platinum group (5d subshell), and the uranium group (5f subshell).

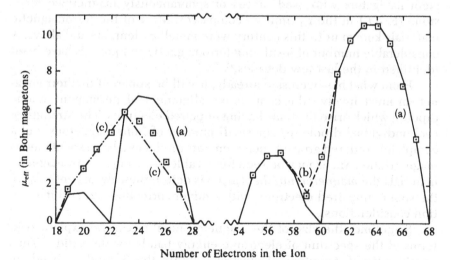

Figure 5–13 Magnetic moment as a function of electron number for paramagnetic ions of transition and rare earth elements, after L. F. Bates, *Modern Magnetism* (Cambridge, Fourth Edition, 1961). The points are experimental data. These are compared with (a), a solid curve for $g \sqrt{J(J+1)}$ Bohr magnetons, as expected for Russell-Saunders coupling; (b), a dashed curve for a more complex model for rare earth ions developed by van Vleck (see bibliography); and (c), a dashed curve of $2\sqrt{S(S+1)}$ Bohr magnetons for the iron group of elements, as expected if the orbital momentum is completely quenched.

The magnetic moment per atom deduced from these experiments conforms quite well with the moment calculated from Equations 5-66 and 5-67 for the rare earth group; but not for the iron group, as demonstrated in the calculated curves and experimental points in the two ranges of Figure 5-13. With the elements of partially filled 3d subshell, the *actual* moments are rather closer to the values expected from spin alone, the orbital part of the angular momentum being "quenched." The curves for Gd^{3+} and Fe^{3+} are uninformative on this subject, since for each of these ions L happens to be zero, which makes J = S and the Landé factor g = 2. However, the curve for Cr^{3+} in Figure 5-12 is clearly compatible with quenching of the orbital momentum to make $J \approx \frac{3}{2}$ and $g \approx 2$, rather than the J = 3, g = 0.4 consequence of Russell-Saunders coupling.

The departure from Russell-Saunders coupling in the iron group of elements is explained as a consequence of the intense inhomogeneous local electric field produced by neighboring atoms, since the 3d electrons involved in this group of elements spend much of their time in the outer parts of the atom, relatively exposed to external influences. The 4f electrons which cause paramagnetism of the rare earth elements are shielded from neighboring atoms by the surrounding 5s and 5p electrons, which makes the *crystal field splitting* much less severe and the Russell-Saunders coupling model more tenable.

FERROMAGNETISM

The existence of materials with a permanent magnetization at room temperature in the absence of an applied magnetic field has been known for many years, and no history is complete without a mention of the ancient navigators who used pieces of spontaneously magnetized lodestone (Fe_3O_4) in their primitive compasses. Most of the ferromagnetic materials known until this century were metallic elements and alloys. A considerable number of insulating ferromagnetic compounds have been identified in the last few decades.[11]

From what has been said already, it will be apparent that ferromagnetism must involve the cooperative alignment of permanent atomic dipoles, which arise in atoms having unpaired electrons. The strength of each individual dipole, μ, is a small number of Bohr magnetons, but a completely ordered array of such moments produces a large spontaneous magnetization M_s, as may be seen from Table 5-4. In most ferromagnetic materials the *magneton number* (μ/μ_B) is determined almost entirely by the spin of unpaired electrons, with a minor correction from orbital motion considerations.

Sometimes the size of the magneton number can be explained in terms of the spectrum of electronic energy bands for the solid.[12] Thus consider the distribution of electrons within the 3d and 4s bands of

[11] A table of ferromagnetic materials which includes some insulating ferromagnetic compounds is given by J. S. Smart, *Effective Field Theories of Magnetism* (Saunders, 1966). Additional tabular information appears in the *American Institute of Physics Handbook* (McGraw-Hill, Third Edition, 1971).

[12] E. C. Stoner, Repts. Progr. Phys. 11, 43 (1948).

TABLE 5-4 THE FERROMAGNETIC ELEMENTS*

Material	Crystal Structure	Curie Temperature	Saturation Magnetization at T = 0		Bohr Magnetons Per Atom
		$T_c(K)$	M_s (A/m)	$B_r = \mu_o M_s$ (T)	(μ/μ_B)
Fe	B.C.C.	1043	1.74×10^6	2.18	2.22
Co	H.C.P.	1395	1.45×10^6	1.82	1.72
Ni	F.C.C.	631	5.11×10^5	0.643	0.606
Gd	H.C.P.	289	2.11×10^6	2.65	7.1
Dy	H.C.P.	85	2.92×10^6	3.67	10.0

* Data from J. S. Smart, *Effective Field Theories of Magnetism* (Saunders, 1966).

nickel. An isolated Ni atom has a $3d^8 4s^2$ configuration, but the energy bands which result in metallic nickel are filled up to the Fermi energy with 9.46 electrons per atom in the 3d band and only 0.54 electrons per atom in the overlapping 4s band. When nickel is in the paramagnetic condition (for a temperature above its Curie point T_c), the occupancy of each band is divided equally between *spin up* and *spin down* states, as illustrated in Figure 5-14(a). Any applied magnetic field will then cause a very small relative shift of the two sets of states, as discussed in connection with Pauli paramagnetism in Section 3.3. For nickel at a temperature low enough to produce ferromagnetic ordering, the very large effective molecular field produces a major energy shift of the spin up states with respect to the spin down states. Figure 5-14(b) shows that for the 3d band one spin configuration becomes completely filled (5 electrons per atom), while the other spin choice has only 4.46 electrons per atom. Thus

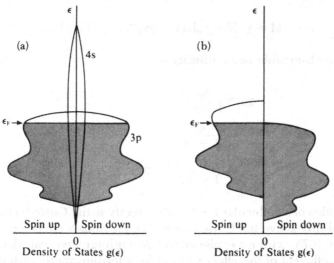

Figure 5-14 (a), Density of states versus energy for the two spin components of the 3d and 4s bands of nickel in a paramagnetic condition. Up to the Fermi energy, the 3d band contains 9.46 electrons per atom and the 4s band 0.54 electrons per atom. (b), The 3d band split by the molecular field, which makes spin down (magnetic moment up) electrons dominate for the ferromagnetic condition of nickel.

the ferromagnetic ordering produces 0.54 3d-band electrons of unpaired spin per atom. Allowance for the g-factor of these, for the slight readjustment of occupancies among the 4s-band electrons, and for the small magnetic contribution of orbital motion results in a saturation magnetization of 0.606 μ_B per atom in this solid.

In the simplest kind of approach to ferromagnetic ordering, it is supposed that the interaction of a magnetic atom with the crystal can be described by an effective field, or *molecular field*. This approach was proposed by Pierre Weiss,[13] who observed that the magnitude H_e of the effective molecular field should be proportional to the magnetic moment per unit volume; that is,

$$H_e = \gamma M \qquad (5\text{-}87)$$

The constant of proportionality, γ, is often called the Weiss coefficient; it should be a function of the specific atoms and the crystal structure but not the temperature.

Now in the ferromagnetic phase of a solid, M and H_e are large quantities without any applied field. Above the Curie temperature T_c, an external field H must be applied to produce any magnetization, but the magnetization is then self-assisted by the molecular field it generates. We can write the total magnetization in a weak applied field as

$$M = \chi_0(H + H_e), \qquad T > T_c \qquad (5\text{-}88)$$

where $\chi_0 = (C/T)$ is a weak-field susceptibility which obeys the Curie law (Equation 5-73) of a paramagnetic material. However, χ_0 will be smaller than the *observable* susceptibility $\chi_m = (M/H)$, the magnetization per unit *applied* field. Inserting Equation 5-87 into Equation 5-88, the total magnetization can be expressed as

$$M = \chi_m H = \chi_0 H(1 - \gamma\chi_0)^{-1}, \qquad T > T_c \qquad (5\text{-}89)$$

Then the observable susceptibility is

$$\chi_m = \frac{\chi_0}{1 - \gamma\chi_0}$$

$$= \frac{C}{T - \gamma C}$$

$$= \frac{C}{T - T_c}, \qquad T > T_c = \gamma C \qquad (5\text{-}90)$$

and the idea of a molecular field leads directly to the *Curie-Weiss* law for susceptibility above the Curie point. (We previously cited this "law" as Equation 5-75, without explanation.) Accordingly, we can deduce the Weiss coefficient of the effective field for a ferromagnetic material if we know the Curie constant and Curie temperature for its magnetic behav-

[13] P. Weiss, J. Phys. Radium 4, 661 (1907).

ior at high temperatures. Typically, $\gamma = (T_c/C) \sim 10^4$, which means that the molecular field $H_e \sim 10^{10}$ amp/meter [$\mu_0 H_e \sim 10^4$ tesla] at low temperatures for a ferromagnetic material, such as the elements listed in Table 5-4.

The Weiss molecular field approach also permits us to create an equation for the temperature-dependence of the spontaneous magnetization M_s in the ferromagnetic phase. Suppose we have N magnetic atoms per unit volume, each of magnetic moment μ_1. If ferromagnetism is to be regarded as paramagnetism with a massive molecular field which creates self-ordering, then the Brillouin function of Equation 5-81 can be used in writing the magnetization as

$$M = N\mu_1 B_J(\mu_1 B/k_0 T) \qquad (5\text{-}91)$$

for magnetic induction B. Consider now a temperature lower than T_c, and with no superimposed external field. Then $B = \mu_0 H_e = \gamma\mu_0 M_s$, and the spontaneous magnetization M_s must satisfy the condition

$$M_s = N\mu_1 B_J(\gamma\mu_0\mu_1 M_s/k_0 T) \qquad (5\text{-}92)$$

For any temperature from absolute zero to T_c, and for a Brillouin function of given order J, there is one value for M_s which satisfies Equation 5-92. Since any Brillouin function is completely saturated at absolute zero, then $M_{s0} = N\mu_1$. Accordingly,

$$(M_s/M_{s0}) = B_J(\gamma\mu_0\mu_1 M_s/k_0 T) \qquad (5\text{-}93)$$

The solution for (M_s/M_{s0}) versus (T/T_c) can be determined graphically, as suggested in Problem 5.9. The decrease of (M_s/M_{s0}) with rising temperature is exemplified by the data for nickel in Figure 5-15, and by the accompanying curve calculated from Equation 5-93 with $J = S = \frac{1}{2}$. The quality of the fit between the model and experiment is obviously good enough to retain our interest in the model.

It was clear to Weiss from the beginning that his "molecular field" model could not arise from a *magnetostatic* coupling of the permanent magnetic dipoles,[14] since the magnitudes he tried to predict were quite impossible. He could deduce from the value of M_s that each dipole in a ferromagnetic solid has a magnetic moment μ_1 of about the magnitude we now call the Bohr magneton. Thus he could calculate the dipole field (μ_1/r_{ab}^3) created by one dipole at the location r_{ab} of the next. He could see that each of an ordered array of dipoles should feel an effective field

$$H_e = \gamma_1\mu_1 r_{ab}^{-3} = \gamma_2\mu_1 N \qquad (5\text{-}94)$$

---

[14] This is perhaps a good time to emphasize the major differences between the ordering processes in a *ferroelectric* and a *ferromagnetic* solid. It is really rather striking that two such dramatic phenomena should arise for such different reasons. In the ferroelectric case, the dipoles are *induced* ones, and the energy which binds a dipole into its ordered place can be accounted for from the strength of dipole-dipole interactions. In contrast, the magnetic dipoles of a ferromagnetic solid are *permanent* ones, and we find that the strength of a direct magnetostatic dipole-dipole coupling is four orders of magnitude smaller than the exchange mechanism which actually produces the ferromagnetic state.

where N is the number of magnetic atoms per unit volume and γ_1 and γ_2 are dimensionless factors. From arguments similar to those used in deducing the effective field within a polarized dielectric, one can justify values for γ_1 or γ_2 which are slightly larger than unity. In this way, Weiss could account for an effective field H_e of no more than about 10^6 A/m. Each dipole would be bound with an energy of only $\mu_0\mu_1 H_e \approx 10^{-23}$ joule (10^{-4} eV), and such a feebly bound array of dipoles would break up above a Curie temperature $T_c \approx (\mu_0\mu_1 H_e/k_0) \approx 1K$. This stood in radical contrast to the Curie temperatures of more than 1000K shown for iron and cobalt in Table 5-4.

This was a gloomy conclusion for Weiss, particularly since his model did account in a tolerable fashion for the form of the spontaneous magnetization below T_c (Figure 5-15) and for the form of the suscepti-bility above this temperature (Equation 5-90). Yet the typical 1000K value for the Curie temperature requires an ordering energy of 0.1 eV or so (which is then consistent with an "effective field" $H_e \sim 10^{10}$ A/m, and an "effective field factor" or *Weiss coefficient* $\gamma \sim 10^4$). Not until the golden era of quantum mechanics could a logical explanation be found for the very strong interaction which is actually responsible for ferromag-netism.

An explanation was then suggested by Heisenberg,[15] who showed that the large correlation energy between adjacent magnetic atoms could be understood on an electrostatic basis, arising from an application of the

---

[15] W. Heisenberg, Z. Physik **49**, 619 (1928) is the original paper on ferromagnetic exchange. The exchange mechanism for insulating ferromagnetics is discussed by P. W. Anderson in *Magnetism, Vol. 1*, edited by G. T. Rado and H. Suhl (Academic Press, 1963).

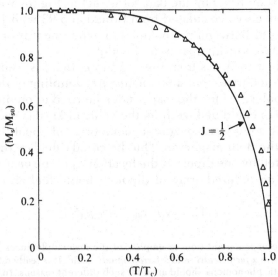

Figure 5-15 Temperature dependence of the spontaneous magnetization M_s (as a fraction of the saturated spontaneous magnetization at absolute zero) for a ferromagnetic solid. The curve conforms with Equation 5-93 for $J = \frac{1}{2}$. The points are the experimental data of P. Weiss and R. Forrer [Ann. Physique **15**, 153 (1926)] for metallic nickel, in which $T_c = 631K$, and $M_{s0} = 5.11 \times 10^5$ amp/meter [$\mu_0 M_{s0} = 0.643$ tesla].

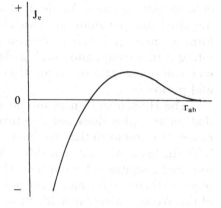

Figure 5-16 A schematic of the variation of the exchange integral with distance r_{ab} between one magnetic atom and the next.

Pauli exclusion principle. Heisenberg noted that there is some overlap of the electron clouds associated with neighboring magnetic atoms, and that this influences the manner in which electrons can be allocated among the various spin states of partially filled inner sub-shells. If two neighboring magnetic atoms have spin S_i and S_j, then their potential energy can be expressed in the form

$$U = \text{constant} - 2J_e S_i \cdot S_j \qquad (5\text{-}95)$$

where J_e is an *exchange integral* performed with respect to the electrostatic interaction potential between the two atoms. The second term on the right of Equation 5-95 is known as the *exchange energy*, and is said to be the consequence of an *exchange force*.

The manner in which the exchange energy is written in Equation 5-95 may make it appear that the spin on one atom is directly coupled with the spin on its neighbor, which is not the case. The quantum-mechanical result is that the exchange integral is a *negative* quantity for a *small* distance r_{ab} between neighboring magnetic atoms. This makes parallel spins energetically unfavorable (just as we found in the Heitler-London model for a hydrogen molecule in Figure 1-5 that the σ_u configuration of parallel spins was very unfavorable compared with the energy of a σ_g configuration with anti-parallel spins). However, when r_{ab} is larger than a critical distance, the exchange integral can be positive (Figure 5-16), and parallel spins are then energetically favored. Ferromagnetism thus places requirements on the minimum spacing of magnetic neighbors, which accounts for the ferromagnetic behavior of Fe, Co, and Ni, while the transition elements preceding them in the Periodic Table do not show the phenomenon. Ferromagnetism also requires that the spacing of magnetic neighbors not be too large, since as sketched in Figure 5-16 the exchange integral declines to zero when r_{ab} is very large.

Thus, given atoms with electrons of unpaired spins, the important consideration is that states be occupied in accordance with the Pauli principle to give a minimum *electrostatic* energy. This just happens to be a situation of ordered parallel spins (and of spontaneous magnetization) if the spacing r_{ab} is appropriate; the ferromagnetic behavior is an accidental consequence. An element such as manganese, which has a large unpaired spin per atom but which is not ferromagnetic in the elemental form because r_{ab} is too small, can be made ferromagnetic by incorporating it in a compound (such as MnSb or MnAs) which increases the manganese-manganese distance far enough to make the exchange integral positive.

The Heisenberg quantum-mechanical picture of an exchange force that orders spins does not overturn the results of the molecular field model of ferromagnetism we have discussed in Equations 5-87 through 5-93. Instead, we may consider the exchange force as justifying the required magnitude for the field H_e which is effective in ordering. The physical theory of ferromagnetism is still often discussed in the language of the Weiss molecular field, though more sophisticated effective field theories have been developed.[16] Such theories attempt to account for the departure of the spontaneous magnetization in any real solid from the family of curves permitted by Equation 5-93. The more complicated theories also deal with the paramagnetic behavior above T_c, which for any real solid always departs to some extent from the Curie-Weiss law of Equation 5-90.

---

[16] J. S. Smart, *Effective Field Theories of Magnetism* (Saunders, 1966), discusses a series of these more complicated models, including the Oguchi model of exchange-coupled pairs in an effective field, and the Bethe-Peierls-Weiss cluster model.

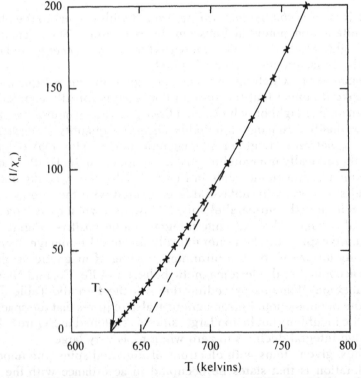

Figure 5-17 Reciprocal of the paramagnetic susceptibility for nickel as a function of temperature above the Curie point. The data are from P. Weiss and R. Forrer, Ann. Physique **15**, 153 (1926). A departure from the Curie-Weiss law of the same general character is found for other ferromagnetic solids.

One such theory was proposed by P. R. Weiss[17] (who is to be distinguished from P. Weiss of the first molecular field theory). P. R. Weiss abandoned the idea that the molecular field must be proportional to the average magnetization, and assumed instead that each moment would tend to lie parallel to the neighboring spins if this latter direction departed at all from the direction of average magnetization. Thus for temperatures close to T_c, Weiss expected to find many large "clusters" of magnetic atoms, with the spin aligned inside each cluster. Such clusters would disperse into separate paramagnetic atoms as the temperature increased, but for temperatures not much larger than T_c, the response to an applied field would be smaller than that given by the Curie-Weiss law of Equation 5-90. As Figure 5-17 shows, $(1/\chi_m)$ increases with temperature for a real solid in a manner distorted from the Curie-Weiss law in the sense to be expected if cluster formation is significant.

SPIN WAVES

At absolute zero the array of spins in a ferromagnetic solid is presumed to be completely ordered. We might regard this as the ground state of the spin system; a state in which one or more spins are reversed is then an excited state of the system.

Now in terms of the Heisenberg formulation for the effective field coupling the charge configuration of one magnetic atom to that of its neighbors, the reversal of a single spin appears to require an extra energy $\Delta U \approx 8 J_e S^2$. However, a much smaller additional energy is necessary if the crystal creates a *spin wave*. This term denotes a partial excitation of a number of spins, an excitation which moves in a wave-like manner through the system of spins. The *magnon* is the quantized unit of spin wave energy, and each magnon excited decreases the total magnetization by one unit of spin. The dispersion laws for magnons in the spin system of a ferromagnetic material at low temperatures may be compared crudely with the phonon dispersion laws for describing the displacements of atoms in a wave-like manner.

Spin waves were postulated in 1930 by Felix Bloch.[18] He showed from spin-wave arguments that the spontaneous magnetization should decline from its saturated value M_{so} at absolute zero in accordance with

$$M_s = M_{so}[1 - A(k_0 T/J_e)^{3/2}], \qquad T \ll T_c \qquad (5\text{-}96)$$

where A is a numerical constant which depends on the crystal structure. This "Bloch $T^{3/2}$ law" is in good agreement with the manner in which M_s starts to decrease (see Problem 5.10).

Despite this success, disussions of ferromagnetism in terms of spin waves were relatively unfashionable for many years because spin waves had not been observed directly. The subject has become a much more

[17] P. R. Weiss, Phys. Rev. **74**, 1493 (1948). See also footnote 16 on this subject.
[18] F. Bloch, Z. Physik **61**, 206 (1930).

lively one since spin wave resonance phenomena have been observed in thin ferromagnetic films. In one such type of experiment, Seavey and Tannenwald[19] applied a microwave magnetic field parallel to the plane of a thin ferromagnetic film and a static magnetic field perpendicular to this plane. When the static field was swept, a series of values were found for which the microwave absorption increased abruptly due to the creation of standing wave modes with an integral number of spin wave wavelengths in the film thickness. The standing wave forms of spin waves appear to have the spin configuration at the surfaces constrained by the magnetic anisotropy of the surface plane.[20]

Another experiment involving spin waves is the conversion of magnon energy into phonons when the speeds of spin waves and elastic waves are made to coincide. This can be done by the choice of a suitable magnetic field. Within a ferromagnetic sample which is subjected to an inhomogeneous magnetic field (such as the demagnetizing field of a sample in the shape of a solid cylinder), spin waves can be excited in a high-field part of the sample and converted into lattice vibrations when the energy is transported into a lower-field region in which phonons travel faster than magnons. Such experiments can be carried out most effectively in a ferromagnetic material in which acoustic losses are small.[21]

In Chapter 2 we discussed the dispersion laws for quantized lattice vibrations, and described how the relationship between phonon energy and phonon wave-vector could be deduced from the inelastic scattering of neutrons. At that time, Equations 2-18 and 2-19 were used as the conservation laws for phonon creation or annihilation. An entirely similar procedure can be used for magnon creation and annihilation through the inelastic scattering of slow neutrons in a ferromagnetic solid. A dispersion curve obtained by this procedure is shown in Figure 5-18. As a con-

---

[19] A procedure by which spin wave resonances should be measurable was postulated by C. Kittel, Phys. Rev. **110**, 1295 (1958), and the first successful results with the procedure were reported by M. H. Seavey, Jr. and P. E. Tannenwald, Phys. Rev. Letters **1**, 168 (1958).

[20] The "pinning" of the spin orientations at the film surfaces is discussed by P. E. Wigen et al., J. Appl. Phys. **34**, 1137 (1963).

[21] Experiments involving the interaction of spin waves and acoustic waves in "YIG" (yttrium iron garnet), and time-delayed "echo" responses of a crystal, are described for example by R. W. Damon and H. van de Vaart, Proc. IEEE **53**, 348 (1965). YIG is actually a *ferrimagnetic* material, a term defined on page 453.

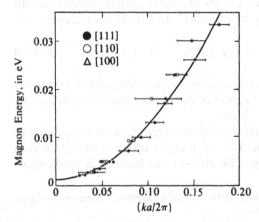

Figure 5-18 Spectrum of magnon energy versus the dimensionless equivalent of wave-vector for spin waves in an alloy of 92 per cent cobalt and 8 per cent iron. Data obtained at room temperature from inelastic neutron scattering experiments by R. N. Sinclair and B. N. Brockhouse, Phys. Rev. **120**, 1638 (1960).

trast to acoustic branch phonons, we see that the magnon energy in Figure 5-18 increases approximately as the square of the reduced wavevector. Consistency of this behavior with the "Bloch $T^{3/2}$ law" is verified in Problem 5.10.

We may observe also from Figure 5-18 that the data indicate a small but finite energy for a magnon of infinite wavelength. This is at least partly due to the limitations of instrumental resolution for the data displayed in the figure, though a finite minimum magnon energy is expected in any ferromagnetic solid because of long-range anisotropy effects.

MAGNETIC DOMAINS

It is quite possible for a piece of a ferromagnetic solid to appear *nonmagnetized* on a macroscopic scale. This behavior is accomplished by an array of *magnetic domains* with random orientations, even though the magnetization within each domain has the large value M_s. Each domain is separated from neighboring domains by a domain wall or *Bloch wall*, in which the direction of spins rotates as a relatively smooth function of distance through the wall, as pictured in Figure 5-19.

The number of domains within a crystal, and the shapes and relationships of these with respect to one another, are determined by the crystal directions in which spin alignment can occur and by the requirement that the sum of Bloch wall energy and sample demagnetization energy must be a minimum. Figure 5-20 shows the arrangements of domains with respect to each other in two kinds of ferromagnetic solid, one encouraging alignment of spins along cubic directions and the other requiring spin alignment along cube diagonal directions.

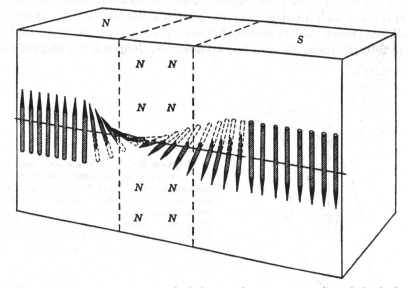

Figure 5-19 The manner in which the spin direction rotates through the thickness of a Bloch wall separating magnetic domains, after C. Kittel, *Introduction to Solid State Physics* (Wiley, 1971). The transition is some 300 lattice constants thick in iron.

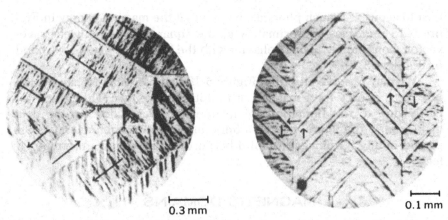

0.3 mm   0.1 mm

Figure 5-20 Two examples of magnetic domain structure visible under the microscope, after S. Chikazumi, *Physics of Magnetism* (Krieger, 1978). The illustration at the right is for iron containing 4 per cent silicon; the easy directions of magnetization are the cubic [100] directions, and domains form 180° and 90° walls as viewed from a surface which is almost a (100) plane. The domains at the left are for Permalloy (78 per cent Ni, 22 per cent Fe), in which the stable magnetic directions are the [111] ones; the resulting domains are separated by 180°, 109°, and 71° walls.

In Figure 5-21, the condition of an apparently "nonmagnetized" sample can be represented by the origin of the coordinate system. When an external field is applied, Bloch walls move in a manner that enlarges domains in which M_s is fairly close to the direction of H. This motion of the Bloch walls can be reversible if the applied field is small but becomes irreversible for a larger field. Once irreversible motion has been produced, the sample has a remanent magnetization if the applied field is removed.

Figure 5-21 shows the magnetization increased to about the saturation value, which involves *domain rotation* at the higher fields as well as wall movement. An alternation between positive and negative applied field then produces the well-known *hysteresis loop* relating magnetic in-

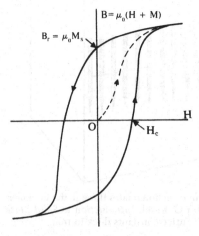

Figure 5-21 A schematic magnetization loop for a multi-domain sample of a ferromagnetic solid. H_c is the *coercivity*, and B_r the *remanence* of the sample. The dashed curve shows what happens when a nonmagnetized sample (domains of random orientations) is first magnetized. The arrows on the solid curves show the course of a subsequent hysteresis loop.

duction and driving field. The area enclosed by this curve is the work done per unit volume in executing the loop path.

The technical aspects of ferromagnetism are very much concerned with the magnetization loops for real solids; with "hard" magnetic materials of great coercive strength, with materials of large permeability, and so forth. The immense literature on this subject can be reached via the bibliography cited at the end of the chapter.

ANTIFERROMAGNETISM AND FERRIMAGNETISM

A form of magnetically ordered solid which would display no gross magnetization because neighboring dipoles were *anti-parallel* to each other was suggested as a possibility by Néel.[22] This type of ordering is called *antiferromagnetism*, and numerous solids have been found which behave this way.[23]

A lowest energy state for two adjacent magnetic ions which requires their spins to be antiparallel evidently corresponds with a negative value for the exchange integral J_e in the Heisenberg model. But the existence of a negative value for J_e is by no means a sufficient condition for the appearance of ordered antiferromagnetism in a real three-dimensional crystal. One must also have a three-dimensional array of the magnetic ions which is favorably disposed to the creation of *magnetic sub-lattices* that oppose the magnetization of each other. For some crystal structures, it is necessary to describe the magnetic ions as a distribution over four or more sub-lattices. However, we can illustrate the manner in which antiferromagnetism comes about by studying the simplest possible arrangement, one in which there are just two opposed magnetic sub-lattices. This model is exemplified by the Mn^{2+} ions of $RbMnF_3$ in Figure 5-22.

The crystal lattice of $RbMnF_3$ permits us to divide the totality of Mn^{2+} ions into an "A" sub-lattice (of which each member has six nearest *magnetic* neighbors, all from the "B" sub-lattice), and a "B" sub-lattice (of which each member has only members of the "A" family as its nearest magnetic neighbors). Thus if there is a negative exchange interaction between nearest magnetic neighbors, and all J_e for more remote magnetic neighbors are zero, the A and B sub-lattices will become spontaneously magnetized in opposite directions. The antiferromagnetic ordering should be complete at absolute zero, but will break up with rising temperature, falling to zero at the Néel temperature T_N.

Once again we can use "molecular field" language to express the effect of the exchange interaction upon the spin orientation of a magnetic

---

[22] The theory of antiferromagnetism proposed by L. Néel, Ann. Physique 17, 5 (1932) was modified into the form customarily used today by J. H. van Vleck, J. Chem. Phys. 9, 85 (1941).

[23] The sources cited in footnote 11 give tabular information on numerous antiferromagnetic solids.

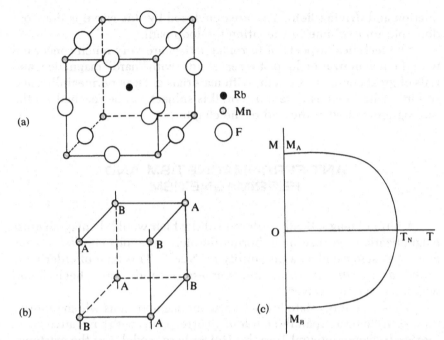

Figure 5-22 Structure and ordering in an antiferromagnetic solid, exemplified by RbMnF$_3$. (a) The crystal structure of RbMnF$_3$. (b) The arrangement of the Mn^{2+} ions into two sub-lattices of opposing spin orientation. (c) The temperature dependences of the equal and opposite magnetizations of the two sub-lattices. After J. S. Smart, *Effective Field Theories of Magnetism* (Saunders, 1966).

ion. The molecular field experienced by a Mn^{2+} ion in the A sub-lattice of Figure 5-22 can be expressed as

$$H_A = -\gamma M_B \qquad (5\text{-}97)$$

by analogy with Equation 5-87 for a ferromagnet. Then, as an analogy with Equation 5-88, the magnetization of the A sub-lattice is

$$M_A = (C/2T)[H - \gamma M_B] \qquad (5\text{-}98)$$

where H is any non-local field and (C/2) is the Curie constant appropriate for one-half of the magnetic ions in the crystal. Similarly, the "B" sub-lattice magnetization is

$$M_B = (C/2T)[H - \gamma M_A] \qquad (5\text{-}99)$$

The simultaneous solution of Equations 5-98 and 5-99 requires that M$_A$ and M$_B$ be non-zero in the absence of an applied field for temperatures equal to or less than T$_N$ = (γC/2).

An observable susceptibility can be derived from Equations 5-98 and 5-99 in the manner previously used for a ferromagnetic solid in Equation 5-90. The two equations give a response per unit of applied field

$$\chi_m = \frac{M_A + M_B}{H} = \frac{C}{T + T_N}, \qquad T \geqslant T_N \qquad (5\text{-}100)$$

which does not entirely accord with experimental observations. The behavior of an antiferromagnetic solid at high temperatures must in practice be described by

$$\chi_m = \frac{C}{T + \theta}, \qquad T \geqslant T_N \qquad (5\text{-}101)$$

where the ratio (θ/T_N) is customarily larger than unity. This discrepancy can be accounted for in more complicated models of magnetic exchange which include the influences of more remote magnetic neighbors. A simple elaboration is suggested in Problem 5.11.

As sketched in part (b) of Figure 5-11, the magnetic susceptibility of an antiferromagnetic solid decreases with falling temperature below the Néel temperature. This behavior is controlled by the exchange interactions which "lock" the magnetic moments into their antiparallel arrays and hinder the response to an applied field. The magnetic susceptibility of a single crystal sample usually displays considerable anisotropy below T_N, since it is much easier to tilt all spins slightly by applying a field perpendicular to the opposing spin systems than it is to induce "spin flips" with a field along the magnetic axis.

In 1948, Néel[24] considered what would happen if the magnetic ions in a crystal could be divided into two groups or sub-lattices of *unequal* magnetic moment. If the molecular field experienced by an ion of sub-lattice A is controlled by a negative exchange integral with respect to its magnetic neighbors of the B sub-lattice (and vice versa), then the solid is *ferrimagnetic*, with a spontaneous magnetization at low temperatures given by the difference between the magnetizations of the two sub-lattices.

The concept of ferrimagnetic ordering originated by Néel was immediately applicable to the magnetic properties of the *ferrite* group of solids. The composition of a ferrite has the form $MO \cdot Fe_2O_3$, where M denotes a divalent metallic cation such as Fe, Ni, Co, Cu, Mg, Zn, or Cd. Thus the ferrite group includes Fe_3O_4, the lodestone material of the ancient mariners, which has a magnetic moment (only $4\mu_B$ per molecule) much smaller than would be expected from ordinary ferromagnetic ordering of all spins. Néel showed that this magnetic moment comes exclusively from the single Fe^{2+} ion per molecule, since the two Fe^{3+} ions per molecule are in *non-equivalent crystallographic locations* and lie on two magnetically opposed sub-lattices.

The spontaneous magnetization of a ferrimagnetic solid decreases from a maximum at very low temperatures to zero magnetization at the upper limit T_f in a manner generally similar to that of a ferromagnetic material. The procedure described in connection with Equation 5-93 is appropriate. For temperatures larger than T_f, the susceptibility behavior is less straightforward than the Curie-Weiss law, as indicated by the dotted curve in part (d) of Figure 5-11. Néel[24] showed that the recipro-

[24] L. Néel, Ann. Physique **3**, 137 (1948).

cal susceptibility could be fitted (see Problem 5.12) with an expression of the form

$$(1/\chi_m) = \left(\frac{T - \theta}{C}\right) - \left(\frac{W}{T - \theta_0}\right), \qquad T > T_f \qquad (5\text{-}102)$$

where C is the Curie constant as before, W and θ_0 are constants that can be related to the various molecular field coefficients (or exchange integrals), and θ is a parameter which is numerically appreciably smaller than T_f.

There has been considerable practical interest in the magnetic properties of several members of the ferrite family which crystallize in the spinel structure. These materials are electrical insulators, a most useful property for many practical uses of a magnetic solid. They can be formed in a way which results in an almost perfectly rectangular hysteresis loop, and toroidal ferrite cores so formed provided the primary information storage for most of the electronic computers built in the 1960's and early 1970's.

Another group of ferrimagnetic solids of intense interest from a practical point of view are the iron garnets, solids with the composition $M_3Fe_5O_{12}$ where M is a trivalent metallic cation such as yttrium or one of the rare earth elements. The garnets have been used in magnetic resonance experiments and in studies of spin wave interactions.

The terms *ferromagnetism*, *ferrimagnetism*, and *antiferromagnetism* describe only the simpler kinds of spontaneous magnetic ordering of which crystalline solids are capable. There are also magnetically interesting solids in which a configuration of minimum energy requires a helical (screw-like) arrangement of spins, and others in which the spins of sub-lattices are canted rather than directly opposed. These exotic systems are discussed in some detail in several of the books cited in the bibliography.

Magnetic Resonance   5.3

The magnetic moment of a charged spinning particle (electron or nucleus) interacts with a macroscopic magnetic induction **B** to cause an energy level to split into two or more states. Transitions between states can be stimulated by photon absorption (subject to appropriate selection rules), and *magnetic resonance* occurs when the energy $\hbar\omega$ of radio-frequency photons coincides with the magnetically induced energy separation of two states of an electron or a nucleus.

The idea of electron spin and of an accompanying magnetic moment was proposed by Goudsmid and Uhlenbeck (1925) as an explanation for the spatial quantization demonstrated in the 1922 Stern-Gerlach experiment and for the observed complexities of atomic spectra, including the "anomalous" Zeeman effect. Pauli (1924) suggested that the hyperfine structure in atomic spectra arose from an interaction between orbital electrons and the small magnetic moment of a spinning nucleus.

We can gain a crude idea of the anticipated sizes for magnetic moments by visualizing a spherical object with mass M uniformly distributed through the volume, and charge Q uniformly distributed over the surface. If this rotates to have angular momentum L_s, the charge circulation creates a magnetic moment μ which is coaxial with and linearly proportional to L_s. As may readily be verified (Problem 5.13),

$$\mu = \gamma L_s = (5Q/6M)L_s \tag{5-103}$$

regardless of the radius of the sphere. The proportionality constant γ is the *gyromagnetic ratio* of the spinning object. When the required quantization $L_s = [\hbar^2 J(J + 1)]^{1/2}$ of angular momentum is taken into account, then the magnetic moment is

$$\mu = (5Q\hbar/6M)\sqrt{J(J + 1)} \tag{5-104}$$

The actual magnetic moment for an electron or a nucleus differs from the expectations of Equation 5-104 because of quantum electrodynamics and the complex distribution of mass and charge. The Bohr magneton and nuclear magneton:

$$\left.\begin{array}{l} \mu_B = (e\hbar/2m) = 9.274 \times 10^{-24} \text{ joule/tesla} \\ \mu_n = (e\hbar/2M_p) = 5.051 \times 10^{-27} \text{ joule/tesla} \end{array}\right\} \tag{5-105}$$

were cited at the beginning of Section 5.2 as the quantities in terms of which magnetic moments of spinning electrons and nuclei are customarily quoted.

THE SIMPLEST FORM FOR
ELECTRON SPIN RESONANCE

The ESR spectra of unpaired electrons in solids are often very complicated, and indeed the contribution of ESR to an improved knowledge of the physics of solids usually depends on these complexities. However, first let us consider the basic condition for ESR absorption. For this, we consider the spin of isolated electrons in a homogeneous environment.

An electron has angular momentum $L_s = \hbar\sqrt{3/4}$, for a spin quantum number choice $m_s = \pm\frac{1}{2}$. Part (a) of Figure 5-23 indicates the two possible angles of inclination between the spin axis and a magnetic field **B**. The corresponding magnetic moment resolved along the direction of **B** is usually written in the form

$$\mu = -g\mu_B m_s \qquad (5\text{-}106)$$

Here the dimensionless quantity g is the spectroscopic splitting factor, or g-factor. (It was noted in Section 5.2 that $g = 2.0023$ for an isolated free electron, to give a magnetic moment of almost exactly one Bohr magneton. However, the g-factor can be radically different from 2.0 in the environment of a solid.)

An unpaired electron with energy U_0 in the absence of a field has an interaction energy $\Delta U = -\mu B$ with a magnetic field. Then, as indicated in Figure 5-23(b), the electron can have either of the two energy levels

$$U = U_0 + \Delta U = U_0 - \mu B = U_0 \pm \tfrac{1}{2}g\mu_B B \qquad (5\text{-}107)$$

corresponding with the spin up/spin down possibilities of $m_s = \pm\frac{1}{2}$.

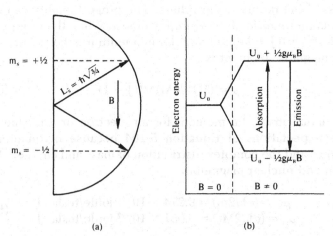

(a)   (b)

Figure 5-23 (a) The possible orientations of an electron's spin axis relative to the direction of a magnetic field. The resolved magnetic moment is proportional to m_s. (b) Splitting of an electronic energy level in a magnetic field. Transitions between the upper and lower levels satisfy both energy and angular momentum conservation if a photon of energy $g\mu_B B$ is absorbed or emitted.

The occupancies N_U and N_L for the upper and lower energy levels are related at thermal equilibrium by

$$(N_U/N_L)_{eq} = \exp\,(-g\mu_B B/k_0 T) \qquad (5\text{-}108)$$

Let ΔN_0 denote the equilibrium value for the difference $\Delta N = (N_L - N_U)$ of occupancies. Since the temperature of an ESR experiment is usually appreciably larger than $(g\mu_B B/k_0)$, then

$$\Delta N_0 \simeq N(g\mu_B B/2k_0 T) \qquad (5\text{-}109)$$

is usually quite small compared with the total spin density $N = (N_L + N_U)$. Nevertheless, ΔN can be perturbed to a value smaller than ΔN_0 by an external influence. For an ESR experiment, this comprises resonant absorption of radio frequency photons with energy

$$h\nu = \hbar\omega = g\mu_B B \qquad (5\text{-}110)$$

since these photons can satisfy both energy and angular momentum conservation in inducing transitions between the two levels of Equation 5-107.

As will be seen from Figure 5-24 and subsequent figures, ESR (and NMR) are usually observed by maintaining a fixed radio frequency ν while B is caused to scan through the range including resonance. The resonant combination of B (in teslas) and ν (in GHz) is

$$\left.\begin{array}{l} \nu = 14.00\ gB \\ g = 0.07145\ (\nu/B) \end{array}\right\} \qquad (5\text{-}111)$$

Thus for free electrons with g = 2.0023, ESR can be carried out with microwaves of frequency ~ 10 GHz (wavelength of 3 cm) in a field of some 0.4 T, or in a smaller field at a lower frequency.

The first reported ESR measurements were those of Zavoisky[25] with a radio frequency of a few MHz and a field of less than 0.01 T. This gave a barely resolvable resonance. Zavoisky subsequently obtained much better resolution by increasing ν and B, and Figure 5-24 gives an example of his later results. The advantages of high frequency/high field resonance were further demonstrated in early ESR work in the United States[26] and Britain,[27] and the microwave spectral range is almost universally used for resonance today.

Use of a high field/high frequency combination not only improves the resolution, but it also increases the absorption probability. Equation 5-109 implies the desirability of both a low temperature and a high frequency, since

$$\Delta N_0 \simeq N(g\mu_B B/2k_0 T) = N(h\nu/2k_0 T) \qquad (5\text{-}112)$$

[25] E. Zavoisky, J. Phys. USSR **9**, 211, 245 (1945).
[26] R. L. Cummerow and D. Halliday, Phys. Rev. **70**, 433 (1946).
[27] D. M. S. Bagguley and J. H. E. Griffiths, Nature **160**, 532 (1947).

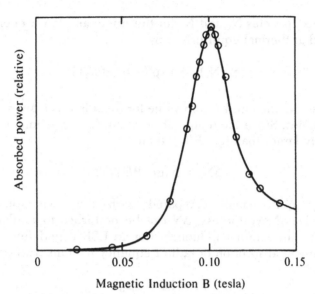

Figure 5–24 Electron spin resonance of the unpaired spins of Mn^{++} ions in $MnSO_4$ at room temperature, taken with $\nu = 2.75$ GHz. [Data of E. Zavoisky, J. Phys. USSR **10**, 197 (1946).] As is usual for resonance experiments, ν is kept constant while **B** is varied to scan through the resonance. The location of the absorption peak is consistent with Equation 5-108 for $g = 1.95$.

The approximate equality sign here reminds us that Equation 5-109 was derived for a temperature that is large compared with $(h\nu/2k_0)$. This is ordinarily true, for $(h\nu/2k_0) \simeq 0.2$K when $\nu \simeq 10$ GHz. While Equation 5-112 indicates that ΔN_0 is proportional to ν, other experimental considerations often make the ESR sensitivity more nearly proportional to ν^2. A modern highly sensitive ESR spectrometer can detect and study as few as 10^{11} electron spins.

Spin-Lattice and Spin-Spin Relaxation

Photons will induce both upward and downward transitions at rates proportional to N_L and N_U respectively. Thus the absorption of microwave power is dependent on the quantity $\Delta N = (N_L - N_U)$ remaining positive throughout an ESR experiment. Ideally, the microwave power input should be small enough so that ΔN remains essentially at the equilibrium value of Equation 5-112 throughout a sweep of the resonant range of B. How much power can be incident depends on the mechanisms for returning spins to the lower energy state, accompanied, of course, by an angular momentum transfer of $\hbar$.

The so-called *direct spin-lattice relaxation* process depends on spin-orbit coupling to connect the spin of an electron with the lattice vibrational spectrum of the solid. In this way, the electron relaxes to the lower-energy spin state and a long-wave acoustic mode phonon is created.[28] The time constant of this direct process is anticipated to vary approximately as $(1/T)$.

---

[28] If $\nu \sim 10$ GHz, the phonons are created with an energy of some 4×10^{-5} eV, and a wavelength of a few hundred interatomic spacings.

For some solids, effective competition to the direct process at higher temperatures is provided by the *Raman spin-lattice relaxation* process, whereby an existing phonon is inelastically scattered by an atom whose unpaired electron is in the upper energy state. The time constant for the Raman process varies as $(1/T)^5$.

The above combination of spin-lattice relaxation processes is characterized by a time constant called τ_1. Thus if ΔN is perturbed from ΔN_0, and allowed to return toward equilibrium after time t_0, we expect that

$$\Delta N = \Delta N_0 \{1 - A \exp[-(t - t_0)/\tau_1]\} \qquad (5\text{-}113)$$

Then τ_1 can be measured from the transient response of a set of spins to an abrupt termination of or reversal of a static magnetic field (as well as in an ESR experiment). Much pioneering work on this topic of *paramagnetic relaxation* was done[29] prior to the successful development of ESR.

In an ESR experiment, τ_1 is clearly connected with the rate at which microwave power can be absorbed and dissipated by a set of spins without collapsing ΔN to zero. It might be expected that τ_1 would also control the width of any ESR line. We shall see that this is so in some cases, but not in others.

A second important relaxation time, τ_2, is set by the probability of *spin-spin relaxation*. This is concerned with the influence (i.e., the local magnetic field contribution) exerted by one magnetic atom on others. $(1/\tau_2)$ is the rate constant for loss of phase coherence of a set of spins with an applied electromagnetic field; as a set of spins loses phase coherence, there is an increased fluctuation in the magnetic field that spins exert on each other. This causes spins to reach the resonance condition for a variety of values of B, and *can* be the principal reason for a resonance of finite width. The combined relaxation time

$$\tau_r = \tau_1 \tau_2/(\tau_1 + \tau_2) \qquad (5\text{-}114)$$

is determined by the shorter of τ_1 and τ_2.

Figure 5-25 plots the spin relaxation time τ_r as deduced from the width of ESR lines for electrons in small particles of sodium metal. The superimposed line for $\tau_r = 3 \times 10^{-8}/T$ suggests that τ_r is controlled by the *direct* spin-lattice relaxation mechanism over the observed range.

It is reasonable to expect that τ_1 alone will control the linewidth when a few magnetic atoms are dispersed in a matrix of nonmagnetic atoms, for then spin-spin correlations are minimized by the large separations. An increase in the density of magnetic atoms may then eventually permit τ_2 to be smaller than τ_1, and a broadened ESR line may then be seen. This does happen for many systems of magnetic atoms.

However, relaxation mechanisms do not dictate the linewidth for *some* systems of interacting spins on nearby atoms. When the electrostatic interaction energy of these electrons (as represented by J_e of Equa-

[29] Much of the early paramagnetic relaxation work was carried out in the Netherlands, and is summarized by C. J. Gorter, *Paramagnetic Relaxation* (Elsevier, 1947); also Physica **14**, 504 (1948).

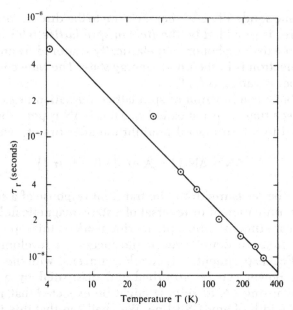

Figure 5-25 Temperature dependence of the relaxation time for electrons in microscopic particles of sodium metal. These particles have a diameter of 5 μm or less so that the radius will be less than the radiofrequency skin depth. τ_r is deduced from the ESR linewidth at each temperature, of which Figure 5-27 shows the 296K data. [From G. Feher and A. F. Kip, Phys. Rev. **98**, 337 (1955).]

tion 5-95) is large compared with the magnetic dipole-dipole energy, then a narrow ESR line is observed. The small width is said to be the beneficiary of *exchange narrowing*.[30] The DPPH radical illustrated by Figure 5-26 has a very narrow line because of this mechanism.

---

[30] J. H. Van Vleck, Phys. Rev. **74**, 1168 (1948).

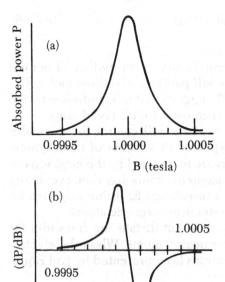

Figure 5-26 (a) The ESR absorption spectrum, and (b), the corresponding trace of the derivative of the line, for DPPH (solid 1,1-diphenyl-2-picryl-hydrazyl) at room temperature and $\nu = 28$ GHz. The very narrow linewidth, $\Delta B_{1/2} \simeq 3 \times 10^{-4}$ T, is a consequence of exchange narrowing. DPPH is often used for ESR spectrometer calibration since the line occurs at $g = 2.0036$, very close to the free electron value.

ESR Linewidth and Line Shape

The position of the ESR line (ratio of B to ν, or effective g-factor) is one useful piece of information in an experiment, but usually much more information is desired than this. Information about the environment of the spins can come from the width and shape of the resonant line (as well as from the fine structure and hyperfine structure complications to be mentioned again later). The last few paragraphs of the preceding subsection have brought us to a concern with linewidth.

ESR measurements are usually made by scanning B (at constant ν) through the range of interest. It is feasible to plot absorbed power P versus B, but greater sensitivity is possible with a phase-sensitive detection method which plots (dP/dB). Figure 5-26 compares traces of P and (dP/dB) for a simple absorption line, and either trace can be used to determine the location B_r of the absorption maximum and the separation $\Delta B_{1/2}$ of half-power points.

The dynamics of stimulated upward and downward transitions for a set of spins when B is swept linearly with time through the resonance condition were considered by Bloch.[31] His discussion was aimed at the conditions for NMR, but is equally valid in the present context. Bloch's result shows that when the incident power level is small enough, and sweep rate fast enough, so that ΔN is barely perturbed from ΔN_0 throughout the sweep, then a Lorentzian line shape

$$P = P_{max}[1 + (B - B_r)^2/\Gamma^2]^{-1} \qquad (5\text{-}115)$$

may be anticipated. The maximum of P and the zero of (dP/dB) occur for $B = B_r$. The half-power points of P are separated by $\Delta B_{1/2} = 2\Gamma$, while the positive and negative peaks of (dP/dB) are separated by $\Delta B_{pp} = (2/\sqrt{3})\Gamma = (\Delta B_{1/2}/\sqrt{3})$. As examined in Problem 5.14, the linewidth in teslas can be converted to an equivalent linewidth in frequency which will often be $(1/\tau_r)$. The result for the spin relaxation lifetime is then

$$\tau_r \simeq 1.1 \times 10^{-11}/g\Gamma \quad \text{seconds} \qquad (5\text{-}116)$$

with Γ expressed in teslas.[32]

When ΔN is forced well away from ΔN_0 during the sweep of resonance, the line shape is apt to be changed toward the Gaussian form

$$P = P_{max} \exp[-(B - B_r)^2/\Gamma^2 \ln2] \qquad (5\text{-}117)$$

This is still written in such a fashion that the half-power points are separated by $\Delta B_{1/2} = 2\Gamma$, but the relationship of Γ to τ_r is not quite as expressed in Equation 5-116, and ΔB_{pp} is now a larger fraction of $\Delta B_{1/2}$.

Figure 5-27 compares an ESR derivative trace with the Lorentzian

---

[31] F. Bloch, Phys. Rev. **70**, 460 (1946).

[32] This is a good time to remind the reader that the literature of ESR and NMR is virtually all written in cgs units, with the terms "gauss" and "oersted" applied indiscriminately for the field strength.

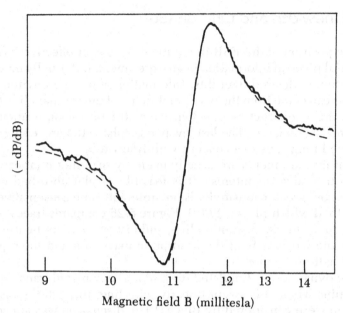

Figure 5–27 Derivative curve for the ESR line of small sodium particles (diameter less than 5 μm), measured at 296K with $\nu = 320$ MHz. [Data of G. Feher and A. F. Kip, Phys. Rev. **98**, 337 (1955).] The dashed curve has the Lorentzian form, the derivative of Equation 5-115. The line shape and corresponding lifetime are examined in Problem 5.14.

form, and Problem 5.14 asks the reader to find out how different this is from a Gaussian shape. The relaxation time deduced from the linewidth in Figure 5-27 can be compared directly with the set of data points for the same metallic sodium sample in Figure 5-25.

COMPLICATIONS IN ESR SPECTRA

So far the simplest type of ESR line has been illustrated, and some complications should now be mentioned which affect many (indeed, most) experiments. Some of the complications are helpful in using ESR to probe the structure of a solid or the location and symmetry of a defect.

Fine Structure when S > 1/2

One complication arises when a magnetically active atom or ion has more than one unpaired electron, so that the total spin quantum number for the atom is greater than 1/2. [This is the case, for example, with the Mn^{++} ions of Figure 5-24. Each ion has five unpaired 3d electrons; granted that the orbital part of the angular momentum is "quenched" for these electrons, as was concluded in connection with Figure 5-13, we must still reckon that S = 5/2 for Mn^{++}.] The spin system of the atom or ion must be capable of alignment with a magnetic field to have the $(2S + 1)$ values $m_s = S, (S - 1), \ldots, (1 - S), -S$ for the magnetic

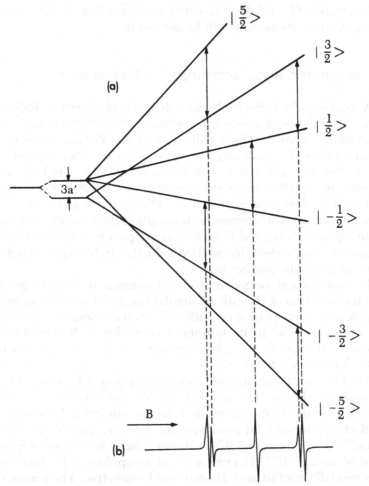

Figure 5–28 (a) Energy level diagram as a function of field B for a 3d⁵ ion, in a crystal field of octahedral symmetry and with B parallel to a major axis of the octahedron. (b) The allowed ESR spectrum when hν >> 3a'. From J. E. Wertz and J. R. Bolton, *Electron Spin Resonance* (McGraw-Hill, 1972).

quantum number. A magnetic field now dictates a different energy for each value of m_s, and the two-level picture of Figure 5-23(b) must be generalized to a $(2S + 1)$-level picture, as exemplified by Figure 5-28(a) for $S = 5/2$. The $(2S + 1)$ levels are usually unequally spaced in energy for a given B and, as illustrated in the figure, usually have the degeneracy for $B = 0$ partly lifted by asymmetry of the crystal field.

The usual selection rule[33] for photon-stimulated transitions among the levels is $\Delta m_s = \pm 1$. For a given microwave frequency this corresponds with the vertical arrows drawn in Figure 5-28(a) at five different values of B. The resulting ESR spectrum sketched in derivative form in Figure 5-28(b) has a *fine structure* of five lines, and in general 2S lines

---

[33] The selection rule is a consequence of the angular momentum ℏ of the photon. Transitions with $\Delta m_s = \pm 2$ are sometimes observed, as for example in the triplet state of naphthalene [J. H. van der Waals and M. S. de Groot, Molecular Phys. **2**, 333 (1959)].

are anticipated. Should the linewidths be wider than the fine structure spacing, a single broad line will be observed.

Ferromagnetic and Ferrimagnetic Resonance

A solid for which the exchange integral J_e of Equation 5-95 is positive has the bulk spontaneous magnetization of the ferromagnetic state at temperatures below the Curie point $T_c \sim (J_e/k_0)$. For a temperature substantially below T_c, relatively few spin waves are excited, and essentially all the unpaired spins of an entire sample are coupled together. This may be regarded as the extreme case of a set of coupled spins (as discussed for a few spins on one atom in the preceding subsection). The magnetic moment of a ferromagnetic sample is the sum of moments from all participating atoms, and the resolved component of this parallel to the direction of some externally applied field H_0 can be represented by an enormous quantum number m_s.

The *total* magnetic moment of a ferromagnetic sample precesses about the direction of $\mathbf{H_0}$ with an angular frequency $\omega = 2\pi\nu$ dependent on H_0. A microwave photon can suffer the strong absorption of *ferromagnetic resonance*[34] when its frequency coincides with this precession frequency. Absorption of a photon changes m_s in accordance with the usual selection rule $\Delta m_s = \pm 1$.

The previous paragraph was couched in terms of $\mathbf{H_0}$, since it is $\mathbf{H_0}$ or $\mathbf{B_0} = \mu_0 \mathbf{H_0}$ which can be measured outside the sample. The effective magnetic field inside the sample is different, but it should be noted that the effective internal field cannot include the very large Weiss "molecular field" $\mathbf{H_e}$ of Equation 5-87, or ferromagnetic resonance frequencies would be around 10^{13} Hz even without an applied field. Since this is experimentally not the case, $\mathbf{H_e}$ must not be effective. The reason is that the R.F. photons used have a wavelength that is large compared with the sample size, and act in phase on all spins together. Then there is never a finite angle between one spin and its neighbors, and $\mathbf{H_e}$ is unable to exert a torque.

What, then, is the effective field to be used in the resonance condition? Only for a spherical sample does $h\nu = g\mu_B B_0$ yield g-values which sound at all like the free electron value. As shown by Kittel,[35] this discrepancy between the applied and effective fields for most sample shapes arises from the strong demagnetizing character of a ferromagnetic sample. In general,

$$h\nu = g\mu_B[(B_0 + \alpha\mu_0 M)(B_0 + \beta\mu_0 M)]^{1/2} \qquad (5\text{-}118)$$

where the numbers α and β depend on the field orientation and sample shape; $\alpha = \beta = 0$ for the simple geometry of a spherical sample.

Ferromagnetic resonance data tend to yield g values slightly larger than the free electron value once Equation 5-118 is applied with values

[34] First reported by J. H. E. Griffiths, Nature **158**, 670 (1946).
[35] C. Kittel, Phys. Rev. **71**, 270 (1947).

for α and β appropriate to the sample and experimental geometry. Thus $g = 2.10$, 2.18, and 2.21 respectively for the 3d transition group of ferromagnetic elements iron, cobalt, and nickel. A comparable resonance is found for *ferrimagnetic* compounds, and $g = 2.11$ for yttrium iron garnet ($Y_3Fe_5O_{12}$) while $g = 2.18$ for nickel ferrite ($NiFe_2O_4$).

The linewidth in ferromagnetic or ferrimagnetic resonance depends in a complicated way on the sample geometry, the temperature, and the character of the particular material. Figure 5-29 shows a relatively broad resonance line for the ferrimagnetic solid $NiFe_2O_4$, but lines as narrow as 10^{-4} T have been reported for some solids. The theory of linewidth in ferromagnetic solids has not yet been fully worked out.[36]

The large electrical conductivity of metallic ferromagnetic elements and alloys creates an experimental problem by virtue of the very small penetration depth (skin depth) of microwave radiation. This problem does not arise with a nonconducting ferromagnetic or ferrimagnetic compound, as exemplified by the sample of Figure 5-29.

Antiferromagnetic Resonance

Consider a solid which is paramagnetic at high temperature and which assumes antiferromagnetic ordering of its spins below a Néel temperature T_N. Early ESR experimenters observed an ESR resonance above T_N, but nothing below. The puzzle was solved when Keffer and Kittel[37] showed that *antiferromagnetic resonance* would require absorp-

---

[36] See, for example, R. E. Prange and V. Korenman, J. Magn. Resonance **6**, 274 (1972).
[37] F. Keffer and C. Kittel, Phys. Rev. **85**, 329 (1952).

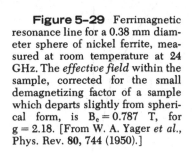

Figure 5–29 Ferrimagnetic resonance line for a 0.38 mm diameter sphere of nickel ferrite, measured at room temperature at 24 GHz. The *effective field* within the sample, corrected for the small demagnetizing factor of a sample which departs slightly from spherical form, is $B_e = 0.787$ T, for $g = 2.18$. [From W. A. Yager *et al.*, Phys. Rev. **80**, 744 (1950).]

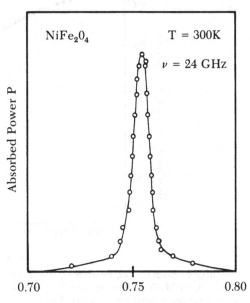

NiFe$_2$O$_4$ T = 300K

ν = 24 GHz

Absorbed Power P

0.70   0.75   0.80

Applied Magnetic Field B (T)

tion of considerably more energetic photons (ν in the range of 10^{11} to 10^{12} Hz).

The two factors taken into account by Keffer and Kittel were:

(a) The large energy stored in an antiferromagnetic solid as a consequence of magnetic anisotropy. This can be represented by a parameter B_A.

(b) The molecular field (exchange field) exerted by each magnetic sub-lattice on the other. In contrast to a ferromagnetic material, this exchange field makes a non-zero (indeed, overwhelming) contribution B_E to the effective field.

As a consequence, the antiferromagnetic resonance condition is

$$h\nu = g\mu_B\{[B_A(2B_E + B_A)]^{1/2} \pm (1 - \alpha)B_{ext}\} \qquad (5\text{-}119)$$

when a field B_{ext} is externally applied. The number α is concerned with the demagnetization factor appropriate to an external field, and declines to zero as the temperature approaches T_N.

Figure 5-30 illustrates how the resonant frequency appropriate to zero applied field varies with temperature for MnF_2. The frequency for the lowest temperature corresponds with $B_A = 0.88$ T and $B_E = 56$ T, so that $[B_A(2B_E + B_A)]^{1/2}$ is almost 10 teslas. This will indicate that a very powerful magnet must be used if it is desired to modify the resonant frequency with an external field. With the superconducting solenoids now available, it has become possible to use a large B_{ext} in the direction appropriate to move the resonant frequency down into the conventional

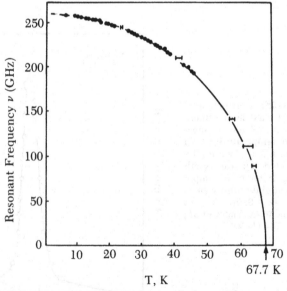

Figure 5-30 Temperature dependence of the antiferromagnetic resonance frequency for MnF_2 with no applied magnetic field. Points actually measured in a finite applied field have been converted via Equation 5-119 to the frequency appropriate for zero field. [After F. M. Johnson and A. H. Nethercot, Phys. Rev. 114, 705 (1959).]

microwave range[38] (taking advantage of the negative sign option in Equation 5-119). This was not done, however, for the earlier work illustrated in Figure 5-30. This pioneering work in antiferromagnetic resonance was carried out with millimeter waves, using a small external field up to 0.5 T to sweep through the resonance line at each temperature. The data points so obtained were then converted via Equation 5-119 to the equivalent resonance frequency for zero applied field, which contains information about B_E and B_A as functions of temperature. The curve of Figure 5-30 is related to, but not identical with, the Brillouin function curve for the magnetization of each opposing magnetic sub-lattice. Both the magnetization and the antiferromagnetic resonance frequency at zero field tend to zero as $T_N = 67.7K$ is approached from below.

ESR in an Electron-Nuclear System

The topic to be discussed next does not have the complications of correlations between spins in a large assembly of atoms. Instead, we shall have to conclude that there can be great complexity in the relation between the electronic spin in an atom and the spin of the nucleus.

The angular momentum of a nucleus is conventionally written as $L_s = [\hbar^2 I(I + 1)]^{1/2}$ in terms of a spin quantum number I. This number takes a variety of values for different isotopes, of which Table 5-5 lists a

---

[38] H. W. de Wijn et al., Phys. Rev. **B.8**, 299 (1973).

TABLE 5-5   SOME EXAMPLES OF NUCLEAR MOMENTS°

Isotope	Spin Quantum Number I	Nuclear Magnetic Moment μ (in units of μ_n)	Nuclear g-Factor $g_N = (\mu/I\mu_n)$	Electric Quadrupole Moment Q (units of $10^{-28}m^2$)
$^1n^0$	1/2	−1.913	−3.826	0
^1H	1/2	2.793	5.586	0
^2H	1	0.857	0.857	0.003
^7Li	3/2	3.256	2.171	−0.042
^9Be	3/2	−1.177	−0.785	0.049
^{14}N	1	0.404	0.404	0.071
^{19}F	1/2	2.627	5.254	0
^{23}Na	3/2	2.216	1.473	0.1
^{25}Mg	5/2	−0.855	−0.342	0.22
^{27}Al	5/2	3.639	1.456	0.149
^{31}P	1/2	1.131	2.262	0
^{35}Cl	3/2	0.821	0.547	−0.080
^{45}Sc	7/2	4.749	1.357	0.22
^{50}V	6	3.341	0.557	−
^{55}Mn	5/2	3.461	1.384	0.3

° A much more extensive table is given by E. R. Andrew, *Nuclear Magnetic Resonance* (Cambridge, 1958). I = 0, so that $\mu = 0$, for any nucleus with an even number of protons and an even number of neutrons.

few examples. There is zero spin $(I = 0)$ for any nucleus containing an even number of both protons and neutrons, and this includes such common nuclei as ^{12}C, ^{16}O, ^{24}Mg, ^{28}Si, and ^{32}S.

Any nucleus with non-zero spin also has a finite magnetic moment μ, and values for this property are similarly listed in Table 5-5. Nuclear magnetic moments are usually expressed as a multiple of the nuclear magneton μ_n of Equation 5-105, and this is in all cases a non-integral multiple. The nuclear g-factor $g_N = (\mu/I\mu_n)$ is useful in expressing the resolved component of nuclear magnetic moment along a field direction as $g_N\mu_n m_I$, where the magnetic quantum number m_I can have any of the $(2I + 1)$ values I, $(I - 1)$, . . . , $(1 - I)$, $-I$. This nuclear magnetic moment has several consequences that we must take into account.

The first and most obvious is that a nucleus of non-zero spin has $(2I + 1)$ possible angles of inclination to the direction of an applied magnetic field. A nucleus with energy u_0 in the absence of a field will have one of the set of energies $u = u_0 + \Delta u = u_0 + g_N\mu_n m_I B$ when the field is turned on. The magnetic quantum number can change by one unit $(\Delta m_I = \pm 1$ corresponds to a change of $\hbar$ in resolved angular momentum) if a photon of energy $\hbar\omega = h\nu = g_N\mu_n B$ is absorbed. If this sounds similar to the introduction to the story of ESR, then it should; for this is the NMR process to be examined later in this section. Since $g_N\mu_n$ is 10^3 to 10^4 times smaller than the electronic $g\mu_B$, the photons used for NMR are much smaller in frequency than those employed in ESR experiments.

The existence of a nuclear magnetic moment does affect the energy of unpaired electrons of the same atom in two ways, known respectively as the *Fermi contact interaction* and the *dipolar interaction*. We shall shortly examine the mechanism of these two interactions, but it is useful first to note that their consequences can be expressed in terms of the effective fields ΔB_{fc} and ΔB_{di} that they superimpose on any external field in affecting the behavior of any unpaired electrons on the atom. The quantities ΔB_{fc} and ΔB_{di} depend on m_I, and may still exist at virtually full strength when there is zero applied field. Thus the energy of a set of unpaired electrons on an atom can have up to $(2I + 1)$ values in the absence of an applied field. Application of an external field completely removes the degeneracy, and then there are $(2I + 1)(2S + 1)$ different energies. Resonant absorption of photons in an ESR experiment can occur for $2S(2I + 1)$ different field strengths at a given photon energy, subject to the angular momentum rules $\Delta m_s = \pm 1$, $\Delta m_I = 0$. The second of these selection rules is necessary since, while flipping a nuclear spin simultaneously with an electron spin takes very little additional energy, it would be inconsistent with the angular momentum $\hbar$ of the absorbed photon (except for the transitions $\Delta m_s = +1$, $\Delta m_I = -2$, which some writers refer to as "forbidden" transitions).

Figure 5-31 illustrates the energy levels as a function of magnetic field for the simplest possible electron spin/nuclear spin combination of $S = \frac{1}{2}$, $I = \frac{1}{2}$. Several of the books cited in the bibliography at the end of this chapter solve the matrix problem for more complicated sets of spins. Even for the case of Figure 5-31, the single ESR line of our original simplified discussion must be replaced with the two lines marked as allowed transitions for a given phtoton energy. The difference $(B_2 - B_1)$

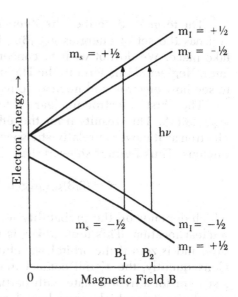

Figure 5-31 Electron energy versus magnetic field for a system of electron spin $S = \frac{1}{2}$, and nuclear spin $I = \frac{1}{2}$. The arrows mark two transitions of equal photon energy, consistent with the selection rule $\Delta m_s = \pm 1$, $\Delta m_I = 0$. This problem was first solved [G. Breit and I. Rabi, Phys. Rev. 38, 2082 (1931)] in connection with atomic spectroscopy, and the resulting levels are sometimes referred to as the Breit-Rabi levels. It will be noted that there is appreciable curvature of the lines at low fields, when there is mixing of m_s, m_I states.

in magnetic field between one resonance line and the other is generally referred to as the *hyperfine splitting*. This name suggests that the hyperfine splitting must always be very small, smaller than the splittings noted as "fine structure" in the discussion of Figure 5-28. However, the hyperfine splitting is small for some atoms (10^{-3} tesla or less) but quite large for other atoms. In atomic hydrogen, the nucleus causes a splitting of 0.0508 tesla, the largest of any atom.

The Spin Hamiltonian

Throughout most of this book, the formalism of quantum mechanics has been avoided as far as possible. However, it is difficult to sum up the interactions of spins on electrons and nuclei without talking about the spin Hamiltonian $\mathcal{H}$. The wavefunction ψ of an electron (or set of unpaired electrons in an atom) must be consistent with the Schrödinger equation for an appropriate Hamiltonian operator, and the eigenvalues of this equation are the permitted energies. For spin purposes, we can omit the large terms in the Hamiltonian concerned with orbital energy and spin-orbit splitting, and write only the terms relating to electron spin, nuclear spin, and their interaction. Thus

$$\mathcal{H} = \mathcal{H}_1 + \mathcal{H}_2 + \mathcal{H}_3 + \mathcal{H}_4$$
$$= g\mu_B \mathbf{B} \cdot \mathbf{S} + g_N \mu_n \mathbf{B} \cdot \mathbf{I} + \mathcal{A} \mathbf{I} \cdot \mathbf{S} + \mathcal{B}[\mathbf{I} \cdot \mathbf{S} - (3/r^2)(\mathbf{I} \cdot \mathbf{r})(\mathbf{S} \cdot \mathbf{r})]$$
$$(5\text{-}120)$$

The first term on the right of Equation 5-120 is the interaction of the external field $\mathbf{B}$ with electron spin, and this leads to the Zeeman splitting of levels which was illustrated in Figures 5-23 and 5-28. When $\mathcal{H}_1$ is the *only* term in the spin Hamiltonian, the energies are the set $g\mu_B B m_s$. Thus we know what $\mathcal{H}_1$ does by itself.

The term $\mathcal{H}_2$ describes the Zeeman splitting of nuclear spin states, and has the set of energies $g_N \mu_n B m_I$ between which NMR transitions take place. We do not wish to concern ourselves with this term until later. Neglect of this term in the Hamiltonian does not affect our ability to see how electron and nuclear spins interact with each other.

The first electron-nuclear interaction term is the operator $\mathcal{H}_3 = \mathcal{A} \mathbf{I} \cdot \mathbf{S}$. This results from the effect of nuclear magnetism on an electron as it moves at relativistic speed in the immediate vicinity of the nucleus. Thus Fermi[39] showed that

$$\mathcal{A} = (2/3)\mu_0 g \mu_B g_N \mu_n |\psi(0)|^2 \quad \text{joule} \quad (5\text{-}121)$$

which depends on the probability $|\psi(0)|^2$ of finding the electron at the nuclear location. This term $\mathcal{A} \mathbf{I} \cdot \mathbf{S}$ is called the *Fermi contact* interaction, and is zero if the orbital wavefunction has a node at the nucleus. Consequently, the Fermi contact interaction is important only for an s-state or for a hybrid state with partially s-like character. It can be ignored for a state which is purely p, d, or f in character. In a similar simplification, the term $\mathcal{H}_4$ is zero for any pure s-state.

When $\mathcal{H}_1$ and $\mathcal{H}_3$ are the only terms to be considered in the spin Hamiltonian, then *to a first order approximation* the electron energy is

$$U = U_0 + g\mu_B B m_s + \mathcal{A} m_s m_I \quad (5\text{-}122)$$

This can be regarded as

$$U = U_0 + g\mu_B m_s (B + \Delta B_{fc}) \quad (5\text{-}123)$$

where ΔB_{fc} is the magnetic field equivalent of the Fermi contact interaction. From Equations 5-121 through 5-123 we can see that

$$\Delta B_{fc} = (\mathcal{A}/g\mu_B) m_I = (2/3)\mu_0 g_N \mu_n |\psi(0)|^2 m_I \quad (5\text{-}124)$$

A change of m_I by unity should then cause the required external field for ESR to change by $(\mathcal{A}/g\mu_B)$. This is the quantity called the *hyperfine splitting*, which was shown as $(B_2 - B_1)$ in Figure 5-31. As examined in Problem 5.16, the hyperfine splitting in teslas can readily be calculated from a knowledge of g_N and $\psi(0)$.

Figures 5-32 and 5-33 illustrate the hyperfine splitting for localized electrons in one semiconductor and localized holes in another. The material of Figure 5-32 is silicon $(I = 0)$ containing substitutional ^{31}P donors, atoms for which $I = \frac{1}{2}$. A Group V donor in silicon uses four outer shell electrons in satisfying the tetrahedral bonding requirements of the lattice, and one outer electron remains, moving in an s-like configuration when bound to its parent nucleus. The weakly-doped sample (b) shows the ESR hyperfine doublet, with a splitting from which $\psi(0)$ can be calculated. Note that the hyperfine spectrum disappears for the strongly

[39] E. Fermi, Z. Physik. **60**, 320 (1930).

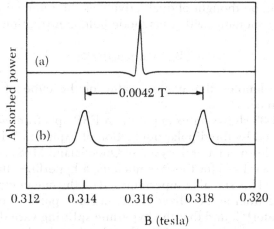

Figure 5-32 ESR spectra of two samples of silicon doped with phosphorus shallow donors, after G. Feher, Phys. Rev. 114, 1219 (1959). The spectra are observed at 8.845 GHz, at a temperature of 1.25K. Sample (a) has $N_d = 3 \times 10^{24}$ m^{-3} of phosphorus donors, and the impurity activation energy has collapsed to zero in a manner comparable to that shown in Figure 4-33. Sample (b) is much less strongly doped, with $N_d = 7 \times 10^{21}$ m^{-3}, and each donor electron is bound in a 1s type of orbit around the parent ^{31}P nucleus. Thus sample (b) shows hyperfine splitting due to the Fermi contact interaction.

doped sample (a), in which the electrons are uncorrelated with their parent nuclei. The single line seen for sample (a) is exchange-narrowed.

The Fermi contact interaction for s-states is *isotropic,* and rotation of a sample showing this interaction will not change the hyperfine structure. This is not so for the dipolar interaction of a spinning nucleus with electronic p-, d-, or f-states, and the latter interaction creates a hyperfine structure which changes with orientation. The quantity $\mathscr{B}$ in Equation 5-120 which relates the nuclear dipole to the electronic dipole is

$$\mathscr{B} = -\mu_0 g \mu_B g_N \mu_n / r^3 \qquad (5\text{-}125)$$

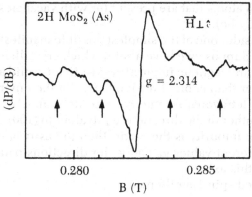

Figure 5-33 Hyperfine structure in the ESR spectrum of a natural crystal of MoS$_2$, containing ^{75}As acceptors. [From R. S. Title and M. W. Shafer, Phys. Rev. **B.8**, 615 (1973).] The measurements were made at 77K, with photons of frequency 9 GHz. The central line is absorption by non-localized holes in some portions of the crystal for which N_a is large, while the four lines split by the dipolar interaction (arrowed) arise from holes localized on the ^{75}As acceptors. The d-like character of the bound hole wavefunction is demonstrated by anisotropy of the hyperfine structure.

The result can be thought of qualitatively as the interaction of the electron magnetic moment with a magnetic field contribution

$$\Delta B_{di} = \mu_0 g_N \mu_n / <r^3> \tag{5-126}$$

where $<r^3>$ denotes the average value of the cube of the electron-nuclear distance.

Figure 5-33 shows an example of an ESR spectrum with hyperfine splitting caused by the dipolar interaction of Equation 5-126. The spectrum is for holes in a natural crystal of Queensland MoS_2 containing ^{75}As acceptors. Since $I = \frac{3}{2}$ for the ^{75}As nucleus, a hyperfine structure of four lines is seen. (This signal is superimposed on the one central "free hole" line resulting from nonlocalized holes in some portions of the crystal.) Title and Shafer[40] found that the hyperfine splitting varied with orientation in a manner consistent with a d_{z^2} orbital, and values of $<r^3>$ could be obtained from the magnitude of ΔB_{di}.

In a great many instances, electrons examined by ESR have a hybrid orbital character, and the hyperfine structure has the complications of both the $\mathscr{A}$ and $\mathscr{B}$ terms in the Hamiltonian and the energy spectrum. These complications are welcomed by the ESR experimentalist, for the angular dependence of hyperfine splitting can reveal just where an atom with unpaired spins is located in a crystal lattice.

Crystal Field Splitting

The *fine structure* splitting of an ESR line because of an anisotropic crystal field was mentioned very briefly in connection with Figure 5-28, but the subject deserves rather more attention. A major fraction of current experimental work in ESR is concerned with the effects of crystal fields, since the anisotropy of the fine structure in an ESR spectrum can be correlated with the locations and bonding arrangements of electrons associated with atoms that are present (or with vacancies where an atom should normally be).

Let us consider one of the simplest possible manifestations of crystal field splitting. Suppose we have a solid which crystallizes in a cubic lattice, and that paramagnetic impurities exist in this solid at locations with symmetry lower than cubic. For simplicity, let the environment of each impurity have tetragonal symmetry, with its symmetry axis coincident with one or another of the directions equivalent to (100). If the symmetry axis for a given impurity is the z-axis, then the associated electron spin system has different values $g_\parallel$ and $g_\perp$ for directions parallel to and perpendicular to this axis.

A simplified spin Hamiltonian

$$\mathscr{H} = g_\parallel \mu_B B_z S_z + g_\perp \mu_B (B_x S_x + B_y S_y) + D S_z^2 \tag{5-127}$$

can then be written, where D is the zero-field splitting. For a center with

[40] R. S. Title and M. W. Shafer, Phys. Rev. **B.8**, 615 (1973).

two unpaired electrons, $S = 1$. The energies U permitted by Equation 5-127 for $S = 1$ are given by the roots of the cubic equation

$$U(U - D)^2 - U(g_{\parallel}{}^2\mu_B{}^2B^2 \cos^2\theta) - (U - D)(g_{\perp}{}^2\mu_B{}^2B^2 \sin^2\theta) = 0 \quad (5\text{-}128)$$

where θ is the angle between the **B** field and the z-axis.

Thus if the crystal is aligned so that **B** is directed along a (100) crystal direction, then one-third of all the centers in the crystal will have their z-axes parallel to the field, and the other two-thirds will have z-axes perpendicular to the field. An ESR experiment will have a spectrum showing a superposition of "parallel" and "perpendicular" spectra, the latter having twice the amplitude of the former. [If **B** is not parallel to a (100) direction, the ESR spectrum is a more complicated sum of the solutions for θ and for $(\pi/2 - \theta)$.]

Figure 5-34 sketches the $B \perp$ z-axis solutions of Equation 5-128 as functions of B, and it is seen that the "perpendicular" component of the ESR spectrum for $h\nu > D$ consists of two lines. These will be quite widely spaced in magnetic field unless measurements are made at a microwave frequency much larger than (D/h). Problem 5.17 asks that the solutions for the "perpendicular" and "parallel" situations be written out, and it will be found from this problem that the "parallel" solution provides two more ESR lines when $h\nu > D$. The quantities D, $g_{\parallel}$, and $g_{\perp}$ can be determined from the magnetic field values of the various lines for a given photon energy, and it is worth noting that this can be carried out *either* for $h\nu > D$ or for $h\nu < D$. [See Problem 5.18 in this regard.]

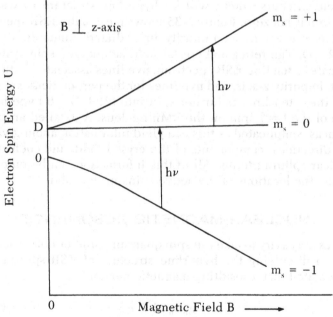

Figure 5-34 Solutions of Equation 5-128 for the three possible energies as a function of magnetic field when **B** is perpendicular to the tetragonal z-axis of the crystal field experienced by an electron spin system. The arrows show the two fields at which allowed transitions can take place for a photon energy $h\nu > D$. Problem 5.17 examines this solution and the solution for **B** parallel to the c-axis, while Problem 5.18 compares the ESR information when $h\nu$ is respectively larger than or smaller than D.

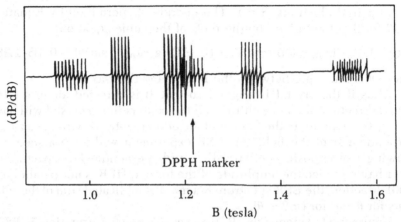

Figure 5-35 Seven of the ten major lines in the ESR spectrum of Mn^{++} ions in $NH_4Cl \cdot 2H_2O$, as measured at 35 GHz by T. J. Seed, J. Chem. Phys. **41**, 1486 (1964). Each line is split into six hyperfine components by the nuclear spin of ^{55}Mn. One additional low-field line at 0.56 T is omitted from this figure, and Seed was unable to observe the two highest field lines. The positions of these high field lines are predicted by the analysis of A. Forman and J. A. van Wyck [Canad. J. Phys. **45**, 3381 (1967)], and they have more recently been successfully measured [J. A. Kennewell, J. R. Pilbrow, and J. H. Price, Phys. Lett. **27A**, 228 (1968)] with a magnetic field of 1.85 T. Note in the figure that the two lines at about 1.2 T are in an approximate intensity ratio of 2:1, as expected for "perpendicular" and "parallel" conformations.

The situation of Equation 5-128 is much simpler than often obtains with paramagnetic impurities in crystals. The effective field is apt to be less symmetrical than tetragonal, the spectrum is more complicated if $S > 1$, and of course there will be hyperfine structure as well if the nuclear spin is not zero. Figure 5-35 shows most of the ESR spectrum for Mn^{++} ions present as an impurity in hydrated ammonium chloride, $NH_4Cl \cdot 2H_2O$. The tetragonal crystal field acting on a spin system with $S = \frac{5}{2}$ creates a ten line ESR spectrum, five lines associated with **B** parallel to the impurity z-axis, and five lines for the perpendicular alignment. Each of these ten lines is further split into $(2I + 1) = 6$ hyperfine lines because of the $I = \frac{5}{2}$ spin on the ^{55}Mn nucleus. A detailed analysis of a spectrum as complicated as this can yield information about g-values for various directions, components of the crystal field, and coefficients for the nuclear splitting terms. All of this information can be employed to determine the locations of the active spins in the solid.

NUCLEAR MAGNETIC RESONANCE

It was necessary to take the spin quantum number I of a nucleus into account in discussing the hyperfine structure of ESR spectra. At that time, we noted that a resulting magnetic moment

$$\mu = g_N\mu_n \sqrt{I(I + 1)} \tag{5-129}$$

can be expressed in terms of I, the nuclear magneton μ_n of Equation 5-105, and the nuclear g-factor g_N. Table 5-5 shows values for several nuclei of I, g_N, and the maximum resolvable magnetic moment $g_N\mu_nI$ along a field direction.

The term $\mathcal{H}_2$ in the Hamiltonian operator of Equation 5-120 has already been noted as describing the interaction of a magnetic field **B** with the nuclear magnetic moment. A nucleus in field **B** has one of the set of energies

$$u = u_0 - g_N \mu_n B m_I \qquad (5\text{-}130)$$

compared with energy u_0 at zero field. Here, m_I can have any of the values I, $(I - 1)$, . . . , $(1 - I)$, $-I$. These correspond to the $(2I + 1)$ possible angles between the field direction and the nuclear dipole axis.

A transition from one of the above levels to another may be induced by a photon provided that $\Delta m_I = \pm 1$ (the angular momentum conservation condition) and that

$$\hbar \omega = h\nu = g_N \mu_n B \qquad (5\text{-}131)$$

This is nuclear magnetic resonance (NMR), a much lower frequency phenomenon than ESR because of the small size of nuclear magnetic moments. Expressing B in tesla, and ν in MHz, we have

$$\nu = 7.623 \, g_N B \qquad \text{(MHz)} \qquad (5\text{-}132)$$

so that a frequency of a few MHz is very convenient for NMR work. The NMR response of water is often used for calibration of magnetic fields, using the relationship $B(\text{tesla}) = [\nu(\text{MHz})/42.58]$ for protons.

As with ESR, measurements of NMR are usually made at a fixed and highly stable radio-frequency, and the magnetic field is varied to pass through resonance. Typically, B is the sum of a steady field from a large electromagnet and a small component $\Delta B \sin(\omega_0 t)$ from an auxiliary coil which causes the total field to pass through the actual resonance point (ω_0 / π) times per second. Phase sensitive detection then permits a display either of absorbed power or of (dP/dB).

At thermal equilibrium, the ratio of occupancies of nuclear spin states separated by $\Delta m_I = 1$ is

$$\exp(-g_N \mu_n B/k_0 T) = \exp(-h\nu/k_0 T) \simeq [1 - (h\nu/k_0 T)] \quad (5\text{-}133)$$

and this is apt to be very close to unity unless measurements are made at an extremely low temperature. For room temperature, and $\nu \simeq 60$ MHz, then $(h\nu/k_0 T) \simeq 10^{-5}$. The small fractional difference between the populations of the various spin states means that the nuclear spin system can easily be saturated if excessive R.F. power is incident.

Nuclear magnetic moments were first measured in the 1930's through refinements of the Stern-Gerlach experiment, and Lasarew and Schubnikow[8] demonstrated the proton paramagnetism of molecular hydrogen in 1937. Precise measurement of nuclear moments became possible with the resonance absorption technique introduced to molecular beams by Rabi,[41] and further refined by Ramsay.[42] These were the

[41] I. Rabi, S. Millman, P. Kusch and J. Zacharias, Phys. Rev. **55**, 526 (1939).
[42] N. F. Ramsay, *Molecular Beams* (Oxford, 1955).

first NMR measurements, though NMR in solids remained elusive for some additional years. The molecular beam experiments provided information on values of I and g_N for various nuclei which are needed for an understanding of the nucleus. Solid state physicists who work with NMR need to know the values of I and g_N for the nuclei in their samples, but use this information merely as a means to an end.

Attempts were made in 1942 by Gorter and Broer[43] to detect nuclear magnetic resonant absorption in crystalline solids ([7]Li in LiCl and ^{19}F in KF) by looking for a weak dispersion of the dielectric constant as the resonance condition was scanned. This was not successful, but the resonant absorption of R.F. energy in a magnetic field was demonstrated only four years later by Purcell et al.[44] for protons in paraffin wax, and by Bloch et al.[45] for protons in water. Figure 5-36 shows a derivative trace of the resonance condition for [63]Cu and [65]Cu in cuprous chloride, as reported by Pound in 1948. The abcissa here is a scale of 15 minutes per division, while the frequency ν was changed (at constant B for this particular experiment) linearly with time; early NMR runs often extended over many hours. Faster data acquisition techniques are now available, though the long relaxation times involved often still necessitate a slow scan procedure in NMR measurements.

In addition to measuring the location (ratio of ν to B) for an NMR line, an observer can also measure the shape and width of the line, and observe in many cases a splitting into fine structure components. The following brief account will indicate that chemists, biophysicists, and solid state physicists have different priorities concerning the information to be obtained from NMR experiments.

---

[43] C. J. Gorter and L. F. J. Broer, Physica **9**, 591 (1942).
[44] E. M. Purcell, H. C. Torrey, and R. V. Pound, Phys. Rev. **69**, 37 (1946).
[45] F. Bloch, W. W. Hansen, and M. E. Packard, Phys. Rev. **69**, 127 (1946).

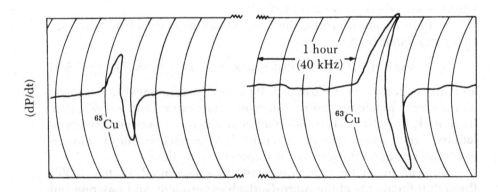

Figure 5-36 A copy of an early NMR trace, that for [63]Cu and [65]Cu nuclei in powdered cuprous chloride, as measured by R. V. Pound, Phys. Rev. **73**, 523 (1948). The value of (dP/dt) was monitored and traced on a chart recorder as resonance for the two copper lines was swept; for this experiment, B was kept constant at 0.3 T while the radio frequency was decreased linearly with time. The amplitude ratio of the two lines is in conformity with the abundance ratio 7:3 for these two copper isotopes. Both [63]Cu and [65]Cu have a nuclear spin I = 3/2, and the figure shown above provided the first accurate measure of the nuclear magnetic moments in copper isotopes: $\mu = 2.22\mu_n$ for [63]Cu, and $\mu = 2.38\mu_n$ for [65]Cu.

High Resolution NMR Spectroscopy

The NMR experiments of the late 1940's placed considerable emphasis on measuring line positions as accurately as possible. This was very important, since for many nuclei the magnetic moments had previously only been approximately estimated from optical spectroscopy. NMR at last gave good values for g_N with these nuclei, information of great interest to nuclear physics.

This stage of investigation is long past. Values for g_N are known with sufficient precision for virtually all stable isotopes and for many radio-isotopes also. Thus g_N values can now be taken for granted by the solid state physicist who uses NMR to study the electrical and magnetic environment in a solid.

It was first noted by Knight[46] that the resonant condition (ν/B) for a given nucleus can depend on the chemical environment. Knight's principal interest was in the rather large *Knight shift* of the resonance condition that will be discussed later for metals, but he did report also a small effect in nonmetals. This *chemical shift* of (ν/B) for a nucleus from one bonding arrangement to another (both in nonmetallic environments) is the result of small contributions to the local magnetic field from atomic and molecular orbital electrons. A chemical shift is usually quoted in terms of the fractional displacement ($\Delta B/B$) from the believed "true" resonance field of the nucleus at a given radio-frequency. For molecules dissolved in a liquid solvent, values for ($\Delta B/B$) are sometimes found to be as large as 10^{-3}. A chemical shift of much smaller size is shown by the triple resonance in Figure 5-37. This figure shows one of the first reported examples of a chemical shift, the three proton NMR peaks cor-

---

[46] W. D. Knight, Phys. Rev. **76**, 1259 (1949).

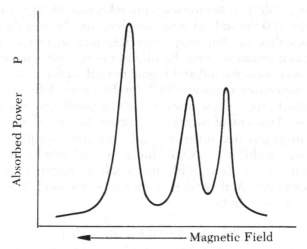

Absorbed Power P

Magnetic Field

Figure 5–37 Trace of the NMR spectrum of protons in ethyl alcohol, after J. T. Arnold, S. S. Dharmetti, and M. E. Packard, J. Chem. Phys. 19, 507 (1951). This was carried out with B = 0.76 T and ν = 32.4 MHz. The three lines correspond to protons from CH_3, CH_2, and OH groups respectively, and the two outer lines are separated by ($\Delta B/B$) = 5 × 10^{-6}.

responding to the three bonding locations of a hydrogen atom in ethyl alcohol.

Both chemistry and molecular biophysics have benefited greatly from high resolution NMR spectroscopy, in such topics as structural organic chemistry[47] and the locations of hydrogen bonds in biological macromolecules. However, high resolution NMR spectroscopy is feasible only if linewidths are narrow compared with the line separations. For a strange reason, NMR lines are narrow in gases and liquids, yet are broad in solid samples. The reason is that the resonance condition for a nucleus in a rigid lattice is broadened by dipole-dipole interaction with its neighbors. In contrast, the rapid motion and rotation of molecules in a fluid smooths out the local field produced by neighboring dipoles to permit a very narrow line!

Thus materials of chemical and biological interest which exist in liquid form are well-suited for high resolution NMR spectroscopy. Unfortunately, solid state physics must manage as best it can with broad resonance lines.

Relaxation and Linewidth for NMR in Solids

The time constants τ_1 for spin-lattice relaxation and τ_2 for spin-spin relaxation were discussed earlier in this section in connection with ESR, and these concepts are equally valid in the NMR context. Both τ_1 and τ_2 affect the solutions of the Bloch equations[31] for the response of a set of spins when (B/ν) is swept at a linear rate through the resonance condition.

Typically, τ_1 has no influence on the NMR linewidth, though it is important in determining the rate at which spin transitions can be induced without saturating the system. τ_1 can become very long for nuclear spin-lattice relaxation, particularly at low temperatures. When Pound discovered in 1951 that the nuclear spin relaxation of ^{7}Li in single crystal LiF was $\tau_1 = 300$ seconds at room temperature, he and Purcell[48] were able to capitalize on this very slow response to create an inverted nuclear spin population, with the higher energy states more populated than the lower ones. Purcell and Pound described this as attainment of a *negative temperature* of some -350K for the nuclei, followed by a decay through infinite spin temperature to normal conditions again. The subsequent development of masers and lasers has made us much more familiar with population inversion and "negative" temperatures.

The large width of an NMR line in a solid results from magnetic dipole-dipole interactions with the nuclear magnetic moments on nearest neighbors. A neighbor of magnetic moment μ at distance $\mathbf{r}$ creates a field contribution

$$\delta \mathbf{B} = (\mu_0/4\pi r^5)[3(\boldsymbol{\mu} \cdot \mathbf{r})\mathbf{r} - r^2\boldsymbol{\mu}] \qquad (5\text{-}134)$$

[47] Numerous books concentrate on the chemical aspects of NMR, including W. W. Paudler, *Nuclear Magnetic Resonance* (Allyn & Bacon, 1971).

[48] E. M. Purcell and R. V. Pound, Phys. Rev. **81**, 279 (1951).

as may be verified by comparison with the corresponding Equation 5-6 for an electric dipole. From Equation 5-134, a moment of magnitude $g_N\mu_n$ at the nearest neighbor distance "a" has an effect with magnitude $|\delta B| \sim (\mu_0 g_N \mu_n / 4\pi a^3)$, typically some 10^{-3} tesla. The effects of various nearest neighbors add vectorially to modify the resonance condition; the effect increases the effective field for some nuclei and decreases it for others at a given moment in time. Thus it is easy to reconcile dipole-dipole interaction arguments with the observed NMR linewidths, which are typically $\Delta B_{pp} = 5$ to 50×10^{-4} tesla.

Motional Narrowing, and Diffusion in Solids

The much narrower NMR lines in fluids occur because rapid motion and rotation of the molecules and atoms averages the sum of all δB to zero over one period of the R.F. radiation.

This *motional narrowing* can also occur in a solid if the temperature is high enough to permit rapid migration of atoms from one lattice or interstitial site to another by the mechanism of solid-state diffusion. The phenomenon of *diffusional narrowing* was first found for solid hydrogen[49] in 1949, and Figure 5-38 illustrates an example of more recent data in a glassy solid.

---

[49] J. Hatton and B. V. Rollin, Proc. Roy. Soc. A.**199**, 222 (1949).

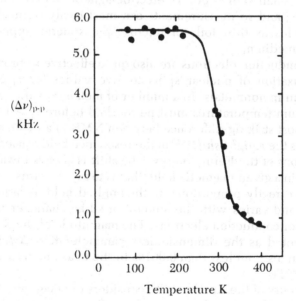

Figure 5-38 Peak-to-peak width of the NMR line for ^{7}Li in a glass of composition 20% Li_2O, 80% SiO_2. The linewidth data has been converted from ΔB_{pp} to $\Delta \nu_{pp}$, and is plotted versus temperature. The temperature dependence for T > 260K is consistent with a jump frequency for diffusion which varies as $\exp(-0.48/k_0 T)$. [See Problem 5.20.] These data are from P. J. Bray, in *Magnetic Resonance*, edited by C. K. Coogan *et al.* (Plenum, 1970), p. 11.

The diffusion coefficient for atoms in a solid is usually expressed in the form

$$D = \xi a^2 \nu_j = \xi a^2 \nu_D \exp(-E/k_0 T) \qquad (5\text{-}135)$$

Here, $\nu_D = (k_0 \theta_D / h)$ is the maximum vibrational frequency of the solid, "a" is the nearest neighbor distance, and E is the activation energy barrier which must be overcome in moving an atom from one site to another. The dimensionless factor ξ depends on the particular lattice, but in all cases is of the order of unity. The Debye frequency $\nu_D \sim 10^{12}$ to 10^{13} Hz, but the magnitude of the jump frequency $\nu_j = \nu_D \exp(-E/k_0 T)$ will be very small if the activation energy is large.

Thus a dipole-dipole broadened NMR line in a solid will start to narrow at a temperature for which the jump frequency ν_j becomes commensurate with the linewidth $\Delta\nu_{pp}$. This is the case just below room temperature for lithium atoms in the glass sample of Figure 5-38, and the diffusion activation energy can be determined from the temperature at which $\Delta\nu_{pp}$ starts to decrease. NMR is thus a useful technique in the study of solid-state diffusion, for a knowledge of the diffusion activation energy tells us what mechanisms are likely to be responsible for diffusion.

NMR in Metals; the Knight Shift

The presence of a large free electron gas affects observation of NMR in a metallic sample in several ways. One complication is that a metal has a quite small skin depth for electromagnetic radiation at high radio frequencies; and so measurements can most easily be made with metallic samples as thin foils or microscopic spheres supported in an insulating medium.

The conduction electrons are also quite effective in promoting spin-lattice relaxation of nuclear spins to give values for τ_1 considerably smaller than in nonmetals. In a number of metals, τ_1 varies as $(1/T)$, and liquid helium temperatures must be reached before $\tau_1 > 1$ sec.

The most striking difference between NMR in a metal and that in a nonmetal is the *Knight shift*[46,50] of the resonance field caused by conduction electrons of the Fermi energy. The shift is always towards a higher frequency in a given magnetic field (that is to say, towards a higher effective g_N), is directly proportional to the applied field, is larger for heavy elements, and varies with the amount of s-like character in the wavefunctions for conduction electrons. The magnitude of the Knight shift is usually quoted as the dimensionless parameter $K = (\Delta\nu/\nu) = (-\Delta B/B)$, and this ranges from less than 0.001% in ^{9}Be to as much as 1.5% in both ^{133}Cs and ^{205}Tl.

The theory of the Knight shift[51] considers the magnetic field created at the nuclear location due to the hyperfine interaction between the

[50] Footnote 46 cites Knight's first paper on this topic, but a fuller account is given by W. D. Knight in *Solid State Physics, Vol. 2* (edited by F. Seitz and D. Turnbull) (Academic Press, 1956), p. 93.

[51] C. H. Townes, C. Herring, and W. D. Knight, Phys. Rev. **77**, 852 (1950).

nuclear moment and the moments of unpaired conduction electrons in s-states. It will be seen that this has much in common with the Fermi contact interaction[39] for the hyperfine splitting of an ESR line. The contribution of each electron depends on how much of its time is spent at the nuclear location, and hence upon $|\psi(0)|^2$.

Electrons of the Fermi energy are the important ones, since all electrons of lower energy are in paired states. Thus the Knight shift depends on the density of states $g(\epsilon_F)$ for the Fermi energy,[52] and on the resulting Pauli paramagnetic susceptibility $\chi_P = \mu_0\mu_B^2 g(\epsilon_F)$ [Equation 3-68] provided by conduction electrons of the Fermi energy. Accordingly, the coefficient of the Knight shift has the functional form

$$K = C\chi_P\langle|\psi_F(0)|^2\rangle_{av} \tag{5-136}$$

with the averaging taken over all electronic states on the Fermi surface. Equation 5-136 can alternatively be written in terms of the hyperfine coupling coefficient, but the above form shows the important dependencies. It can be seen why K is larger for the heavier elements, and is dependent on s-like wavefunction character for the conduction electrons.

For numerous metals, the Knight shift coefficient K varies rather little with temperature, perhaps by less than one part in ten over the range from absolute zero to the melting point. An interesting exception to this trend is provided by the data for ^{113}Cd in Figure 5-39. This curve

---

[52] For a *free electron* gas, $g(\epsilon_F) = (3n/2\epsilon_F)$ as quoted in Equation 3-69, but the periodic potential introduced in the band theory can make $g(\epsilon_F)$ larger or smaller than the free electron value.

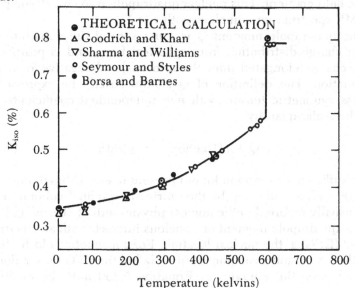

Figure 5-39 The isotropic Knight shift [expressed as a percentage of $(\Delta\nu/\nu)$] versus temperature for ^{113}Cd. Taken from R. V. Kasowski and L. M. Falicov, Phys. Rev. Letters 22, 1001 (1969), and showing their calculated points for the solid (at 0, 298 and 462K) and for the liquid at 600K, as well as various experimental results. Kasowski and Falicov concluded that both $g(\epsilon_F)$ and $|\psi_F(0)|^2$ are appropriate for a free electron gas in liquid cadmium, but that $|\psi_F(0)|^2$ is only 0.74 of the free electron value at absolute zero, and $g(\epsilon_F)$ a still smaller fraction.

has been explained by Kasowski and Falicov[53] as the consequence of both $g(\epsilon_F)$ and $|\psi_F(0)|^2$ varying with temperature, because of the complications of band overlap in this solid. Thus the two conduction electrons per atom for Cd result in a first Brillouin zone which is mostly full, a moderate occupancy of the second zone, and tiny pockets of electrons in the third and fourth zones. The important states in Zone 2 are s-p hybrids, but they become more s-like as temperature rises. Moreover, the change in periodic potential causes increased occupancy of s-like states in Zones 3 and 4 upon heating. Both of these factors contribute to a rising curve for K. Upon melting, the electronic distribution is that of an unrestricted free electron gas.

In this way, Knight shift data can be combined with other information about a metal to learn about $g(\epsilon_F)$ and about the fraction of s-like character for states of the Fermi energy. Problem 5.21 takes a highly simplified approach to how $g(\epsilon_F)$ may vary with temperature for cadmium, using the data of Figure 5-39.

Nuclear Quadrupole Resonance

When the energy levels of a nucleus are split only by a magnetic field, then the $(2I + 1)$ levels of Equation 5-130 form a linear progression. The selection rule $\Delta m_I = \pm 1$ permits 2I transitions among these levels, but each has the identical energy $h\nu = g_N\mu_n B$. However, these transitions occur at 2I slightly different energies if the energy spectrum is modified by interaction of a nuclear quadrupole moment Q with an electric field gradient. This nuclear quadrupole resonance[54] complicates the NMR spectrum for a nucleus with $I \geqslant 1$.

The quadrupole moment Q is a measure of the departure of the nuclear charge distribution from spherical form, and is positive for a prolate charge (elongated along the spin axis) and negative for an oblate configuration. The definition of Q should strictly be expressed as a traceless symmetric tensor, with five independent coefficients. However, the scalar quantity

$$Q = \int (1/2e)(3z^2 - r^2)\rho(r)d^3r \qquad (5\text{-}137)$$

gives a sufficient description for our present needs. Q has dimensions of (length)^2, and, as indicated by the entries in the final column of Table 5-5, is usually reported in the nuclear physics unit of "barns" (10^{-28} m^2).

The quadrupole moment of a nucleus interacts with an electric field gradient (EFG) at the nuclear location. For a magnetic field B_z directed along the z-axis, and a component (d^2V/dz^2) of the EFG tensor along this direction, then the energies of Equation 5-130 must be modified to

$$u = u_0 - g_N\mu_n B_z m_I + eQ\left(\frac{d^2V}{dz^2}\right)\left[\frac{3m_I^2 - I(I+1)}{4I(2I-1)}\right] \qquad (5\text{-}138)$$

[53] R. V. Kasowski and L. M. Falicov, Phys. Rev. Letters **22**, 1001 (1969).

[54] First reported in crystalline solids by R. V. Pound, Phys. Rev. **79**, 685 (1950).

A photon will now be absorbed in stimulating a transition $m_I \rightarrow (m_I - 1)$ if its energy is

$$hv = g_N\mu_n B_z - eQ \left(\frac{d^2V}{dz^2}\right)\left[\frac{3(2m_I - 1)}{4I(2I - 1)}\right] \qquad (5\text{-}139)$$

This still coincides with the "ordinary" NMR transition energy for the transition from $m_I = \frac{1}{2}$ to $m_I = -\frac{1}{2}$, but the other NMR transitions will take place at a different frequency (for given B_z) or different B_z (for given frequency) if the nucleus has a finite Q and has a location in the solid for which (d^2V/dz^2) is non-vanishing. The "satellite" lines at larger and smaller fields can thus be used to *measure* the electric field gradient at the location of a spinning nucleus. In this way, nuclear quadrupole resonance has an important role to play in solid state physics.

Figure 5-40 shows a typical example of the quadrupole splitting of an NMR line. The ^{11}B nuclei have a spin quantum number $I = \frac{3}{2}$, and hence a three-line spectrum would be seen if all ^{11}B experienced the same environment and alignment with respect to the direction of B_z. The figure actually shows a strong central line (corresponding to the transition from $m_I = \frac{1}{2}$ to $m_I = -\frac{1}{2}$) and *two* sets of satellites, for crystallites of differing orientations. Detailed information on the components of the electric field gradient tensor must be obtained in such a case by analysis of the quadrupole spectrum as a function of angle.

Both ESR and NMR lead to very complicated spectra, and problems of analysis, for paramagnetic ions and spinning nuclei in solid environments. It is hoped that the reader who has been interested by the examples presented in this section will seek out some of the much more comprehensive accounts cited in the bibliography which ends this chapter.

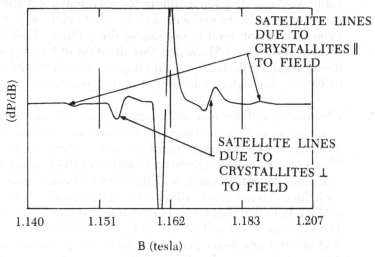

Figure 5–40 Derivative NMR spectrum for ^{11}B in powdered CrB_2 at a frequency of 16 MHz. Satellites of the main line arise from the interaction of nuclear quadrupole moment with the electric field gradient (EFG) in this non-cubic solid. Some crystallites are oriented such that the major axis of the EFG tensor is parallel to B, and approximately twice as many have this axis perpendicular to B. From R. G. Barnes, *Magnetic Resonance*, edited by C. K. Coogan *et al.* (Plenum, 1970), p. 63.

Problems

5.1   Add vectorially the components of electric field resulting from the charges $+q$ and $-q$ in Figure 5-1 and show that the dipole field is indeed given by Equation 5-6.

5.2   Consider a spherical cavity within a uniform dielectric solid, in which the dipole moment per unit volume is **P**. Let the z-axis denote the direction of this polarization and let θ be the angle between the z-axis and a vector **r** which extends from the center of the cavity to a location on the cavity wall. Describe the field $\mathbf{E}_{surf}$ at the center of the cavity caused by the distribution of surface charge over the cavity wall as an integral with respect to θ, and show that this integral reduces to $(P/3\varepsilon_0)$ as quoted in Equation 5-11. How does $\mathbf{E}_{surf}$ change if we wish to express it for a location displaced from the center of the cavity along the direction of **P**?

5.3   Expand the exponentials of Equation 5-25 as far as terms in w^5 to show that the Langevin function for low field conditions is a converging series with

$$L(w) \approx (w/3)[1 - (w^2/15) + \cdots], \qquad w \ll 1$$

as the leading terms. Using the full form of Equation 5-25, show that when the polarization is 50 per cent of the saturation value, you can develop a simple transcendental equation for $w_{0.5}$ which can be solved with a slide rule to yield $w_{0.5} = 1.80$. [You should not need to use any of the published tables for $L(w)$ to show this.] Show that this situation of 50 per cent of the saturation polarization would require the implausible field of 10^9 V/m for water molecules at room temperature.

5.4   Observations of the real and imaginary parts of the dielectric constant are made as a function of temperature for a polar solid in which the Debye temperature for lattice vibrations is 153K. The dielectric measurements are made at 110 kHz, and show a maximum loss factor at $T = 270$K. Show that this is consistent with the existence of two dipole orientations with a barrier 0.4 eV high between them. Plot curves of κ_1 and κ_2 for this solid at 110 kHz (let $A = 5$ and $B = 15$ in Equation 5-33). You should find that the imaginary part of the dielectric constant exceeds 50 per cent of its peak value over the temperature range 250K to about 290K.

5.5   Show that the equation of motion, Equation 5-41, does have the solution given by Equation 5-42, with a polarizability given by Equation 5-43. Use the latter in the Clausius-Mosotti equation to show that the real and imaginary parts of the dielectric constant are given by Equations 5-46 and 5-47. Locate the turning points of κ_1 in Equation 5-46, and use this to determine the values of γ, ω_1, ω_0, $(e^2N_e/m\varepsilon_0)$, and $\beta = \omega_0^2/m$ for a solid in which the real part of the dielectric constant is supposed to go through a maximum and minimum in the visible part of the spectrum. Let the maximum of κ_1 occur for $\omega = 1.67 \times 10^{15}$ rad/s, and the subsequent minimum be $\kappa_1 = 0.25$ for $\omega = 3.6 \times 10^{15}$ rad/s. You should conclude that the restoring force constant for an electron in this solid is $\beta = 10$ N/m. Plot curves for the real and imaginary parts of the dielectric constant versus frequency (using a logarithmic scale for frequency). You should find that $\kappa = \kappa_1 = 2.25$ at the low frequency end of the range.

5.6 Two atoms with a large ionic polarizability are brought to within a distance D of each other, far away from any other matter and isolated from external fields. How large must α_i be in order that it can be energetically favorable for the atoms to spontaneously polarize?

5.7 Consider an electron in a circular orbit around a hydrogen atom, with radius r and angular frequency ω. Describe ω in terms of e, m, and r without worrying about quantum restrictions. Now let magnetic induction B exist in a direction perpendicular to the orbital plane, thus changing the balance of centrifugal and centripetal forces because of the Lorentz force. Show that to first order in B the angular frequency is shifted by an amount $\Delta\omega = (eB/2m)$. This result is known as the *Larmor theorem*, and is central to the classical theory of diamagnetism. How large a magnetic field would be necessary to make $\Delta\omega$ as large as 1 per cent of the orbital frequency of an electron in a hydrogen atom?

5.8 A set of paramagnetic atoms (N per unit volume) has all electrons except one per atom in paired states, and this one is in an s-like state. What are the possible values for m_J? Draw a sketch of the directions of these with respect to a magnetic field, indicating their potential energies in the field. If the occupancies of these states are determined by a Boltzmann distribution, show that the *net* occupancies give a magnetization

$$M = N\mu_B \tanh(B\mu_B/k_0T)$$

per unit volume. Demonstrate that this is consistent with Equation 5-80 for the set of conditions described.

5.9 In this problem you wish to solve Equation 5-93 of ferromagnetic magnetization for the simple case of $J = \frac{1}{2}$. Show that

$B_{1/2}(y)$ reduces to a simple hyperbolic function of the dimensionless variable $y = (\gamma\mu_0\mu_1 M_s/k_0 T)$, and plot a curve of this function for positive y. On the same piece of graph paper, construct a series of straight lines from the origin which intersects the curve. These straight lines must represent (M_s/M_{s0}) as linear functions of y (i.e., for various temperatures). You can use the asymptotic weak-field form of $B_{1/2}(y)$ at T_c in order to express T_c in terms of $\gamma\mu_0\mu_1 M_{s0}$, and hence to calibrate the slopes of your straight lines. Deduce from your intersection points some combinations of (M_s/M_{s0}) and (T/T_c), and compare them with the curve and data of Figure 5-15.

5.10 You may observe from Figure 5-18 that the energy of a magnon varies approximately as the square of the wave-vector. From this piece of information, describe the density of magnon states as a power law function of energy. Given that magnon states are occupied in the manner of a Bose-Einstein distribution law, show that the total number of magnons excited at any temperature T is in accordance with the "Bloch $T^{3/2}$" law for the reduction of spontaneous magnetization with temperature. Show in the same way that the magnon population makes a contribution towards the specific heat which varies as $T^{3/2}$. Discuss how these results are modified when we take into account that long-range anisotropic effects make the minimum magnon energy a little larger than zero for infinite spin-wave wavelength.

5.11 Kittel suggests (in his *Introduction to Solid State Physics*) that the molecular field of Equation 5-97 for an antiferromagnetic material be elaborated to $H_A = (-\gamma M_B - \eta M_A)$; $H_B = (-\gamma M_A - \eta M_B)$. Show that for this situation the Néel temperature T_N and the negative temperature intercept of the reciprocal susceptibility are related by $(\theta/T_N) = (\gamma + \eta)/(\gamma - \eta)$.

5.12 Apply the formalism of Problem 5.11 to a *ferrimagnetic* material in which the magnetic moments associated with the two sub-lattices are of differing strengths. Show how this leads to the high temperature susceptibility behavior described in Equation 5-102.

5.13 Evaluate the angular momentum L_s and magnetic moment μ for a sphere of mass M (mass uniformly distributed through the volume) and charge Q (uniformly distributed over the surface), assuming a radius r and an angular velocity ω. Thereby, show that the ratio of magnetic moment to angular momentum is as quoted in Equation 5-103.

5.14 Differentiate Equations 5-115 and 5-117 to locate the maxima and minima of (dP/dB) when an ESR line is respectively Lorentzian or Gaussian. How does ΔB_{pp} compare with 2Γ for the Gaussian case? Make a photocopy of Figure 5-27, and draw

Gaussian curves on that copy for ΔB_{pp} the same as in the Lorentzian and for $\Delta B_{1/2} = 2\Gamma$ the same as the Lorentzian. Determine the location of P_{max} for the experimental curve, and hence evaluate the effective g-factor. Also determine the half-width Γ, and convert this to an equivalent frequency width. Application of the Heisenberg uncertainty principle will then allow you to calculate the corresponding relaxation time τ_r, for comparison with Figure 5-25.

5.15 Plot a curve of the necessary *applied* magnetic field as a function of temperature if it is desired to observe antiferromagnetic resonance in the MnF_2 sample of Figure 5-30 at a frequency of 10 GHz.

5.16 Any modern physics text will tell you that the ground state wavefunction for a hydrogen atom has a radial dependence $\psi(r) = (\pi a_0{}^3)^{-1/2} \exp(-r/a_0)$. Evaluate $|\psi(0)|^2$ in m^{-3}, and thereby obtain the Fermi contact interaction coefficient $\mathscr{A}$ for this atom. The result in joules can be expressed alternatively as the hyperfine splitting $[\mathscr{A}/g\mu_B]$ in teslas, which you know should be 0.0508 T for a hydrogen atom.

5.17 Write down the solutions of Equation 5-128 for $\theta = 0$, and $\theta = \pi/2$. There are three solutions for each choice of θ. Verify that the solutions for $\theta = \pi/2$ are as sketched in Figure 5-34, and draw a sketch of the three solutions for $\theta = 0$. Mark on this sketch the two transitions which can be photon stimulated for $h\nu > D$.

5.18 Show that the energy level schemes of Problem 5.17 yield enough allowed ESR lines subject to the selection rule $\Delta m_s = \pm 1$ so that $g_\parallel$, $g_\perp$, and D can be determined from the combined parallel/perpendicular spectrum, whether one uses a photon energy $h\nu > D$, or a smaller photon energy $h\nu < D$.

5.19 Express the linewidth of the ^{63}Cu resonance in cuprous chloride (Figure 5-36) in terms of $\Delta\nu_{pp}$ and ΔB_{pp}. By what ratio is this resonance wider than that illustrated in Figure 5-37 for protons associated with OH groups in liquid ethyl alcohol?

5.20 Consider the temperature at which the linewidth shown in Figure 5-38 starts to decrease. Show that this is consistent with a 0.48 eV activation energy for diffusion, as noted in the figure caption. Based on the elastic data of Table 2-2, you may assume that $\theta_D \simeq 600K$ for a glass which is mostly silica.

5.21 This problem attempts a highly simplified analysis of the cadmium Knight shift data of Figure 5-39. Find the density of this metal (assume the same for both solid and liquid), and determine the electron density, Fermi energy ϵ_F, Fermi density of states $g(\epsilon_F)$ and Pauli susceptibility χ_P for a free electron gas of two electrons per atom. This is supposed to be a good assumption for liquid Cd. From the Knight shift shown for the liquid, determine the value of the parameter C in Equation 5-136. Now note from the figure caption that $|\psi_F(0)|^2$ is less than unity at absolute zero, and assume for simplicity that it has the same reduced value at all temperatures in solid Cd. Upon this (incorrect) assumption, evaluate $g(\epsilon_F)$ as a function of temperature for the solid, plotting your results with an ordinate scale of $g(\epsilon_F)$ as a fraction of the free electron value.

5.22 You observe NMR for ^{23}Na nuclei in an anisotropic crystalline environment, and think that you can see satellite lines caused by interaction of the nuclear quadrupole moment with an electric field gradient. These satellites are barely resolved because with operation at 30 MHz, the NMR linewidth $\Delta B_{pp} = 10^{-3}$ tesla. What can you set as the upper limit for (d^2V/dz^2) at the nuclear locations? How could you change the experiment to improve your conditions of observation?

Bibliography

Dielectrics
W. F. Brown, in *Handbuch der Physik, Vol. 17* (Springer, 1956).
I. Bungat and M. Popescu, *Physics of Solid Dielectrics* (North-Holland, 1984).
H. Fröhlich, *Theory of Dielectrics* (Oxford Univ. Press, 2nd ed., 1958).
B. K. P. Scaife, *Complex Permittivity* (Crane, Russak, 1971).
I. S. Zheludev, *Physics of Crystalline Dielectrics* (Plenum Press, 1971), 2 vols.

Piezoelectrics and Ferroelectrics
J. C. Burfoot, *Ferroelectrics* (Van Nostrand, 1967).
W. G. Cady, *Piezoelectricity* (Dover, 1964).
P. W. Forsbergh, in *Handbuch der Physik, Vol. 17* (Springer, 1956).
V. K. Fridkin, *Ferroelectric Semiconductors* (Plenum Press, 1980).
F. Jona and G. Shirane, *Ferroelectric Crystals* (Pergamon Press, 1962).
S. B. Lang, *Source Book of Pyroelectricity* (Gordon and Breach, 1974).
M. E. Lines and A. M. Glass, *Principles and Applications of Ferroelectrics and Related Materials* (Oxford Univ. Press, 1977).

Diamagnetism and Paramagnetism
B. D. Cullity, *Introduction to Magnetic Materials* (Addison-Wesley, 1972).
J. B. Goodenough, *Magnetism and the Chemical Bond* (Krieger, 1976).
G. M. Kalvius and R. S. Tebble, *Experimental Magnetism* (Wiley, 1979).
D. H. Martin, *Magnetism in Solids* (MIT Press, 1967).
J. H. Van Vleck, *Theory of Electric and Magnetic Susceptibilities* (Oxford Univ. Press, 1932).

D. Wagner, *Introduction to the Theory of Magnetism* (Pergamon Press, 1972).
R. M. White, *Quantum Theory of Magnetism* (McGraw-Hill, 1970).

Ferro-, Antiferro-, and Ferri-magnetism
R. Boll, *Soft Magnetic Materials* (Wiley, 1979).
R. M. Bozorth, *Ferromagnetism* (Van Nostrand, 1951).
S. Chikazumi, *Physics of Magnetism* (Krieger, 1978).
D. J. Craik (ed.), *Magnetic Oxides* (Wiley, 1975).
M. Cyrot (ed.), *Magnetism of Metals and Alloys* (North-Holland, 1982).
F. Keffer, in *Handbuch der Physik, Vol. 18* (Springer, 1966). [Spin waves.]
A. H. Morrish, *Physical Principles of Magnetism* (Krieger, 1980).
T. H. O'Dell, *Magnetic Bubbles* (Halsted, 1974).
G. T. Rado and H. Suhl (eds.), *Magnetism* (Academic Press), 5 vols., 1963–1973.
K. J. Standley, *Oxide Magnetic Materials* (Oxford Univ. Press, 2nd ed., 1972).
E. P. Wohlfarth (ed.), *Handbook on Ferromagnetic Materials* (North-Holland, 1980), 3 vols.
H. J. Zeiger, *Magnetic Interactions in Solids* (Oxford Univ. Press, 1973).
G. Winkler, *Magnetic Garnets* (Vieweg, 1981).

Magnetic Resonance
A. Abragam and M. Goldman, *Nuclear Magnetism: Order and Disorder* (Oxford Univ. Press, 1982).
E. Andrew, *Nuclear Magnetic Resonance* (Cambridge Univ. Press, 1958).
A. Ault and G. Dudek, *An Introduction to Proton NMR Spectroscopy* (Holden-Day, 1976).
C. K. Coogan *et al.* (eds.), *Magnetic Resonance* (Plenum Press, 1970).
U. Habeberlen, *High Resolution NMR in Solids* (Academic Press, 1976).
J. E. Hamman (ed.), *Theoretical Foundations of Electron Spin Resonance* (Academic Press, 1978).
W. Low, *Paramagnetic Resonance in Solids* (Academic Press, 1960).
G. E. Pake and T. L. Estle, *Physical Principles of Paramagnetic Resonance* (Benjamin, 2nd ed., 1972).
C. P. Poole, *Electron Spin Resonance* (Wiley, 2nd ed., 1983).
C. P. Slichter, *Principles of Magnetic Resonance* (Springer, 2nd ed., 1980).
M. Sparks, *Ferromagnetic Relaxation Theory* (McGraw-Hill, 1964).
J. E. Wertz and J. R. Bolton, *Electron Spin Resonance* (McGraw-Hill, 1972).
J. Winter, *Magnetic Resonance in Metals* (Oxford Univ. Press, 1971).

SOME USEFUL NUMERICAL CONSTANTS*

Quantity	Symbol	Value
Speed of light in vacuum	c	2.9979×10^8 m/s
Permittivity of free space	$\varepsilon_o = (10^7/4\pi c^2)$	8.8542×10^{-12} F/m
Permeability of free space	$\mu_o = 4\pi \times 10^{-7}$	1.2566×10^{-6} H/m
Planck's constant	$\begin{cases} h \\ \hbar = (h/2\pi) \end{cases}$	6.6261×10^{-34} J s 1.0546×10^{-34} J s
Elementary charge	e	1.6022×10^{-19} C
Electron volt	eV	1.6022×10^{-19} J
Atomic mass unit	u	1.6605×10^{-27} kg
Electron rest mass	m	9.1094×10^{-31} kg
Proton rest mass	M_p	1.6726×10^{-27} kg
Neutron rest mass	M_n	1.6749×10^{-27} kg
Electron Compton wavelength	$\lambdabar_c = (\hbar/mc)$	3.8616×10^{-13} m
Ångstrom unit	Å	1.0000×10^{-10} m
Bohr radius	$a_0 = (4\pi\,\varepsilon_o\hbar^2/me^2)$	5.2918×10^{-11} m
Fine structure constant	$\alpha = (\lambdabar_c/a_o)$	$(1/137.036)$
Rydberg energy	$R_\infty = (\hbar^2/2ma_o^2)$	$\begin{cases} 2.1799 \times 10^{-18} \text{ J} \\ 13.606 \text{ eV} \end{cases}$
Bohr magneton	$\mu_B = (e\hbar/2m)$	9.2740×10^{-24} J/T
Nuclear magneton	$\mu_n = (e\hbar/2M_p)$	5.0508×10^{-27} J/T
Boltzmann's constant	k_o	1.3807×10^{-23} J/K
Avogadro's number	N_A	6.0221×10^{23} mole^{-1}
Ideal Gas constant	$R = k_0 N_A$	8.3145 J/mol·K
Temperature for $k_o T = 1$ eV		1.1604×10^4 K
Frequency for $h\upsilon = 1$ eV		2.4180×10^{14} Hz
Wavelength for $hc/\lambda = 1$ eV		1.2398×10^{-6} m

* Using numbers for the fundamental physical constants from the 1986 CODATA evaluation, as reported by E. R. Cohen and B. N. Taylor, Rev. Mod. Phys. **59**, 1121 (1987).

AUTHOR INDEX

SUBJECT INDEX

497